T0155842

Texts in Applied Mathematics

Volume 64

More information about this series at http://www.springer.com/series/1214

Sören Bartels

Numerical Approximation of Partial Differential Equations

 Springer

Sören Bartels
Angewandte Mathematik
Albert-Ludwigs-Universitaet
Freiburg, Germany

Additional material to this book can be downloaded from http://extras.springer.com.

ISSN 0939-2475   ISSN 2196-9949   (electronic)
Texts in Applied Mathematics
ISBN 978-3-319-81265-6   ISBN 978-3-319-32354-1   (eBook)
DOI 10.1007/978-3-319-32354-1

Printed on acid-free paper

This Springer imprint is published by Springer Nature
The registered company is Springer International Publishing AG Switzerland

Preface

> *Musicians are like mathematicians. Every part has to be right for it to work, ...*
>
> – Jeff Beck, 2015

This textbook is meant to serve as an introduction to the numerical analysis and practical treatment of linear partial differential equations. It introduces the main concepts for discretizing prototypical partial differential equations and discusses the application to model problems in continuum mechanics and electromagnetism. Short MATLAB implementations illustrate the practicality of the numerical methods.

The first part of the textbook covers the development and analysis of elementary finite difference methods, the mathematical theory for elliptic partial differential equations, and the construction and numerical analysis of finite element methods. The second part is devoted to iterative solution methods either by refining a triangulation locally to efficiently resolve corner singularities, or by solving the resulting linear system of equations via multilevel or domain decomposition techniques. In the third part of the text, linearly constrained and singularly perturbed boundary value problems are investigated. Their numerical treatment is based on a mathematical framework for saddle-point problems and requires using nonstandard finite element spaces. This leads to accurate numerical schemes for simulating nearly incompressible materials, thin elastic objects, electromagnetic fields, and turbulent flows.

The reader is assumed to be familiar with basic numerical techniques such as interpolation, quadrature, sparse matrices, and iterative solution techniques for linear systems of equations, although this is not mandatory for understanding the main concepts. It is also helpful to have experience with linear functional analysis, properties of Lebesgue and Sobolev spaces, and theories of existence and regularity for linear partial differential equations. The corresponding results needed in this text are explained but not proved in detail.

Each of the three parts of the book can be used for a one-term lecture accompanied by theoretical and practical tutorials. The problems and projects included in the text may require additional comments and some specific supervision depending

v

on the prerequisites of the students. Their main purpose is to provide ideas for individual work and experiments with the included MATLAB codes that are available at

http://extras.springer.com/2016/978-3-319-32353-4

In Chap. 1 we discuss the mathematical description of transport, diffusion, and wave phenomena and their numerical simulation with finite difference methods. We investigate the accuracy of the methods via stability and consistency properties assuming the existence of regular solutions. Optimal order convergence rates for general boundary conditions will be addressed and the practicality of the methods illustrated with short implementations.

Chapter 2 is concerned with general existence theories for solutions of partial differential equations using concepts from functional analysis and considering generalizations of classical derivatives based on a multidimensional integration-by-parts formula. The chapter introduces Sobolev spaces, discusses their main properties, states existence theories for elliptic second order linear partial differential equations, and sketches regularity results for solutions.

In Chap. 3 we show how finite element methods provide an abstract framework for interpolating functions or vector fields in multidimensional domains. They allow for specifying Galerkin methods for approximating partial differential equations. In combination with regularity results, error estimates in various norms can be proved. We discuss the efficient implementation of low order and isoparametric methods in the case of stationary and evolutionary model problems.

The starting point for Chap. 4 is the observation that convergence rates of standard numerical methods are suboptimal when the solution has corner singularities as in the case of elliptic equations on nonconvex domains. Optimal rates can be obtained by using locally refined triangulations which are either specifically constructed for particular domains or generated automatically via adaptive mesh-refinement algorithms. We introduce both approaches, analyze their convergence, and illustrate their implementation.

In Chap. 5 we make use of the fact that linear systems of equations resulting from finite element discretizations of partial differential equations are typically large, sparse, and ill-conditioned. Their efficient numerical solution exploits properties of the underlying continuous problem or a sequence of discretizations. The chapter discusses multigrid, domain decomposition, and preconditioning methods.

The goal of Chap. 6 is to provide alternatives to standard numerical methods that fail to provide accurate approximations when partial differential equations involve constraints defined by a differential operator or when they contain terms weighted by a larger parameter. Generalizations of the Lax–Milgram and Céa lemmas provide a concise framework for the development and analysis of appropriate numerical methods. Central to the construction is the validity of an inf-sup condition that defines a compatibility requirement on involved finite element spaces.

In Chap. 7 we introduce and analyze stable finite element methods for discretizing a saddle-point formulation of the Poisson problem and the Stokes system. Moreover, we investigate characteristic properties of convection-dominated equations and their numerical approximation via introducing stabilizing terms. Flexible discontinuous Galerkin methods are derived and analyzed for a model Poisson equation. Simple implementations illustrate the performance of the numerical methods.

The final Chap. 8 discusses the development, analysis, and implementation of numerical methods for boundary value problems in elasticity, electromagnetism, and fluid mechanics. Each of the considered problems requires a suitable numerical treatment to capture relevant effects with a low number of degrees of freedom.

This textbook results from several courses which I have taught at the University of Maryland at College Park, the Humboldt University of Berlin, the University of Bonn, and the Albert Ludwig University Freiburg. Besides many colleagues, assistants, and students that have contributed to the development of this text, I would particularly like to thank Marijo Milicevic, Alexis Papathanassopoulos, Dirk Pauly, and Patrick Schön for their help and support.

Freiburg, Germany   Sören Bartels
March 2015

Contents

Part III   Constrained and Singularly Perturbed Problems

Part I
Finite Differences and Finite Elements

Chapter 1
Finite Difference Method

1.1 Transport Equation

1.1.1 Mathematical Model

We consider a fluid, e.g., water, flowing at constant speed $a > 0$ through an infinitely long tube of constant cross-section. At time $t = 0$ we inject a substance, e.g., ink, at arbitrary positions. Omitting the molecular structure of the fluid, this is described by a function u_0 that specifies the concentration, i.e., the number of particles per unit volume of the substance at $t = 0$, cf. Fig. 1.1.

We want to derive a mathematical model that allows us to predict the concentration of the substance at arbitrary positions x and times $t > 0$. For this we omit diffusion effects, and assume that the concentration only depends on the horizontal position x, i.e., the concentration at time $t \geq 0$ and position $x \in \mathbb{R}$ is given by a function $u(t, x)$. These assumptions are justified if the speed a is sufficiently large and the radius of the tube sufficiently small. The amount M of the substance, i.e., the number of particles, which is at time t contained in the section $[x_1, x_2]$ is at a later time $t' = t + \tau$ contained in the section $[x_1 + a\tau, x_2 + a\tau]$, which leads to the identity

$$M = \pi r^2 \int_{x_1}^{x_2} u(t, x)\, dx = \pi r^2 \int_{x_1 + a\tau}^{x_2 + a\tau} u(t + \tau, x)\, dx.$$

Letting U be a primitive of u with respect to x, i.e., we have $\partial_x U(t, x) = u(t, x)$, we deduce that

$$U(t, x_2) - U(t, x_1) = U(t + \tau, x_2 + a\tau) - U(t + \tau, x_1 + a\tau).$$

We differentiate this identity with respect to x_2 and obtain

$$u(t, x_2) = u(t + \tau, x_2 + a\tau).$$

© Springer International Publishing Switzerland 2016
S. Bartels, *Numerical Approximation of Partial Differential Equations*,
Texts in Applied Mathematics 64, DOI 10.1007/978-3-319-32354-1_1

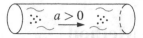

Fig. 1.1 Transport of a substance by a constantly moving fluid through a tube

Differentiating this identity with respect to τ gives

$$0 = \partial_t u(t + \tau, x_2 + a\tau) + a\partial_x u(t + \tau, x_2 + a\tau).$$

Considering the limit $\tau \to 0$ and writing x instead of x_2, we obtain the *transport equation*

$$\partial_t u(t, x) + a\partial_x u(t, x) = 0$$

which holds for all $t > 0$ and all $x \in \mathbb{R}$. This equation defines a relation between the partial derivatives of the function u and is called a *partial differential equation*. To predict the concentration at a time $t > 0$, we need to solve this equation subject to the *initial condition*

$$u(0, x) = u_0(x)$$

for all $x \in \mathbb{R}$. If we are only interested in a finite spatial interval $[\alpha, \beta] \subset \mathbb{R}$, then we have to know the concentration at the left end of the interval called an *inflow boundary* at all times $t > 0$, i.e., we have to include the *boundary condition*

$$u(t, \alpha) = u_\ell(t)$$

to describe the transport process from left to right. If the fluid is flowing in the opposite direction, i.e., in case $a < 0$, the boundary condition has to be imposed at the right end of the interval, i.e., at $x = \beta$.

1.1.2   Explicit Solution

If we interpret the solution $u(t, x)$ of the transport process as a function on $\mathbb{R}_{\geq 0} \times \mathbb{R}$, we see that its gradient

$$\nabla u = [\partial_t u, \partial_x u]^\top$$

satisfies

$$\nabla u \cdot [1, a]^\top = \partial_t u + a\partial_x u = 0,$$

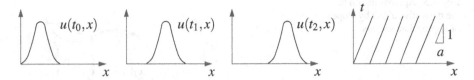

Fig. 1.2 A solution of the transport equation at times $t_0 < t_1 < t_2$ (*left*). Some characteristics $\{(t,x) : x - at = c\}$ for the transport equation along which solutions are constant (*right*)

i.e., the function u is constant in the direction of the vector $[1,a]^{\top}$. Equivalently, u is constant on lines given by $\{(t,x) : x - at = c\}$ for every fixed constant $c \in \mathbb{R}$. Therefore the solution u is given by

$$u(t,x) = f(x - at)$$

for some function f. Incorporating the initial condition $u(0,x) = u_0(x)$, we find that $f = u_0$, i.e.,

$$u(t,x) = u_0(x - at).$$

A typical solution is shown in Fig. 1.2. The lines $L_c = \{(t,x) : x - at = c\}$ for $c \in \mathbb{R}$ are called the *characteristics* of the partial differential equation $\partial_t u + a \partial_x u = 0$. In the (t,x)-plane the characteristics are straight lines and information is transported along them, cf. Fig. 1.2.

1.1.3 Difference Quotients

In general it is not possible to find a closed formula for the solution of a partial differential equation and it is desirable to approximate it numerically. For this, we need to replace the partial derivatives by computable quantities, e.g., by appropriate slopes of secants. For simplicity, we consider the interval $[0, 1]$. The points on the interval $[0, 1]$ that are used to define the secants are defined through a *step-size* $\Delta x = 1/J > 0$ and its integer multiples $x_j = j\Delta x$, $j = 0, 1, \ldots, J$, called *grid points*, as indicated in Fig. 1.3. For a function $u \in C^1([0, 1])$, we then consider the approximations

$$u'(x_j) \approx \frac{u(x_{j+1}) - u(x_j)}{\Delta x},$$

$$u'(x_j) \approx \frac{u(x_j) - u(x_{j-1})}{\Delta x},$$

$$u'(x_j) \approx \frac{u(x_{j+1}) - u(x_{j-1})}{2\,\Delta x}.$$

Fig. 1.3 Approximation of
the derivative $u'(x_j)$ by slopes
of appropriate secants

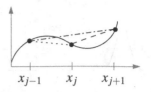

Note that the first quotient is only defined for $j = 0, 1, \ldots, J - 1$, the second for $j = 1, 2, \ldots, J$, and the third for $j = 1, 2, \ldots, J - 1$. The unknown solution u of a partial differential equation is in general not at our disposal and therefore we approximate its values $u(x_j)$ by quantities $U_j, j = 0, 1, \ldots, J$.

Definition 1.1 Given a step-size $\Delta x = 1/J$ for $J \geq 1$ and a sequence $(U_j)_{j=0,\ldots,J}$ the quantities

$$\partial_x^+ U_j = \frac{U_{j+1} - U_j}{\Delta x}, \quad \partial_x^- U_j = \frac{U_j - U_{j-1}}{\Delta x}, \quad \widehat{\partial}_x U_j = \frac{U_{j+1} - U_{j-1}}{2\Delta x}$$

for $j = 0, 1, \ldots, J - 1, j = 1, 2, \ldots, J$, and $j = 1, 2, \ldots, J - 1$ are called *forward*, *backward*, and *central difference quotients*, respectively.

To estimate the approximation error of a difference quotient, we assume that $u \in C^2([0, 1])$ and employ a Taylor expansion about a point $x_j = j\Delta x \in (0, 1)$, i.e., we have

$$u(x_{j+1}) = u(x_j + \Delta x) = u(x_j) + u'(x_j)\Delta x + \frac{1}{2} u''(\xi)\Delta x^2$$

with a point $\xi \in [0, 1]$. This implies that

$$\frac{u(x + \Delta x) - u(x)}{\Delta x} = u'(x) + \frac{1}{2} u''(\xi)\Delta x$$

and hence

$$\left|\partial^+ u(x_j) - u'(x_j)\right| \leq \frac{\Delta x}{2} \sup_{\xi \in [0,1]} |u''(\xi)|.$$

Difference quotients can be applied repeatedly to a sequence to construct approximations of higher order derivatives, e.g., we have

$$\partial^+ \partial^- U_j = \partial^- \partial^+ U_j = \frac{U_{j+1} - 2U_j + U_{j-1}}{\Delta x^2}.$$

Definition 1.2 The quantity

$$\partial^+ \partial^- U_j = \frac{U_{j+1} - 2U_j + U_{j-1}}{\Delta x^2}$$

for $j = 1, 2, \ldots, J - 1$ is called a *second-order central difference quotient*.

Generalizing the above argument leads to the following estimates.

Proposition 1.1 (Difference Quotients) *We have*

$$|\partial^\pm u(x_j) - u'(x_j)| \leq \frac{\Delta x}{2} \|u''\|_{C([0,1])},$$

$$|\widehat{\partial} u(x_j) - u'(x_j)| \leq \frac{\Delta x^2}{6} \|u'''\|_{C([0,1])},$$

$$|\partial^+ \partial^- u(x_j) - u''(x_j)| \leq \frac{\Delta x^2}{12} \|u^{(4)}\|_{C([0,1])}.$$

Proof Exercise. □

1.1.4 Approximation Scheme

We consider the transport problem on the spatial interval $[0, 1]$ and temporal interval $[0, T]$ for some *time horizon* $T > 0$ with positive speed $a > 0$, initial concentration u_0, and vanishing boundary concentration $u_\ell(t) = 0$ for all $t > 0$ at the left end of the interval. We thus consider the *initial boundary value problem*

$$\begin{cases} \partial_t u(t, x) + a\partial_x u(t, x) = 0 & \text{for all } (t, x) \in (0, T] \times (0, 1), \\ u(t, 0) = 0 & \text{for all } t \in (0, T], \\ u(0, x) = u_0(x) & \text{for all } x \in [0, 1]. \end{cases}$$

To discretize this problem we employ spatial and temporal step-sizes $\Delta x = 1/J$ and $\Delta t = T/K$ defined by positive integers J, K, respectively. We then consider a point $(t_k, x_j) = (k\Delta t, j\Delta x)$ for $0 \leq k \leq K$ and $0 \leq j \leq J$ and replace the partial derivatives in the partial differential equation evaluated at (t_k, x_j) by difference quotients, e.g., a forward difference quotient in time and a backward difference quotient in space, i.e.,

$$\partial_t^+ U_j^k + a\partial_x^- U_j^k = 0.$$

Here, the coefficients $(U_j^k)_{0 \leq j \leq J, 0 \leq k \leq K}$ are approximations of the unknown function values $u(t_k, x_j)$. Replacing the difference quotients by their definitions, we can equivalently write

$$\frac{U_j^{k+1} - U_j^k}{\Delta t} + a\frac{U_j^k - U_{j-1}^k}{\Delta x} = 0$$

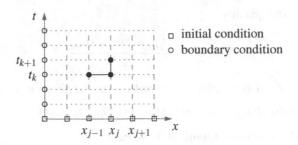

Fig. 1.4 Schematic description of the numerical method for the transport equation

or

$$U_j^{k+1} = U_j^k - a\frac{\Delta t}{\Delta x}\left(U_j^k - U_{j-1}^k\right).$$

This relation shows that we can compute the coefficients U_j^{k+1}, $j = 1, 2, \ldots, J$, provided we know the coefficients U_j^k, $j = 0, 1, \ldots, J$. It is therefore called an *explicit scheme*. For $k = 0$ they are obtained from the initial condition, i.e., $U_j^0 = u_0(x_j)$, $j = 0, 1, \ldots, J$, and for $j = 0$ we employ the boundary condition, i.e, $U_0^k = 0$ for $k = 1, 2, \ldots, K$. The numerical method is schematically depicted in Fig. 1.4 and realized in the following algorithm.

Algorithm 1.1 *Given positive integers J, K, set $\Delta x = 1/J$ and $\Delta t = T/K$.*

(1) Define $U_j^0 := u_0(x_j)$ for $j = 0, 1, \ldots, J$, and set $k = 0$.

(2) Set $U_0^{k+1} = 0$.

(3) For $j = 1, 2, \ldots, J$, compute

$$U_j^{k+1} = U_j^k - a\frac{\Delta t}{\Delta x}\left(U_j^k - U_{j-1}^k\right).$$

(4) Stop if $t_{k+1} = T$, i.e., if $k + 1 = K$; increase $k \to k + 1$ and continue with (2) otherwise.

Remark 1.1 Clearly, note that in general we have $U_j^k \neq u(t_k, x_j)$.

A MATLAB implementation of the algorithm is shown in Fig. 1.5. It has to be taken into account that MATLAB does not allow for the use of the index 0 for arrays. Therefore, all indices are increased by 1 when an array is accessed. The results of two experiments with different discretization parameters, leading to stable and unstable approximations, are shown in Fig. 1.6.

```
function transport
a = 1; T = 1;
J = 20; K = 20;
Delta_x = 1/J; Delta_t = T/K;
U = zeros(K+1,J+1);
for j = 0:J
    U(1,j+1) = u_0(j*Delta_x);
end
plot(Delta_x*(0:J),U(1,:)); axis([0,1,0,1]); pause
for k = 0:K-1
    U(k+1,1) = 0;
    for j = 1:J
        U(k+2,j+1) = U(k+1,j+1)...
            -a*(Delta_t/Delta_x)*(U(k+1,j+1)-U(k+1,j));
    end
    plot(Delta_x*(0:J),U(k+2,:)); axis([0,1,0,1]); pause(.1)
end
clf; mesh(Delta_x*(0:J),Delta_t*(0:K),U);

function val = u_0(x)
if x <= .5
    val = .5*(1+sin(4*pi*x-pi/2));
else
    val = 0;
end
```

Fig. 1.5 Implementation of an approximation scheme for the transport equation. The access of arrays requires increasing indices by one

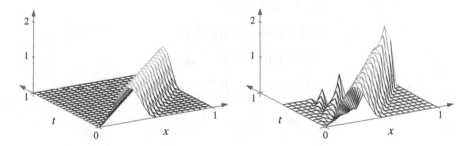

Fig. 1.6 Stable (*left*) and unstable (*right*) approximation of the transport equation

1.1.5   Stability

Testing different parameters Δt and Δx in the numerical scheme for the transport equation shows that approximations may become very large which contradicts the physical behavior of the process. In particular, letting

$$\tilde{u}_0(x) = \begin{cases} u_0(x) & \text{for } x \in [0, 1], \\ 0 & \text{for } x \notin [0, 1], \end{cases}$$

be the trivial extension of u_0 by zero to the entire real line, we have that the exact solution is given by $u(t, x) = \tilde{u}_0(x - at)$. This implies that for all $t \in [0, T]$ we have

$$\sup_{x \in [0,1]} |u(t, x)| = \sup_{x \in [0,1]} |\tilde{u}_0(x - at)| \leq \sup_{x \in [0,1]} |u_0(x)|,$$

i.e., the concentration is at all times bounded by the maximum of the initial concentration. We want to find conditions under which such a result also holds for the approximations.

Proposition 1.2 (Stability) *Set $\mu = a\Delta t / \Delta x$. If $0 \leq \mu \leq 1$, then the numerical solution (U_j^k) of the difference scheme satisfies*

$$\sup_{j=0,\ldots,J} |U_j^k| \leq \sup_{j=0,\ldots,J} |U_j^0|$$

for all $k = 0, 1, \ldots, K$.

Proof For $0 \leq k \leq K - 1$ and $1 \leq j \leq J$, we have

$$U_j^{k+1} = U_j^k - \mu(U_j^k - U_{j-1}^k)$$

and since $1 - \mu \geq 0$ and $\mu \geq 0$ we deduce that

$$\begin{aligned}
|U_j^{k+1}| &= |(1 - \mu)U_j^k + \mu U_{j-1}^k| \\
&\leq |(1 - \mu)U_j^k| + |\mu U_{j-1}^k| \\
&= |1 - \mu||U_j^k| + |\mu||U_{j-1}^k| \\
&= (1 - \mu)|U_j^k| + \mu|U_{j-1}^k| \\
&\leq (1 - \mu) \sup_{j'=0,\ldots,J} |U_{j'}^k| + \mu \sup_{j'=0,\ldots,J} |U_{j'}^k| \\
&= \sup_{j'=0,\ldots,J} |U_{j'}^k|.
\end{aligned}$$

Noting $U_0^{k+1} = 0$, we thus have

$$\sup_{j=0,\ldots,J} |U_j^{k+1}| \leq \sup_{j=0,\ldots,J} |U_j^k|.$$

An inductive argument implies the assertion.   □

Remark 1.2 The conditions of the theorem cannot be improved.

The uniform boundedness of approximations is called *stability* of a numerical scheme. We have proved this under the condition $0 \leq \mu \leq 1$, i.e., if $a \geq 0$ and $\Delta t \leq \Delta x / a$.

1.1.6 Convergence

The boundedness of approximations, i.e., the stability of the scheme, is a necessary condition to obtain meaningful approximations. We next want to address the question of how close the approximations are to the exact solution.

Proposition 1.3 (Convergence) *Set* $\mu = a\Delta t/\Delta x$. *If* $0 \leq \mu \leq 1$ *and* $u \in C^2([0, T] \times [0, 1])$, *then we have*

$$\sup_{j=0,\ldots,J} |u(t_k, x_j) - U_j^k| \leq \frac{t_k}{2}(\Delta t + a\Delta x)\|u\|_{C^2([0,T]\times[0,1])}$$

for all $k = 0, 1, \ldots, K$ *and with*

$$\|u\|_{C^2([0,T]\times[0,1])} = \sup_{0\leq\ell+m\leq 2} \sup_{(t,x)\in[0,T]\times[0,1]} |\partial_t^\ell \partial_x^m u(t, x)|.$$

Proof

(i) We define the consistency term

$$\mathscr{C}_j^k = \partial_t^+ u(t_k, x_j) + a\partial_x^- u(t_k, x_j)$$

which measures how much the exact solution fails to satisfy the discrete scheme. Since $\partial_t u + a\partial_x u = 0$ in $(0, 1) \times (0, T)$ it follows with estimates for the difference quotients that

$$|\mathscr{C}_j^k| = |\partial_t^+ u(t_k, x_j) - \partial_t u(t_k, x_j) + a\partial_x^- u(t_k, x_j) - a\partial_x u_x(t_k, x_j)|$$

$$\leq |\partial_t^+ u(t_k, x_j) - \partial_t u(t_k, x_j)| + a|\partial_x^- u(t_k, x_j) - \partial_x u(t_k, x_j)|$$

$$\leq \frac{\Delta t}{2} \sup_{t\in[0,T]} |\partial_t^2 u(\cdot, x_j)| + a\frac{\Delta x}{2} \sup_{x\in[0,1]} |\partial_x^2 u(t_k, \cdot)|.$$

(ii) We subtract the equations

$$u(t_{k+1}, x_j) = u(t_k, x_j) - a\Delta t\partial_x^- u(t_k, x_j) + \Delta t\mathscr{C}_j^k,$$

$$U_j^{k+1} = U_j^k - a\Delta t\,\partial_x^- U_j^k,$$

to deduce that for the error $Z_j^k = u(t_k, x_j) - U_j^k$, we have

$$Z_j^{k+1} = Z_j^k - a\,\Delta t\,\partial_x^- Z_j^k + \Delta t\,\mathscr{C}_j^k.$$

Using the triangle inequality and arguing as in the proof of the stability result, the condition $0 \leq \mu \leq 1$ implies that

$$|Z_j^{k+1}| \leq |Z_j^k - \mu Z_j^k + \mu Z_{j-1}^k| + |\Delta t \, \mathscr{C}_j^k|$$

$$\leq \sup_{j'=0,...,J} |Z_{j'}^k| + \Delta t \, |\mathscr{C}_j^k|,$$

i.e., using $Z_0^{k+1} = 0$,

$$\sup_{j=0,...,J} |Z_j^{k+1}| \leq \sup_{j'=0,...,J} |Z_{j'}^k| + \Delta t \sup_{j'=0,...,J} |\mathscr{C}_{j'}^k|.$$

An induction over $k = 0, 1, \ldots, K-1$ with $Z_j^0 = 0, j = 0, 1, \ldots, J$, proves the asserted estimate. $\square$

Remarks 1.3

(i) The condition $u \in C^2([0, T] \times [0, 1])$ is satisfied if $\tilde{u}_0 \in C^2([0, 1])$.
(ii) In the proof of the error estimate we combined the consistency and the stability of the finite difference scheme.
(iii) The error estimate of the proposition implies the *convergence* of approximations as $(\Delta t, \Delta x) \to 0$.

1.1.7 CFL Condition

If $a < 0$, i.e., if the flow is from right to left, then the approximation scheme $\partial_t^+ U_j^k + a\partial_x^- U_j^k = 0$ is unstable, i.e., numerical solutions always become unbounded. In this case the discretization $\partial_t^+ U_j^k + a\partial_x^+ U_j^k = 0$ is stable whenever $-1 \leq \mu \leq 0$. The choice of the approximation scheme according to the sign of a is related to a condition introduced by Courant, Friedrichs, and Levy.

Definition 1.3 An explicit numerical scheme

$$U_j^{k+1} = \Phi(t_k, x_j, U_{j-m_\ell}^k, \ldots, U_{j+m_r}^k, \Delta t, \Delta x)$$

is said to satisfy the *CFL condition* if the characteristic through the point (t_{k+1}, x_j) intersects the line $\{(x, t) : t = t_k, x \in \mathbb{R}\}$ within the convex hull of all grid points $(t_k, x_{j-m_\ell}), \ldots, (t_k, x_{j+m_r})$ that are involved in the computation of U_j^{k+1}, i.e., within the segment $t_k \times [x_{j-m_\ell}, x_{j+m_r}]$, cf. Fig. 1.7.

The CFL condition guarantees that the numerical scheme can reflect the physical property that information is transported along characteristics. This is a minimal condition on a numerical scheme to provide meaningful approximations, and it is necessary for the stability of a numerical scheme.

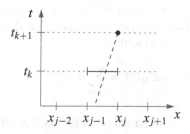

Fig. 1.7 The CFL condition requires that the characteristic (*dashed line*) through the point (t_{k+1}, x_j) intersects the convex hull of points involved in the computation of U_j^{k+1} (*solid line*)

Examples 1.1

(i) For the scheme $U_j^{k+1} = U_j^k - \mu(U_j^k - U_{j-1}^k)$ with $\mu = a\Delta t/\Delta x$, the points involved in computing U_j^{k+1} are (t_k, x_j) and (t_k, x_{j-1}). Since the slope of the characteristic equals $1/a$, the CFL condition is satisfied if and only if $a^{-1} \geq \Delta t/\Delta x$ and $a^{-1} \geq 0$, i.e, if $0 \leq \mu \leq 1$.

(ii) The finite difference scheme $\partial_t^+ U_j^k + a\widehat{\partial_x} U_j^k = 0$ satisfies the CFL condition if $|\mu| \leq 1$, but the scheme is always unstable, i.e., the CFL condition is not a sufficient criterion for stability.

For nonconstant coefficients $a : [0, T] \times [0, 1] \to \mathbb{R}$ of varying sign, the *upwinding scheme*

$$U_j^{k+1} = U_j^k - \begin{cases} \mu_j^k(U_j^k - U_{j-1}^k) & \text{if } \mu_j^k \geq 0, \\ \mu_j^k(U_{j+1}^k - U_j^k) & \text{if } \mu_j^k < 0, \end{cases}$$

with $\mu_j^k = a(t_k, x_j)\Delta t/\Delta x$ is stable if $\sup_{(t,x)\in[0,T]\times[0,1]} |a(t,x)|(\Delta t/\Delta x) \leq 1$.

1.1.8 Fourier Stability Analysis

The *Fourier series* or *von Neumann stability analysis* for a difference scheme of the form

$$U_j^{k+1} = (E_{\Delta t} U^k)_j = \sum_{m\in\mathbb{Z}} a_m U_{j-m}^k$$

yields necessary and sufficient conditions for stability with respect to the norm

$$\|V\|_{\ell^2(\mathbb{Z})} = \Big(\sum_{j\in\mathbb{Z}} V_j^2\Big)^{1/2}$$

for $V = (V_j)_{j \in \mathbb{Z}}$. Here finite vectors $(V_j)_{j=0,\dots,J}$ are extended trivially by zero to sequences $(V_j)_{j \in \mathbb{Z}}$. Given a sequence $(a_m)_{m \in \mathbb{Z}}$ that defines $E_{\Delta t}$, we define its *(Fourier) symbol* for $\xi \in [-\pi, \pi]$ through

$$\widetilde{E}_{\Delta t}(\xi) = \sum_{m \in \mathbb{Z}} a_m e^{-im\xi}.$$

The function $\widetilde{E}_{\Delta t}$ contains information about the stability of the scheme defined by $E_{\Delta t}$. The proof of the following result requires elementary results about the Fourier transform which are recommended as exercises.

Proposition 1.4 (Fourier Stability) *The operator $E_{\Delta t}$ is stable with respect to the norm $\| \cdot \|_{\ell^2(\mathbb{Z})}$, i.e., we have*

$$\|E_{\Delta t}V\|_{\ell^2(\mathbb{Z})} \leq \|V\|_{\ell^2(\mathbb{Z})}$$

for all $V = (V_j)_{j \in \mathbb{Z}}$, if and only if $|\widetilde{E}(\xi)| \leq 1$ for all $\xi \in [-\pi, \pi]$.

Proof Parseval's identity states that for all $V = (V_j)_{j \in \mathbb{Z}}$ with $\|V\|_{\ell^2(\mathbb{Z})} < \infty$, we have

$$\|V\|_{\ell^2(\mathbb{Z})}^2 = \frac{1}{2\pi \Delta x} \int_{-\pi}^{\pi} |\widehat{V}(\xi)|^2 \, d\xi,$$

where $\widehat{V}(\xi) = \sum_{j \in \mathbb{Z}} V_j e^{-ij\xi}$ is the Fourier transform of $V = (V_j)_{j \in \mathbb{Z}}$. The Fourier transform of $E_{\Delta t}V$ is given by

$$\widehat{(E_{\Delta t}V)}(\xi) = \sum_{j \in \mathbb{Z}} (E_{\Delta t}V)_j \, e^{-ij\xi}$$

$$= \sum_{j \in \mathbb{Z}} \left(\sum_{m \in \mathbb{Z}} a_m V_{j-m} \right) e^{-ij\xi}$$

$$= \sum_{m \in \mathbb{Z}} a_m e^{-im\xi} \sum_{j \in \mathbb{Z}} V_{j-m} \, e^{-i(j-m)\xi}$$

$$= \sum_{m \in \mathbb{Z}} a_m e^{-im\xi} \sum_{\ell \in \mathbb{Z}} V_\ell \, e^{-i\ell\xi}$$

$$= \sum_{m \in \mathbb{Z}} a_m e^{-im\xi} \widehat{V}(\xi)$$

$$= \widetilde{E}_{\Delta t}(\xi) \, \widehat{V}(\xi).$$

Therefore the stability of $E_{\Delta t}$ holds if and only if

$$\int_{-\pi}^{\pi} |\widetilde{E}_{\Delta t}(\xi)|^2 |\widehat{V}(\xi)|^2 \, d\xi \leq \int_{-\pi}^{\pi} |\widehat{V}(\xi)|^2 \, d\xi$$

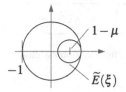

Fig. 1.8 The values of the symbol $\widetilde{E}(\xi) \in \mathbb{C}$ for $\xi \in [-\pi, \pi]$ lie on the circle with midpoint $1 - \mu$ and radius μ

is satisfied for all possible $\widehat{V}$. Since the Fourier transform is an isomorphism from $\{V = (V_j)_{j \in \mathbb{Z}} : \|V\|_{\ell^2(\mathbb{Z})} < \infty\}$ onto $L^2(-\pi, \pi)$, one can show that the estimate holds if and only if $|\widetilde{E}_{\Delta t}(\xi)| \leq 1$ for all $\xi \in [-\pi, \pi]$. $\square$

We illustrate the application of the stability criterion with the difference scheme for the transport equation.

Example 1.2 For the numerical scheme

$$U_j^{k+1} = (1 - \mu)U_j^k + \mu U_{j-1}^k,$$

we have $U_j^{k+1} = \sum_{m \in \mathbb{Z}} a_m U_{j-m}^k$ if we set $a_0 = (1 - \mu)$, $a_1 = \mu$, and $a_m = 0$ for $m \in \mathbb{Z} \setminus \{0, 1\}$. Then

$$\widetilde{E}_{\Delta t}(\xi) = a_0 + a_1 e^{-i\xi} = (1 - \mu) + \mu e^{-i\xi}.$$

The complex function values $\widetilde{E}_{\Delta t}(\xi)$ lie on the circle of radius $|\mu|$ with midpoint $1 - \mu$. If $a \geq 0$ and $\mu \leq 1$, we have $|\widetilde{E}_{\Delta t}(\xi)| \leq 1$ for all $\xi \in \mathbb{R}$, cf. Fig. 1.8.

1.2 Heat Equation

1.2.1 Mathematical Model

We consider a motionless fluid in a straight thin tube of finite length and a substance that is diffusing through the liquid, cf. Fig. 1.9. The substance moves from regions of higher concentrations to regions of lower concentrations. According to *Fick's law* of diffusion, the rate of motion is proportional to the concentration gradient.

We let $u(t, x)$ be the concentration of the substance, i.e., the number of particles per unit volume at time t and position x. The amount of substance in the section $[x_1, x_2]$ is

$$M(t) = \pi r^2 \int_{x_1}^{x_2} u(t, x) \, dx,$$

Fig. 1.9 Diffusion of a substance within a motionless fluid

and the rate of change of mass in $[x_1, x_2]$ is

$$\frac{dM}{dt}(t) = \pi r^2 \int_{x_1}^{x_2} \partial_t u(t, x) \, dx.$$

The mass in the section cannot change except by flowing in or flowing out at its ends. According to Fick's law, the quantity $-\kappa \partial_x u(t, x)$ determines how much substance is passing the point x from left to right. Hence we have

$$\frac{dM}{dt}(t) = -\kappa \partial_x u(t, x_1) + \kappa \partial_x u(t, x_2)$$

with a proportionality constant $\kappa > 0$. We thus have that

$$\pi r^2 \int_{x_1}^{x_2} \partial_t u(t, x) dx = \kappa \partial_x(t, x_2) - \kappa \partial_x u(t, x_1).$$

Differentiating with respect to x_2 leads to the identity

$$\partial_t u(t, x_2) = \frac{\kappa}{\pi r^2} \partial_x^2 u(t, x_2),$$

which is called the *diffusion equation*. If we want to describe that no substance can enter or escape at the ends $x = \alpha$ and $x = \beta$ of the tube, then by Fick's law the concentration gradient has to vanish there, i.e., for all $t \in (0, T)$ we impose that

$$\kappa \partial_x u(t, \alpha) = \kappa \partial_x u(t, \beta) = 0.$$

These boundary conditions are called *(homogeneous) Neumann boundary conditions*. If instead the concentration itself is prescribed at the ends $x = \alpha$ and $x = \beta$ by functions u_ℓ and u_r, then we have to impose for all $t \in (0, T)$ the conditions

$$u(t, \alpha) = u_\ell(t), \quad u(t, \beta) = u_r(t)$$

which are called *Dirichlet boundary conditions*.

Remark 1.4 The partial differential equation $\partial_t u = \kappa \partial_x^2 u$ also provides a description of the heat distribution in a thin wire and is therefore also called a *heat equation*. Homogeneous Neumann boundary conditions model the wire being insulated at its ends while Dirichlet boundary conditions describe a situation in which the ends of the wire are connected to a body of fixed temperature, e.g., a water basin.

1.2.2 Explicit Solution

On the entire real line, i.e., in the absence of boundary conditions, an explicit representation formula for solving the heat equation can be derived. In particular, we have that the function

$$u(t, x) = \frac{1}{(4\pi t)^{1/2}} \int_{\mathbb{R}} e^{-|x-y|^2/(4t)} u_0(y) \, dy$$

solves $\partial_t u - \partial_x^2 u = 0$ in $(0, T) \times \mathbb{R}$ and $u(0, x) = u_0(x)$ for all $x \in \mathbb{R}$.

Example 1.3 For u_0 given by

$$u_0(x) = \begin{cases} 1 & \text{if } |x| \le 1, \\ 0 & \text{if } |x| > 1, \end{cases}$$

the solution for $t = 1/10, 1/2, 1$ is depicted in Fig. 1.10. For every positive $t > 0$ the function $x \mapsto u(t, x)$ is a smooth and positive function. Theoretically, this corresponds to an infinite speed of propagation of information which is not physical. The function values are however negligible for large $|x|$ and the heat equation provides accurate descriptions of heat distribution processes.

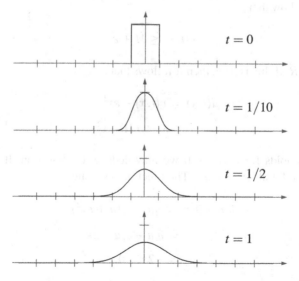

Fig. 1.10 The solution of the heat equation is smooth and positive for times $t > 0$

1.2.3   Properties of Solutions

For simplicity, we assume $\kappa/(\pi r^2) = 1$ and consider the initial boundary value problem

$$
\begin{cases}
\partial_t u(t,x) - \partial_x^2 u(t,x) = 0 & \text{for all } (t,x) \in (0,T] \times (0,1), \\
u(t,0) = u(t,1) = 0 & \text{for all } t \in (0,T], \\
u(0,x) = u_0(x) & \text{for all } x \in [0,1].
\end{cases}
$$

The existence of a solution will be established in an exercise by a *separation of variables*.

Proposition 1.5 (Maximum Principle) *If $u \in C^2([0,T] \times [0,1])$ satisfies $\partial_t u - \partial_x^2 u = 0$ in $R = \{(t,x) \in \mathbb{R}^2 : 0 < t < T, \ 0 < x < 1\}$, then the maximum value of u is attained on the* parabolic boundary

$$
\Gamma_p = \{(t,x) \in \partial R : t = 0, \ x = 0, \ or \ x = 1\}.
$$

Proof Let M be the maximum value of u on Γ_p. We want to show that $u(t,x) \leq M$ for all $(t,x) \in \overline{R}$. For this let $\varepsilon > 0$ and define

$$
v(t,x) = u(t,x) + \varepsilon x^2.
$$

Our goal is to show that

$$
v(t,x) \leq M + \varepsilon
$$

for all $(t,x) \in \overline{R}$. If this is true, then it follows that

$$
u(t,x) = v(t,x) - \varepsilon x^2
$$

$$
\leq M + \varepsilon(1 - x^2),
$$

and since this holds for all $\varepsilon > 0$ we may deduce the assertion. It is clear that $v(t,x) \leq M + \varepsilon$ for all $(t,x) \in \Gamma_p$. The function v satisfies

$$
\partial_t v - \partial_x^2 v = \partial_t u - \partial_x^2(u + \varepsilon x^2)
$$

$$
= \partial_t u - \partial_x^2 u - 2\varepsilon
$$

$$
= -2\varepsilon < 0.
$$

Since v is continuous, it must attain its maximum in $\overline{R}$. Suppose that v attains its maximum in the open set R at the point (t_0, x_0). Then, necessarily

$$
\partial_t v(t_0, x_0) = 0, \quad \partial_x^2 v(t_0, x_0) \leq 0,
$$

but this contradicts $\partial_t v - \partial_x^2 v < 0$. Therefore, v has its maximum on the boundary ∂R. Suppose that v attains its maximum on $\partial R \setminus \Gamma_p$, i.e., at a point (T, x_0) with $x_0 \in (0, 1)$. We then have

$$\partial_x^2 v(T, x_0) \leq 0, \quad \partial_t v(T, x_0) = \lim_{\delta \to 0} \frac{v(T, x_0) - v(T - \delta, x_0)}{\delta} \geq 0,$$

since $v(T, x_0) \geq v(T - \delta, x_0)$ for all $\delta > 0$. But this again contradicts $\partial_t v - \partial_x^2 v < 0$. Hence the maximum has to be on Γ_p. Therefore,

$$v(t, x) \leq M + \varepsilon$$

for all $(t, x) \in \bar{R}$ and this implies the theorem. □

The maximum principle has the interpretation that there exist *no interior hot spots* in a heat conducting wire.

Remarks 1.5

(i) The *strong maximum principle* states that the maximum is only on Γ_p unless the solution is constant.

(ii) A minimum principle follows from considering $-u(t, x)$.

Proposition 1.6 (Energy Decay) *Suppose that u solves the heat equation. Then for all $t \in [0, T]$ we have*

$$\frac{d}{dt} \int_0^1 u(t, x)^2 \, dx \leq 0, \quad \frac{d}{dt} \int_0^1 \left(\partial_x u(t, x)\right)^2 dx \leq 0.$$

Proof We multiply the equation $\partial_t u - \partial_x^2 u = 0$ with u and integrate with respect to x over the interval $(0, 1)$ to verify that

$$\int_0^1 \left(u \partial_t u - u \partial_x^2 u\right) dx = 0.$$

The identity $\partial_t(u^2/2) = u \partial_t u$ allows us to rewrite the first term as

$$\int_0^1 u \partial_t u \, dx = \frac{1}{2} \frac{d}{dt} \int_0^1 u^2 \, dx.$$

Integrating by parts in the second term shows that

$$\int_0^1 u \partial_x^2 u \, dx = -\int_0^1 (\partial_x u)^2 \, dx + [u \partial_x u]_0^1$$

and the boundary term vanishes due to the boundary conditions $u(t, 0) = u(t, 1) = 0$. A combination of the identities yields that

$$\frac{1}{2}\frac{d}{dt}\int_0^1 u^2\,dx = -\int_0^1 (\partial_x u)^2\,dx \le 0.$$

Defining $F(t) = (1/2)\int_0^1 u(t, x)^2\,dx$, we see that $F'(t) \le 0$ for all $t \in (0, T)$ which is the first decay estimate. The second estimate follows analogously from multiplying the heat equation with $\partial_t u$. $\square$

The energy estimate implies the uniqueness of solutions.

Corollary 1.1 (Uniqueness) *The initial boundary value problem for the heat equation has a unique solution.*

Proof Exercise. $\square$

1.2.4 Explicit Scheme

We choose a mesh-size $\Delta x = 1/J$ and a step-size $\Delta t = T/K$ and replace the partial derivatives in the heat equation by appropriate difference quotients. With the forward difference quotient for the time derivative this leads to the *forward Euler scheme*:

$$\begin{cases} \partial_t^+ U_j^k - \partial_x^+ \partial_x^- U_j^k = 0, & j = 1, 2, \ldots, J-1,\ k = 0, 1, \ldots, K-1, \\ U_0^k = U_J^k = 0, & k = 0, 1, \ldots, K, \\ U_j^0 = u_0(x_j),\ j = 0, 1, 2, \ldots, J. \end{cases}$$

Equivalently, we have

$$\frac{1}{\Delta t}(U_j^{k+1} - U_j^k) = \frac{1}{\Delta x^2}(U_{j-1}^k - 2U_j^k + U_{j+1}^k)$$

or with $\lambda = \Delta t/\Delta x^2$

$$U_j^{k+1} = (1 - 2\lambda)U_j^k + \lambda U_{j-1}^k + \lambda U_{j+1}^k.$$

This is an *explicit scheme* since U_j^{k+1} can be directly computed from the coefficients $(U_j^k)_{j=0,\ldots,J}$ at the previous time step. The scheme is also called the *explicit Euler scheme*.

Proposition 1.7 (Stability and Convergence) *If $\lambda \le 1/2$, then the solution (U_j^k) of the forward Euler scheme satisfies*

$$\sup_{j=0,\ldots,J} |U_j^k| \le \sup_{j=0,\ldots,J} |U_j^0|.$$

If in addition we have $u \in C^4([0, T] \times [0, 1])$, then

$$\sup_{j=0,...,J} |u(t_k, x_j) - U_j^k| \leq \frac{t_k}{2}(\Delta t + \Delta x^2)\left(\|\partial_x^4 u\|_{C([0,T]\times[0,1])} + \|\partial_t^2 u\|_{C([0,T]\times[0,1])}\right)$$

for all $k = 0, 1, \ldots, K$.

Proof Since $1 - 2\lambda \geq 0$, we have that

$$|U_j^{k+1}| \leq (1 - 2\lambda)|U_j^k| + \lambda|U_{j-1}^k| + \lambda|U_{j+1}^k|$$

$$\leq (1 - 2\lambda) \sup_{j'=0,...,J} |U_{j'}^k| + 2\lambda \sup_{j'=0,...,J} |U_{j'}^k|$$

$$\leq \sup_{j'=0,...,J} |U_{j'}^k|,$$

which implies the first assertion. The consistency terms

$$\mathscr{C}_j^k = \partial_t^+ u(t_k, x_j) - \partial_x^+ \partial_x^- u(t_k, x_j)$$

satisfy

$$|\mathscr{C}_j^k| = |\partial_t^+ u(t_k, x_j) - \partial_t u(t_k, x_j) + \partial_x^2 u(t_k, x_j) - \partial_x^+ \partial_x^- u(t_k, x_j)|$$

$$\leq |\partial_t^+ u(t_k, x_j) - \partial_t u(t_k, x_j)| + |\partial_x^+ \partial_x^- u(t_k, x_j) - \partial_x^2 u(t_k, x_j)|.$$

With the estimates for the difference quotients, cf. Proposition 1.1, we deduce that

$$|\mathscr{C}_j^k| \leq \frac{1}{2}(\Delta t + \Delta x^2)\left(\|\partial_x^4 u\|_{C([0,T]\times[0,1])} + \|\partial_t^2 u\|_{C([0,T]\times[0,1])}\right).$$

The error $Z_j^k = u(t_k, x_j) - U_j^k$ satisfies

$$\partial_t^+ Z_j^k - \partial_x^+ \partial_x^- Z_j^k = \mathscr{C}_j^k,$$

and hence

$$Z_j^{k+1} = Z_j^k + \lambda(Z_{j-1}^k - 2Z_j^k + Z_{j+1}^k) + \Delta t \, \mathscr{C}_j^k$$

$$= (1 - 2\lambda)Z_j^k + \lambda Z_{j-1}^k + \lambda Z_{j+1}^k + \Delta t \, \mathscr{C}_j^k.$$

Arguing as above, this leads to

$$\sup_{j=0,...,J} |Z_j^{k+1}| \leq \sup_{j=0,...,J} |Z_j^k| + \Delta t \sup_{j=0,...,J} |\mathscr{C}_j^k|.$$

An inductive argument with $Z_j^0 = 0, j = 0, 1, \ldots J$, implies the error estimate. $\square$

```
function explicit_euler
T = 1;
J = 20; K = 400;
Delta_x = 1/J; Delta_t = T/K;
lambda = Delta_t/Delta_x^2;
U = zeros(K+1,J+1);
for j = 0:J
    U(1,j+1) = u_0(j*Delta_x);
end
plot(Delta_x*(0:J),U(1,:)); axis([0,1,0,1]); pause
for k = 0:K-1
    U(k+2,1) = 0;
    for j = 1:J-1
        U(k+2,j+1) = (1-2*lambda)*U(k+1,j+1)...
            +lambda*(U(k+1,j+2)+U(k+1,j));
    end
    U(k+1,J+1) = 0;
    plot(Delta_x*(0:J),U(k+2,:)); axis([0,1,0,1]); pause(.1)
end
clf; mesh(Delta_x*(0:J),Delta_t*(0:K),U);

function val = u_0(x)
val = sin(pi*x);
```

Fig. 1.11 Explicit Euler scheme for approximating the heat equation

Remarks 1.6

(i) The condition $\lambda \leq 1/2$ requires that we have $\Delta t \leq \Delta x^2/2$, i.e., the step-size has to be very small compared to the mesh-size. This condition cannot be improved in general.

(ii) The condition $u \in C^4([0,T] \times [0,1])$ requires certain regularity properties of u_0. It suffices that its trivial extension to $\mathbb{R}$ satisfies $\tilde{u}_0 \in C^4(\mathbb{R})$.

The implementation of the explicit Euler scheme is straightforward and a realization in MATLAB is shown in Fig. 1.11.

1.2.5 Implicit Scheme

We obtain a different numerical scheme if instead of the forward difference quotient in the approximation of the time derivative we employ the backward difference quotient. The resulting scheme is called the *backward Euler scheme*:

$$\begin{cases} \partial_t^- U_j^k - \partial_x^+ \partial_x^- U_j^k = 0, & j = 1, 2, \ldots, J-1, \ k = 1, 2, \ldots, K \\ U_0^k = U_J^k = 0, & k = 1, 2, \ldots, K, \\ U_j^0 = u_0(x_j), \ j = 0, 1, 2, \ldots, J. \end{cases}$$

Rewriting the first equation with the abbreviation $\lambda = \Delta t/\Delta x^2$, we have

$$U_j^k - \lambda(U_{j-1}^k - 2U_j^k + U_{j+1}^k) = U_j^{k-1}$$

for $k = 1, 2, \ldots, K$. Equivalently, we may replace k by $k + 1$ so that

$$U_j^{k+1} - \lambda(U_{j-1}^{k+1} - 2U_j^{k+1} + U_{j+1}^{k+1}) = U_j^k$$

for $k = 0, 1, \ldots, K - 1$. From this equation we see that we cannot directly compute U_j^{k+1} from the approximations $(U_j^k)_{j=0,\ldots,J}$ at the previous time-step. Instead, we have to solve a linear system of equations to determine the new approximations $(U_j^{k+1})_{j=0,\ldots,J}$. The scheme is therefore said to be *implicit* and is also called the *implicit Euler scheme*. By writing down the equations for $j = 1, 2, \ldots, J - 1$, we find that the linear system of equations reads as follows:

$$\begin{bmatrix} 1+2\lambda & -\lambda & & & \\ -\lambda & 1+2\lambda & -\lambda & & \\ & \ddots & \ddots & \ddots & \\ & & -\lambda & 1+2\lambda & -\lambda \\ & & & -\lambda & 1+2\lambda \end{bmatrix} \begin{bmatrix} U_1^{k+1} \\ U_2^{k+1} \\ \vdots \\ U_{J-2}^{k+1} \\ U_{J-1}^{k+1} \end{bmatrix} = \begin{bmatrix} U_1^k \\ U_2^k \\ \vdots \\ U_{J-2}^k \\ U_{J-1}^k \end{bmatrix}.$$

Here we eliminated U_0^{k+1} and U_J^{k+1} from the system using the boundary conditions $U_0^{k+1} = U_J^{k+1} = 0$.

Proposition 1.8 (Existence, Stability, and Convergence) *There exist unique coefficients (U_j^k) that solve the implicit Euler scheme and that satisfy*

$$\sup_{j=0,\ldots,J} |U_j^k| \leq \sup_{j=0,\ldots,J} |U_j^0|$$

for all $k = 0, 1, \ldots, K$, independently of $\lambda = \Delta t/\Delta x^2$. If $u \in C^4([0, T] \times [0, 1])$, then we have

$$\sup_{j=0,\ldots,J} |u(t_k, x_j) - U_j^k| \leq \frac{t_k}{2}(\Delta t + \Delta x^2)(\|\partial_x^4 u\|_{C([0,T]\times[0,1])} + \|\partial_t^2 u\|_{C([0,T]\times[0,1])})$$

for $k = 0, 1, \ldots, K$.

Proof

(i) It follows from the Gerschgorin theorem that the matrix

$$A = \begin{bmatrix} 1+2\lambda & -\lambda & & \\ -\lambda & \ddots & \ddots & \\ & \ddots & \ddots & -\lambda \\ & & -\lambda & 1+2\lambda \end{bmatrix}$$

is regular and hence, given $(U_j^k)_{j=1,\dots,J-1}$, there exists a uniquely defined vector $(U_j^{k+1})_{j=1,\dots,J-1}$ that solves the linear system of equations related to the backward Euler scheme.

(ii) Let $j' \in \{1, 2, \dots, J-1\}$ be such that $|U_{j'}^{k+1}| = \sup_{j=0,\dots,J} |U_j^{k+1}|$. We have

$$(1 + 2\lambda)U_{j'}^{k+1} = U_{j'}^k + \lambda U_{j'-1}^{k+1} + \lambda U_{j'+1}^{k+1},$$

and hence

$$(1 + 2\lambda)|U_{j'}^{k+1}| \leq |U_{j'}^k| + \lambda|U_{j'-1}^{k+1}| + \lambda\,|U_{j'+1}^{k+1}|$$

$$\leq \sup_{j=0,\dots,J} |U_j^k| + 2\lambda \sup_{j=0,\dots,J} |U_j^{k+1}|.$$

Due to the choice of j' we thus have

$$(1 + 2\lambda) \sup_{j=0,\dots,J} |U_j^{k+1}| \leq \sup_{j=0,\dots,J} |U_j^k| + 2\lambda \sup_{j=0,\dots,J} |U_j^{k+1}|$$

which proves the first estimate.

(iii) The consistency error

$$\mathscr{C}_j^{k+1} = \partial_t^- u(t_{k+1}, x_j) - \partial_x^+ \partial_x^- u(t_{k+1}, x_j)$$

is bounded, as in the explicit case by

$$|\mathscr{C}_j^{k+1}| \leq \frac{1}{2}(\Delta t + \Delta x^2)(\|\partial_x^4 u\|_{C([0,T]\times[0,1])} + \|\partial_t^2 u\|_{C([0,T]\times[0,T])}).$$

The error $Z_j^k = u(t_k, x_j) - U_j^k$ satisfies the equation

$$(1 + 2\lambda)Z_j^{k+1} - \lambda(Z_{j-1}^{k+1} + Z_{j+1}^{k+1}) = Z_j^k + \Delta t\,\mathscr{C}_j^{k+1}.$$

We choose j' such that

$$|Z_{j'}^{k+1}| = \sup_{j=0,\dots,J} |Z_j^{k+1}|$$

and estimate that

$$(1 + 2\lambda) \sup_{j=0,\dots,J} |Z_j^{k+1}| = (1 + 2\lambda)|Z_{j'}^{k+1}|$$

$$\leq \lambda|Z_{j'-1}^{k+1}| + \lambda|Z_{j'+1}^{k+1}| + \Delta t\,|\mathscr{C}_{j'}^{k+1}| + |Z_{j'}^k|$$

$$\leq 2\lambda \sup_{j=0,\dots,J} |Z_j^{k+1}| + \Delta t\,|\mathscr{C}_{j'}^{k+1}| + \sup_{j=0,\dots,J} |Z_j^k|.$$

```
function implicit_euler
T = 1;
J = 20; K = 10;
Delta_x = 1/J; Delta_t = T/K;
lambda = Delta_t/Delta_x^2;
U = zeros(K+1,J+1);
e = ones(J-1,1);
A = speye(J-1)+lambda*spdiags([-e,2*e,-e],[-1,0,1],J-1,J-1);
for j = 0:J
    U(1,j+1) = u_0(j*Delta_x);
end
plot(Delta_x*(0:J),U(1,:)); axis([0,1,0,1]); pause
for k = 0:K-1
    U(k+2,1) = 0;
    x = A\U(k+1,2:J)';
    U(k+2,2:J) = x';
    U(k+2,J+1) = 0;
    plot(Delta_x*(0:J),U(k+2,:)); axis([0,1,0,1]); pause(.05)
end
mesh(Delta_x*(0:J),Delta_t*(0:K),U);

function val = u_0(x)
val = sin(pi*x);
```

Fig. 1.12 Implicit Euler scheme for approximating the heat equation

This implies that

$$\sup_{j=0,\dots,J} |Z_j^{k+1}| \le \sup_{j=0,\dots,J} |Z_j^k| + \Delta t\, |\mathscr{C}_{j'}^k|.$$

The bound on $|\mathscr{C}_j^{k+1}|$ and an induction over $k = 0, 1, \dots, K$ imply the error estimate.   □

Remark 1.7 The theorem shows that the implicit Euler scheme is *unconditionally stable*. The error estimate is the same as the one for the explicit Euler scheme but holds unconditionally.

A realization of the implicit Euler scheme requires the solution of linear systems of equations in every time step. In the MATLAB realization, shown in Fig. 1.12, these are solved with the backslash operator. Since the system matrix is the same in every time step, a very efficient approach would be to compute an *LU* or Cholesky factorization once. The involved factors are sparse matrices with limited bandwidth which allows for directly solving the resulting linear systems with linear computational complexity. For efficiency, it is important that the matrices be defined as sparse matrices in the code. The result of an experiment is shown in Fig. 1.13.

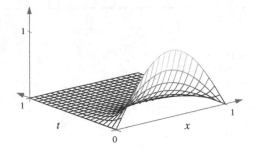

Fig. 1.13 Approximation of the heat equation with the implicit Euler scheme

1.2.6   Midpoint Scheme

A family of approximation schemes results from combining the explicit and implicit
Euler schemes. This is done by choosing a parameter $\theta \in [0, 1]$ and considering the
difference scheme:

$$\begin{cases} \partial_t^- U_j^{k+1} - (1 - \theta)\partial_x^+ \partial_x^- U_j^k - \theta \partial_x^+ \partial_x^- U_j^{k+1} = 0, & j = 1, 2, \ldots, J - 1, \\ & k = 0, 1, \ldots, K - 1, \\ U_0^{k+1} = U_J^{k+1} = 0, & k = 0, 1, \ldots, K - 1, \\ U_j^0 = u_0(x_j), j = 0, 1, 2, \ldots, J. \end{cases}$$

The scheme is called the θ-*method* or *midpoint scheme*. For $\theta = 1$, it coincides with
the implicit Euler scheme, and for $\theta = 0$, it reduces to the explicit Euler scheme as
can be seen from the relation

$$U_j^{k+1} - \theta\lambda(U_{j-1}^{k+1} - 2U_j^{k+1} + U_{j+1}^{k+1}) = U_j^k + (1 - \theta)\lambda(U_{j-1}^k - 2U_j^k + U_{j+1}^k).$$

The combination of the schemes is illustrated in Fig. 1.14.

The θ-method is implicit whenever $\theta > 0$ and leads to the following system
of linear equations, in which the vanishing components U_0^{k+1} and U_J^{k+1} have been
eliminated:

$$A \begin{bmatrix} U_1^{k+1} \\ \vdots \\ U_{J-1}^{k+1} \end{bmatrix} = B \begin{bmatrix} U_1^k \\ \vdots \\ U_{J-1}^k \end{bmatrix}$$

with

$$A = \begin{bmatrix} 1 + 2\theta\lambda & -\theta\lambda & & \\ -\theta\lambda & \ddots & \ddots & \\ & \ddots & \ddots & -\theta\lambda \\ & & -\theta\lambda & 1 + 2\theta\lambda \end{bmatrix}$$

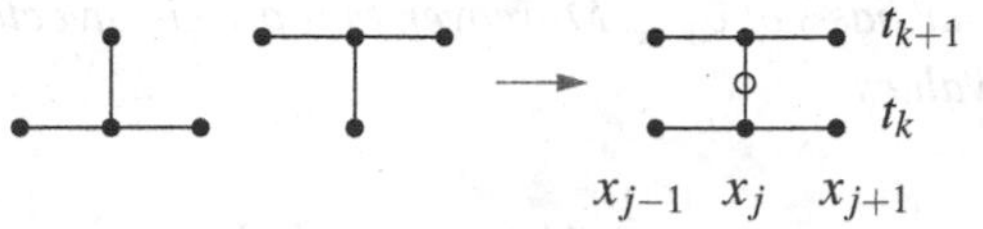

Fig. 1.14 The convex combination with weights $1 - \theta$ and θ of the explicit and implicit Euler schemes results in the θ-method

and

$$B = \begin{bmatrix} 1 - 2(1 - \theta)\lambda & (1 - \theta)\lambda & & \\ (1 - \theta)\lambda & \ddots & & \ddots \\ & \ddots & \ddots & (1 - \theta)\lambda \\ & & (1 - \theta)\lambda & 1 - 2(1 - \theta)\lambda \end{bmatrix}.$$

Proposition 1.9 (Well-Posedness) *The midpoint scheme admits a unique solution* (U_j^k) *for every choice of* Δt *and* Δx.

Proof Exercise. □

In order to analyze the stability of the midpoint scheme, we employ the linear space

$$\ell_{0,\Delta x}^2 = \{V \in \mathbb{R}^{J+1} : V_0 = V_J = 0\}$$

having the inner product

$$(V, W)_{\Delta x} = \Delta x \sum_{j=0}^{J} V_j W_j$$

which induces the norm $\|V\|_{2,\Delta x} = (V, V)_{\Delta x}^{1/2}$. We regard the difference quotient $-\partial_x^+ \partial_x^-$ as an operator on $\ell_{0,\Delta x}^2$ by defining

$$\left(-\partial_x^+ \partial_x^- V\right)_j = \begin{cases} -\partial_x^+ \partial_x^- V_j & \text{for } j = 1, 2, \ldots, J - 1, \\ 0 & \text{for } j = 0, J. \end{cases}$$

The choice of an appropriate basis for the space $\ell_{0,\Delta x}^2$ will lead to simplifying the scheme.

Lemma 1.1 (Eigenvectors for $-\partial_x^+ \partial_x^-$) *The vectors* $(\varphi_p : p = 1, \ldots, J - 1) \subset \ell_{\Delta x}^2$ *defined for* $j = 0, 1, \ldots, J$ *by*

$$\varphi_{p,j} = \sqrt{2} \sin(\pi p j \Delta x)$$

define an orthonormal basis of $\ell^2_{0,\Delta x}$. Moreover, they are eigenvectors of the operator $-\partial^+_x \partial^-_x$ with eigenvalues

$$\sigma_p = \frac{2}{\Delta x^2}\left(1 - \cos(\pi p \Delta x)\right).$$

Proof Exercise.   □

Since $(\varphi_p : p = 1, \ldots, J-1)$ is an orthonormal basis, we have

$$V = \sum_{p=1}^{J-1} \widehat{V}_p \varphi_p, \quad \widehat{V}_p = (V, \varphi_p)_{\Delta x}.$$

for every $V \in \ell^2_{0,\Delta x}$. We substitute the corresponding representations for $U^{k+1} = (U^{k+1}_j)_{j=0,\ldots,J}$ and $U^k = (U^k_j)_{j=0,\ldots,J}$ into the midpoint scheme and obtain the relation

$$\left(\frac{1}{\Delta t} - \theta\, \partial^+_x \partial^-_x\right) \sum_{p=1}^{J-1} \widehat{U}^{k+1}_p \varphi_p = \left(\frac{1}{\Delta t} + (1-\theta)\partial^+_x \partial^-_x\right) \sum_{p=1}^{J-1} \widehat{U}^k_p \varphi_p.$$

Incorporating the fact that the vectors φ_p are eigenvectors of $-\partial^+_x \partial^-_x$ with eigenvalues σ_p, we deduce that

$$\sum_{p=1}^{J-1} (1 + \theta \Delta t\, \sigma_p)\widehat{U}^{k+1}_p \varphi_p = \sum_{p=1}^{J-1} \left(1 - (1-\theta)\Delta t\, \sigma_p\right)\widehat{U}^k_p \varphi_p.$$

The coefficients in the linear combinations have to coincide, i.e., we have

$$(1 + \theta \Delta t\, \sigma_p)\widehat{U}^{k+1}_p = \left(1 - (1-\theta)\Delta t\, \sigma_p\right)\widehat{U}^k_p.$$

With respect to the basis of eigenvectors the numerical scheme thus becomes explicit, i.e., given the coefficients $(\widehat{U}^k_p)_{p=1,\ldots,J-1}$ that define U^k we can directly compute the coefficients $(\widehat{U}^{k+1}_p)_{p=1,\ldots,J-1})$ that define U^{k+1}. Sufficient for stability is that the modulus of the factors $\widetilde{E}(p)$ in $\widehat{U}^{k+1}_p = \widetilde{E}(p)\widehat{U}^k_p$ be bounded by one. We have

$$\widetilde{E}(p) = \frac{1 - (1-\theta)\Delta t\, \sigma_p}{1 + \theta \Delta t\, \sigma_p}.$$

With $\lambda = \Delta t/\Delta x^2$ and $z = 1 - \cos(\pi p \Delta x)$ we have $\widetilde{E}(p) = f(z)$ for

$$f(z) = \frac{1 - 2(1-\theta)\lambda z}{1 + 2\theta\lambda z} = \frac{1 + 2\theta\lambda z - 2\lambda z}{1 + 2\theta\lambda z} = 1 - \frac{2\lambda z}{1 + 2\theta\lambda z}.$$

The function $f(z)$ is monotonically decreasing for $z \in [0, 2]$ with $f(0) = 1$ so that it thus suffices to guarantee that $f(2) \geq -1$, i.e.,

$$f(2) = 1 - \frac{4\lambda}{1 + 4\theta\lambda} \geq -1,$$

which is satisfied if $(1 - 2\theta)\lambda \leq 1/2$. This leads to the following proposition.

Proposition 1.10 (Stability) *Let (U_j^k) be the solution of the midpoint scheme. If $(1 - 2\theta)\lambda \leq 1/2$, then*

$$\|U^k\|_{2,\Delta x} \leq \|U^0\|_{2,\Delta x}$$

for all $k = 0, 1, \ldots, K$.

Proof Since $(\varphi_p : p = 1, 2, \ldots, J - 1)$ is an orthonormal basis, we have

$$\|V\|_{2,\Delta x}^2 = \Delta x \sum_{p=1}^{J-1} |\widehat{V}_p|^2$$

for every $V \in \ell_{0,\Delta x}^2$. Noting that under the condition $(1 - 2\theta)\lambda \leq 1/2$ we have $|\widehat{U}_p^{k+1}| \leq |\widehat{U}_p^k|$ for $p = 1, 2, \ldots, J - 1$ and $k = 0, 1, \ldots, K - 1$, we deduce the assertion. □

Remarks 1.8

(i) The θ-method is unconditionally stable if $\theta \geq 1/2$. For $\theta < 1/2$ we have to guarantee that $\lambda \leq 1/2$ as in the case of the explicit Euler scheme.
(ii) We have $U^{k+1} = \widetilde{A}^{-1}\widetilde{B}U^k$ with the matrices $\widetilde{A}, \widetilde{B} \in \mathbb{R}^{(J+1)\times(J+1)}$ defined by

$$\widetilde{A} = \begin{bmatrix} 1 & 0 & 0 \\ 0 & A & 0 \\ 0 & 0 & 1 \end{bmatrix}, \quad \widetilde{B} = \begin{bmatrix} 0 & 0 & 0 \\ 0 & B & 0 \\ 0 & 0 & 0 \end{bmatrix}$$

and the proof of the proposition shows that $\|\widetilde{A}^{-1}\widetilde{B}V\|_{2,\Delta x} \leq \|V\|_{2,\Delta x}$ for every vector $V \in \ell_{\Delta x}^2$ provided that $(1 - 2\theta)\lambda \leq 1/2$.

1.2.7 Crank–Nicolson Scheme

The Euler schemes are particular cases of the θ-method and lead to an approximation error of the order $\mathcal{O}(\Delta t + \Delta x^2)$. We want to address the question whether this error bound can be improved for other choices of θ. The essential quantity which determines the accuracy of the scheme is the consistency term

$$\mathscr{C}_j^{k+1} = \partial_t^- u(t_{k+1}, x_j) - \partial_x^+ \partial_x^- \big(\theta u(t_{k+1}, x_j) + (1 - \theta)u(t_k, x_j)\big).$$

The second term on the right-hand side may be regarded as an approximation of $\partial_x^2 u$ evaluated at the intermediate point $(t_{k+\theta}, x_j)$, where $t_{k+\theta} = (k + \theta)\Delta t$. Taylor expansions provide the identities

$$u(t_{k+1}, x_j) = u(t_{k+\theta}, x_j) + (1 - \theta)\Delta t\, \partial_t u(t_{k+\theta}, x_j)$$

$$+ (1 - \theta)^2 \frac{\Delta t^2}{2} \partial_t^2 u(t_{k+\theta}, x_j) + \mathcal{O}(\Delta t^3),$$

$$u(t_k, x_j) = u(t_{k+\theta}, x_j) - \theta\Delta t\, \partial_t(t_{k+\theta}, x_j) + \theta^2 \frac{\Delta t^2}{2} \partial_t^2 u(t_{k+\theta}, x_j) + \mathcal{O}(\Delta t^3).$$

Subtracting the equations and dividing by Δt lead to

$$\partial_t^- u(t_{k+1}, x_j) = \partial_t u(t_{k+\theta}, x_j) + \left((1 - \theta)^2 - \theta^2\right) \frac{\Delta t}{2} \partial_t^2 u(t_{k+\theta}, x_j) + \mathcal{O}(\Delta t^2).$$

For $\theta = 1/2$ the second term on the right-hand side disappears and we obtain

$$\partial_t^- u(t_{k+1}, x_j) - \partial_t u(t_{k+1/2}, x_j) = \mathcal{O}(\Delta t^2).$$

The θ-method with $\theta = 1/2$ is called the *Crank–Nicolson scheme*. It is unconditionally stable and approximates the exact solution with an error $\mathcal{O}(\Delta t^2 + \Delta x^2)$. The reason for this improved accuracy is a cancellation of terms related to the symmetry properties of the method.

Proposition 1.11 (Crank–Nicolson Scheme) *Let (U_j^k) solve the midpoint scheme with $\theta = 1/2$. For $k = 0, 1, \ldots, K$, we have*

$$\|U^k - u^k\|_{2,\Delta x} \leq c t_k (\Delta t^2 + \Delta x^2) \|u\|_{C^4([0,T] \times [0,1])},$$

where $U^k = (U_j^k)_{j=0,\ldots,J}$ and $u^k = (u(t_k, x_j))_{j=0,\ldots,J}$, $c \geq 0$ is a constant that is independent of Δx and Δt, and

$$\|u\|_{C^r([0,T] \times [0,1])} = \max_{0 \leq \ell + m \leq r}\, \sup_{(t,x) \in [0,T] \times [0,1]}\, |\partial_t^\ell \partial_x^m u(t, x)|.$$

Proof

(i) Inserting the vanishing expression $\partial_t u(t_{k+1/2}, x_j) - \partial_x^2 u(t_{k+1/2}, x_j)$ in the consistency term leads to

$$|\mathscr{C}_j^{k+1}| = |\partial_t^- u(t_{k+1}, x_j) - \partial_x^+ \partial_x^- \left(u(t_{k+1}, x_j) + u(t_k, x_j)\right)/2|$$

$$\leq |\partial_t^- u(t_{k+1}, x_j) - \partial_t u(t_{k+1/2}, x_j)|$$

$$+ |\partial_x^+ \partial_x^- \left(u(t_{k+1}, x_j) + u(t_k, x_j)\right)/2 - \partial_x^2(t_{k+1/2}, x_j)|$$

$$= I + II.$$

We proved above that

$$I \leq c\Delta t^2 \|u\|_{C^3([0,T]\times[0,1])}.$$

To derive a bound for the second term we employ Taylor expansions in t, i.e.,

$$\partial_x^2 u(t_k, x_j) = \partial_x^2 u(t_{k+1/2}, x_j) - \frac{\Delta t}{2}\partial_x^2\partial_t u(t_{k+1/2}, x_j) + \mathcal{O}(\Delta t^2)$$

$$\partial_x^2 u(t_{k+1}, x_j) = \partial_x^2 u(t_{k+1/2}, x_j) + \frac{\Delta t}{2}\partial_x^2\partial_t u(t_{k+1/2}, x_j) + \mathcal{O}(\Delta t^2).$$

Adding these equations and noting that

$$\partial_x^2 u(t_k, x_j) = \partial_x^+ \partial_x^- u(t_k, x_j) + \mathcal{O}(\Delta x^2),$$
$$\partial_x^2 (t_{k+1}, x_j) = \partial_x^+ \partial_x^- u(t_{k+1}, x_j) + \mathcal{O}(\Delta x^2),$$

leads to

$$\partial_x^2 u(t_{k+1/2}, x_j) - \partial_x^+ \partial_x^- \big(u(t_k, x_j) + u(t_{k+1}, x_j)\big)/2 = \mathcal{O}(\Delta x^2) + \mathcal{O}(\Delta t^2).$$

We thus have

$$II \leq c(\Delta x^2 + \Delta t^2)\,\|u\|_{C^4([0,T]\times[0,1])}$$

which shows that

$$|\mathscr{C}_j^{k+1}| \leq c(\Delta x^2 + \Delta t^2)\,\|u\|_{C^4([0,T]\times[0,1])}.$$

(ii) The error $Z^{k+1} = u^{k+1} - U^{k+1}$ satisfies

$$Z^{k+1} = \widetilde{A}^{-1}\widetilde{B}u^k + \Delta t\widetilde{A}^{-1}\mathscr{C}^{k+1} - \widetilde{A}^{-1}\widetilde{B}U^k$$
$$= \widetilde{A}^{-1}\widetilde{B}Z^k + \Delta t\widetilde{A}^{-1}\mathscr{C}^{k+1}.$$

The stability analysis for the midpoint scheme, cf. Remarks 1.8, showed that for all $V \in \ell_{0,\Delta x}^2$ we have

$$\|\widetilde{A}^{-1}\widetilde{B}V\|_{2,\Delta x} \leq \|V\|_{2,\Delta x}.$$

Therefore we get

$$\|Z^{k+1}\|_{2,\Delta x} \leq \|\widetilde{A}^{-1}\widetilde{B}Z^k\|_{2,\Delta x} + \Delta t\,\|\widetilde{A}^{-1}\mathscr{C}^{k+1}\|_{2,\Delta x}$$
$$\leq \|Z^k\|_{2,\Delta x} + \Delta t\,\|\mathscr{C}^{k+1}\|_{2,\Delta x}$$
$$\leq \|Z^k\|_{2,\Delta x} + c\Delta t\,(\Delta t^2 + \Delta x^2)\|u\|_{C^4([0,T]\times[0,1])},$$

```
function theta_method
T = 1;
theta = .5;
J = 20; K = 10;
Delta_x = 1/J; Delta_t = T/K;
lambda = Delta_t/Delta_x^2;
U = zeros(K+1,J+1);
e = ones(J-1,1);
X = spdiags([-e,2*e,-e],[-1,0,1],J-1,J-1);
A = speye(J-1)+theta*lambda*X;
B = speye(J-1)-(1-theta)*lambda*X;
for j = 0:J
    U(1,j+1) = u_0(j*Delta_x);
end
plot(Delta_x*(0:J),U(1,:)); axis([0,1,0,1]); pause
for k = 0:K-1
    U(k+2,1) = 0;
    x = A\(B*U(k+1,2:J)');
    U(k+2,2:J) = x';
    U(k+2,J+1) = 0;
    plot(Delta_x*(0:J),U(k+2,:)); axis([0,1,0,1]); pause(.05)
end
mesh(Delta_x*(0:J),Delta_t*(0:K),U);

function val = u_0(x)
val = sin(pi*x);
```

Fig. 1.15 Midpoint scheme which reduces to the explicit and implicit Euler schemes for $\theta = 0$ and $\theta = 1$, respectively, and realizes the Crank–Nicolson scheme for $\theta = 1/2$

where we used that $\|\widetilde{A}^{-1}\|_{2,\Delta x} \leq 1$ which follows from the fact that the eigenvalues of $\widetilde{A}$ are given by $\lambda_0 = 1$, $\lambda_\ell = 1 + \lambda\big(1 - \cos(\ell\pi/J)\big)$, $\ell = 1, 2, \ldots, J$, and $\lambda_{J+1} = 1$. An inductive argument with $Z^0 = 0$ implies the error estimate.   □

Remark 1.9 In general, the Crank–Nicolson scheme does not satisfy a maximum principle. In particular, in general we do not have that

$$\sup_{j=0,\ldots,J} |U_j^{k+1}| \leq \sup_{j=0,\ldots,J} |U_j^k|$$

for $k = 0, 1, \ldots, K - 1$.

A MATLAB implementation of the midpoint scheme is shown in Fig. 1.15. It is essential that the generation of the matrices A and B be as sparse matrices.

1.2.8 Source Terms and Boundary Conditions

The finite difference schemes discussed in the previous sections are easily modified to incorporate possible source terms in the partial differential equation, i.e., to approximately solve the equation

$$\partial_t u(t, x) - \partial_x^2 u(t, x) = f(t, x)$$

together with initial and boundary conditions. The θ-method is then based on the discretization

$$\partial_t^- U_j^{k+1} - \partial_x^+ \partial_x^- \left(\theta U_j^{k+1} + (1 - \theta)U_j^k\right) = f(t_{k+\theta}, x_j).$$

The incorporation of inhomogeneous Dirichlet conditions, i.e., boundary conditions of the form

$$u(t, 0) = u_{\text{D},\ell}(t), \quad u(t, 1) = u_{\text{D},r}(t)$$

with given functions $u_{\text{D},\ell}$, $u_{\text{D},r} : [0, T] \to \mathbb{R}$, can be reduced to homogeneous Dirichlet conditions by modifying the right-hand side. For this, one defines an extension of the boundary data, e.g.,

$$\tilde{u}_{\text{D}}(t, x) = (1 - x)u_{\text{D},\ell}(t) + xu_{\text{D},r}(t),$$

and decomposes the unknown solution u as

$$u = \widehat{u} + \tilde{u}_{\text{D}}.$$

The function $\widehat{u}$ then satisfies the homogeneous Dirichlet boundary conditions $\widehat{u}(t, 0) = \widehat{u}(t, 1) = 0$, the partial differential equation

$$\partial_t \widehat{u} - \partial_x^2 \widehat{u} = f - \partial_t \tilde{u}_{\text{D}} + \partial_x^2 \tilde{u}_{\text{D}},$$

where in fact $\partial_x^2 \tilde{u}_{\text{D}} = 0$, and the initial condition $\widehat{u}(0, x) = u_0(x) - \tilde{u}_{\text{D}}(0, x)$. The incorporation of Neumann boundary conditions, i.e., boundary conditions of the form

$$\partial_x u(t, 0) = g_\ell(t), \quad \partial_x u(t, 1) = g_r(t)$$

with given functions g_ℓ, $g_r : [0, T] \to \mathbb{R}$ requires a careful treatment since a derivative has to be approximated. The approximation $\partial_x u(t_{k+1}, 0) \approx \partial_x^+ U_0^{k+1}$ introduces a consistency error $\mathcal{O}(\Delta x)$ and could contaminate the spatial discretization error $\mathcal{O}(\Delta x^2)$. In order to maintain the quadratic convergence behavior with respect to Δx, the *ghost points*

$$x_{-1} = -\Delta x, \quad x_{J+1} = 1 + \Delta x$$

are introduced and the boundary conditions are approximated with central difference quotients, i.e.,

$$\widehat{\partial}_x U_0^{k+1} = \frac{U_1^{k+1} - U_{-1}^{k+1}}{2\Delta x} = g_\ell(t_{k+1})$$

and analogously $\widehat{\partial}_x U_J^{k+1} = g_r(t_{k+1})$. Since this introduces two additional unknowns at every time step, two additional equations are required. These are obtained by considering the discretized partial differential equations also at the boundary nodes x_0 and x_J. The θ-method in the case of Neumann boundary conditions then reads as follows:

$$\begin{cases} \partial_t^- U_j^{k+1} = f(t_{k+\theta}, x_j) + \partial_x^+ \partial_x^- (\theta U_j^{k+1} + (1-\theta)U_j^k), & j = 0, 1, \ldots, J, \\ & k = 0, 1, \ldots, K-1, \\ \widehat{\partial}_x U_0^{k+1} = g_\ell(t_{k+1}), \quad \widehat{\partial}_x U_J^{k+1} = g_r(t_{k+1}), & k = 0, 1, \ldots, K-1, \\ U_j^0 = u_0(x_j), & j = 0, 1, \ldots, J. \end{cases}$$

The approximations at the ghost points are initialized with the equations

$$\widehat{\partial}_x U_0^0 = g_\ell(0), \quad \widehat{\partial}_x U_J^0 = g_r(0).$$

This scheme defines numerical approximations that approximate the exact solution with the orders proved in the previous sections.

1.3   Wave Equation

1.3.1   Mathematical Model

We consider a flexible elastic string of finite length, which undergoes small transverse vibrations. We assume that the movement of the string is confined to a plane and let $u(t, x)$ denote the displacement from the equilibrium position at time t and position x, cf. Fig. 1.16.

Imagining the string as a chain of elastic springs suggests assuming that the restoring force or tension F acting on the string due to a displacement be directed

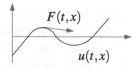

Fig. 1.16   Flexible elastic string with tangential tension vector

tangentially along the string. We let $\sigma(t, x)$ denote the length of the vector $F(t, x)$. Assuming that $|\partial_x u| \ll 1$, we have

$$F(t, x) = \frac{\sigma(t, x)}{\left(1 + \partial_x u(t, x)^2\right)^{1/2}} \begin{bmatrix} 1 \\ \partial_x u(t, x) \end{bmatrix} \approx \sigma(t, x) \begin{bmatrix} 1 \\ \partial_x u(t, x) \end{bmatrix}.$$

The net force acting on a segment $[x_1, x_2]$ of the string is given by the difference of the forces at the endpoints. Newton's second law $F = ma$ for a segment $[x_1, x_2]$ thus reads

$$F(t, x_2) - F(t, x_1) = \int_{x_1}^{x_2} \varrho \begin{bmatrix} 0 \\ \partial_t^2 u(t, x) \end{bmatrix} dx,$$

where ϱ denotes the constant density, i.e., mass per unit length, of the string, and where we used that the motion is only transversal. Equivalently, we have

$$\sigma(t, x_2) - \sigma(t, x_1) = 0,$$

$$\sigma(t, x_2) \partial_x u(t, x_2) - \sigma(t, x_1) \partial_x u(t, x_1) = \int_{x_1}^{x_2} \varrho \partial_t^2 u(t, x) \, dx.$$

The first equation implies that σ is constant and the second equation becomes

$$\sigma \big(\partial_x u(t, x_2) - \partial_x u(t, x_1) \big) = \int_{x_1}^{x_2} \varrho \partial_t^2 u(t, x) \, dx.$$

Differentiating this identity with respect to x_2 results in

$$\sigma \partial_x^2 u(t, x_2) = \varrho \partial_t^2 u(t, x_2)$$

which is called the *wave equation*. We have to complement this equation by boundary and initial conditions. If the string occupies the interval $[\alpha, \beta]$ in its reference configuration and is fixed at both ends, then we impose the Dirichlet boundary conditions

$$u(t, \alpha) = u(t, \beta) = 0$$

for all $t \in [0, T]$. It is clear that it is impossible to predict the motion of a string only from its displacement at some time $t = 0$. In addition we also have to know the initial velocity. If this information is given by functions u_0 and v_0, then we impose the initial conditions

$$u(0, x) = u_0(x), \quad \partial_t u(0, x) = v_0(x)$$

for all $x \in [\alpha, \beta]$. These complete the initial boundary value problem for the description of a vibrating string.

Remarks 1.10

(i) Instead of being fixed, the displacement at an end, e.g., at $x = \alpha$, could be prescribed by some function $u_\ell(t)$. In this case we impose the inhomogeneous Dirichlet condition $u(t, \alpha) = u_\ell(t)$.

(ii) If an end of the string is free to move, then there is no tension, e.g., $F(t, \alpha) = 0$, and we have to impose the Neumann boundary condition

$$\partial_x u(t, \alpha) = 0$$

at that end instead of the Dirichlet boundary condition $u(t, \alpha) = u_\ell(t)$.

1.3.2 Explicit Solution

The factorization

$$\partial_t^2 u - c^2 \partial_x^2 u = \left(\frac{\partial}{\partial t} - c \frac{\partial}{\partial x} \right) \left(\frac{\partial}{\partial t} + c \frac{\partial}{\partial x} \right) u$$

and an appropriate change of coordinates imply that solutions of the wave equation $\partial_t^2 u - c^2 \partial_x^2 u = 0$ are given by

$$u(t, x) = f(x + ct) + g(x - ct),$$

with functions $f, g \in C^2(\mathbb{R})$. Incorporating the initial conditions $u(0, x) = u_0(x)$ and $\partial_t u(0, x) = v_0(x)$, we find that the wave equation on the whole real line, i.e., in the case of an infinitely long string, is solved by the function

$$u(t, x) = \frac{1}{2} \left[u_0(x + ct) + u_0(x - ct) \right] + \frac{1}{2c} \int_{x-ct}^{x+ct} v_0(s) \, ds.$$

This identity is known as *d'Alembert's formula*.

Example 1.4 Let $c = 1$, $v_0(x) = 0$, and

$$u_0(x) = \begin{cases} 1 - |x| & \text{for } |x| \leq 1, \\ 0 & \text{for } |x| \geq 1. \end{cases}$$

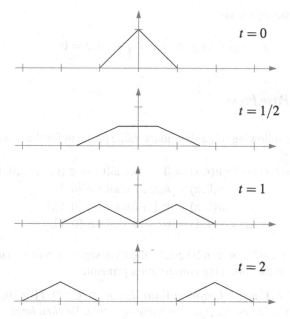

Fig. 1.17 Solution of the wave equation on the real line at times $t = 0, 1/2, 1, 2$

For these initial data d'Alembert's formula provides for $t = 1/2$ the solution

$$u(1/2, x) = \frac{1}{2}\left(u_0(x + 1/2) + u_0(x - 1/2)\right)$$

$$= \begin{cases} 0 & \text{for } |x| \geq 3/2, \\ 3/4 + |x|/2 & \text{for } 1/2 \leq |x| \leq 3/2, \\ 1/2 & \text{for } |x| \leq 1/2. \end{cases}$$

Similar expressions are found for $t = 1/2$, $t = 1$, and $t = 2$. The solution is for $t = 0, 1/2, 1, 2$ depicted in Fig. 1.17. We see that it splits into two similar parts travelling with speed $c = 1$ symmetrically to the left and right.

The solution of the wave equation given by d'Alembert's formula reveals a close connection to the transport equation. In fact, one can write the wave equation as a system of transport equations.

Example 1.5 Assume that $u \in C^2([0, T] \times [0, 1])$ solves the wave equation $\partial_t^2 u - c^2 \partial_x^2 u = 0$. Then the functions $p, q \in C^1([0, T] \times [0, 1])$ defined by

$$\begin{bmatrix} p \\ q \end{bmatrix} = \frac{1}{\sqrt{2}} \begin{bmatrix} 1 & 1 \\ 1 & -1 \end{bmatrix} \begin{bmatrix} c^{-1}\partial_t u \\ \partial_x u \end{bmatrix}$$

solve the transport equations

$$\partial_t p - c\partial_x p = 0, \quad \partial_t q + c\partial_x q = 0.$$

1.3.3 Well-Posedness

We consider the following initial boundary value problem for the wave equation:

$$\begin{cases} \partial_t^2 u(t,x) - c^2\partial_x^2 u(t,x) = 0 & \text{for all}\, (t,x) \in (0,T] \times (0,1), \\ u(0,x) = u_0(x) \,\text{for all}\, x \in [0,1], \\ \partial_t u(0,x) = v_0(x) \,\text{for all}\, x \in [0,1], \\ u(t,0) = u(t,1) = 0 & \text{for all}\, t \in (0,T]. \end{cases}$$

The existence of a solution can be established via separation of variables. Uniqueness follows from the following conservation principle.

Proposition 1.12 (Energy Conservation) *Let* $u \in C^2([0,T] \times [0,1])$ *solve the initial boundary value problem for the wave equation. We then have*

$$\frac{1}{2}\int_0^1 (\partial_t u(t,x))^2 + c^2(\partial_x u(t,x))^2\, dx = \frac{1}{2}\int_0^1 (v_0(x))^2 + c^2(\partial_x u_0(x))^2\, dx$$

for all $t \in [0,T]$.

Proof We multiply the partial differential equation by $\partial_t u$, integrate the resulting identity over $x \in [0,1]$, and use integration-by-parts to verify that

$$0 = \int_0^1 \partial_t^2 u\, \partial_t u - c^2\partial_x^2 u\, \partial_t u\, dx = \int_0^1 \partial_t^2 u\, \partial_t u + c^2\partial_x u\, \partial_t\partial_x u\, dx,$$

where we incorporated the boundary conditions to eliminate boundary terms. Since

$$\partial_t(\partial_t u)^2 = 2(\partial_t^2 u)(\partial_t u), \quad \partial_t(\partial_x u)^2 = 2(\partial_t\partial_x u)(\partial_x u),$$

we can rewrite the integral identity as

$$0 = \frac{d}{dt}\frac{1}{2}\int_0^1 (\partial_t u)^2 + c^2(\partial_x u)^2\, dx.$$

The value of the integral is thus independent of $t \in [0,T]$ and this proves the proposition. □

Remark 1.11 The proposition states that the sum of total kinetic and total elastic energy is constant as a function of $t \in [0,T]$.

The estimate of the proposition implies the uniqueness of solutions.

Corollary 1.2 *The initial boundary value problem for the wave equation admits a unique solution.*

Proof Exercise.   □

From practical experience and d'Alembert's formula, we deduce that we cannot expect a maximum principle for the wave equation.

Example 1.6 For $u_0(x) = 0$ and $v_0(x) = \cos(x)$ we obtain

$$u(t, x) = \frac{1}{2c}[\sin(x + ct) - \sin(x - ct)] = \frac{1}{c}\sin(ct)\cos(x)$$

as a solution of the wave equation on the entire real line.

1.3.4 Explicit Scheme

We replace the second-order partial derivatives in the wave equation by second-order difference quotients, i.e.,

$$\partial_t^+ \partial_t^- U_j^k = c^2 \partial_x^+ \partial_x^- U_j^k,$$

or, with $\mu = c\Delta t/\Delta x$,

$$U_j^{k+1} - 2U_j^k + U_j^{k-1} = \mu^2(U_{j+1}^k - 2U_j^k + U_{j-1}^k).$$

This identity for $j = 1, 2, \ldots, J - 1$ and the boundary conditions $U_0^{k+1} = U_J^{k+1} = 0$ allow us to compute $(U_j^{k+1})_{j=0,\ldots,J}$ if the vectors $(U_j^k)_{j=0,\ldots,J}$ and $(U_j^{k-1})_{j=0,\ldots,J}$ are known, i.e., we need the approximations from two previous time-steps. We set $U_j^0 = u_0(x_j)$ for $j = 0, 1, \ldots, J$. We could use the approximation

$$\partial_t^+ U_j^0 \approx \partial_t u(0, x_j) = v_0(x_j)$$

to define

$$U_j^1 = U_j^0 + \Delta t\, v_0(x_j).$$

This, however, introduces an approximation error $\mathscr{O}(\Delta t)$ that does not match the consistency error $\mathscr{O}(\Delta t^2)$ of the discretization of the differential equation. Similar to the introduction of ghost points for the discretization of Neumann boundary conditions in the heat equation, we introduce the auxiliary time-step $t_{-1} = -\Delta t$ and use an approximation with a central difference quotient, i.e.,

$$\widehat{\partial}_t U_j^0 \approx \partial_t u(0, x_j) = v_0(x_j).$$

We use the resulting identity

$$U_j^{-1} = U_j^1 - 2\Delta t\, v_0(x_j)$$

in the discretized partial differential equation for $k = 0$ and obtain

$$U_j^1 = 2(1 - \mu^2)U_j^0 + \mu^2(U_{j+1}^0 + U_{j-1}^0) - U_j^{-1}$$
$$= 2(1 - \mu^2)U_j^0 + \mu^2(U_{j+1}^0 + U_{j-1}^0) + 2\Delta t\, v_0(x_j) - U_j^1$$

for $j = 1, 2, \ldots, J - 1$. This together with the boundary conditions $U_0^1 = U_J^1 = 0$ defines the approximation $(U_j^1)_{j=0,\ldots,J}$. The complete difference scheme is summarized as follows:

$$\begin{cases} U_j^{k+1} = 2(1 - \mu^2)U_j^k - U_j^{k-1} + \mu^2(U_{j+1}^k + U_{j-1}^k), & j = 1, \ldots, J-1, \\ & k = 1, \ldots, K-1, \\ U_0^{k+1} = U_J^{k+1} = 0, & k = 1, \ldots, K-1, \\ U_j^0 = u_0(x_j), & j = 0, \ldots, J, \\ U_j^1 = (1 - \mu^2)U_j^0 + \mu^2(U_{j+1}^0 + U_{j-1}^0)/2 + \Delta t\, v_0(x_j), & j = 1, \ldots, J-1. \end{cases}$$

An implementation of the scheme is displayed in Fig. 1.18. A finite difference approximation obtained with the code is shown in Fig. 1.19.

In order to determine the stability of the difference scheme we write the solution vectors $U^k = (U_j^k)_{j=0,\ldots,J}$ of the difference scheme in terms of the eigenvectors $(\varphi_p : p = 1, \ldots, J - 1)$ for $-\partial_x^+\partial_x^-$, cf. Lemma 1.1, i.e.,

$$U^k = \sum_{p=1}^{J-1} \xi_p^k \varphi_p$$

with the coefficients $\xi_p^k = (\varphi_p, U^k)_{\Delta x}$ for $p = 1, 2, \ldots, J - 1$ and $k = 0, 1, \ldots, K$. The identity

$$\partial_t^+\partial_t^- U^k = c^2\partial_x^+\partial_x^- U^k$$

then becomes

$$\partial_t^+\partial_t^- \sum_{p=1}^{J-1} \xi_p^k \varphi_p = c^2 \sum_{p=1}^{J-1} \xi_p^k(-\sigma_p)\varphi_p.$$

Since the vectors $(\varphi_p)_{p=1,\ldots,J-1}$ define a basis we deduce that the coefficients of the linear combinations coincide, i.e.,

$$\partial_t^+\partial_t^- \xi_p^k = -c^2\sigma_p\xi_p^k,$$

```
function wave_explicit
T = 2; c = 1;
J = 20; K = 40;
Delta_x = 1/J;
Delta_t = T/K;
mu = c*Delta_t/Delta_x;
U = zeros(K+1,J+1);
for j = 0:J
    U(1,j+1) = u_0(j*Delta_x);
end
plot(Delta_x*(0:J),U(1,:)); axis([0,1,-1,1]); pause
U(2,1) = 0;
for j = 1:J-1
    U(2,j+1) = (1-mu^2)*U(1,j+1)+(mu^2/2)*(U(1,j)+U(1,j+2))...
               +Delta_t*v_0(j*Delta_x);
end
U(2,J+1) = 0;
for k = 1:K-1
    U(k+2,1) = 0;
    for j = 1:J-1
        U(k+2,j+1) = 2*(1-mu^2)*U(k+1,j+1)-U(k,j+1)...
                     +mu^2*(U(k+1,j)+U(k+1,j+2));
    end
    U(k+2,J+1) = 0;
    plot(Delta_x*(0:J),U(k+2,:)); axis([0 1 -1 1]); pause(.1)
end
mesh(Delta_x*(0:J),Delta_t*(0:K),U);

function val = u_0(x)
% val = 2*max(min(x-1/4,3/4-x),0);
val = sin(pi*x);

function val = v_0(x)
val = 0;
```

Fig. 1.18 Explicit finite difference scheme for the wave equation

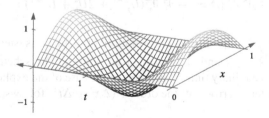

Fig. 1.19 Approximation of the wave equation with an explicit scheme

for $p = 1, 2, \ldots, J - 1$, which is equivalent to

$$
\begin{bmatrix} \xi_p^k \\ \xi_p^{k+1} \end{bmatrix} = A_p \begin{bmatrix} \xi_p^{k-1} \\ \xi_p^k \end{bmatrix} = \begin{bmatrix} 0 & 1 \\ -1 & 2(1 - c^2 \Delta t^2 \sigma_p/2) \end{bmatrix} \begin{bmatrix} \xi_p^{k-1} \\ \xi_p^k \end{bmatrix}.
$$

The sequences $(\xi_p^k)_{k=0,1,...}$ remain bounded for every $p = 1, 2, \ldots, J-1$ if the complex eigenvalues of the matrices A_p are strictly bounded by one. The eigenvalues are the roots of the characteristic polynomials

$$z^2 - 2\gamma_p z + 1 = 0$$

with $\gamma_p = 1 - c^2 \Delta t^2 \sigma_p / 2$, i.e., $z_{1,2} = \gamma_p \pm (\gamma_p^2 - 1)^{1/2}$. A sufficient condition for $|z_{1,2}| < 1$ is that $|\gamma_p| < 1$. Recalling that $\sigma_p = (2/\Delta x^2)(1 - \cos(\pi p \Delta x))$, we find that

$$\gamma_p = 1 - \mu^2 \big(1 - \cos(\pi p \Delta x)\big).$$

Since $0 < \cos(\pi p \Delta x) < 1$ for $p = 1, 2, \ldots, J-1$ and $\Delta x = 1/J$, a sufficient condition for $|z_{1,2}| < 1$ is that $\mu \leq 1$. This implies the following proposition.

Proposition 1.13 (Stability) *If $\mu \leq 1$, then the solution of the explicit finite difference scheme for the wave equation remains bounded.*

Remark 1.12 The condition of the proposition coincides with the stability criterion for the transport equation and cannot be improved in general.

1.3.5 Implicit Scheme

In view of its connection to the transport equation, the stability condition $c\Delta t \leq \Delta x$ appears natural for the wave equation. Following [6] we will see that a Crank–Nicolson type discretization leads to an implicit scheme that is unconditionally stable and has the same consistency error as the explicit scheme. In particular, we consider the discretization scheme

$$\partial_t^+ \partial_t^- U_j^k = \frac{c^2}{4} \partial_x^+ \partial_x^- (U_j^{k+1} + 2U_j^k + U_j^{k-1})$$

which defines $(U_j^{k+1})_{j=1,\ldots,J-1}$ as the solution of a linear system of equations provided that $(U_j^k)_{j=0,\ldots,J}$ and $(U_j^{k-1})_{j=0,\ldots,J}$ are known. The scheme is initialized with the help of an auxiliary time level t_{-1} as in the case of the explicit scheme. The fact that its consistency error is of the order $\mathcal{O}(\Delta x^2 + \Delta t^2)$ follows from the Taylor expansions

$$\partial_x^2 u(t_{k+1}, x_j) = \partial_x^2 u(t_k, x_j) + \Delta t \, \partial_x^2 \partial_t u(t_k, x_j) + \mathcal{O}(\Delta t^2),$$

$$\partial_x^2 u(t_{k-1}, x_j) = \partial_x^2 u(t_k, x_j) - \Delta t \, \partial_x^2 \partial_t u(t_k, x_j) + \mathcal{O}(\Delta t^2)$$

which imply that

$$\partial_x^2 u(t_{k+1}, x_j) + \partial_x^2 u(t_{k-1}, x_j) = 2\partial_x^2 u(t_k, x_j) + \mathcal{O}(\Delta t^2).$$

The stability proof mimics the derivation of the energy conservation principle and is based on a discrete version of the integration-by-parts formula.

Proposition 1.14 (Summation-by-Parts) *For sequences* $(V_j)_{j=0,\ldots,J}$ *and* $(W_j)_{j=0,\ldots,J}$ *and* $\Delta x > 0$, *we have*

$$\Delta x \sum_{j=0}^{J-1} (\partial_x^+ W_j) V_j + \Delta x \sum_{j=1}^{J} W_j (\partial_x^- V_j) = W_J V_J - W_0 V_0,$$

Proof Exercise. □

This leads to the following stability result.

Theorem 1.1 (Discrete Energy Conservation) *Let* (U_j^k) *be the solution of the implicit difference scheme with boundary conditions* $U_0^k = U_J^k = 0$ *for* $k = 0, 1, \ldots, K$. *Then the expression*

$$\Gamma^k = \frac{\Delta x}{2} \sum_{j=1}^{J-1} |\partial_t^+ U_j^k|^2 + c^2 \frac{\Delta x}{2} \sum_{j=1}^{J} |\partial_x^- U_j^{k+1/2}|^2,$$

where $U_j^{k+1/2} = (U_j^{k+1} + U_j^k)/2$, *is independent of* $k = 0, 1, \ldots, K - 1$.

Proof We multiply the difference scheme by $\widehat{\partial}_t U_j^k$ and use the elementary identities

$$LHS_j^k = (\partial_t^+ \partial_t^- U_j^k)(\widehat{\partial}_t U_j^k) = \frac{1}{2} \partial_t^- (\partial_t^+ U_j^k)^2$$

and

$$RHS_j^k = \frac{c^2}{4} \partial_x^+ \partial_x^- (U_j^{k+1} + 2U_j^k + U_j^{k-1}) \widehat{\partial}_t U_j^k$$

$$= \frac{c^2}{2\Delta t} (\partial_x^+ \partial_x^- (U_j^{k+1/2} + U_j^{k-1/2}))(U_j^{k+1/2} - U_j^{k-1/2}).$$

With the abbreviations $W_j^k = \partial_x^- (U_j^{k+1/2} + U_j^{k-1/2})$ and $V_j^k = U_j^{k+1/2} - U_j^{k-1/2}$, we have, noting $V_0^k = V_J^k = 0$ and using the summation-by-parts formula, that

$$\Delta x \sum_{j=1}^{J-1} RHS_j^k = \Delta x \frac{c^2}{2\Delta t} \sum_{j=0}^{J-1} (\partial_x^+ W_j^k) V_j^k$$

$$= -\Delta x \frac{c^2}{2\Delta t} \sum_{j=1}^{J} W_j^k (\partial_x^- V_j^k)$$

$$= -\Delta x \frac{c^2}{2\Delta t} \sum_{j=1}^{J} \partial_x^- (U_j^{k+1/2} + U_j^{k-1/2}) \partial_x^- (U_j^{k+1/2} - U_j^{k-1/2})$$

$$= -\Delta x \frac{c^2}{2\Delta t} \sum_{j=1}^{J} ((\partial_x^- U_j^{k+1/2})^2 - (\partial_x^- U_j^{k-1/2})^2)$$

$$= -\Delta x \frac{c^2}{2} \partial_t^- \sum_{j=1}^{J} (\partial_x^- U_j^{k+1/2})^2.$$

Since $LHS_j^k = RHS_j^k$, after summation over $j = 0, 1, \ldots, J-1$, we get

$$-\Delta x \frac{c^2}{2} \partial_t^- \sum_{j=1}^{J} (\partial_x^- U_j^{k+1/2})^2 = \frac{\Delta x}{2} \partial_t^- \sum_{j=0}^{J-1} (\partial_t^+ U_j^k)^2.$$

Equivalently, we have $\partial_t^- \Gamma^k = 0$ which implies the proposition. $\square$

1.4   Poisson Equation

1.4.1   Mathematical Model

We consider a heat conducting object that occupies the domain $\Omega \subset \mathbb{R}^d$, $d = 2, 3$, and assume that at any point $x \in \Omega$ and any time $t \in [0, T]$, we can measure a temperature $\theta(t, x)$. The heat $H(t)$ contained in a subdomain $\omega \subset \Omega$ is given by

$$H_\omega(t) = \int_\omega c\varrho\theta(t, x)\, dx,$$

where c is the specific heat capacity and ϱ the density of the medium. The specific heat capacity relates the temperature to the internal energy which is a conserved quantity. The heat flux $q(t, x)$ specifies how much internal energy passes the point x at time t. Possible sources or sinks of heat are described by a function $f(t, x)$. For the domain ω, the quantity

$$-\int_{\partial\omega} q(t, s) \cdot n(s)\, ds + \int_\omega f(t, x)\, dx$$

with the outer unit normal $n : \partial\omega \to \mathbb{R}^d$ thus measures how much heat enters or escapes from ω at time t, cf. Fig. 1.20. Therefore we have

$$\frac{d}{dt} H_\omega(t) = -\int_{\partial\omega} q(t, s) \cdot n(s)\, ds + \int_\omega f(t, x)\, dx.$$

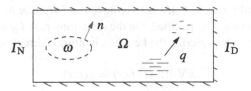

Fig. 1.20 Heat balance for a subdomain $\omega \subset \Omega$ in a heat conducting object $\Omega \subset \mathbb{R}^d$; heat flows from regions of higher temperatures into those of lower temperatures

Fourier's law of heat conductivity states that the heat flux q is proportional to the negative temperature gradient, i.e.,

$$q(t,x) = -\kappa \nabla \theta(t,x)$$

with the heat conductivity κ. This relation states that heat flows from regions of higher temperatures to those of lower temperatures. We thus have

$$\int_\omega c\varrho \partial_t \theta(t,x)\,dx = \int_{\partial\omega} \kappa \nabla \theta(t,s) \cdot n(s)\,ds + \int_\omega f(t,x)\,dx.$$

Gauss's theorem guarantees that

$$\int_{\partial\omega} \big(\kappa \nabla \theta(t,s)\big) \cdot n(s)\,ds = \int_\omega \operatorname{div}\big(\kappa \nabla \theta(t,x)\big)\,dx$$

which yields that

$$\int_\omega \big(c\varrho \partial_t \theta(t,x) - \operatorname{div}\big(\kappa \nabla \theta(t,x)\big) - f(t,x)\big)\,dx = 0.$$

Assuming that the integrand is continuous, since the identity holds for arbitrary subdomains $\omega \subset \Omega$ with C^1 boundary, we have that

$$c\varrho \partial_t \theta(t,x) = \operatorname{div}\big(\kappa \nabla \theta(t,x)\big) + f(t,x)$$

holds for all $x \in \Omega$ and $t \in [0,T]$. For constant material parameters c, ϱ, κ, and noting that $\operatorname{div} \nabla \theta = \Delta \theta = \partial_{x_1}^2 \theta + \cdots + \partial_{x_d}^2 \theta$, the equation simplifies to

$$c\varrho \partial_t \theta(t,x) = \kappa \Delta \theta(t,x) + f(t,x).$$

We impose the Dirichlet boundary conditions

$$\theta(t,x) = \theta_D(t,x)$$

for all $t \in [0, T]$ and points $x \in \partial\Omega$ belonging to the *Dirichlet boundary* $\Gamma_D \subset \partial\Omega$ on which the temperature is specified. On the remaining part $\Gamma_N = \partial\Omega \setminus \Gamma_D$, called *Neumann boundary*, we prescribe the heat flux in normal direction, i.e.,

$$\kappa \nabla \theta(t, x) \cdot n(x) = g(t, x)$$

for all $t \in [0, T]$ and $x \in \Gamma_N$. Initial conditions $\theta(0, x) = \theta_0(x)$ for all $x \in \Omega$ complete the initial boundary value problem that describes the heat distribution in Ω. Assuming that the given functions f, g, and θ_D are independent of $t \in [0, T]$, we expect that for sufficiently large times, the temperature distribution becomes *stationary*, i.e., that

$$\partial_t \theta(t, x) \approx 0$$

for all $x \in \Omega$. In this case the initial boundary value problem simplifies to the *boundary value problem*

$$-\kappa \Delta\theta = f \text{ in } \Omega, \quad \theta = \theta_D \text{ on } \Gamma_D, \quad \kappa \nabla \theta \cdot n = g \text{ on } \Gamma_N.$$

This stationary problem is called the *Poisson problem* or *Poisson equation*. If f vanishes, then the problem is referred to as the *Laplace equation* and solutions are *harmonic functions*. The operator Δ is called the *Laplace operator*.

1.4.2 Poisson Problem

Given $\Omega \subset \mathbb{R}^d, d = 2, 3$, and functions $f : \Omega \to \mathbb{R}, u_D : \Gamma_D \to \mathbb{R}$, and $g : \Gamma_N \to \mathbb{R}$, where Γ_D and Γ_N satisfy $\Gamma_D \cap \Gamma_N = \emptyset$ and $\Gamma_D \cup \Gamma_N = \partial\Omega$, we aim to determine $u \in C^2(\overline{\Omega})$ such that

$$\begin{cases} -\Delta u = f & \text{in } \Omega, \\ u = u_D & \text{on } \Gamma_D, \\ \nabla u \cdot n = g & \text{on } \Gamma_N. \end{cases}$$

For simple domains such as rectangles or paralellepipeds and disks or balls, the existence of a solution can be established with a separation of variables or the introduction of polar coordinates. The following maximum principle for the Laplace equation implies the uniqueness of solutions.

Proposition 1.15 (Maximum Principle) *Let Ω be a connected bounded open set in $\mathbb{R}^d$, $d = 2, 3$. Suppose that $u \in C^2(\overline{\Omega})$ satisfies $-\Delta u = 0$ in Ω. Then the maximum and the minimum value of u are attained on $\partial\Omega$.*

Proof Let $\varepsilon > 0$ and set $v(x) = u(x) + \varepsilon|x|^2$. Then

$$\Delta v(x) = \Delta u(x) + \varepsilon \Delta(x_1^2 + \cdots + x_d^2) = 0 + 2d\varepsilon > 0.$$

But

$$\Delta v(x_M) = \partial_{x_1}^2 v(x_M) + \cdots + \partial_{x_d}^2 v(x_M) \leq 0$$

at an interior point $x_M \in \Omega$ corresponding to a maximum. Therefore, v has no interior maximum in Ω. Since the continuous function v has a maximum in $\overline{\Omega}$, it has to be on $\partial\Omega$, say at $x_M \in \partial\Omega$. Then for all $x \in \Omega$, we have

$$u(x) \leq v(x) \leq v(x_M) = u(x_M) + \varepsilon|x_M|^2 \leq \max_{y \in \partial\Omega} u(y) + \varepsilon R^2,$$

where $R = \max_{x \in \overline{\Omega}} |x|$. Since the estimate holds for all $\varepsilon > 0$, we deduce that

$$u(x) \leq \max_{y \in \partial\Omega} u(y).$$

The existence of a minimum point on $\partial\Omega$ is similarly demonstrated. $\square$

The uniqueness of solutions for the Poisson problem is an immediate consequence.

Corollary 1.3 (Uniqueness) *The Poisson problem with $\Gamma_D = \partial\Omega$ admits at most one solution.*

Proof If $u_1, u_2 \in C^2(\overline{\Omega})$ solve the Poisson problem with $\Gamma_D = \partial\Omega$, then their difference $u = u_1 - u_2$ satisfies $-\Delta u = 0$ in Ω and $u = 0$ on $\partial\Omega$. The maximum principle implies $u = 0$ in Ω, i.e., $u_1 = u_2$. $\square$

An important tool for analyzing the Poisson problem is Gauss's theorem which states that

$$\int_U \operatorname{div} F(x) \, dx = \int_{\partial U} F(s) \cdot n(s) \, ds,$$

for C^1 vector fields $F : U \to \mathbb{R}^d$ and domains $U \subset \mathbb{R}^d$ with C^1 boundary. Noting that $\Delta = \operatorname{div} \nabla$, an immediate consequence is *Green's identity*

$$\int_{\partial U} v \nabla u \cdot n \, ds = \int_U \nabla u \cdot \nabla v \, dx + \int_U v \Delta u \, dx$$

which holds for $u \in C^2(U)$ and $v \in C^1(U)$. If u is harmonic, i.e., if $-\Delta u = 0$, then the identity has important consequences.

Proposition 1.16 (Mean Value Property) *Suppose that $u \in C^2(\overline{\Omega})$ satisfies $-\Delta u = 0$ in $\Omega \subset \mathbb{R}^d$. For any $x_0 \in \Omega$ and $a > 0$ such that the open ball $B_a(x_0)$*

satisfies $B_a(x_0) \subset \Omega$, we have

$$u(x_0) = \frac{1}{|\partial B_a(x_0)|} \int_{\partial B_a(x_0)} u(s)\, ds,$$

where $|\partial B_a(x_0)|$ is the surface area of $\partial B_a(x_0)$.

Proof Without loss of generality we assume $x_0 = 0$. We apply Green's identity with u and $v = 1$ to verify that

$$\int_{\partial B_a(x_0)} \nabla u \cdot n\, ds = 0.$$

We have $\nabla u \cdot n = \partial_r u$ in polar coordinates (r, θ) or (r, θ, φ). In particular, for $d = 3$, we have

$$\int_0^{2\pi} \int_0^\pi \partial_r u(a, \theta, \varphi) a^2 \sin(\theta)\, d\theta\, d\varphi = 0.$$

We divide by the surface area $|\partial B_a(x_0)| = 4\pi a^2$ and pull the derivative with respect to the radius out of the integral to obtain

$$\frac{d}{dr}\bigg|_{r=a} \frac{1}{4\pi a^2} \int_0^{2\pi} \int_0^\pi u(r, \theta, \varphi) a^2 \sin(\theta)\, d\theta\, d\varphi = 0.$$

This holds for all $0 < a \le \bar{a}$ and therefore the function

$$r \mapsto \frac{1}{4\pi r^2} \int_0^{2\pi} \int_0^\pi u(r, \theta, \varphi) r^2 \sin(\theta)\, d\theta\, d\varphi = \frac{1}{|\partial B_r(x_0)|} \int_{\partial B_r(x_0)} u(s)\, ds$$

is independent of $0 < r < \bar{a}$. The expression on the right is the average of u on the sphere $\{|x| = r\}$. By continuity of u, we have

$$u(0) = \lim_{r \to 0} \frac{1}{4\pi r^2} \int_0^{2\pi} \int_0^\pi u(r, \theta, \varphi) r^2 \sin(\theta)\, d\theta\, d\varphi,$$

which proves the assertion for $d = 3$. The case $d = 2$ follows analogously.   □

The mean value property implies the strong version of the maximum principle which asserts that a harmonic function attains its extrema only on the boundary unless u is constant. To see this, assume that $u \in C^2(\overline{\Omega})$ is harmonic, i.e., $-\Delta u = 0$ in Ω, and suppose that u attains its maximum M at the interior point $x_M \in \Omega$. For $\bar{a} > 0$ such that $B_{\bar{a}}(x_M) \subset \Omega$, the mean value property for every $0 < a \le \bar{a}$ shows that

$$u(x_M) = \frac{1}{|\partial B_a(x_M)|} \int_{\partial B_a(x_M)} u(s)\, ds \le u(x_M).$$

This can only be true if $u(x) = u(x_M)$ for all $x \in B_{\overline{a}}(x_M)$. By covering Ω with overlapping balls, we find that u is constant.

1.4.3 Finite Difference Scheme

For simplicity we restrict to the case that $\Omega = (0, 1)^2$, $\Gamma_D = \partial\Omega$, and $u_D = 0$. Generalizations will be discussed below. The model problem thus seeks a solution of the following boundary value problem:

$$\begin{cases} -\Delta u = f & \text{in } \Omega = (0, 1)^2, \\ u = 0 & \text{on } \partial\Omega. \end{cases}$$

For an integer $J \geq 1$, we set $\Delta x = 1/J$ and define grid points $x_{j,m} = (j\Delta x, m\Delta x)$ for $0 \leq j, m \leq J$ and replace the partial derivatives by central difference quotients. The discretized Poisson problem thus consists in finding the coefficients of a *grid function* $(U_{j,m} : j, m = 0, \ldots, J) \subset \mathbb{R}$ such that

$$\begin{cases} -\partial_{x_1}^+ \partial_{x_1}^- U_{j,m} - \partial_{x_2}^+ \partial_{x_2}^- U_{j,m} = f(x_{j,m}) & \text{for } 1 \leq j, m \leq J, \\ U_{0,m} = U_{J,m} = U_{j,0} = U_{j,J} = 0 & \text{for } j, m = 0, 1, \ldots, J. \end{cases}$$

In the following we denote by $-\Delta_h$ the discretized negative Laplace operator, i.e.,

$$\begin{aligned} -\Delta_h U_{j,m} &= -\partial_{x_1}^+ \partial_{x_1}^- U_{j,m} - \partial_{x_2}^+ \partial_{x_2}^- U_{j,m} \\ &= -\frac{U_{j+1,m} - 2U_{j,m} + U_{j-1,m}}{\Delta x^2} - \frac{U_{j,m+1} - 2U_{j,m} + U_{j,m-1}}{\Delta x^2} \\ &= -\frac{1}{\Delta x^2} \left(U_{j+1,m} + U_{j,m+1} - 4U_{j,m} + U_{j-1,m} + U_{j,m-1} \right). \end{aligned}$$

Its evaluation at a grid point $x_{j,m}$ involves the grid point $x_{j,m}$ and its four neighbors $x_{j-1,m}$, $x_{j+1,m}$, $x_{j,m-1}$, and $x_{j,m+1}$. The corresponding stencil is referred to as a *5-point-stencil*, cf. Fig. 1.21.

We remark that if $-\Delta_h U_{j,m} = 0$, then we have

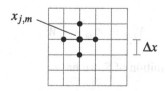

Fig. 1.21 Uniform grid on the domain $[0, 1]^2$ and 5-point-stencil

$$U_{j,m} = \frac{1}{4}\left(U_{j-1,m} + U_{j+1,m} + U_{j,m-1} + U_{j,m+1}\right)$$

which is a discrete version of the mean value property for harmonic functions. It implies the following discrete maximum principle.

Lemma 1.2 (Discrete Maximum Principle) *If $U = (U_{j,m} : j,m = 0,\ldots,J)$ satisfies $-\Delta_h U_{j,m} \le 0$ for all $j,m = 1,2,\ldots,J-1$, then U attains its maximum for $j = 0, j = J, m = 0$ or $m = J$.*

Proof The condition $-\Delta_h U_{j,m} \le 0$ implies that

$$U_{j,m} \le \frac{1}{4}\left(U_{j-1,m} + U_{j+1,m} + U_{j,m-1} + U_{j,m+1}\right)$$

for $1 \le j,m \le J-1$. If $U_{j,m}$ is an interior maximum, then this holds with equality and then the maximum is also attained at the neighboring points, i.e.,

$$U_{j,m} = U_{j-1,m} = U_{j+1,m} = U_{j,m-1} = U_{j,m+1}.$$

Repeating the argument with the neighboring grid points shows that U is constant except for the corner points $x_{0,0}, x_{0,J}, x_{J,0}$ and $x_{J,J}$ which could only increase the maximum on the boundary. □

The discrete maximum principle leads to the following boundedness result; we follow [6].

Lemma 1.3 (Discrete Boundedness) *For all $(Z_{j,m} : j,m = 0,\ldots,J)$ with $Z_{j,m} = 0$ for $j = 0, j = J, m = 0$, or $m = J$, we have*

$$\max_{j,m=0,\ldots,J} |Z_{j,m}| \le \frac{1}{2} \sup_{j,m=1,\ldots,J-1} |\Delta_h Z_{j,m}|.$$

Proof We abbreviate $S = \max_{j,m=1,\ldots,J-1} |\Delta_h Z_{j,m}|$ and define the grid function

$$W_{j,m} = (j\,\Delta x)^2 + (m\Delta x)^2$$

which coincides at the grid points with the function $w(x_1,x_2) = x_1^2 + x_2^2$. Then $W_{j,m} \ge 0$ for $j,m = 0,1,\ldots,J$ and $\Delta_h W_{j,m} = 4$ for $j,m = 1,2,\ldots,J-1$. We set

$$V_{j,m} = Z_{j,m} + \frac{S}{4}W_{j,m},$$

and note that due to the definition of S, we have

$$-\Delta_h V_{j,m} = -\Delta_h Z_{j,m} - S \le 0.$$

The discrete maximum principle implies that $V_{j,m}$ attains its maximum for $j = 0$, $j = J$, $m = 0$, or $m = J$. For these indices we have $Z_{j,m} = 0$ and $0 \leq W_{j,m} \leq 2$. We thus have $V_{j,m} \leq S/2$ for all $j, m = 0, 1, \ldots, J$, which implies that

$$Z_{j,m} = V_{j,m} - \frac{S}{4} W_{j,m} \leq \frac{S}{2}.$$

Repeating the argument with $V_{j,m} = -Z_{j,m} + (S/4) W_{j,m}$ proves $-Z_{j,m} \leq S/2$ and completes the proof of the lemma. □

The discrete boundedness implies the injectivity of the discretized Laplace operator as a mapping on grid functions with zero boundary values. Since this is an endomorphism we also deduce the surjectivity and hence the existence of a unique solution for the discrete problem. Furthermore, we deduce the following error estimate.

Proposition 1.17 (Error Estimate) *Let $u \in C^2(\overline{\Omega})$ and $U = (U_{j,m} : j, m = 0, \ldots, J)$ be the solutions of the Poisson problem and its discretization, respectively. We then have*

$$\sup_{j,m=0,\ldots,J} |u(x_{j,m}) - U_{j,m}| \leq \frac{\Delta x^2}{24} \big(\|\partial_{x_1}^4 u\|_{C([0,1]^2)} + \|\partial_{x_2}^4 u\|_{C([0,1]^2)} \big).$$

Proof Since $-\Delta u(x_{j,m}) = f(x_{j,m})$ for all $0 \leq j, m \leq J$, the error $Z_{j,m} = u(x_{j,m}) - U_{j,m}$ satisfies

$$
\begin{aligned}
-\Delta_h Z_{j,m} &= -\Delta_h u_{j,m} + \Delta_h U(x_{j,m}) \\
&= f(x_{j,m}) - f(x_{j,m}) + \Delta u(x_{j,m}) - \Delta_h u(x_{j,m}) \\
&= \partial_{x_1}^2 u(x_{j,m}) - \partial_{x_1}^+ \partial_{x_1}^- u(x_{j,m}) + \partial_{x_2}^2 u(x_{j,m}) - \partial_{x_2}^+ \partial_{x_2}^- u(x_{j,m}).
\end{aligned}
$$

From the estimates for difference quotients, we get

$$|-\Delta_h Z_{j,m}| \leq \frac{\Delta x^2}{12} \big(\|\partial_{x_1}^4 u\|_{C([0,1]^2)} + \|\partial_{x_2}^4 u\|_{C([0,1]^2)} \big).$$

Since $Z_{j,m} = 0$ whenever $j = 0$, $j = J$, $m = 0$, or $m = J$, an application of the discrete boundedness lemma yields that

$$\sup_{j,m=0,\ldots,J} |Z_{j,m}| \leq \frac{\Delta x^2}{24} \big(\|\partial_{x_1}^4 u\|_{C([0,1]^2)} + \|\partial_{x_2}^4 u\|_{C([0,1]^2)} \big).$$

This proves the estimate. □

1.4.4 Implementation

The finite difference discretization of the Poisson problem leads to a linear system of equations that requires determining the coefficients $(U_{j,m} : j, m = 0, \ldots, J)$ such that, with $f_{j,m} = f(x_{j,m})$,

$$-(U_{j-1,m} + U_{j,m-1} - 4U_{j,m} + U_{j+1,m} + U_{j,m+1}) = \Delta x^2 f_{j,m}$$

for $j, m = 1, 2, \ldots, J - 1$ and

$$U_{0,m} = U_{J,m} = U_{j,0} = U_{j,J} = 0$$

for $j, m = 0, 1, \ldots, J$. The vanishing coefficients associated with the boundary nodes can be eliminated from the system. To simplify the equations further, we introduce a *lexicographic enumeration* of the interior nodes by identifying

$$(j, m) \equiv j + (m - 1)(J - 1) = \ell$$

for $j, m = 1, 2, \ldots, J - 1$ and $\ell = 1, 2, \ldots, L$ with $L = (J - 1)^2$. With the matrix $X \in \mathbb{R}^{(J-1) \times (J-1)}$ defined by

$$X = \begin{bmatrix} 4 & -1 & & \\ -1 & \ddots & \ddots & \\ & \ddots & \ddots & -1 \\ & & -1 & 4 \end{bmatrix}$$

and the identity matrix $I \in \mathbb{R}^{(J-1) \times (J-1)}$, the reduced system of equations can be rewritten as $AU = b$, i.e.,

$$\begin{bmatrix} X & -I & & \\ -I & \ddots & \ddots & \\ & \ddots & \ddots & -I \\ & & -I & X \end{bmatrix} \begin{bmatrix} U_1 \\ U_2 \\ \vdots \\ U_L \end{bmatrix} = \Delta x^2 \begin{bmatrix} f_1 \\ f_2 \\ \vdots \\ f_L \end{bmatrix}.$$

The regularity of the matrix A follows from the unique existence of the discrete solution U. Alternatively, it follows from its diagonal dominance and irreducibility. Note that every block matrix corresponds to one row in the grid. Figure 1.22 displays a MATLAB implementation of the difference scheme. For this particular linear system of equations, an iterative solution procedure that avoids the generation of the sparse matrix A, e.g., a Gauss–Seidel iteration, would be more efficient. The result of a numerical experiment is shown in Fig. 1.23.

```
function poisson_2d
J = 20;
Delta_x = 1/J;
L = (J-1)^2;
e = ones(J-1,1); E = ones(L,1);
X = spdiags([-e,4*e,-e],[-1,0,1],J-1,J-1);
A = sparse(L,L);
for j = 1:(J-1):L
    A(j:j+J-2,j:j+J-2) = X;
end
A = A+spdiags([-E,-E],[-J+1,J-1],L,L);
for j = 1:J-1
    for m = 1:J-1
        b(j+(m-1)*(J-1),1) = Delta_x^2*f([j,m]*Delta_x);
    end
end
U = A\b;
show(U,J,Delta_x);

function val = f(x)
val = 1;

function show(U,J,Delta_x)
U_mat = zeros(J+1,J+1); U_mat(2:J,2:J) = reshape(U,J-1,J-1)';
mesh(Delta_x*(0:J),Delta_x*(0:J),U_mat); axis([0,1,0,1,0,.1]);
```

Fig. 1.22 Finite difference scheme for the Poisson problem on the square $\Omega = (0,1)^2$ with homogeneous Dirichlet boundary conditions on $\Gamma_D = \partial\Omega$

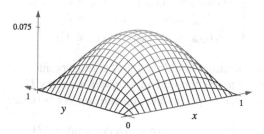

Fig. 1.23 Numerical approximation of a two-dimensional Poisson problem

1.4.5   Boundary Conditions

The numerical treatment of the Poisson problem with general boundary conditions or in a curved domain requires certain modifications of the techniques discussed above. For an open domain $\Omega \subset \mathbb{R}^2$, subsets $\Gamma_D \subset \partial\Omega$ and $\Gamma_N = \partial\Omega \setminus \Gamma_D$, and functions $f \in C(\overline{\Omega})$, $u_D \in C(\Gamma_D)$, and $g \in C(\Gamma_N)$, we consider the following problem:

$$\begin{cases} -\Delta u = f & \text{in } \Omega, \\ u = u_D & \text{on } \Gamma_D, \\ \nabla u \cdot n = g & \text{on } \Gamma_N. \end{cases}$$

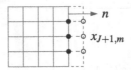

Fig. 1.24 Treatment of Neumann boundary conditions on the side $\{1\} \times (0,1)$ with central difference quotients and ghost points $x_{J+1,m}$

The inhomogeneous Dirichlet boundary condition defined by u_D can be reduced to a homogeneous Dirichlet condition, provided that there exists a function $\tilde{u}_D \in C^2(\overline{\Omega})$ such that $\tilde{u}_D|_{\Gamma_D} = u_D$. For this, the unknown solution $u \in C^2(\overline{\Omega})$ is written as $u = \widehat{u} + \tilde{u}_D$ with an unknown function $\widehat{u} \in C^2(\overline{\Omega})$ that satisfies a modified Poisson problem with the homogeneous Dirichlet boundary condition:

$$\begin{cases} -\Delta\widehat{u} = f + \Delta\tilde{u}_D & \text{in } \Omega, \\ \widehat{u} = 0 & \text{on } \Gamma_D, \\ \nabla\widehat{u} \cdot n = g - \nabla\tilde{u}_D \cdot n & \text{on } \Gamma_N. \end{cases}$$

Neumann boundary conditions are best discretized with a central difference quotient and ghost points. If, e.g., $\Omega = (0,1)^2$ and $\Gamma_N = \{1\} \times (0,1)$, then we have $n = [1,0]^\top$ on Γ_N, and by introducing the ghost points $x_{J+1,m}$, $m = 1, 2, \ldots, J-1$, cf. Fig. 1.24, a discretization of the Poisson problem reads as follows:

$$\begin{cases} -\Delta_h U_{j,m} = f(x_{j,m}) & \text{for } x_{j,m} \in \Omega \cup \Gamma_N, \\ U_{j,m} = 0 & \text{for } x_{j,m} \in \Gamma_D, \\ \widehat{\partial_x U_{J,m}} = g(x_{J,m}) & \text{for } x_{J,m} \in \Gamma_N. \end{cases}$$

A lexicographic enumeration should include the ghost points.

If Dirichlet conditions are imposed on a curved boundary $\Gamma_D = \partial\Omega$, then we use a uniform grid on $\mathbb{R}^2$ that is specified by a parameter $\Delta x > 0$, i.e.,

$$\mathcal{N}_h = \{x_{j,m} = (j\Delta x, m\Delta x), \; (j,m) \in \mathbb{Z}^2\},$$

and define sets of *interior nodes* Ω_h and *boundary nodes* $\Gamma_{D,h}$ by

$$\Omega_h = \{x_{j,m} \in \mathcal{N}_h \cap \Omega : \text{all neighbors of } x_{j,m} \text{ belong to } \overline{\Omega}\},$$

$$\Gamma_{D,h} = \{x_{j,m} \in \mathcal{N}_h \cap \overline{\Omega} : \text{a neighbor of } x_{j,m} \text{ does not belong to } \overline{\Omega}\}.$$

With every boundary node $x_{j,m} \in \Gamma_{D,h}$ we may associate a neighboring node in the set $\{x_{j-1,m}, x_{j+1,m}, x_{j,m-1}, x_{j,m+1}\}$ which does not belong to $\overline{\Omega}$, e.g., $x_{j+1,m}$. We assume that Δx is sufficiently small so that there exists a unique point $\widetilde{x}_{j,m} \in \partial\Omega$ that is on the segment of length Δx connecting $x_{j,m}$ and $x_{j+1,m}$, and that the neighbor in the opposite direction, i.e., $x_{j-1,m}$, belongs to Ω_h, cf. Fig. 1.25. We then impose

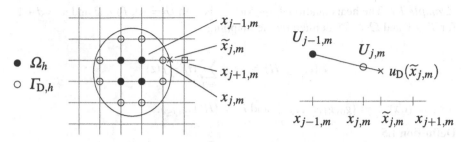

Fig. 1.25 Treatment of Dirichlet boundary conditions on a curved boundary; the coefficient $U_{j,m}$ is obtained by linearly interpolating $U_{j-1,m}$ and $u_D(\tilde{x}_{j,m})$

the condition that the coefficient $U_{j,m}$ associated with the boundary node $x_{j,m}$ is given by the linear interpolation of $U_{j-1,m}$ and $u_D(\tilde{x}_{j,m})$, i.e.,

$$U_{j,m} = \frac{|x_{j,m} - \tilde{x}_{j,m}|}{\Delta x + |x_{j,m} - \tilde{x}_{j,m}|} U_{j-1,m} + \frac{\Delta x}{\Delta x + |x_{j,m} - \tilde{x}_{j,m}|} u_D(\tilde{x}_{j,m}).$$

For all interior nodes we use the discretized partial differential equation, i.e.,

$$-\Delta_h U_{j,m} = f(x_{j,m}).$$

for all $x_{j,m} \in \Omega_h$.

1.5 General Concepts

1.5.1 Abstract Boundary Value Problems

The previous sections showed that differential equations have the potential to accurately describe important physical processes. We aim for a general definition of boundary value problems and their well-posedness.

Definition 1.4 A *partial differential equation* is a mapping

$$F : U \times \mathbb{R} \times \mathbb{R}^n \times \mathbb{R}^{n^2} \times \cdots \times \mathbb{R}^{n^k} \to \mathbb{R}$$

on an open domain $U \subset \mathbb{R}^n$ that defines a relation between the partial derivatives of a function $u \in C^k(U)$ via the equation

$$F\big(z, u(z), Du(z), D^2u(z), \ldots, D^ku(z)\big) = 0$$

for all $z \in U$, abbreviated by $F(u) = 0$. Functions that satisfy this relation are called the *solutions of the partial differential equation*.

Example 1.7 The heat equation $\partial_t u - \Delta u = f$ is with $U = (0, T) \times \Omega$ and $n = d+1$
for $T > 0$ and $\Omega \subset \mathbb{R}^d$ defined by the mapping

$$F(z, s, g, H) = g_0 - \sum_{i=1}^{d} H_{ii} - f(z)$$

for $z = (t, x)$, $g = (g_0, g_1, \ldots, g_d)$, and $H = (H_{ij})_{0 \le i,j \le d}$.

Definition 1.5

(i) The *order* of a partial differential equation is the order of the highest partial
derivative on which F depends.
(ii) A partial differential equation of order k is called *linear*, if there exists a linear
operator $L : C^k(U) \to C(U)$ and a function $f \in C(U)$ such that

$$F\big(z, u(z), Du(z), \ldots, D^k u(z)\big) = Lu(z) - f(z)$$

for all $z \in U$.

Example 1.8 The Poisson equation $-\Delta u = f$ defines a linear partial differential
equation of second order with linear operator $Lu = -\Delta u = \partial_{x_1}^2 u + \ldots \partial_{x_d}^2 u$ called
the *Laplace operator*.

Partial differential equations are completed with boundary or initial conditions.
Initial conditions will be interpreted as boundary conditions in the following
definition.

Definition 1.6 A *boundary value problem* seeks a *solution* $u \in C^k(\overline{U})$ of a partial
differential equation $F(u) = 0$ that satisfies the *boundary condition*

$$G\big(z, u(z), Du(z), \ldots, D^{k-1} u(z)\big) = 0$$

for all $z \in \partial U$ with a mapping $G : \partial U \times \mathbb{R} \times \mathbb{R}^n \times \cdots \times \mathbb{R}^{n^{k-1}}$. It is called *linear* if
the partial differential equation is linear and the boundary condition is linear in the
sense that $G(u) = Mu - \ell$ with an operator $M : C^k(\overline{U}) \to C(\partial U)$ and a function
$\ell \in C(\partial U)$.

Example 1.9 The initial condition $u(0, x) = u_0(x)$ for all $x \in \Omega$ and the Dirichlet
boundary condition $u(t, x) = u_D(x)$ for all $x \in \partial \Omega$ and $t \in (0, T)$ are realized by the
mapping

$$G\big((t, x), u(t, x), Du(t, x)\big) = \begin{cases} u(0, x) - u_0(x) & \text{if } t = 0, \\ u(t, x) - u_D(x) & \text{if } t \in (0, T), \, x \in \partial \Omega, \\ 0 & \text{if } t = T, \end{cases}$$

for all $(t, x) \in \big(\{0, T\} \times \Omega\big) \cup \big((0, T) \times \partial \Omega\big)$.

The existence of a solution for a boundary value problem is in general not
sufficient to justify it as a meaningful mathematical model.

Definition 1.7 A boundary value problem is *well-posed* if there exists a unique solution and if small perturbations of data lead to small changes in the solution. Otherwise it is called *ill-posed*.

Examples 1.10

(i) The Laplace equation $-\Delta u = 0$ in Ω subject to the Dirichlet boundary condition $u = u_D$ on $\partial\Omega$ is a well-posed problem in the sense that it admits a unique solution, and if $\tilde{u}$ is the solution subject to the perturbed boundary condition $\tilde{u} = \tilde{u}_D$ on $\partial\Omega$, then the maximum principle proves

$$\|u - \tilde{u}\|_{C(\Omega)} \le \|u_D - \tilde{u}_D\|_{C(\partial\Omega)},$$

i.e., if the perturbation of the boundary data $u_D - \tilde{u}_D$ is small, then also the difference in the solutions $u - \tilde{u}$ is small.

(ii) Examples of ill-posed boundary value problems are $-u'' - u = 0$ in $(0, 1)$ subject to $u(0) = u(1) = 0$ since solutions are nonunique, $u' = 1$ in $(0, 1)$ subject to $u(0) = u(1) = 0$ since no solution exists, and $u' = u^2$ in $(0, 1)$ subject to $u(0) = \alpha$ since the solution $u(x) = \alpha/(1 - \alpha x)$ is unbounded and has a singularity at $x = 1/\alpha$.

Remark 1.13 In the context of numerical mathematics a well-posed mathematical problem is also called well-conditioned.

1.5.2 Classification of Second-Order Equations

Second-order linear partial differential equations with constant coefficients are given by

$$\sum_{i,j=1}^{n} a_{ij}\partial_{z_i}\partial_{z_j}u + \sum_{i=1}^{n} b_i\partial_{z_i}u + cu = f$$

with coefficients $A = (a_{ij}) \in \mathbb{R}^{n \times n}$, $b = (b_i)_{i=1,\dots,n} \in \mathbb{R}^n$, and $c \in \mathbb{R}$. Since $\partial_{z_i}\partial_{z_j}u = \partial_{z_j}\partial_{z_i}u$ for $u \in C^2(U)$ we assume that A is symmetric. Hence, there exist a diagonal matrix $\Lambda \in \mathbb{R}^{n \times n}$ whose entries are the eigenvalues $\lambda_i \in \mathbb{R}$, $i = 1, 2, \dots, n$, of A, and an orthogonal matrix $Q \in \mathbb{R}^{n \times n}$, such that

$$QAQ^\mathsf{T} = \Lambda.$$

The change of coordinates $\xi = Qz$ transforms the main part of the partial differential equation according to

$$\sum_{i,j=1}^{n} a_{ij}\partial_{z_i}\partial_{z_j}u = \sum_{\ell=1}^{n} \lambda_\ell\partial_{\xi_\ell}^2\tilde{u},$$

where $\bar{u}(\xi) = u(Q^{\mathsf{T}}\xi)$. This leads to the following nonexhaustive classification.

Definition 1.8 A second-order linear partial differential equation is called:

- *elliptic*, if all eigenvalues of A are nonzero and have the same sign;
- *hyperbolic*, if all eigenvalues of A are nonzero and have the same sign except for one;
- *parabolic*, if exactly one eigenvalue of A vanishes and all other eigenvalues have the same sign.

The Poisson, wave, and heat equation are prototypical examples of their equivalence classes.

Examples 1.11

(i) The Poisson equation $-\Delta u = f$ is elliptic.
(ii) The wave equation $\partial_t^2 u - \Delta u = f$ is hyperbolic.
(iii) The heat equation $\partial_t u - \Delta u = f$ is parabolic.

Remark 1.14 The transport equation $\partial_t u - a \cdot \nabla u = f$ is called a *first-order hyperbolic equation*.

The discussion of the wave and heat equations revealed characteristic properties of problems of the same class which are listed in Table 1.1 and explained in the following remark.

Remarks 1.15

(i) We have verified with d'Alembert's formula that waves travel at a finite speed; the representation formula for solutions of the heat equation on the real line shows that information propagates at an infinite speed.
(ii) It follows from d'Alembert's formula that solutions of the wave equation have at times $t > 0$ the same differentiability properties as the initial data; the representation formula for the heat equation shows that the solutions are smooth for positive times.
(iii) The energy conservation principle and the maximum principle imply the well-posedness of the wave and heat equations, respectively.

Table 1.1 Characteristic properties of hyperbolic and parabolic partial differential equations

Property	Wave equation	Heat equation
Speed of propagation	Finite	Infinite
Discontinuities	Transported	Lost
Well-posedness	Yes	Yes
Reversibility	Yes	No
Maximum principle	No	Yes
Behavior for $t \to \infty$	Energy conservation	Energy dissipation

(iv) The reverse wave equation $\partial_t^2 u - \partial_x^2 u = 0$ for $t \in (-T, 0)$ with initial condition at $t = 0$ has the same properties as the forward equation. This is not the case for the heat or diffusion equation which is a fundamental irreversibility principle of thermodynamics.

(v) A maximum principle does not hold for the wave equation but is true for the heat equation.

(vi) We have proved the conservation of the sum of total kinetic and stored energy for the wave equation. For the heat equation, the temperature distribution tends to an equilibrium configuration and the energy decays during this process.

1.5.3 Abstract Convergence Theory

We consider the following abstract k-th order linear boundary value problem:

$$\begin{cases} F(u) = Lu - f = 0 \text{ in } U, \\ G(u) = Mu - \ell = 0 \text{ on } \partial U. \end{cases}$$

A finite difference discretization with positive step-sizes $h = (\Delta z_1, \ldots, \Delta z_n)$ then leads to a finite-dimensional problem that seeks a vector $U_h \in \mathbb{R}^{N_h}$ that satisfies the following equations:

$$\begin{cases} F_h(U_h) = L_h U_h - f_h = 0, \\ G_h(U_h) = M_h U_h - \ell_h = 0. \end{cases}$$

We consider a sequence of discretizations indexed by a sequence of step-size vectors $h \to 0$ and we are interested in the behavior of the approximation error.

Definition 1.9 The sequence of discretizations is *stable under the condition* $\mathscr{D}$ if there exists a constant $c_1 > 0$ that does not depend on $h > 0$ such that the solution U_h of the discretized problem satisfies

$$\|U_h\|_{\ell, N_h} \le c_1 \|(f_h, \ell_h)\|_{r, N_h}$$

for all $h > 0$ with $\mathscr{D}(h) \le 0$ and with appropriate norms $\| \cdot \|_{\ell, N_h}$ and $\|(\cdot, \cdot)\|_{r, N_h}$. If the condition $\mathscr{D}(h) \le 0$ is satisfied for all $h > 0$, then the discretization is called *unconditionally stable*.

Requirements on the norms will be discussed below.

Definition 1.10 The sequence of discretizations is *consistent of order* $\alpha \in \mathbb{N}^n$, if there exist an interpolation operator $\mathscr{I}_h : C^{k+s}(\overline{U}) \to \mathbb{R}^{N_h}$ and a constant $c_2 > 0$, such that

$$\left\| \left(F_h(\mathscr{I}_h u), G_h(\mathscr{I}_h u) \right) \right\|_{r, N_h} \le c_2 h^\alpha \|u\|_{C^{k+s}(\overline{U})}$$

for all $h > 0$ and every solution $u \in C^{k+s}(\overline{U})$, where $h^\alpha = \Delta z_1^{\alpha_1} + \cdots + \Delta z_n^{\alpha_n}$.

Remark 1.16 The interpolation operator is typically defined by the evaluation of u at the grid points. A minimal requirement on the discrete norms is that $\|\mathscr{I}_h u\|_{\ell,N_h} \to \|u\|$ as $h \to 0$ for some norm $\|\cdot\|$ on $C(\overline{U})$.

The stability and consistency imply the following error estimate.

Theorem 1.2 (Lax–Richtmyer Theorem) *Assume that the sequence of discretizations is stable under the condition $\mathscr{D}$, and consistent of order $\alpha \in \mathbb{N}^n$. If the continuous boundary value problem is well-posed and the solution satisfies $u \in C^{k+s}(\overline{U})$, then the discretization is convergent of order α, i.e.,*

$$\|\mathscr{I}_h u - U_h\|_{\ell,N_h} \le c_1 c_2 h^\alpha \|u\|_{C^{k+s}(\overline{U})}$$

for all $h > 0$ with $\mathscr{D}(h) \le 0$.

Proof The error $Z_h = \mathscr{I}_h u - U_h$ satisfies the equations

$$L_h Z_h = L_h \mathscr{I}_h u - L_h U_h = L_h \mathscr{I}_h u - f_h = F_h(\mathscr{I}_h u),$$

$$M_h Z_h = M_h \mathscr{I}_h u - M_h U_h = M_h \mathscr{I}_h u - \ell_h = G_h(\mathscr{I}_h u).$$

The stability of the discrete problems implies that

$$\|Z_h\|_{\ell,N_h} \le c_1 \big\| \big(F_h(\mathscr{I}_h u), G_h(\mathscr{I}_h u)\big) \big\|_{r,N_h}.$$

The consistency estimate leads to

$$\|Z_h\|_{\ell,N_h} \le c_1 c_2 h^\alpha \|u\|_{C^{k+s}(\overline{U})}$$

and proves the error estimate.   □

Remarks 1.17

 (i) The converse implication, i.e., that convergence of stable discretizations implies consistency, can also be proved.
 (ii) The theorem establishes the concept that stability and consistency imply convergence.

Example 1.12 The explicit Euler scheme is stable under the condition $\mathscr{D}(\Delta t, \Delta x) = \Delta t/\Delta x^2 - 1/2 \le 0$, while the implicit Euler scheme is unconditionally stable. Under these conditions we proved the stability estimate

$$\|U\|_\infty = \sup_{0 \le k \le K} \sup_{0 \le j \le J} |U_j^k| \le \sup_{0 \le j \le J} |U_j^0| = \|U^0\|_\infty.$$

The explicit and implicit Euler schemes are linearly consistent with respect to Δt and quadratically consistent with respect to Δx, i.e., we have

$$L_h u(t_k, x_j) = \partial_t^\pm u(t_k, x_j) - \partial_x^+ \partial_x^- u(t_k, x_j) = \mathscr{O}(\Delta t + \Delta x^2),$$

i.e., $F_h(\mathscr{I}_h u) = \mathcal{O}(\Delta t + \Delta x^2)$, and $M_h u(0, x_j) - u_0(x_j) = 0$, i.e., $G_h(\mathscr{I}_h u) = 0$. Thus, with $\alpha = (1, 2)$ and $h = (\Delta t, \Delta x)$,

$$\left\| \left(F_h(\mathscr{I}_h u), G_h(\mathscr{I}_h u) \right) \right\|_\infty \leq c h^\alpha \|u\|_{C^4([0,T]\times[0,1])}.$$

The equivalence theorem implies the error estimates.

1.5.4 Two-Dimensional Heat and Wave Equation

Combining the methods developed for the one-dimensional heat and wave equations with the method developed for the two-dimensional Poisson problem leads to difference schemes for the two-dimensional heat and wave equations. We let A denote the matrix representing the discretized negative Laplace operator $-\Delta_h$ subject to homogeneous Dirichlet boundary conditions on $\Omega = (0, 1)^2$, i.e.,

$$\frac{1}{\Delta x^2} A U = -\Delta_h U,$$

where $U \in \mathbb{R}^L$ is a vector that contains the coefficients of a grid function associated with interior grid points x_ℓ, $\ell = 1, 2, \ldots, L$, in lexicographic enumeration.

Example 1.13 For the approximation of the two-dimensional heat equation $\partial_t u - \Delta u = f$ in $(0, T) \times \Omega$ subject to the initial condition $u(0, x) = u_0(x)$ and homogeneous Dirichlet boundary condition $u(t, x) = 0$ for $(t, x) \in (0, T) \times \partial\Omega$, the θ-method reads, with $\lambda = \Delta t / \Delta x^2$,

$$U^{k+1} - U^k = -\lambda \left(\theta A U^{k+1} + (1 - \theta) A U^k \right) + \Delta t F^{k+\theta},$$

$$U_\ell^0 = u_0(x_\ell),$$

where $F^{k+\theta} = \left(f(t_{k+\theta}, x_\ell) \right)_{\ell=1,\ldots,L}$. For $\theta = 1, 1/2, 0$, this is the implicit Euler scheme, the Crank–Nicolson scheme, and the explicit Euler scheme, respectively. The explicit Euler scheme is stable if $\lambda \leq 1/2^d = 1/4$, while the Crank–Nicolson and implicit Euler scheme are unconditionally stable.

A MATLAB implementation of the θ-method for the two-dimensional heat equation is shown in Fig. 1.26.

Example 1.14 The two-dimensional wave equation $\partial_t^2 u - c^2 \Delta u = f$ in $(0, T) \times \Omega$ with initial conditions $u(0, x) = u_0(x)$ and $\partial_t u(0, x) = v_0(x)$ for all $x \in \Omega$ and the homogeneous Dirichlet boundary condition $u(t, x) = 0$ for all $(t, x) \in (0, T) \times \partial\Omega$ can be approximated with the *implicit scheme*

$$U^{k+1} - 2U^k + U^{k-1} = \frac{\mu^2}{4} A(U^{k+1} + 2U^k + U^{k-1}) + \Delta t^2 F^k.$$

```
function theta_method_2d
T = 1;
theta = .5;
J = 20; K = 10; L = (J-1)^2;
Delta_x = 1/J; Delta_t = T/K;
lambda = Delta_t/Delta_x^2;
U = zeros(K+1,L);
e = ones(J-1,1); E = ones(L,1);
X = spdiags([-e,4*e,-e],[-1,0,1],J-1,J-1);
A = sparse(L,L);
for j = 1:(J-1):L
    A(j:j+J-2,j:j+J-2) = X;
end
A = A+spdiags([-E,-E],[-J+1,J-1],L,L);
B = speye(L)+lambda*theta*A;
C = speye(L)-lambda*(1-theta)*A;
for j = 1:J-1
    for m =1:J-1
        U(1,j+(m-1)*(J-1)) = u_0([j,m]*Delta_x);
    end
end
show(U(1,:),J,Delta_x); pause
for k = 0:K-1
    for j = 1:J-1
        for m = 1:J-1
            F(j+(m-1)*(J-1)) = f((k+theta)*Delta_t,[j,m]*Delta_x);
        end
    end
    U(k+2,:) = B\(C*U(k+1,:)'+Delta_t*F');
    show(U(k+2,:),J,Delta_x); pause(.05);
end

function val = u_0(x)
val = 16*x(1)*(1-x(1))*x(2)*(1-x(2));

function val = f(t,x)
val = 8;

function show(U,J,Delta_x)
U_mat = zeros(J+1,J+1); U_mat(2:J,2:J) = reshape(U,J-1,J-1)';
mesh(Delta_x*(0:J),Delta_x*(0:J),U_mat); axis([0,1,0,1,0,1,0,1]);
```

Fig. 1.26 Finite difference scheme for the heat equation on the square $\Omega = (0, 1)^2$ with homogeneous Dirichlet boundary conditions on $\Gamma_D = \partial\Omega$

The scheme is unconditionally stable and consistent of order 2. The discretized equation for $k = 0$, the initial condition $U_\ell^0 = u_0(x_\ell)$, and the approximation $U_\ell^1 - U_\ell^{-1} = 2\Delta t v_0(x_\ell)$ provide the initial vectors U^0 and U^1.

A MATLAB implementation of the implicit scheme for the two-dimensional wave equation is shown in Fig. 1.27.

```
function wave_implicit_2d
T = 100; c = 1;
J = 20; K = 20; L = (J-1)^2;
Delta_x = 1/J; Delta_t = T/K;
mu = c*Delta_t/Delta_x;
U = zeros(K+1,L);
e = ones(J-1,1); E = ones(L,1);
X = spdiags([-e,4*e,-e],[-1,0,1],J-1,J-1);
A = sparse(L,L);
for j = 1:(J-1):L
    A(j:j+J-2,j:j+J-2) = X;
end
A = A+spdiags([-E,-E],[-J+1,J-1],L,L);
B = speye(L)+(mu/4)*A;
C = speye(L)-(mu/4)*A;
for j = 1:J-1
    for m = 1:J-1
        U(1,j+(m-1)*(J-1)) = u_0([j,m]*Delta_x);
        V(j+(m-1)*(J-1)) = v_0([j,m]*Delta_x);
        F(j+(m-1)*(J-1)) = f(0,[j,m]*Delta_x);
    end
end
U(2,:) = B\(C*U(1,:)'+B*Delta_t*V'+(1/2)*Delta_t^2*F');
show(U(1,:),J,Delta_x); pause
for k = 1:K-1
    for j = 1:J-1
        for m = 1:J-1
            F((m-1)*(J-1)+j) = f(k*Delta_t,[j,m]*Delta_x);
        end
    end
    U(k+2,:) = B\(2*C*U(k+1,:)'-B*U(k,:)'+Delta_t^2*F');
    show(U(k+2,:),J,Delta_x); pause(.05);
end

function val = u_0(x)
val = 16*x(1)*(1-x(1))*x(2)*(1-x(2));

function val = v_0(x)
val = 0;

function val = f(t,x)
val = 0;

function show(U,J,Delta_x)
U_mat = zeros(J+1,J+1); U_mat(2:J,2:J) = reshape(U,J-1,J-1)';
mesh(Delta_x*(0:J),Delta_x*(0:J),U_mat);axis([0,1,0,1,-1,1,-1,1]);
```

Fig. 1.27 Finite difference scheme for the wave equation on the square $\Omega = (0,1)^2$ with homogeneous Dirichlet boundary conditions on $\Gamma_D = \partial\Omega$

References

Important contributions to the analysis and development of finite difference methods are the articles [3, 7]. Specialized textbooks on finite difference methods are the references [5, 9, 13]. Elementary properties of partial differential equations and numerical methods are discussed in [2, 6, 8, 10, 12]. The derivation of partial differential equations from physical principles is the subject of [4, 11, 14]; see [1] for a careful derivation of the equations of a vibrating string.

1. Antman, S.S.: The equations for large vibrations of strings. Am. Math. Monthly **87**(5), 359–370 (1980). URL http://dx.doi.org/10.2307/2321203
2. Carstensen, C.: Wissenschaftliches Rechnen (1997). Lecture Notes, University of Kiel, Germany
3. Crank, J., Nicolson, P.: A practical method for numerical evaluation of solutions of partial differential equations of the heat-conduction type. Proc. Cambridge Philos. Soc. **43**, 50–67 (1947)
4. Eck, C., Garcke, H., Knabner, P.: Mathematische Modellierung. Springer-Lehrbuch. Springer, Berlin-Heidelberg-New York (2011)
5. Jovanović, B.S., Süli, E.: Analysis of finite difference schemes. Springer Series in Computational Mathematics, vol. 46. Springer, London (2014). URL http://dx.doi.org/10.1007/978-1-4471-5460-0
6. Larsson, S., Thomée, V.: Partial differential equations with numerical methods. Texts in Applied Mathematics, vol. 45. Springer, Berlin (2009)
7. Lax, P.D., Richtmyer, R.D.: Survey of the stability of linear finite difference equations. Comm. Pure Appl. Math. **9**, 267–293 (1956)
8. Plato, R.: Concise numerical mathematics. Graduate Studies in Mathematics, vol. 57. American Mathematical Society, Providence, RI (2003)
9. Richtmyer, R.D., Morton, K.W.: Difference methods for initial-value problems. Second edition. Interscience Tracts in Pure and Applied Mathematics, No. 4. Interscience Publishers John Wiley & Sons, New York-London-Sydney (1967)
10. Salsa, S., Vegni, F.M.G., Zaretti, A., Zunino, P.: A primer on PDEs. Unitext, vol. 65, Italian edn. Springer, Milan (2013). URL http://dx.doi.org/10.1007/978-88-470-2862-3
11. Schweizer, B.: Partielle Differentialgleichungen. Springer-Lehrbuch Masterclass. Springer, New York (2013)
12. Strauss, W.A.: Partial Differential Equations, 2nd edn. Wiley, Chichester (2008)
13. Strikwerda, J.C.: Finite Difference Schemes and Partial Differential Equations, 2nd edn. Society for Industrial and Applied Mathematics (SIAM), Philadelphia, PA (2004). URL http://dx.doi.org/10.1137/1.9780898717938
14. Temam, R., Miranville, A.: Mathematical Modeling in Continuum Mechanics, 2nd edn. Cambridge University Press, Cambridge (2005). URL http://dx.doi.org/10.1017/CBO9780511755422

Chapter 2
Elliptic Partial Differential Equations

2.1 Weak Formulation of the Poisson Problem

2.1.1 Classical Solutions

Due to its occurrence in describing many physical processes, the Laplace operator plays a central role in the analysis of partial differential equations. In particular, the existence of solutions for the Poisson problem enables establishing the existence of solutions for many initial boundary value problems. For an open domain $\Omega \subset \mathbb{R}^d$ with a partitioning of the boundary $\partial\Omega = \Gamma_D \cup \Gamma_N$, and functions $f \in C(\overline{\Omega})$, $g \in C(\Gamma_N)$, and $u_D \in C(\Gamma_D)$, we consider the following boundary value problem:

$$\begin{cases} -\Delta u = f & \text{in } \Omega, \\ u = u_D & \text{on } \Gamma_D, \\ \nabla u \cdot n = g & \text{on } \Gamma_N. \end{cases}$$

Elementary methods prove the existence of solutions for simple domains such as rectangular or circular ones.

Definition 2.1 A *classical solution* of the Poisson problem is a function $u \in C^1(\overline{\Omega}) \cap C^2(\Omega)$ that solves the boundary value problem.

The concept of classical solutions is too restrictive in many situations.

Proposition 2.1 (Nonclassical Solution) *For $\gamma \in (0, 2\pi)$ let*

$$\Omega = \{r(\cos\phi, \sin\phi) : 0 < r < 1, 0 < \phi < \gamma\},$$

S. Bartels, *Numerical Approximation of Partial Differential Equations*,
Texts in Applied Mathematics 64, DOI 10.1007/978-3-319-32354-1_2

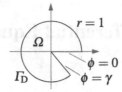

Fig. 2.1 Domain Ω that leads to solutions with unbounded gradients for $\gamma \in (\pi, 2\pi)$

cf. Fig. 2.1, $\Gamma_D = \partial\Omega$, *and* $\Gamma_N = \emptyset$, *and define* $f = 0$ *in* Ω, *and*

$$u_D(r, \phi) = \begin{cases} 0 & \text{for } \phi \in \{0, \gamma\}, \\ \sin(\phi\pi/\gamma) & \text{for } r = 1. \end{cases}$$

Then

$$u(r, \phi) = r^{\pi/\gamma} \sin(\phi\pi/\gamma)$$

satisfies $u \in C^\infty(\Omega)$ *and solves the Poisson problem with* $f = 0$. *It is a classical solution if and only if* $\gamma \in (0, \pi]$, *i.e., we have* $u \notin C^1(\overline{\Omega})$ *if* $\gamma \in (\pi, 2\pi)$.

Proof Obviously, $u = u_D$ on $\Gamma_D = \partial\Omega$. The Laplace operator is in polar coordinates given by

$$\Delta u = \partial_r^2 u + r^{-1}\partial_r u + r^{-2}\partial_\phi^2 u$$

and this implies that u solves the Poisson problem with $f = 0$. The function u is infinitely often differentiable in the open domain Ω. Its gradient is in polar coordinates given by

$$\nabla u = \begin{bmatrix} \partial_r u \\ r^{-1}\partial_\phi u \end{bmatrix} = (\pi/\gamma)r^{\pi/\gamma - 1} \begin{bmatrix} \sin(\phi\pi/\gamma) \\ \cos(\phi\pi/\gamma) \end{bmatrix}.$$

Hence the gradient is bounded if and only if $\gamma \in (0, \pi]$.   □

2.1.2 Weak Formulation

To weaken the notion of solutions we need to find an appropriate reformulation of the Poisson problem. A promising approach is to reduce the order of the derivatives in the equation. This is achieved with the following steps:

Step 1:   We multiply the partial differential equation by a function $v \in C^1(\overline{\Omega})$ and integrate the resulting identity over Ω, i.e., we have

$$-\int_\Omega v \Delta u \, dx = \int_\Omega vf \, dx.$$

Step 2:   We apply Gauss's theorem to the vector field $F = v \nabla u$ and use $\operatorname{div} F = v \Delta u + \nabla v \cdot \nabla u$ to obtain with the outer unit normal n on $\partial \Omega$ that

$$\int_{\partial\Omega} (v \nabla u) \cdot n \, ds = \int_\Omega \operatorname{div}(v \nabla u) \, dx = \int_\Omega v \Delta u \, dx + \int_\Omega \nabla u \cdot \nabla v \, dx.$$

With the identity from Step 1 this leads to

$$\int_\Omega \nabla u \cdot \nabla v \, dx = \int_\Omega fv \, dx + \int_{\partial\Omega} (\nabla u \cdot n) v \, ds.$$

Step 3:   We incorporate the Neumann boundary condition $\partial_n u = \nabla u \cdot n = g$ on Γ_N and impose the restriction $v = 0$ on Γ_D so that we have

$$\int_\Omega \nabla u \cdot \nabla v \, dx = \int_\Omega fv \, dx + \int_{\Gamma_N} gv \, ds.$$

This motivates us to define the following notion of a solution.

Definition 2.2   A function $u \in C^1(\overline{\Omega})$ solves the *weak formulation* of the Poisson problem if it satisfies $u = u_D$ on Γ_D and if the identity

$$\int_\Omega \nabla u \cdot \nabla v \, dx = \int_\Omega fv \, dx + \int_{\Gamma_N} gv \, ds$$

holds for all functions $v \in C^1(\overline{\Omega})$ with $v = 0$ on Γ_D.

The derivation of the weak formulation implies that it is satisfied for every classical solution. The converse implication is true if a solution of the weak formulation satisfies $u \in C^2(\Omega)$. To reverse the derivation, the following lemma is needed.

Lemma 2.1 (Fundamental Lemma)   *Assume that $h \in C(\Omega)$ is such that*

$$\int_\Omega hv \, dx = 0$$

for every function $v \in C(\overline{\Omega})$ with $v = 0$ on $\partial\Omega$. Then $h = 0$.

Proof Exercise.   □

For simplicity we restrict to the case $\Gamma_N = \emptyset$. We assume that $\Omega \subset \mathbb{R}^d$ is open and bounded, and that $\partial\Omega$ is the union of finitely many C^1-submanifolds. This implies that Gauss's theorem holds for all vector fields $F \in C^1(\overline{\Omega}; \mathbb{R}^d)$.

Proposition 2.2 (Formal Equivalence) *Assume that* $\Gamma_D = \partial\Omega$ *and that there exists a solution* $u \in C^1(\overline{\Omega})$ *of the weak formulation that satisfies* $u \in C^2(\Omega)$. *Then* u *is also a classical solution.*

Proof Let $v \in C^1(\overline{\Omega})$ with $v = 0$ on $\partial\Omega$. Using that

$$\int_\Omega \nabla u \cdot \nabla v \, dx = - \int_\Omega (\Delta u) v \, dx,$$

we find that

$$\int_\Omega (\Delta u + f) v \, dx = 0.$$

The fundamental lemma implies that $h = \Delta u + f = 0$. Hence u solves the Poisson problem in its classical formulation. □

The weak formulation of the Poisson problem has an important structure. Real valued mappings will also be referred to as *forms*.

Proposition 2.3 (Bilinear Form) *Assume without loss of generality that* $u_D = 0$ *and set*

$$\widehat{V} = \{v \in C^1(\overline{\Omega}) : v|_{\Gamma_D} = 0\}.$$

Define $a : \widehat{V} \times \widehat{V} \to \mathbb{R}$ *and* $b : \widehat{V} \to \mathbb{R}$ *by*

$$a(u, v) = \int_\Omega \nabla u \cdot \nabla v \, dx, \quad b(v) = \int_\Omega f v \, dx + \int_{\Gamma_N} g v \, ds$$

for all $u, v \in \widehat{V}$. *Then* $\widehat{V}$ *is a linear space, a is a bilinear form, b is a linear form, and $u \in \widehat{V}$ solves the weak formulation of the Poisson problem if and only if for all* $v \in \widehat{V}$ *we have*

$$a(u, v) = b(v).$$

Proof The result follows directly from the definitions. □

2.1.3   Minimization Problem

Establishing the existence of a solution for the weak formulation of the Poisson problem is a nontrivial task. One approach is based on the observation that the weak formulation defines optimality conditions for a minimization problem.

Proposition 2.4 (Dirichlet's Principle) *Assume* $u_D = 0$ *and let*

$$\widehat{V} = \{v \in C^1(\overline{\Omega}) : v|_{\Gamma_D} = 0\}.$$

The function $u \in \widehat{V}$ *solves the weak formulation of the Poisson problem if and only if it is minimal for the* Dirichlet energy

$$I(v) = \frac{1}{2} \int_\Omega |\nabla u|^2 \, dx - \int_\Omega fv \, dx - \int_{\Gamma_N} gv \, ds$$

in the set of all $v \in \widehat{V}$.

Proof

(i) Assume that $u \in \widehat{V}$ is minimal. Then, for every $v \in \widehat{V}$ the function

$$\varphi(t) = I(u + tv)$$

is minimal for $t = 0$. We have

$$\varphi'(0) = \lim_{t \to 0} t^{-1}\big(I(u + tv) - I(u)\big)$$

$$= \lim_{t \to 0} t^{-1}\left(\frac{1}{2} \int_\Omega |\nabla(u + tv)|^2 - |\nabla u|^2 \, dx - t \int_\Omega fv \, dx - t \int_{\Gamma_N} gv \, ds\right)$$

$$= \int_\Omega \nabla u \cdot \nabla v \, dx - \int_\Omega fv \, dx - \int_{\Gamma_N} gv \, ds.$$

This shows that φ is differentiable at $t = 0$, and that $\varphi'(0) = 0$ is equivalent to the weak formulation.

(ii) Conversely, assume that $u \in \widehat{V}$ solves the weak formulation. Using the binomial identity

$$|b|^2 - |a|^2 = 2a \cdot (b - a) + |a - b|^2,$$

which holds for all $a, b \in \mathbb{R}^d$, we find that

$$I(u + v) - I(u) = \frac{1}{2} \int_\Omega |\nabla(u + v)|^2 - |\nabla u|^2 \, dx - \int_\Omega fv \, dx - \int_\Omega gv \, ds$$

$$= \int_\Omega \nabla u \cdot \nabla v \, dx - \int_\Omega fv \, dx - \int_\Omega gv \, ds + \frac{1}{2} \int_\Omega |\nabla v|^2 \, dx$$

$$= \frac{1}{2} \int_\Omega |\nabla v|^2 \, dx \geq 0.$$

Hence $I(u) \leq I(u + v)$ for all $v \in \widehat{V}$ which implies that u is minimal. $\qquad \square$

Remark 2.1 Establishing the existence of a minimizer is a topic of the calculus of variations.

2.2 Elementary Functional Analysis

2.2.1 Riesz Representation Theorem

We want to address the solvability of the problem of determining $u \in V$ such that

$$a(u, v) = b(v)$$

for all $v \in V$, where V is a linear space, a a bilinear form, and b a linear form.

Definition 2.3

(i) A *Hilbert space* is a Banach space V such that its norm is induced by a symmetric bilinear form $\langle \cdot, \cdot \rangle_V : V \times V \to \mathbb{R}$, i.e., for all $v \in V$ we have

$$\|v\|_V^2 = \langle v, v \rangle_V.$$

(ii) An element $v \in V$ in a Hilbert space V is said to be orthogonal to a subset $U \subset V$ if

$$\langle v, u \rangle_V = 0$$

for all $u \in U$.

Remark 2.2 In a Hilbert space V we have the *Cauchy–Schwarz inequality*

$$\langle v, w \rangle_V \le \|v\|_V \|w\|_V$$

for all $v, w \in V$.

Example 2.1 The linear space of square summable sequences, i.e.,

$$\ell^2(\mathbb{N}) = \Big\{ (v_j)_{j \in \mathbb{N}} : \sum_{j \in \mathbb{N}} v_j^2 < \infty \Big\}$$

is a Hilbert space with the scalar product

$$\langle v, w \rangle_{\ell^2(\mathbb{N})} = \sum_{j \in \mathbb{N}} v_j w_j.$$

Fig. 2.2 Orthogonal projection onto a subspace U

In Hilbert spaces, the best approximation of an element in a closed subspace exists and defines an orthogonality relation, which is depicted in Fig. 2.2 and specified in the following lemma.

Lemma 2.2 (Projection Onto Subspaces) *Let V be a Hilbert space and let $U \subset V$ be a closed subspace. Then, for every $v \in V$, there exists a uniquely defined element $u \in U$ such that*

$$\|v - u\|_V = \inf_{r \in U} \|v - r\|_V$$

and, for all $r \in U$,

$$\langle v - u, r \rangle_V = 0.$$

Proof

(i) We define $\gamma = \inf_{r \in U} \|v - r\|_V$, and let $(u_j)_{j \in \mathbb{N}}$ be a sequence in U so that

$$\|v - u_j\|_V^2 \leq \gamma^2 + \frac{1}{j}.$$

We have that

$$\|u_j - u_k\|_V^2 = 2\|v - u_j\|_V^2 + 2\|v - u_k\|_V^2 - \|(v - u_j) + (v - u_k)\|_V^2$$
$$= 2\|v - u_j\|_V^2 + 2\|v - u_k\|_V^2 - 4\|v - (u_j + u_k)/2\|_V^2.$$

Since $(u_j + u_k)/2 \in U$ we have $\|v - (u_j + u_k)/2\|_V \geq \gamma$ and hence

$$\|u_j - u_k\|_V^2 \leq 4\gamma^2 + \frac{2}{j} + \frac{2}{k} - 4\gamma^2 = \frac{2}{j} + \frac{2}{k}.$$

This implies that $(u_j)_{j \in \mathbb{N}}$ is a Cauchy sequence which has a limit $u \in V$. Since U is closed, we have $u \in U$. Moreover, we have $\gamma = \|v - u\|_V$.

(ii) Assume that there exists $r \in U$ with $\beta = \|r\|_V^{-2} \langle v - u, r \rangle_V \neq 0$. Then, defining $z = u + \beta r$, we have

$$\|v - z\|_V^2 = \|(v - u) - \beta r\|_V^2 = \|v - u\|_V^2 - \beta^2 \|r\|_V^2 < \|v - u\|_V^2$$

which contradicts $z \in U$.

(iii) It remains to show that u is unique. For this, assume that there exists $r \in U$ with $\|v - u\|_V = \|v - r\|_V$. We then have $u + r \in U$ and hence

$$\|v - r\|_V^2 = \|(v - u) + (u - r)\|_V^2 = \|v - u\|_V^2 + \|u - r\|_V^2.$$

This implies that $u = r$.   □

Remark 2.3 The orthogonality relation uniquely defines the best approximation.

The following theorem generalizes the fact that linear mappings between finite dimensional spaces are represented by matrices.

Theorem 2.1 (Riesz Representative) *Let V be a Hilbert space and let $b : V \to \mathbb{R}$ be linear and continuous. Then there exists a unique $u_b \in V$ with*

$$\langle u_b, v \rangle_V = b(v)$$

for all $v \in V$. This defines a bijective linear mapping $R : b \mapsto u_b$.

Proof

(i) The set $U = \{v \in V : b(v) = 0\}$ is a closed subspace. If $U = V$, then we choose $u_b = 0$. Otherwise, there exists $w \in V \setminus U$ with $b(w) \neq 0$. We let $u \in U$ be the orthogonal projection of w onto U so that we have

$$\langle w - u, r \rangle_V = 0$$

for all $r \in U$. We set $z = w - u$ and define

$$u_b = \frac{b(z)}{\|z\|_V^2} z.$$

We note that $b(z)v - b(v)z \in U$ and hence $\langle z, b(z)v - b(v)z \rangle_V = 0$ for all $v \in V$. This leads to

$$\langle u_b, v \rangle_V = \frac{b(z)}{\|z\|_V^2} \langle z, v \rangle_V = \frac{b(v)\|z\|_V^2}{\|z\|_V^2} = b(v).$$

(ii) If $u_1 = Rb_1 = Rb_2 = u_2$, then $\langle u_1 - u_2, v \rangle_V = 0$ for all $v \in V$ implies $u_1 = u_2$ and thus $b_1 = b_2$ which proves injectivity. Given $u \in V$ let $b(v) = \langle u, v \rangle_V$ for all $v \in V$ so that $Rb = u$ and R is surjective. To verify the linearity of R, we note that for all linear and continuous functionals $b, d : V \to \mathbb{R}$ and $\alpha, \beta \in \mathbb{R}$ we have

$$\langle R(\alpha b + \beta d), v \rangle_V = (\alpha b + \beta d)(v) = \alpha b(v) + \beta d(v)$$

$$= \alpha \langle Rb, v \rangle_V + \beta \langle Rd, v \rangle_V,$$

which implies that $R(\alpha b + \beta d) = \alpha Rb + \beta Rd$.   □

Corollary 2.1 (Representation Via Bilinear Forms) *Assume that V is a Banach space and the bilinear form $a : V \times V \to \mathbb{R}$ is symmetric and defines an equivalent norm on V, i.e., there exists $c > 0$ such that*

$$c^{-1} \|v\|_V^2 \leq a(v, v) \leq c \|v\|_V^2$$

for all $v \in V$. Then for every continuous and linear mapping $b : V \to \mathbb{R}$ there exists a uniquely defined $u_b \in V$ such that

$$a(u_b, v) = b(v)$$

for all $v \in V$.

Proof The bilinear form *a* defines a scalar product on *V* so that it is a Hilbert space and hence Riesz's representation theorem can be applied. □

2.2.2 Linear Operators and Functionals

We let V, W be Banach spaces. A linear mapping between V and W will be called a *linear operator*.

Definition 2.4 We let $L(V, W)$ be the set of all continuous linear operators $A : V \to W$.

Proposition 2.5 (Bounded Linear Operators) *A linear operator $A : V \to W$ is continuous if and only if it is bounded in the sense that there exists $c > 0$ such that*

$$\|Av\|_W \leq c \|v\|_V$$

for all $v \in V$. The infimum of all such constants $c > 0$ is denoted by $\|A\|_{L(V,W)}$ and called operator norm, *i.e., for all $v \in V$ we have*

$$\|Av\|_W \leq \|A\|_{L(V,W)} \|v\|_V.$$

This defines a norm on $L(V, W)$ such that it is a Banach space.

Proof Exercise. □

Of particular importance is the case $W = \mathbb{R}$.

Definition 2.5 The space $L(V, \mathbb{R})$ is called *dual space* of *V* and is denoted by V'. Its elements are called *(bounded) linear functionals*.

Remark 2.4 The Riesz representation operator $R : V' \to V$ defines a bounded linear operator with $\|R\|_{L(V',V)} = \|R^{-1}\|_{L(V,V')} = 1$. This follows from

$$\|Rb\|_V^2 = \langle Rb, Rb \rangle_V = b(Rb) \le \|b\|_{V'} \|Rb\|_V$$

and $b(v) = \langle Rb, v \rangle_V \le \|Rb\|_V \|v\|_V$, i.e., $\|b\|_{V'} \le \|Rb\|_V$.

Hahn–Banach theorems are fundamental results of functional analysis.

Theorem 2.2 (Hahn–Banach Extension, See [4, Chap. 1]) *Let $U \subset V$ be a closed subspace and $b \in U'$. Then there exists $\tilde{b} \in V'$ such that $\tilde{b}|_U = b$ and $\|\tilde{b}\|_{V'} = \|b\|_{U'}$.*

Proof (Sketched) Assume for simplicity that V is a Hilbert space. If $U \ne V$, then there exists $v \in V \setminus U$ which is orthogonal to U and we define $\hat{b} : \text{span}\{U, v\} \to \mathbb{R}$ by

$$\hat{b}(u + \alpha v) = b(u)$$

for all $u + \alpha v \in \text{span}\{U, v\}$. Then $\hat{b}$ extends b with identical operator norms, denoted $b \preceq \hat{b}$. We may thus consider the set $\mathcal{E}$ of all extensions of b which is equipped with the partial ordering $\preceq$. Zorn's lemma implies the existence of a maximal element and this element has to be defined on the entire space V, since otherwise we could extend it to a larger subspace as above. $\square$

Various important results are related to this theorem.

Remarks 2.5

(i) If $C_1, C_2 \subset V$ are disjoint convex sets and C_1 is open, then there exist $b \in V'$ and $m \in \mathbb{R}$ such that

$$b(w_1) > m \ge b(w_2)$$

for all $w_1 \in C_1$ and $w_2 \in C_2$. The set $\{v \in V : b(v) = m\}$ is called a *separating hyperplane*.

(ii) For every $v \in V$ we have

$$\|v\|_V = \sup_{b \in V' \setminus \{0\}} \frac{b(v)}{\|b\|_{V'}}.$$

(iii) The *open mapping theorem* asserts that if $A : V \to W$ is a surjective bounded linear operator, then A is *open*, i.e., the image of every open set is open. In particular, the inverse of a bijective bounded linear operator is bounded. This implies that if $\|\cdot\|_1$ and $\|\cdot\|_2$ are norms on a Banach space X with $\|v\|_1 \le c\|v\|_2$ for all $v \in V$, then $\|v\|_2 \le c'\|v\|_2$ for some constant $c' > 0$ and all $v \in V$.

2.2.3 Lax–Milgram Lemma

The implication of the Riesz representation theorem can be generalized in the sense of the following theorem.

Theorem 2.3 (Lax–Milgram Lemma) *Let V be a Hilbert space and assume that the bilinear form $a : V \times V \to \mathbb{R}$ is coercive and bounded, i.e., there exist $\alpha, k_a > 0$ such that*

$$a(u, u) \geq \alpha \|u\|_V^2, \quad |a(u, v)| \leq k_a \|u\|_V \|v\|_V$$

for all $u, v \in V$. Then for every $b \in V'$ there exists a unique $u_b \in V$ such that

$$a(u_b, v) = b(v)$$

for all $v \in V$. Moreover, we have $\|u_b\|_V \leq \alpha^{-1} \|b\|_{V'}$.

Proof For every $v \in V$, we have $a(v, \cdot) \in V'$. Hence, by the Riesz representation theorem, there exists $Av = Ra(v, \cdot) \in V$ such that

$$\langle Av, w \rangle_V = a(v, w)$$

for all $w \in V$. Moreover, there exists $f \in V$ such that

$$\langle f, w \rangle_V = b(w)$$

for all $w \in V$. With these definitions it suffices to solve the equation $Au = f$. The mapping $A : v \mapsto Av$ is linear and bounded with

$$\|Av\|_V = \|Ra(v, \cdot)\|_V = \|a(v, \cdot)\|_{V'} = \sup_{w \in V \backslash \{0\}} \frac{a(v, w)}{\|w\|_V} \leq k_a \|v\|_V,$$

where we used that $\|R\|_{L(V', V)} = 1$. We want to show that the mapping

$$T_\delta : V \to V, \quad v \mapsto v - \delta(Av - f)$$

is a contraction for an appropriate $\delta > 0$, i.e., satisfies $\|T_\delta v - T_\delta w\|_V \leq q \|v - w\|_V$ with some $0 \leq q < 1$ for all $v, w \in V$. Writing $e = v - w$ and using $\langle Ae, e \rangle_V = a(e, e) \geq \alpha \|e\|_V^2$, we have

$$\|T_\delta v - T_\delta w\|_V^2 = \|e - \delta Ae\|_V^2$$

$$= \|e\|_V^2 - 2\delta \langle Ae, e \rangle_V + \delta^2 \|Ae\|_V^2$$

$$\leq \|e\|_V^2 - 2\delta\alpha \|e\|_V^2 + \delta^2 k_a^2 \|e\|_V^2$$

$$= (1 - 2\delta\alpha + \delta^2 k_a^2) \|v - w\|_V^2.$$

Choosing, e.g., $\delta = \alpha/k_a^2$ and noting $\alpha \leq k_a$, we find that the operator T_δ is a contraction. Banach's fixed point theorem implies the existence of a unique $u \in V$ with

$$u = T_\delta u = u - \delta(Au - f).$$

This yields that $Au = f$, i.e., $a(u, v) = b(v)$ for all $v \in V$. Finally, choosing $v = u$, we verify that

$$\alpha \|u\|_V^2 \leq a(u, u) = b(u) \leq \|b\|_{V'} \|u\|_V,$$

which completes the proof.   □

2.2.4 Weak Convergence

According to the Bolzano–Weierstraß theorem every bounded sequence in a finite-dimensional space admits a convergent subsequence. This is not true in infinite-dimensional spaces.

Example 2.2 Let $(v_j)_{j \in \mathbb{N}} \subset \ell^2(\mathbb{N})$ be defined by $v_{j,n} = \delta_{jn}$ for all $j, n \in \mathbb{N}$, i.e.,

$$v_j = (0, \ldots, 0, 1, 0, \ldots).$$

Then $\|v_j\|_{\ell^2(\mathbb{N})} = 1$ for all $j \in \mathbb{N}$, but for distinct $j, k \in \mathbb{N}$, we have

$$\|v_j - v_k\|_{\ell^2(\mathbb{N})} = \sqrt{2},$$

i.e., there cannot exist a convergent subsequence.

A meaningful limit can be constructed by resorting to a different notion of convergence.

Definition 2.6 A sequence $(v_j)_{j \in \mathbb{N}} \subset V$ is called *weakly convergent* if there exists $v \in V$, such that for all $b \in V'$, we have for $j \to \infty$ that

$$b(v_j) \to b(v).$$

The element v is called the *weak limit* of the sequence $(v_j)_{j \in \mathbb{N}}$, denoted $v_j \rightharpoonup v$.

To distinguish weak from ordinary convergence, the latter is also called *norm* or *strong convergence*.

Definition 2.7 The Banach space V is called *reflexive* if the mapping $i : V \to (V')' = V''$, $i(v)[b] = b(v)$, is an isomorphism.

In reflexive spaces, e.g., Hilbert spaces, bounded sequences have weakly convergent subsequences.

Theorem 2.4 (Eberlein–Šmulian, See [4, Chap. 3]) *Assume that V is reflexive. Then every bounded sequence has a weakly convergent subsequence.*

An important ingredient in the proof of the theorem is the *separability* of certain subspaces, e.g., that there exists a countable set $\{u_j : j \in \mathbb{N}\} \subset U$ which is dense in U.

Example 2.3 If $(v_j)_{j \in \mathbb{N}}$ is an orthonormal system in a Hilbert space, then the sequence is weakly convergent with weak limit $v = 0$.

Definition 2.8 An operator $A : V \to W$ is called *compact*, if for every bounded sequence $(v_j)_{j \in \mathbb{N}} \subset V$ in V, the sequence $(Av_j)_{j \in \mathbb{N}} \subset W$ has a strongly convergent subsequence in W.

Remarks 2.6

  (i) Compact linear operators are bounded; if A is compact and $v_j \rightharpoonup v$, then $Av_j \to Av$.
 (ii) Every bounded linear mapping with finitedimensional image is compact.
(iii) Due to the Arzelà–Ascoli theorem, Hölder-continuous functions are compactly embedded in the space of continuous functions.
(iv) The composition of a compact and a bounded linear operator defines a compact operator.
 (v) A compact, self-adjoint operator $A : V \to V$ on a Hilbert space V admits a complete orthonormal system of eigenfunctions.

2.3 Sobolev Spaces

2.3.1 Lebesgue Integral and L^p Spaces

The Lebesgue integral is based on constructing the *Lebesgue measure* that associates a volume $\mu(A) \in [0, \infty]$ with sets $A \subset \mathbb{R}^d$ contained in the Borel σ-algebra $\mathscr{E}$ of *measurable sets*, which is the smallest collection of all subsets of $\mathbb{R}^d$ that contains all open sets and is closed under countable union and relative complement. Sets $N \subset \mathbb{R}^d$ with $\mu(N) = 0$ are called *null sets* or *sets of measure zero*. If a boolean expression $b(x)$ is true for all $x \in \mathbb{R}^d \setminus N$ with a null set N, then it is said to hold *almost everywhere*. For parallelepipeds $A = (a_1, b_1) \times \cdots \times (a_d, b_d)$, the Lebesgue measure coincides with its volume, i.e., $\mu(A) = (b_1 - a_1) \cdots (b_d - a_d)$. A function $f : \mathbb{R}^d \to [-\infty, \infty]$ is *measurable* if its sublevel sets are measurable, i.e.,

$$\{x \in \mathbb{R}^d : f(x) < s\} \in \mathscr{E}$$

for all $s \in \mathbb{R}$. The integral of a measurable nonnegative function $f : \mathbb{R}^d \to [0, \infty]$ is defined via lower approximation by simple functions, i.e.,

$$\int_{\mathbb{R}^d} f(x)\, dx = \sup \left\{ \sum_{i=1}^{n} \alpha_i \mu(A_i) : 0 \leq \sum_{i=1}^{n} \alpha_i \chi_{A_i} \leq f \text{ almost everywhere} \right\}$$

and may be infinite. For a general measurable function $f : \mathbb{R}^d \to [-\infty, \infty]$ its *Lebesgue integral* is with the positive and negative parts f^+ and f^-, defined by

$$\int_{\mathbb{R}^d} f(x)\, dx = \int_{\mathbb{R}^d} f^+(x)\, dx - \int_{\mathbb{R}^d} (-f^-(x))\, dx.$$

The integral is a linear and monotone operation. For a function defined on a measurable set $\Omega \subset \mathbb{R}^d$ its Lebesgue integral is defined by extending the function by zero to the entire space. The reader is referred to [6] for details.

Definition 2.9 For a measurable function $f : \Omega \to [-\infty, \infty]$ and $p \in [1, \infty]$, define

$$\|f\|_{L^p(\Omega)} = \begin{cases} \left(\int_{\Omega} |f(x)|^p\, dx \right)^{1/p} & \text{for } 1 \leq p < \infty, \\ \inf_{\mu(N)=0} \sup_{x \in \Omega \setminus N} |f(x)| & \text{for } p = \infty. \end{cases}$$

Let $\mathscr{L}^p(\Omega)$ be the set of all measurable functions with $\|f\|_{L^p(\Omega)} < \infty$ and define an equivalence relation $f \sim g$ on $\mathscr{L}^p(\Omega)$ by identifying f and g if they coincide almost everywhere in Ω. The *Lebesgue space* $L^p(\Omega)$ then consists of the corresponding equivalence classes, i.e.,

$$L^p(\Omega) = \mathscr{L}^p(\Omega)/\sim.$$

The equivalence classes in $L^p(\Omega)$ are called *Lebesgue functions*.

Due to the identification of functions we have that $\|f\|_{L^p(\Omega)} = 0$ if and only if $f = 0$, where f stands for the equivalence class of all functions that are zero outside of a set of measure zero. To show that $L^p(\Omega)$ is a normed space, we need Young's inequality.

Lemma 2.3 (Young's Inequality) *For $1 < p, q < \infty$ with $1/p + 1/q = 1$ and all $a, b \in \mathbb{R}_{\geq 0}$ we have $ab \leq a^p/p + b^q/q$.*

Proof Exercise. $\square$

Young's inequality leads to the triangle inequality in $L^p(\Omega)$. We set $1/p = 0$ if $p = \infty$.

Proposition 2.6 (Hölder and Minkowski Inequalities)
(i) Let $1 \leq p, q \leq \infty$ with $1/p + 1/q = 1$ and $f \in L^p(\Omega)$, and $g \in L^q(\Omega)$. We then have

$$\int_{\Omega} |fg|\, dx \leq \|f\|_{L^p(\Omega)} \|g\|_{L^q(\Omega)}.$$

(ii) Let $1 \leq p \leq \infty$ and $f, g \in L^p(\Omega)$. We then have

$$\|f + g\|_{L^p(\Omega)} \leq \|f\|_{L^p(\Omega)} + \|g\|_{L^p(\Omega)}.$$

Proof Exercise.   □

Fatou's lemma states that if $(f_j)_{j \in \mathbb{N}}$ is a sequence of nonnegative measurable functions that converge pointwise to a function f, then f is measurable and we have

$$\int_\Omega f \, dx \leq \liminf_{j \to \infty} \int_\Omega f_j \, dx.$$

This fact leads to the following result.

Theorem 2.5 (L^p Spaces) *For $1 \leq p \leq \infty$ we have that $L^p(\Omega)$ is a Banach space.*

Remarks 2.7

(i) For $1 < p < \infty$ the space $L^p(\Omega)$ is a reflexive space, whose dual can be identified with $L^q(\Omega)$ if $1/p + 1/q = 1$.
(ii) The space $L^p(\Omega)$ is separable for $1 \leq p < \infty$.

2.3.2 Transformation, Integration-by-Parts, and Regularization

The integral of a measurable function can be approximated by step functions, i.e., linear combinations of characteristic functions of measurable sets. Since for every measurable set and every invertible affine mapping $T(x) = Mx + b$ we have

$$\mu(A) = |\det M| \, \mu(T^{-1}(A)),$$

we can derive the following formula.

Proposition 2.7 (Transformation Formula, See [6, Chap. III]) *Let $\Phi : \mathbb{R}^d \to \mathbb{R}^d$ be a C^1-diffeomorphism, i.e., Φ is bijective and Φ and Φ^{-1} are differentiable, and let $f \in L^1(\widehat{\Omega})$ for $\widehat{\Omega} = \Phi(\Omega)$. Then we have*

$$\int_{\Phi(\Omega)} f(y) \, dy = \int_\Omega f(\Phi(x)) |\det D\Phi(x)| \, dx.$$

Important for the analysis of elliptic partial differential equations is *Gauss's theorem* which is also called the *divergence theorem*. We follow [8] for a sketch of the proof.

Proposition 2.8 (Gauss's Theorem, See [6, Chap. IV]) *Let $\Omega \subset \mathbb{R}^d$ be open and bounded such that $\partial\Omega$ is the finite union of C^1 graphs. For every vector field $F = [F_1, \ldots, F_d]^\top \in C^1(\overline{\Omega}; \mathbb{R}^d)$ we have*

$$\int_\Omega \operatorname{div} F \, dx = \int_{\partial\Omega} F \cdot n \, ds,$$

where $n(s)$ is for every $s \in \partial\Omega$ an outer unit normal.

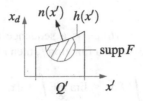

Fig. 2.3 Localization to a ball and parametrization of the boundary in the proof of Gauss's theorem

Proof (Sketched) By introducing a partition of unity and employing the transformation formula with an orthogonal transformation, we may assume that F is supported in a set Q given by

$$Q = \{(x', x_d) \in \mathbb{R}^{d-1} \times \mathbb{R} : x' \in Q', 0 \le x_d \le h(x')\},$$

cf. Fig. 2.3, and $F(x', x_d) = 0$ for $x' \in \partial Q'$ or $x_d = 0$. By Fubini's theorem we have

$$\int_Q \operatorname{div} F(x)\, \mathrm{d}x = \sum_{k=1}^{d} \int_{Q'} \int_0^{h(x')} \partial_k F_k(x', x_d)\, \mathrm{d}x_d\, \mathrm{d}x'.$$

For $k = d$ we use one-dimensional integration-by-parts to verify that

$$\int_0^{h(x')} \partial_d F_d(x', x_d)\, \mathrm{d}x_d = F_d\big(x', h(x')\big) - F_d(x', 0) = F_d\big(x', h(x')\big).$$

For $k = 1, 2, \ldots, d-1$, we define

$$f_k(x') = \int_0^{h(x')} F_k(x', x_d)\, \mathrm{d}x_d$$

and note that we have

$$\partial_k f_k(x') = \int_0^{h(x')} \partial_k F_k(x', x_d)\, \mathrm{d}x_d + \int_0^{h(x')} \partial_d F_k(x', x_d)\, \mathrm{d}x_d\, \partial_k h(x')$$

$$= \int_0^{h(x')} \partial_k F_k(x', x_d)\, \mathrm{d}x_d + F_k\big(x', h(x')\big)\partial_k h(x').$$

The fact that $f_k(x') = 0$ for $x' \in \partial Q'$ implies that

$$0 = \int_{Q'} \partial_k f_k(x')\, \mathrm{d}x'$$

for $k = 1, 2, \ldots, d-1$. This leads to

$$0 = \int_{Q'} \int_0^{h(x')} \partial_k F_k(x', x_d) \, dx_d \, dx' + \int_{Q'} F_k(x', h(x')) \partial_k h(x') \, dx'.$$

Hence we have

$$\int_Q \operatorname{div} F(x) \, dx = -\sum_{k=1}^{d-1} \int_{Q'} F_k(x', h(x')) \partial_k h(x') \, dx' + \int_{Q'} F_d(x', h(x')) \, dx'$$

$$= \int_{Q'} F(x', h(x')) \cdot \left[-\nabla' h(x'), 1 \right]^T dx'$$

$$= \int_S F \cdot n \, ds,$$

where we used $S = \{(x', h(x')) : x' \in Q'\} = \partial Q \cap \partial \Omega$, $\nabla' = [\partial_1, \ldots, \partial_{d-1}]^T$,

$$n(x') = (1 + |\nabla' h(x')|^2)^{-1/2} \left[-\nabla' h(x'), 1 \right]^T,$$

and that the surface integral involves the term $(1 + |\nabla' h(x')|^2)^{1/2}$. □

Lebesgue functions can be approximated by smooth functions. These can be constructed with *convolution kernels*, which are nonnegative compactly supported functions $J \in C^\infty(\mathbb{R}^d)$ with $\|J\|_{L^1(\mathbb{R}^d)} = 1$. An example is

$$J(x) = \begin{cases} c_d e^{-1/(1-|x|^2)} & \text{for } |x| < 1, \\ 0 & \text{for } |x| \geq 1, \end{cases}$$

with an appropriate constant $c_d > 0$. We have the following result.

Proposition 2.9 (Approximation by Smooth Functions, See [1, Chap. II]) *Let $u \in L^p(\Omega)$ and let $\tilde{u}$ be its extension by zero to $\mathbb{R}^d$. For a convolution kernel $J \in C^\infty(\mathbb{R}^d)$ and $\varepsilon > 0$, set $J_\varepsilon(x) = \varepsilon^{-d} J(x/\varepsilon)$, and define*

$$\tilde{u}_\varepsilon(x) = (J_\varepsilon * \tilde{u})(x) = \int_{\mathbb{R}^d} J_\varepsilon(x - y) \tilde{u}(y) \, dy.$$

Then, $\tilde{u}_\varepsilon$ is compactly supported with $\tilde{u}_\varepsilon \in C^\infty(\mathbb{R}^d)$. For $u_\varepsilon = \tilde{u}_\varepsilon|_\Omega \in C^\infty(\overline{\Omega})$, we have $\|u_\varepsilon\|_{L^p(\Omega)} \leq \|u\|_{L^p(\Omega)}$ and $u_\varepsilon \to u$ in $L^p(\Omega)$ as $\varepsilon \to 0$.

Proof (Sketched)

(i) By considering limits of difference quotients, one establishes with the dominated convergence theorem for every $\alpha \in \mathbb{N}_0^d$ that

$$\partial^\alpha \tilde{u}_\varepsilon(x) = \int_{\mathbb{R}^d} \partial^\alpha J_\varepsilon(x - y) \tilde{u}(y) \, dy.$$

(ii) With Hölder's inequality we verify that

$$|J_\varepsilon * \tilde{u}(x)| \leq \left(\int_{\mathbb{R}^d} J_\varepsilon(x - y) \, dy \right)^{1/p'} \left(\int_{\mathbb{R}^d} J_\varepsilon(x - y) |\tilde{u}(y)|^p \, dy \right)^{1/p}$$

and the first factor on the right-hand side equals 1. It follows that

$$\int_{\mathbb{R}^d} |J_\varepsilon * \tilde{u}(x)|^p \, dx \leq \int_{\mathbb{R}^d} \int_{\mathbb{R}^d} J_\varepsilon(x - y) |\tilde{u}(y)|^p \, dy \, dx = \|\tilde{u}\|^p_{L^p(\mathbb{R}^d)}.$$

(iii) The approximation property $u_\varepsilon \to u$ is a consequence of the estimate

$$\left| (J_\varepsilon * \phi)(x) - \phi(x) \right| \leq \int_{\mathbb{R}^d} J_\varepsilon(x - y) |\phi(y) - \phi(x)| \, dy = \sup_{|x-y| \leq c\varepsilon} |\phi(x) - \phi(y)|$$

for a compactly supported continuous function ϕ that approximates u.   □

2.3.3 Sobolev Spaces

As a consequence of Gauss's theorem, we have for every open set $\Omega \subset \mathbb{R}^d$ that

$$\int_\Omega u(\partial_i \phi) \, dx = - \int_\Omega (\partial_i u) \phi \, dx$$

for all $u \in C^1(\overline{\Omega})$ and all $\phi \in C_0^\infty(\Omega)$, where

$$C_0^\infty(\Omega) = \{ \phi \in C^\infty(\mathbb{R}^d) : \operatorname{supp} \phi \subset \Omega \}.$$

Note that $\operatorname{supp} \phi$ is closed so that ϕ vanishes in a neighborhood of $\partial\Omega$. We use the identity above to define derivatives for a class of Lebesgue functions.

Definition 2.10 Let $\Omega \subset \mathbb{R}^d$ be open and bounded. The function $u \in L^1(\Omega)$ is called *weakly differentiable* or a *Sobolev function* if for $i = 1, 2, \ldots, d$, there exists $u_i \in L^1(\Omega)$ such that

$$\int_\Omega u \, \partial_i \phi \, dx = - \int_\Omega u_i \phi \, dx$$

for all $\phi \in C_0^\infty(\Omega)$. We denote $\partial_i u = u_i$, called *weak partial derivatives* of u with respect to the i-th coordinate and define the *weak gradient* by $\nabla u = [\partial_1 u, \ldots, \partial_d u]^\top$.

Every function $u \in C^1(\overline{\Omega})$ is weakly differentiable and its weak and classical partial derivatives coincide. This is based on the following generalization of the fundamental lemma, cf. Lemma 2.1.

Lemma 2.4 (Fundamental Lemma, Refined Version, See [1, Chap. III]) *Let $h \in L^1(\Omega)$ be such that*

$$\int_\Omega h\phi \, dx \geq 0$$

for all $\phi \in C_0^\infty(\Omega)$ with $\phi \geq 0$ in Ω. We then have $h \geq 0$ in Ω.

Proof (Sketched) The proof follows from regularizing h.   □

Remark 2.8 By considering h and $-h$ it follows that if $\int_\Omega h\phi \, dx = 0$ for all $\phi \in C_0^\infty(\Omega)$, then we have that $h = 0$.

The consistency of weak and classical derivatives is an immediate consequence of this result.

Corollary 2.2 (Consistency of Weak Derivatives)

(i) *If $u \in L^1(\Omega)$ is weakly differentiable, then its weak derivatives are uniquely defined.*
(ii) *Every $u \in C^1(\overline{\Omega})$ is weakly differentiable. In particular, weak and classical derivatives coincide.*

Proof

(i) Assume that $u_i, v_i \in L^1(\Omega)$ are weak partial derivatives of u. We then have

$$\int_\Omega (u_i - v_i)\phi \, dx = 0$$

for all $\phi \in C_0^\infty(\Omega)$ and it follows from the fundamental lemma that $u_i = v_i$.
(ii) Let $u \in C^1(\overline{\Omega})$. Gauss's theorem implies that

$$\int_\Omega u \, \partial_i \phi \, dx = -\int_\Omega \partial_i u \, \phi \, dx$$

for all $\phi \in C_0^\infty(\Omega)$. Hence u is weakly differentiable with unique weak derivative $\partial_i u$.   □

Examples 2.4

(i) The function $u(x) = |x|$, $x \in \Omega = (-1, 1)$ is weakly differentiable with weak derivative $u'(x) = \text{sign}(x)$. To prove this, let $\phi \in C_0^\infty(\Omega)$ and use integration-by-parts on the intervals $(-1, 0)$ and $(0, 1)$, i.e.,

$$\int_{(-1,1)} |x|\phi'(x) \, dx = \int_{(-1,0)} (-x)\phi'(x) \, dx + \int_{(0,1)} x\phi'(x) \, dx$$

$$= -\int_{(-1,0)} (-1)\phi(x) \, dx - \int_{(0,1)} 1\phi(x) \, dx$$

$$= -\int_{(-1,1)} \text{sign}(x)\phi(x) \, dx.$$

(ii) The function $u(x) = \text{sign}(x)$, $x \in \Omega = (-1, 1)$, is not weakly differentiable. To verify this, assume that u is weakly differentiable. It then follows that $u'|_{(-1,0)} = 0$ and $u'|_{(0,1)} = 0$, i.e., $u' = 0$. But for $\phi \in C_0^\infty(\Omega)$ with $\phi(0) \neq 0$ we have

$$\int_{(-1,1)} \text{sign}(x)\phi'(x)\,dx = -\int_{(-1,0)} \phi'(x)\,dx + \int_{(0,1)} \phi'(x)\,dx = -2\phi(0),$$

which contradicts $u' = 0$.

The example can be generalized.

Proposition 2.10 (Continuous, Piecewise Differentiable Functions) *Assume that* $(\Omega_j)_{j=1,\ldots,J}$ *is an open partition of* Ω, *i.e.,* $\overline{\Omega} = \overline{\Omega}_1 \cup \cdots \cup \overline{\Omega}_J$, Ω_j *is open for* $j = 1, 2, \ldots, J$, *and* $\Omega_j \cap \Omega_\ell = \emptyset$ *for* $j \neq \ell$. *Let* $u \in C(\overline{\Omega})$ *be such that* $u|_{\Omega_j} \in C^1(\overline{\Omega}_j)$ *for* $j = 1, 2, \ldots, J$. *Then* u *is weakly differentiable.*

Proof Exercise.   □

Remark 2.9 Lipschitz continuous functions are weakly differentiable.

Weak derivatives of higher order are obtained inductively or equivalently by the following characterization in which we use the *multi-index notation* $\partial^\alpha \phi = \partial_1^{\alpha_1} \partial_2^{\alpha_2} \ldots \partial_d^{\alpha_d} \phi$ for $\alpha \in \mathbb{N}_0^d$.

Definition 2.11 The function $u \in L^1(\Omega)$ is *weakly differentiable of order* $k \geq 0$ if for every $\alpha \in \mathbb{N}_0^d$ with $|\alpha| = \alpha_1 + \alpha_2 + \cdots + \alpha_d \leq k$, there exists a function $u_\alpha \in L^1(\Omega)$ such that

$$\int_\Omega u\, \partial^\alpha \phi\, dx = (-1)^{|\alpha|} \int_\Omega u_\alpha \phi\, dx$$

for all $\phi \in C_0^\infty(\Omega)$. In this case we define $\partial^\alpha u = u_\alpha$ and for $0 \leq j \leq k$

$$D^j u = \left(\partial^\alpha u\right)_{|\alpha|=j}.$$

The Frobenius norm of $D^j u(x)$ is for almost every $x \in \Omega$ defined via $|D^j u(x)| = \left(\sum_{|\alpha|=j} |\partial^\alpha u(x)|^2\right)^{1/2}$.

We are now in a position to define Sobolev spaces.

Definition 2.12 For $k \in \mathbb{N}_0$ and $p \in [1, \infty]$ the *Sobolev space* $W^{k,p}(\Omega)$ consists of all $u \in L^p(\Omega)$ such that all weak partial derivatives $\partial^\alpha u$ with $|\alpha| \leq k$ satisfy $\partial^\alpha u \in L^p(\Omega)$. The space is equipped with the norm

$$\|u\|_{W^{k,p}(\Omega)} = \left(\sum_{|\alpha| \leq k} \|\partial^\alpha u\|_{L^p(\Omega)}^p\right)^{1/p}$$

if $1 \leq p < \infty$ and

$$\|u\|_{W^{k,\infty}(\Omega)} = \max_{|\alpha| \leq k} \|\partial^\alpha u\|_{L^\infty(\Omega)}$$

if $p = \infty$. If $p = 2$, we abbreviate $H^k(\Omega) = W^{k,2}(\Omega)$.

Proposition 2.11 (Completeness) *For $k \in \mathbb{N}_0$ and $p \in [1, \infty]$ the space $W^{k,p}(\Omega)$ is a Banach space. For $p = 2$ it is a Hilbert space with scalar product*

$$\langle u, v \rangle_{H^k(\Omega)} = \sum_{|\alpha| \leq k} \int_\Omega \partial^\alpha u \, \partial^\alpha v \, dx.$$

Proof It is straightforward to verify that the space $W^{k,p}(\Omega)$ is a normed space and a pre-Hilbert space if $p = 2$. To show that it is complete, let $(u_j)_{j \in \mathbb{N}}$ be a Cauchy sequence in $W^{k,p}(\Omega)$. Then, due to the definition of the norm on $W^{k,p}(\Omega)$, for every $\alpha \in \mathbb{N}_0^d$ with $|\alpha| \leq k$, the sequence $(\partial^\alpha u_j)_{j \in \mathbb{N}}$ is a Cauchy sequence in the Banach space $L^p(\Omega)$ and has a limit $u_\alpha \in L^p(\Omega)$. It remains to show that $u = u_{(0,\dots,0)}$ belongs to $W^{k,p}(\Omega)$, i.e., that it has weak derivatives up to order k. Let $\alpha \in \mathbb{N}_0^d$ with $|\alpha| \leq k$ and $\phi \in C_0^\infty(\Omega)$. For every $j \in \mathbb{N}$ we have

$$\int_\Omega u_j \partial^\alpha \phi \, dx = (-1)^{|\alpha|} \int_\Omega \partial^\alpha u_j \phi \, dx.$$

Hölder inequalities imply that in the limit for $j \to \infty$ this becomes

$$\int_\Omega u \, \partial^\alpha \phi \, dx = (-1)^{|\alpha|} \int_\Omega u_\alpha \phi \, dx.$$

Since this holds for all $\phi \in C_0^\infty(\Omega)$, we find that u is weakly differentiable with $\partial^\alpha u = u_\alpha$. This implies that $u \in W^{k,p}(\Omega)$ and proves the completeness property. $\square$

Remark 2.10 Sobolev spaces are separable if $1 \leq p < \infty$ and reflexive if $1 < p < \infty$.

As in the case of Lebesgue functions, Sobolev functions can be approximated by smooth functions if $1 \leq p < \infty$. The proof of the following theorem is highly technical.

Theorem 2.6 (Meyers–Serrin, See [1, Chap. III]) *For $k \in \mathbb{N}_0$ and $1 \leq p < \infty$, the space $C^\infty(\Omega) \cap W^{k,p}(\Omega)$ is dense in $W^{k,p}(\Omega)$.*

Some important properties of weak derivatives are listed in the following proposition, see e.g., [6, Chap. IV] for details.

Proposition 2.12 (Weak Derivatives Calculus)

(i) *If $\nabla u = 0$ and Ω is connected, then u is constant.*
(ii) *If $u, v \in W^{1,2}(\Omega)$, then $uv \in W^{1,1}(\Omega)$ and $\nabla(uv) = u\nabla v + v\nabla u$.*

(iii) If $g \in C^1(\mathbb{R})$ with $|g'| \leq C$ and $u \in W^{1,p}(\Omega)$, $1 \leq p < \infty$, then $\tilde{u} = g \circ u \in W^{1,p}(\Omega)$ with $\nabla\tilde{u} = g'(u)\nabla u$.

(iv) If $u \in W^{1,p}(\Omega)$, $1 \leq p < \infty$, then $|u| \in W^{1,p}(\Omega)$ with $\nabla|u| = \text{sign}(u)\nabla u$, where $\text{sign}(0) = 0$.

2.3.4 Traces and the Subspace $W_D^{1,p}(\Omega)$

In order to deal with Dirichlet boundary conditions in boundary value problems, it is necessary to assign a meaning to boundary values of Sobolev functions. For this, a stronger density result is needed.

Definition 2.13 A set $\Omega \subset \mathbb{R}^d$ is called *Lipschitz domain*, if it is open and connected, and if for each $x \in \partial\Omega$ there exists a transformation $\Phi(y) = My + r$ with an orthogonal matrix $M \in \mathbb{R}^{d \times d}$ and a vector $r \in \mathbb{R}^d$, a parameter $\delta > 0$, an open set $Q' \subset \mathbb{R}^{d-1}$, and a Lipschitz continuous function $h : Q' \to \mathbb{R}$ such that

$$\Omega \cap B_\delta(x) = \Phi(\{(y', y_d) \in Q' \times \mathbb{R} : h(y') < y_d\}) \cap B_\delta(x),$$

$$\partial\Omega \cap B_\delta(x) = \Phi(\{(y', y_d) \in Q' \times \mathbb{R} : h(y') = y_d\}) \cap B_\delta(x),$$

$$\overline{\Omega}^c \cap B_\delta(x) = \Phi(\{(y', y_d) \in Q' \times \mathbb{R} : h(y') > y_d\}) \cap B_\delta(x).$$

The transformations and parametrizations are illustrated in Fig. 2.4.

Examples of domains that are not Lipschitz domains are indicated in Fig. 2.5.

Remarks 2.11

(i) Lipschitz domains are open and connected sets, whose boundary is locally parameterized by a Lipschitz continuous function and which lie locally on one side of their boundary.

(ii) On Lipschitz domains we have that the space $W^{1,\infty}(\Omega)$ can be identified with the space of Lipschitz continuous functions on Ω.

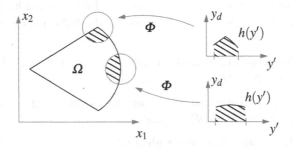

Fig. 2.4 Local Lipschitz continuous parametrizations of the boundary

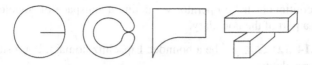

Fig. 2.5 Sets which are not Lipschitz domains

Smooth, bounded functions are dense in Sobolev spaces on Lipschitz domains if $p < \infty$.

Theorem 2.7 (Density of Smooth Functions, See [1, Chap. III]) *Let $\Omega \subset \mathbb{R}^d$ be a bounded Lipschitz domain. Then the set $C^\infty(\overline{\Omega})$ is dense in $W^{k,p}(\Omega)$ for $k \in \mathbb{N}_0$ and $1 \leq p < \infty$.*

Remark 2.12 If Ω is not a Lipschitz domain, then the statement is false in general, as the example $W^{1,2}((-1,1) \setminus \{0\})$ shows.

Since Lebesgue functions can be modified on sets of measure zero, it is in general not meaningful to specify boundary values. For Sobolev functions this is possible. We remark that Lebesgue spaces on submanifolds are defined via local parametrizations.

Proposition 2.13 (Trace Operator, See [1, Chap. IV]) *Let $\Omega \subset \mathbb{R}^d$ be a bounded Lipschitz domain and $1 \leq p \leq \infty$. There exists a uniquely defined bounded linear operator $\gamma : W^{1,p}(\Omega) \to L^p(\partial\Omega)$, such that for all $u \in C^\infty(\overline{\Omega}) \cap W^{1,p}(\Omega)$, we have $\gamma(u) = u|_{\partial\Omega}$.*

Proof (Sketched) We consider the case $p = 1$ and show that the restriction $\gamma(u) = u|_{\partial\Omega}$ satisfies

$$\|\gamma(u)\|_{L^1(\partial\Omega)} \leq c\big(\|u\|_{L^1(\Omega)} + \|\nabla u\|_{L^1(\Omega)}\big)$$

for all $u \in C^\infty(\overline{\Omega})$. By density, the operator can be extended to a bounded linear operator on $W^{1,1}(\Omega)$. The case $p > 1$ then follows by replacing u with $|u|^\gamma$ and a suitable number γ. To indicate the proof of the asserted estimate, we use a partition of unity and assume that u is supported in a neighborhood of a boundary point. After an appropriate reparametrization we may assume that this neighborhood is given by $Q' \times (0,1)$ for some set $Q' \subset \mathbb{R}^{d-1}$ and the boundary part is $Q' \times \{0\}$. We then have, using $u(x', 1) = 0$, that

$$u(x', 0) = -\int_0^1 \partial_d u(x', x_d)\, dx_d.$$

An integration of this identity over $x' \in Q'$ leads to

$$\|u\|_{L^1(\partial\Omega)} \leq \int_{Q'}\int_0^1 |\partial_d u(x', x_d)|\, dx_d\, dx' \leq \|\nabla u\|_{L^1(Q)}.$$

For functions with arbitrary support in Ω, we obtain the asserted estimate. $\square$

With the continuous trace operator we define a subspace of Sobolev functions vanishing on a part of the boundary.

Definition 2.14 Let $\Omega \subset \mathbb{R}^d$ be a bounded Lipschitz domain. For a set $\Gamma_D \subset \partial\Omega$ and $1 \le p \le \infty$, define

$$W_D^{1,p}(\Omega) = \{v \in W^{1,p}(\Omega) : \gamma(v)|_{\Gamma_D} = 0\}.$$

For $p = 2$, we abbreviate $H_D^1(\Omega) = W_D^{1,2}(\Omega)$. For $\Gamma_D = \partial\Omega$ we set $H_0^1(\Omega) = H_D^1(\Omega)$.

Instead of $\gamma(v)$ or $\gamma(v)|_{\Gamma_D}$ we typically write $v|_{\partial\Omega}$ or $v|_{\Gamma_D}$. Due to the continuity of γ, the subspace $W_D^{1,p}(\Omega)$ is closed.

Remark 2.13 The space $W_D^{1,p}(\Omega)$ is a Banach space for $1 \le p \le \infty$ and for $p = 2$ it is a Hilbert space.

Functions in $W_D^{1,p}(\Omega)$ whose gradients vanish are constantly zero. The following proposition is a quantitative version of this implication. It is also known as *Friedrichs inequality*.

Proposition 2.14 (Poincaré Inequality, See [10, Chap. IV]) *Let $\Omega \subset \mathbb{R}^d$ be a bounded Lipschitz domain, assume that $\Gamma_D \subset \partial\Omega$ has positive surface measure, and let $1 \le p \le \infty$. Then there exists a constant $C_P > 0$ such that for all $u \in W_D^{1,p}(\Omega)$ we have*

$$\|u\|_{L^p(\Omega)} \le c_P \|\nabla u\|_{L^p(\Omega)}.$$

Proof (Sketched) We provide a proof for the case that $\Gamma_D = \partial\Omega$ and $p = 2$. In this case the set $C_0^\infty(\Omega)$ is dense in $H_D^1(\Omega)$ and it suffices to prove the estimate for $u \in C_0^\infty(\Omega)$. We let u be such a function and extend it by zero to a function $\tilde{u} \in C_0^\infty(\mathbb{R}^d)$. Since Ω is bounded there exists $r > 0$ such that $\operatorname{supp}\tilde{u} \subset B_r(0)$. Using one-dimensional integration-by-parts and Hölder's inequality, we have for every $x \in \Omega$ that

$$|\tilde{u}(x)|^2 = |\tilde{u}(x_1, x_2, \ldots, x_d) - \tilde{u}(-r, x_2, \ldots, x_d)|^2$$

$$= \left| \int_{-r}^{x_1} \partial_1 \tilde{u}(s, x_2, \ldots, x_d) \, ds \right|^2$$

$$\le \left(\int_{-r}^{x_1} 1 \, ds \right) \left(\int_{-r}^{x_1} |\partial_1 \tilde{u}(s, x_2, \ldots, x_d)|^2 \, ds \right)$$

$$\le 2r \int_{-r}^{r} |\partial_1 \tilde{u}(s, x_2, \ldots, x_d)|^2 \, ds.$$

An integration over Ω implies that

$$\|\tilde{u}\|_{L^2(\Omega)}^2 \leq 2r \int_\Omega \int_{-r}^{r} |\partial_1 \tilde{u}(s, x_2, \ldots, x_d)|^2 \, ds \, dx \leq 4r^2 \|\nabla \tilde{u}\|_{L^2(\Omega)}^2.$$

Noting $u = \tilde{u}$ on Ω, we deduce the estimate. □

2.3.5 Sobolev Embeddings

While Lebesgue functions can have nearly arbitrary jumps and discontinuities, these effects are limited for Sobolev functions. If, e.g., $\Omega = (0, 1)$ and $u \in H^1(\Omega)$, Hölder's inequality shows that

$$|u(x_2) - u(x_1)| = \left| \int_{x_1}^{x_2} u'(x) \, dx \right| \leq (x_2 - x_1)^{1/2} \|u'\|_{L^2(\Omega)}.$$

This implies that we can modify u on a set of measure zero to obtain a continuous and uniformly bounded function. This observation can be generalized.

Definition 2.15 For $1 \leq p \leq \infty$, the *Sobolev (conjugate) exponent* $p^* \in [1, \infty]$ of p is defined by

$$p^* = \begin{cases} dp/(d - p), & \text{if } p < d, \\ 1 \leq q < \infty, & \text{if } p = d, \\ \infty, & \text{if } p > d, \end{cases}$$

where $1 \leq q < \infty$ stands for an arbitrary number $q \in \mathbb{R}$ with $q \geq 1$.

Theorem 2.8 (Sobolev Inequalities, See [1, Chap. IV]) *Let $\Omega \subset \mathbb{R}^d$ be a bounded Lipschitz domain. For $1 \leq p < \infty$, and $1 \leq q \leq p^*$, we have that every $u \in W^{1,p}(\Omega)$ satisfies $u \in L^q(\Omega)$. In particular, there exists a constant $c_S > 0$, such that we have*

$$\|u\|_{L^q(\Omega)} \leq c_S \|u\|_{W^{1,p}(\Omega)}$$

for all $u \in W^{1,p}(\Omega)$, i.e., the embedding $W^{1,p}(\Omega) \to L^q(\Omega)$ is continuous.

The proof of the result is based on an extension of u to a function $\hat{u} \in W^{1,p}(\widehat{\Omega})$, such that $\hat{u} = 0$ on $\partial\widehat{\Omega}$ and $\Omega \subset \widehat{\Omega}$ and an integration argument as in the proof of the Poincaré inequality. If one restricts to functions in $W_0^{1,p}(\Omega)$ then the Lipschitz condition on Ω can be omitted.

Remark 2.14 The condition of the theorem cannot be improved, e.g., the function $u(x) = \log(\log(|x|))$ satisfies $u \in H^1(B_{1/2}(0))$ for $d = 2$ but does not belong to $L^\infty(B_{1/2}(0))$.

If one restricts to $1 \leq q < p^*$, then every bounded sequence in $W^{1,p}(\Omega)$ has a strongly convergent subsequence in $L^q(\Omega)$.

Theorem 2.9 (Rellich–Kandrachov, See [1, Chap. VI]) *Let Ω be a bounded Lipschitz domain, $1 \leq p < \infty$, and $1 \leq q < p^*$. Then the embedding $W^{1,p}(\Omega) \to L^q(\Omega)$ is compact.*

The proof of this result follows from regularizations, applications of the Arzelà–Ascoli characterization of compact subsets of $C^0(\Omega)$, and careful passages to limits. The compactness results allow for indirect proofs of *Poincaré type inequalities*.

Corollary 2.3 (Poincaré Inequality with Vanishing Mean) *Let $\Omega \subset \mathbb{R}^d$ be a bounded Lipschitz domain and $1 \leq p < \infty$. Then there exists a constant $c_P > 0$ such that*

$$\|u\|_{L^p(\Omega)} \leq c_P \|\nabla u\|_{L^p(\Omega)}$$

for all $u \in W^{1,p}(\Omega)$ with $\int_\Omega u \, dx = 0$.

Proof Assume that there exists no such constant $c_P > 0$. This implies that for every $j \in \mathbb{N}$ we can find $\tilde{u}_j \in W^{1,p}(\Omega)$ such that

$$\|\tilde{u}_j\|_{L^p(\Omega)} > j \|\nabla \tilde{u}_j\|_{L^p(\Omega)}$$

and $\int_\Omega \tilde{u}_j \, dx = 0$. We define $u_j = \tilde{u}_j / \|\tilde{u}_j\|_{L^p(\Omega)}$ so that $\|u_j\|_{L^p(\Omega)} = 1$, $\|\nabla u_j\|_{L^p(\Omega)} < 1/j$, and $\int_\Omega u_j \, dx = 0$. We have that $\nabla u_j \to 0$ in $L^p(\Omega)$ as $j \to \infty$. Due to the compact embedding there exists a subsequence $(u_{j_\ell})_{\ell \in \mathbb{N}}$ such that $u_{j_\ell} \to u$ in $L^p(\Omega)$. Hence the sequence $(u_{j_\ell})_{\ell \in \mathbb{N}}$ is a Cauchy sequence in $W^{k,p}(\Omega)$ with limit $u \in W^{k,p}(\Omega)$. Since $\nabla u_{j_\ell} \to 0$ we have $\nabla u = 0$, i.e., that u is constant. Since also $\int_\Omega u \, dx = 0$, it follows that $u = 0$. This contradicts $\|u\|_{L^p(\Omega)} = \lim_{\ell \to \infty} \|u_{j_\ell}\|_{L^p(\Omega)} = 1$ and proves the estimate. $\qquad\square$

Sobolev functions $u \in W^{1,p}(\Omega)$ with $p > d$ can be identified with continuous functions. In general, this is not the case, as the function $u(x) = \log(\log(|x|))$ for $p = d = 2$ shows.

Theorem 2.10 (Morrey, See [1, Chap. IV]) *Let $\Omega \subset \mathbb{R}^d$ be a bounded Lipschitz domain and assume $k > d/p$. Then there exists $c_M > 0$ such that for all $u \in W^{k,p}(\Omega)$ we have $u \in C(\overline{\Omega}) \cap L^\infty(\Omega)$ and*

$$\|u\|_{L^\infty(\Omega)} \leq c_M \|u\|_{W^{k,p}(\Omega)},$$

i.e., the embedding $W^{k,p}(\Omega) \to C(\overline{\Omega}) \cap L^\infty(\Omega)$ is continuous.

Remark 2.15 Morrey's theorem remains true under the weaker *cone condition* on Ω instead of the Lipschitz property. This condition requires that there exists an open cone $C \subset \mathbb{R}^d$ with vertex 0, such that for every $x \in \Omega$, there exists a rotation $Q \in \mathbb{R}^{d \times d}$ so that $\widetilde{C} = QC + x \subset \Omega$.

2.4 Weak Solutions

2.4.1 Existence and Uniqueness

With the identification of the Sobolev space $H_D^1(\Omega)$ as the completion of the set of functions in $C^1(\overline{\Omega})$ that vanish on Γ_D, we are in a position to apply the Lax–Milgram lemma to the weak formulation of the Poisson problem. Key ingredients are the Poincaré and Hölder inequalities.

Definition 2.16 A function $u \in H^1(\Omega)$ is called a *weak solution* of the Poisson problem if it satisfies $u|_{\Gamma_D} = u_D$ in the sense of traces and

$$\int_\Omega \nabla u \cdot \nabla v \, dx = \int_\Omega f v \, dx + \int_{\Gamma_N} g v \, ds$$

for all $v \in H_D^1(\Omega)$.

Theorem 2.11 (Existence and Uniqueness) *Assume that $\Omega \subset \mathbb{R}^d$ is a bounded Lipschitz domain, $\Gamma_D \subset \partial\Omega$ a closed set of positive surface measure, and $\Gamma_N = \partial\Omega \setminus \Gamma_D$. Then for every $\tilde{u}_D \in H^1(\Omega)$, $u_D = \tilde{u}_D|_{\Gamma_D}$, $f \in L^2(\Omega)$, and $g \in L^2(\Gamma_N)$, there exists a unique weak solution of the Poisson problem.*

Proof The function $u \in H_D^1(\Omega)$ is a weak solution of the Poisson problem if and only if $u = \hat{u} + \tilde{u}_D$ and $\hat{u} \in H_D^1(\Omega)$ satisfies

$$\int_\Omega \nabla \hat{u} \cdot \nabla v \, dx = \int_\Omega f v \, dx + \int_{\Gamma_N} g v \, ds - \int_\Omega \nabla \tilde{u}_D \cdot \nabla v \, dx$$

for all $v \in H_D^1(\Omega)$. We let $a(\hat{u}, v)$ and $b(v)$ denote the left- and right-hand side of this formulation, respectively. The bilinear form $a : H_D^1(\Omega) \times H_D^1(\Omega) \to \mathbb{R}$ is coercive, since by the Poincaré inequality we have

$$a(u, u) = \int_\Omega |\nabla u|^2 \, dx = \|\nabla u\|_{L^2(\Omega)}^2 \geq (c_P^2 + 1)^{-1} \|u\|_{H^1(\Omega)}^2.$$

With Hölder's inequality and the trivial estimate $\|\nabla v\|_{L^2(\Omega)} \leq \|v\|_{H^1(\Omega)}$, we find that

$$|a(u, v)| \leq \int_\Omega |\nabla u| |\nabla v| \, dx \leq \|\nabla u\|_{L^2(\Omega)} \|\nabla v\|_{L^2(\Omega)} \leq \|u\|_{H^1(\Omega)} \|v\|_{H^1(\Omega)}.$$

In order to apply the Lax–Milgram lemma it remains to show that the linear mapping b is bounded. Hölder inequalities imply that

$$|b(v)| \leq \|f\|_{L^2(\Omega)}\|v\|_{L^2(\Omega)} + \|g\|_{L^2(\Gamma_N)}\|v\|_{L^2(\Gamma_N)} + \|\nabla \tilde{u}_D\|_{L^2(\Omega)}\|\nabla v\|_{L^2(\Omega)}.$$

With the help of the trace inequality we verify that

$$\|v\|_{L^2(\Gamma_N)} \leq \|v\|_{L^2(\partial\Omega)} \leq c_{\mathrm{Tr}}\|v\|_{H^1(\Omega)}.$$

Hence we have that

$$|b(v)| \leq c\|v\|_{H^1(\Omega)}.$$

Altogether we see that a is a coercive and bounded bilinear mapping on $H_D^1(\Omega)$, and b is a bounded linear functional on $H^1(\Omega)$. Hence the Lax–Milgram lemma implies the existence of a unique weak solution.   □

The result of the theorem can be generalized to a class of elliptic partial differential equations.

Remark 2.16 Assume that f, g, and u_D are as above. Let $K \in C(\overline{\Omega}; \mathbb{R}^{d \times d})$, $b \in C^1(\overline{\Omega}; \mathbb{R}^d)$, and $c \in C(\overline{\Omega})$ such that

$$p^\top K(x)p \geq C_K|p|^2$$

for some $C_K > 0$ and almost every $x \in \Omega$, and all $p \in \mathbb{R}^d$, and

$$c - \frac{1}{2}\operatorname{div} b \geq 0 \text{ in } \Omega, \quad b \cdot n \geq 0 \text{ on } \Gamma_N.$$

Then there exists a unique weak solution of the elliptic boundary value problem

$$\begin{cases} -\operatorname{div}(K\nabla u) + b \cdot \nabla u + cu = f & \text{in } \Omega, \\ u = u_D & \text{on } \Gamma_D, \\ (K\nabla u) \cdot n = g & \text{on } \Gamma_N. \end{cases}$$

2.4.2 Galerkin Approximation

To approximate the weak solution of the Poisson problem, we replace the space $H_D^1(\Omega)$ by a finite-dimensional subspace $V_h \subset H_D^1(\Omega)$.

Proposition 2.15 (Céa's Lemma) *Let V be a Hilbert space, let $a : V \times V \to \mathbb{R}$ be a bounded and coercive bilinear form, and let $b : V \to \mathbb{R}$ be a bounded and linear form. Given any finite-dimensional subspace $V_h \subset V$, there exists a unique Galerkin* approximation $u_h \in V_h$ *that satisfies*

$$a(u_h, v_h) = b(v_h)$$

for all $v_h \in V_h$. *With the weak solution* $u \in V$ *defined by*

$$a(u, v) = b(v)$$

for all $v \in V$, *we have the* Galerkin orthogonality

$$a(u - u_h, v_h) = 0$$

for all $v_h \in V_h$. *Moreover, we have the* quasi-best-approximation property

$$\|u - u_h\|_V \le (k_a/\alpha) \inf_{w_h \in V_h} \|u - w_h\|_V.$$

Proof

(i) The finite-dimensional subspace V_h is complete and we apply the Lax–Milgram lemma with V_h to deduce the existence of a unique solution $u_h \in V_h$ that satisfies

$$a(u_h, v_h) = b(v_h)$$

for all $v_h \in V_h$.

(ii) Due to the inclusion $V_h \subset V$, the exact solution $u \in V$ satisfies $a(u, v_h) = b(v_h)$ for all $v_h \in V_h$. Hence the bilinearity of a implies that

$$a(u - u_h, v_h) = a(u, v_h) - a(u_h, v_h) = b(v_h) - b(v_h) = 0.$$

(iii) The coercivity of a, the Galerkin orthogonality, and the boundedness of a show that

$$\alpha \|u - u_h\|_V^2 \le a(u - u_h, u - u_h)$$
$$= a(u - u_h, u - u_h - v_h)$$
$$\le k_a \|u - u_h\|_V \|u - u_h - v_h\|_V.$$

Replacing $w_h = u_h + v_h$ proves the quasi-best-approximation property. $\quad\square$

By choosing a basis for the subspace V_h, the Galerkin approximation is obtained via the solution of a linear system of equations with positive definite matrix. We use the same notation for the right-hand side vector $b \in \mathbb{R}^n$ and the functional $b \in V'$.

Proposition 2.16 (Stiffness Matrix) *Let* $a : V \times V \to \mathbb{R}$ *and let* $b : V \to \mathbb{R}$ *be as above. Assume that* $(\varphi_1, \varphi_2, \dots, \varphi_n)$ *is a basis for* V_h. *Define* $A \in \mathbb{R}^{n \times n}$ *and* $b \in \mathbb{R}^n$ *by*

$$A_{jk} = a(\varphi_j, \varphi_k), \quad b_j = b(\varphi_j)$$

for $j, k = 1, 2, \ldots, n$. Then A is positive definite, and for the unique solution $U = [U_1, U_2, \ldots, U_n]^\top \in \mathbb{R}^n$ of the linear system of equations

$$A^\top U = b,$$

the Galerkin approximation $u_h \in V_h$ is given by

$$u_h = \sum_{j=1}^n U_j \varphi_j.$$

Proof

(i) Let $U \in \mathbb{R}^n$ and define $u_h = \sum_{j=1}^n U_j \varphi_j$. We then have

$$U^\top A U = \sum_{j,k=1}^n U_j A_{jk} U_k$$

$$= \sum_{j,k=1}^n a(U_j \varphi_j, U_k \varphi_k)$$

$$= a(u_h, u_h) \geq \alpha \|u_h\|_V^2.$$

Since $u_h = 0$ if and only if $U = 0$ we deduce that A is positive definite and regular.

(ii) For the unique solution $U \in \mathbb{R}^n$ of the linear system $A^\top U = b$ and every $V \in \mathbb{R}^n$, we have

$$(A^\top U)^\top V = b^\top V$$

which is equivalent to

$$a(u_h, v_h) = b(v_h)$$

for $v_h = \sum_{j=1}^n V_j \varphi_j$. This shows that u_h is the Galerkin approximation.   □

2.4.3 Regularity

To derive convergence rates for numerical methods, it is important to identify higher differentiability properties of weak solutions.

Definition 2.17 A second-order elliptic boundary value problem is called H^2-*regular*, if for $u_D = 0$, $g \in L^2(\Gamma_N)$, and $f \in L^2(\Omega)$, its weak solution $u \in H_D^1(\Omega)$ satisfies $u \in H^2(\Omega)$ and there exists $c_L > 0$ such that

$$\|D^2 u\|_{L^2(\Omega)} \leq c_L \big(\|f\|_{L^2(\Omega)} + \|g\|_{L^2(\Gamma_N)} \big).$$

We prove H^2-regularity for the Poisson problem in a convex domain Ω with piecewise smooth boundary and $\Gamma_D = \partial\Omega$ following [5].

Theorem 2.12 (H^2-Regularity of the Poisson Problem, See [7, Chap. VIII]) *Assume that Ω is convex and $\partial\Omega$ is piecewise C^2, and let $\Gamma_D = \partial\Omega$ and $u_D = 0$. Then the Poisson problem is H^2-regular with $c_\Delta = 1$.*

Proof An exercise shows that for $u \in C^3(\overline{\Omega})$, we have

$$|\Delta u|^2 - |D^2 u|^2 = \operatorname{div}\left(\nabla u \Delta u - \frac{1}{2}\nabla|\nabla u|^2\right).$$

Integrating this identity over Ω and applying Gauss's theorem leads to

$$\int_\Omega |\Delta u|^2 - |D^2 u|^2 \, dx = -\int_{\partial\Omega} \nabla u \cdot n \, \Delta u - \frac{1}{2}(\nabla|\nabla u|^2) \cdot n \, ds.$$

We will show that if u vanishes on $\partial\Omega$, then the integrand of the boundary term is nonnegative so that

$$\int_\Omega |D^2 u|^2 \, dx \le \int_\Omega |\Delta u|^2 \, dx.$$

By the density of smooth functions, the estimate also holds for $u \in H^2(\Omega)$. In particular, the right-hand side is finite for $u \in H^1(\Omega)$ with $\Delta u \in L^2(\Omega)$, and the estimate implies $u \in H^2(\Omega)$. For the solution of the Poisson problem we have $-\Delta u = f$ and deduce H^2-regularity with $c_\Delta = 1$. To prove the nonnegativity property, we consider a point $z \in \partial\Omega$ at which $\partial\Omega$ is C^2. By choosing an appropriate coordinate system, we may assume that in a neighborhood of z the points on $\partial\Omega$ are given by

$$x = \left(x', h(x')\right)$$

with a concave C^2 function $h : Q' \to \mathbb{R}$ such that $\nabla' h(z') = [\partial_1 h(z'), \dots, \partial_{d-1} h(z')]^\top = 0$, cf. Fig. 2.6.

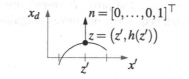

Fig. 2.6 Boundary point $z = (z', h(z')) \in \partial\Omega$ at which the boundary is C^2; after an appropriate choice of coordinates we have $\nabla' h(z') = 0$

Noting that the outer unit normal at z is then given by $n = [0, \ldots, 0, 1]^{\top}$, we have for

$$\psi = \nabla u \cdot n \, \Delta u - \frac{1}{2} (\nabla |\nabla u|^2) \cdot n,$$

that

$$\psi(z) = \partial_d u(z) \Delta u(z) - \frac{1}{2} \partial_d |\nabla u|^2(z)$$

$$= \partial_d u(z) \Delta u(z) - \sum_{i=1}^{d} \partial_i u(z) \partial_i \partial_d u(z)$$

$$= \sum_{i=1}^{d-1} \partial_d u(z) \partial_i^2 u(z) - \sum_{i=1}^{d-1} \partial_i u(z) \partial_i \partial_d u(z).$$

Since $u|_{\partial \Omega} = 0$ and hence $u(x', h(x')) = 0$ for all $x' \in Q'$, we deduce that for $i, j = 1, 2, \ldots, d-1$,

$$0 = \partial_i u + \partial_d u \partial_i h,$$

$$0 = \partial_i \partial_j u + \partial_d \partial_i u \partial_j h + \partial_d \partial_j u \partial_i h + \partial_d^2 u \partial_i h \partial_j h + \partial_d u \partial_i \partial_j h.$$

Using $\nabla' h(z') = 0$, the first identity implies that at the point $z = (z', h(z'))$ we have $\partial_i u(z) = 0$. The second equation then leads to

$$0 = \partial_i \partial_j u(z) + \partial_d u(z) \partial_i \partial_j h(z').$$

Using this with $i = j$ and $\partial_i u(z) = 0$ in the identity for $\psi(z)$ shows that

$$\psi(z) = -(\partial_d u(z))^2 \sum_{i=1}^{d-1} \partial_i^2 h(z').$$

Since h is concave we deduce that $\psi(z) \geq 0$. $\square$

Remark 2.17 More generally, one can prove H^2-regularity for a class of elliptic second-order partial differential equations on convex domains and on domains with C^2-boundary.

References

Specialized textbooks on weakly differentiable functions and Sobolev spaces are the references [1, 6, 11]. Introductions to Lebesgue measure are contained in [6, 8]. Details about results from functional analysis can be found in [4, 10]. Aspects of functional analysis and partial differential equations are discussed in [2–5, 9]. Regularity properties of elliptic partial differential equations can be found in [7].

1. Adams, R.A., Fournier, J.J.F.: Sobolev spaces. Pure and Applied Mathematics (Amsterdam), vol. 140, 2nd edn. Elsevier/Academic Press, Amsterdam (2003)
2. Alt, H.W.: Lineare Funktionalanalysis, 6th edn. Springer-Lehrbuch Masterclass. Springer, Berlin-Heidelberg-New York (2012)
3. Attouch, H., Buttazzo, G., Michaille, G.: Variational analysis in Sobolev and BV spaces. MOS-SIAM Series on Optimization, 2nd edn. Society for Industrial and Applied Mathematics (SIAM), Philadelphia, PA; Mathematical Optimization Society, Philadelphia, PA (2014). URL http://dx.doi.org/10.1137/1.9781611973488
4. Brezis, H.: Functional Analysis, Sobolev Spaces and Partial Differential Equations. Universitext. Springer, New York (2011)
5. Dobrowolski, M.: Angewandte Funktionalanalysis. Springer, Berlin-Heidelberg-New York (2006)
6. Evans, L.C., Gariepy, R.F.: Measure theory and fine properties of functions. Studies in Advanced Mathematics. CRC Press, Boca Raton, FL (1992)
7. Gilbarg, D., Trudinger, N.S.: Elliptic partial differential equations of second order. Classics in Mathematics. Springer, Berlin (2001)
8. Königsberger, K.: Analysis. 1, 6th edn. Springer-Lehrbuch. [Springer Textbook]. Springer, Berlin (2004). URL http://dx.doi.org/10.1007/978-3-642-18490-1
9. Reddy, B.D.: Introductory functional analysis. Texts in Applied Mathematics, vol. 27. Springer, New York (1998). URL http://dx.doi.org/10.1007/978-1-4612-0575-3
10. Yosida, K.: Functional analysis. Grundlehren der Mathematischen Wissenschaften [Fundamental Principles of Mathematical Sciences], vol. 123, 6th edn. Springer, Berlin-New York (1980)
11. Ziemer, W.P.: Weakly differentiable functions. Graduate Texts in Mathematics, vol. 120. Springer, New York (1989). URL http://dx.doi.org/10.1007/978-1-4612-1015-3

Chapter 3
Finite Element Method

3.1 Interpolation with Finite Elements

3.1.1 Abstract Finite Elements

Finite elements generalize the concept of splines, i.e., using functions that are piecewise polynomials, and have certain global continuity or differentiability properties. We follow [16].

Definition 3.1 For a closed set $T \subset \mathbb{R}^d$ and $k \in \mathbb{N}_0$, we define the space of *polynomials of (total) degree k* restricted to T by

$$\mathscr{P}_k(T) = \{ v \in C(T) : v(x) = \sum_{\alpha \in \mathbb{N}_0^d, |\alpha| \le k} a_\alpha x^\alpha, a_\alpha \in \mathbb{R} \},$$

where $x^\alpha = x_1^{\alpha_1} x_2^{\alpha_2} \ldots x_d^{\alpha_d}$.

A finite element is a subspace of polynomials on a subset $T \subset \mathbb{R}^d$. For the assembly of finite elements, additional structures are required.

Definition 3.2 A *finite element* is a triple $(T, \mathscr{P}, \mathscr{K})$, consisting of a closed set $T \subset \mathbb{R}^d$ called *element*, a space of polynomials $\mathscr{P}$ called *ansatz-functions*, with dim $\mathscr{P} = R+1$, and a set $\mathscr{K} = \{\chi_0, \chi_1, \ldots, \chi_R\}$ of *node functionals* $\chi_j : C^\infty(T) \to \mathbb{R}$ such that:

(a) if for $q \in \mathscr{P}$ we have $\chi(q) = 0$ for all $\chi \in \mathscr{K}$, then $q = 0$;
(b) there exists $m \ge 1$ with $\mathscr{P}_{m-1}(T) \subset \mathscr{P}$;
(c) there exists $p \in [1, \infty]$ such that every $\chi \in \mathscr{K}$ extends to a bounded linear operator on $W^{m,p}(T)$. The integer $k = m - 1$ is called the *(complete) polynomial degree*

of the finite element.

© Springer International Publishing Switzerland 2016
S. Bartels, *Numerical Approximation of Partial Differential Equations*,
Texts in Applied Mathematics 64, DOI 10.1007/978-3-319-32354-1_3

Example 3.1 For a closed set $T \subset \mathbb{R}^d$ the triple $(T, \mathscr{P}_0(T), \chi_T)$, with $\chi_T(v) = v(x_T)$ for an arbitrary point $x_T \in T$, is a finite element called a *P0-element* with $m = 1$ and $p > d$. If χ_T is defined by

$$\chi_T(v) = \int_T v \, dx,$$

then the triple is a finite element with $m = 1$ and $p = 1$.

We say in the following that points $z_0, z_1, \ldots, z_d \in \mathbb{R}^d$ are noncollinear, if they do not belong to one hyperplane in $\mathbb{R}^d$.

Proposition 3.1 (P1-Element) *For noncollinear vertices* $z_0, z_1, \ldots, z_d \in \mathbb{R}^d$, $d = 1, 2, 3$, *let* $T \subset \mathbb{R}^d$ *be the line segment, triangle, or tetrahedron*

$$T = \mathrm{conv}\{z_0, z_1, \ldots, z_d\} \subset \mathbb{R}^d,$$

set $\mathscr{P} = \mathscr{P}_1(T)$, *and* $\mathscr{K} = \{\chi_0, \chi_1, \ldots, \chi_d\}$ *with*

$$\chi_j(v) = v(z_j)$$

for $j = 0, 1, \ldots, d$ *and* $v \in C^\infty(T)$. *Then* $(T, \mathscr{P}, \mathscr{K})$ *is a finite element with* $m = 2$ *and* $p = 2$ *called a* (linear) *P1-element, cf. Fig. 3.1*.

Proof Assume that $q \in \mathscr{P}_1(T)$ is such that $\chi_j(q) = 0$ for $j = 0, 1, \ldots, d$, i.e., the affine function $x \mapsto q(x)$ vanishes at the noncollinear points $z_0, z_1, \ldots, z_d$. This can only be the case if $q = 0$. Since for $d \le 3$ we have $W^{2,2}(T) \subset C(T)$ due to Morrey's theorem, cf. Thm. 2.10, the functionals are well defined on $W^{m,p}(T)$ with $p = 2$.   □

For further examples we restrict to $d = 2$ and to reference sets $\widehat{T} \subset \mathbb{R}^2$.

Examples 3.2

(i) Let $\hat{z}_0 = (0,0)$, $\hat{z}_1 = (1,0)$, $\hat{z}_2 = (0,1)$, and set $\widehat{T} = \mathrm{conv}\{\hat{z}_0, \hat{z}_1, \hat{z}_2\}$, and $\widehat{\mathscr{P}} = \mathscr{P}_2(\widehat{T})$. With

$$\hat{z}_3 = (1/2, 0), \quad \hat{z}_4 = (0, 1/2), \quad \hat{z}_5 = (1/2, 1/2),$$

define $\varphi_j(v) = v(\hat{z}_j)$ for $j = 0, 1, \ldots, 5$, and $\widehat{\mathscr{K}} = \{\chi_0, \ldots, \chi_5\}$. Then $(\widehat{T}, \widehat{\mathscr{P}}, \widehat{\mathscr{K}})$ is a finite element called a *(quadratic) P2-element*.

Fig. 3.1 Linear *P1*-elements for $d = 1, 2, 3$. The linear functionals are defined by point evaluation at the vertices

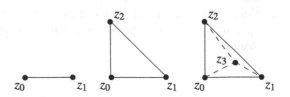

(ii) Let $k \in \mathbb{N}$ and $\widehat{T} = [0, 1]^2$. For $i, j = 0, 1, \ldots, k$ let

$$\hat{z}_{ij} = (i, j)/k, \quad \chi_{ij}(v) = v(\hat{z}_{ij}).$$

Let $\widehat{\mathscr{P}}$ be the set of polynomials of partial degree k on $\widehat{T}$, i.e.,

$$\widehat{\mathscr{P}} = Q_k(\widehat{T}) = \Big\{ \sum_{i,j=0}^{k} a_{ij} x_1^i x_2^j : a_{ij} \in \mathbb{R} \Big\}.$$

Then $(\widehat{T}, \widehat{\mathscr{P}}, \widehat{\mathscr{K}})$ is a finite element called a Q_k-element.

(iii) Let $\hat{z}_0 = (0, 0)$, $\hat{z}_1 = (1, 0)$ and $\hat{z}_2 = (0, 1)$, and set $\widehat{T} = \mathrm{conv}\{\hat{z}_0, \hat{z}_1, \hat{z}_2\}$. Define the barycenter of T by

$$\hat{z}_4 = (\hat{z}_1 + \hat{z}_2 + \hat{z}_3)/3.$$

Then for $j = 0, 1, \ldots, 3$, define the functional

$$\chi_j(v) = v(\hat{z}_j)$$

and the functionals $\chi_4, \chi_5, \ldots, \chi_9$ for $j = 0, 1, 2$ and $i = 1, 2$ by

$$\chi_{3+2j+i}(v) = \partial_i v(\hat{z}_j).$$

Set $\widehat{\mathscr{K}} = \{\chi_0, \ldots, \chi_9\}$ and $\widehat{\mathscr{P}} = \mathscr{P}_3(T)$. Then $(\widehat{T}, \widehat{\mathscr{P}}, \widehat{\mathscr{K}})$ is a finite element called a *cubic Hermite element*, cf. Fig. 3.2.

Remark 3.1 The node functionals contained in the set $\mathscr{K}$ are also called the *(local) degrees of freedom* of the finite element.

Specifying the linear functionals leads to the definition of an interpolant.

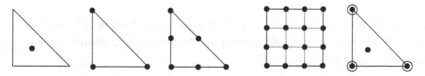

Fig. 3.2 Constant $P0$-element, linear $P1$-element, quadratic $P2$-element, quadrilateral Q_k-element, and cubic Hermite element (*from left to right*). *Filled dots* represent an evaluation of a function and *circles* the evaluation of its gradient

Definition 3.3 Given a finite element $(T, \mathscr{P}, \mathscr{K})$ and $v \in W^{m,p}(T)$, the *(nodal) interpolant* $\mathscr{I}_T v \in \mathscr{P}$ is the uniquely defined function in $\mathscr{P}$ with

$$\chi(\mathscr{I}_T v) = \chi(v)$$

for all $\chi \in \mathscr{K}$.

With the help of an appropriate basis of $\mathscr{P}$, the nodal interpolant has a simple representation.

Proposition 3.2 (Nodal Basis) *Let* $(T, \mathscr{P}, \mathscr{K})$ *be a finite element. There exists a basis* $(\varphi_0, \varphi_1, \ldots, \varphi_R)$ *for* $\mathscr{P}$ *such that*

$$\chi_j(\varphi_k) = \delta_{jk}$$

for all $j, k = 0, 1, \ldots, R$. *In particular, we have*

$$\mathscr{I}_T(v) = \sum_{j=0}^{R} \chi_j(v) \varphi_j$$

for all $v \in W^{m,p}(T)$.

Proof Let $(q_0, q_1, \ldots, q_R)$ be a basis for $\mathscr{P}$, and define the matrix $A \in \mathbb{R}^{(R+1) \times (R+1)}$ by

$$A_{jk} = \chi_j(q_k)$$

for $j, k = 0, 1, \ldots, R$. Then A is regular, and we let $c_j \in \mathbb{R}^{R+1}$ be the solution of $A c_j = e_j$, for $j = 0, 1, \ldots, R$, where $(e_0, e_1, e_2, \ldots, e_R)$ is the canonical basis of $\mathbb{R}^{R+1}$. Then the polynomials

$$\varphi_j = \sum_{\ell=0}^{R} c_{j\ell} q_\ell$$

have the property $\chi_j(\varphi_k) = \delta_{jk}$, $j, k = 0, 1, \ldots, R$, and define a basis for $\mathscr{P}$. If $\mathscr{I}_T v = \sum_{\ell=0}^{R} b_\ell \varphi_\ell$, then it follows with the definition of the interpolant that $b_j = \chi_j(\mathscr{I}_T v) = \chi_j(v), j = 0, 1, \ldots, R$. $\square$

Example 3.3 Let $\widehat{T}$ be the triangle with vertices $\hat{z}_0 = (0, 0)$, $\hat{z}_1 = (1, 0)$ and $\hat{z}_2 = (0, 1)$. The nodal basis of the $P1$-finite element on $\widehat{T}$ is given by the functions

$$\widehat{\varphi}_0(x_1, x_2) = 1 - x_1 - x_2, \quad \widehat{\varphi}_1(x_1, x_2) = x_1, \quad \widehat{\varphi}_2(x_1, x_2) = x_2.$$

3.1.2   Bramble–Hilbert Lemma

The properties of interpolants can be analyzed with the Bramble–Hilbert lemma, which is a consequence of three auxiliary lemmas. We recall the multi-index notation

$$\partial^\alpha v = \partial_1^{\alpha_1} \partial_2^{\alpha_2} \ldots \partial_d^{\alpha_d} v$$

for $\alpha \in \mathbb{N}_0^d$, and $D^m v = \left(\partial^\alpha v\right)_{\alpha \in \mathbb{N}_0^d, |\alpha| = m}$, where $|\alpha| = \alpha_1 + \alpha_2 + \cdots + \alpha_d$. The norm in $W^{m,p}(T)$ is for a set $T \subset \mathbb{R}^d$, whose interior is a bounded Lipschitz domain in $\mathbb{R}^d$, given by

$$\|v\|_{W^{m,p}(T)}^p = \Big(\sum_{k=0}^m \|D^k v\|_{L^p(T)}^p\Big)^{1/p} = \Big(\sum_{k=0}^m \sum_{\alpha \in \mathbb{N}_0^d, |\alpha|=k} \|\partial^\alpha v\|_{L^p(T)}^p\Big)^{1/p}.$$

A related seminorm in $W^{m,p}(T)$ is defined by

$$|v|_{W^{m,p}(T)} = \|D^m v\|_{L^p(T)} = \Big(\sum_{\alpha \in \mathbb{N}_0^d, |\alpha|=m} \|\partial^\alpha v\|_{L^p(T)}^p\Big)^{1/p}.$$

We let $1 \le p < \infty$ and $m \in \mathbb{N}$ in what follows. Equality of functions is understood in the sense of Lebesgue functions.

Lemma 3.1 (Kernel of Differential Operators) *Suppose that* $v \in W^{m,p}(T)$ *satisfies* $\partial^\alpha v = 0$ *for all* $\alpha \in \mathbb{N}_0^d$ *with* $|\alpha| = m$. *Then there exists a polynomial* $q \in \mathscr{P}_{m-1}(T)$ *such that* $v = q$.

Proof If $\nabla v = 0$, then v is constant, which proves the result for $m = 1$. Assume that the implication has been proved for some $m \ge 1$, and let $v \in W^{m+1,p}(T)$ be such that $\partial^\alpha v = 0$ for all $\alpha \in \mathbb{N}_0^d$ with $|\alpha| = m + 1$. For every $\beta \in \mathbb{N}_0^d$ with $|\beta| = m$ and $j = 1, 2, \ldots, d$, we then have, since weak derivatives commute, that

$$\partial_j \partial^\beta v = \partial^\beta (\partial_j v) = 0.$$

Hence $w_j = \partial_j v \in \mathscr{P}_{m-1}(T)$ for $j = 1, 2, \ldots, d$. By integrating $w = (w_1, w_2, \ldots, w_d)$ along appropriate paths, we can construct $\tilde{v} \in \mathscr{P}_m(T)$ with $\nabla \tilde{v} = w$. Hence $\nabla(\tilde{v} - v) = 0$, i.e., $v - \tilde{v}$ is constant, which implies that $v \in \mathscr{P}_m(T)$. $\square$

Remark 3.2 Alternatively, one can use that $\partial^\beta \partial^\alpha v = 0$ for all $\beta \in \mathbb{N}_0^d$, i.e., $v \in H^k(T)$ for every $k \in \mathbb{N}$, and by embedding theorems it follows that $v \in C^\infty(T)$. The result then follows with classical arguments.

Lemma 3.2 (Projection Onto Polynomials) *For all $v \in W^{m,p}(T)$ there exists a uniquely defined polynomial $q \in \mathscr{P}_{m-1}(T)$ such that*

$$\int_T \partial^\alpha (v - q)\, dx = 0$$

for all $\alpha \in \mathbb{N}_0^d$ with $|\alpha| \le m - 1$.

Proof Setting $N = |\{\alpha \in \mathbb{N}_0^d : |\alpha| \le m - 1\}|$, and noting that

$$\mathscr{P}_{m-1}(T) = \Big\{ \sum_{\alpha \in \mathbb{N}_0^d, |\alpha| \le m-1} a_\alpha x^\alpha : a_\alpha \in \mathbb{R} \Big\},$$

it follows that the mapping

$$\mathscr{P}_{m-1}(T) \to \mathbb{R}^N, \quad q \mapsto \Big(\int_T \partial^\alpha q(x)\, dx \Big)_{\alpha \in \mathbb{N}_0^d, |\alpha| \le m-1}$$

is an isomorphism, which implies the result.   □

Lemma 3.3 (Generalized Poincaré Inequality) *There exists $c_P' > 0$ such that for all $v \in W^{m,p}(T)$ satisfying*

$$\int_T \partial^\alpha v\, dx = 0$$

for all $\alpha \in \mathbb{N}_0^d$ with $|\alpha| \le m - 1$, we have

$$\|v\|_{W^{m,p}(T)} \le c_P' |v|_{W^{m,p}(T)}.$$

Proof Assume that the statement is false. Then for every $k \in \mathbb{N}$ there exists $\hat{v}_k \in W^{m,p}(T)$ such that

$$\int_T \partial^\alpha \hat{v}_k\, dx = 0$$

for all $|\alpha| \le m - 1$ and

$$\|\hat{v}_k\|_{W^{m,p}(T)} > k |\hat{v}_k|_{W^{m,p}(T)}.$$

For each $k \in \mathbb{N}$, we define $v_k = \hat{v}_k / \|\hat{v}_k\|_{W^{m,p}(T)}$, so that

$$\|v_k\|_{W^{m,p}(T)} = 1, \quad |v_k|_{W^{m,p}(T)} < k^{-1}.$$

The sequence $(v_k)_{k \in \mathbb{N}}$ is bounded in $W^{m,p}(T)$, and by the Rellich–Kandrachov embedding theorem, cf. Theorem 2.9, there exists $v \in W^{m,p}(T)$ and a subsequence $(v_{k_j})_{j \in \mathbb{N}}$ such that $v_{k_j} \to v$ in $W^{m-1,p}(T)$ as $j \to \infty$. Since we also have that $|v_k|_{W^{m,p}(T)} \to 0$, it follows that $(v_{k_j})_{j \in \mathbb{N}}$ is a Cauchy sequence in $W^{m,p}(T)$ with limit v. But then $|v|_{W^{m,p}(T)} = 0$ and Lemma 3.1 implies that $v = q$ for some $q \in \mathscr{P}_{m-1}(T)$. Since we have

$$\int_T \partial^\alpha v \, dx = \lim_{j \to \infty} \int_T \partial^\alpha v_{k_j} \, dx = 0,$$

for all $\alpha \in \mathbb{N}_0^d$ with $|\alpha| = m - 1$, Lemma 3.2 yields that $q = 0$. Hence $v = 0$, but this contradicts

$$\|v\|_{W^{m,p}(T)} = \lim_{j \to \infty} \|v_{k_j}\|_{W^{m,p}(T)} = 1,$$

and this proves the lemma.   □

The Bramble–Hilbert lemma follows from the lemmas.

Theorem 3.1 (Bramble–Hilbert Lemma) *Let $1 \le p < \infty$ and assume that F : $W^{m,p}(T) \to \mathbb{R}$ is a bounded and quasisublinear functional, i.e., there exist $c_1, c_2 > 0$ such that for all $v, w \in W^{m,p}(T)$, we have*

$$|F(v)| \le c_1 \|v\|_{W^{m,p}(T)}, \quad |F(v + w)| \le c_2(|F(v)| + |F(w)|),$$

and assume that F vanishes on $\mathscr{P}_{m-1}(T)$. Then we have

$$|F(v)| \le c_p' c_1 c_2 \|D^m v\|_{L^p(T)}$$

for all $v \in W^{m,p}(T)$.

Proof Let $v \in W^{m,p}(T)$. For all $q \in \mathscr{P}_{m-1}(T)$ we have that

$$|F(v)| \le c_2 |F(v - q)| \le c_1 c_2 \|v - q\|_{W^{m,p}(T)}.$$

Due to Lemma 3.2 there exists a uniquely defined $q \in \mathscr{P}_{m-1}(T)$ satisfying $\int_T \partial^\alpha (v - q) \, dx = 0$ for all $\alpha \in \mathbb{N}_0^d$ with $|\alpha| \le m - 1$, and the generalized Poincaré inequality implies that $\|v - q\|_{W^{m,p}(T)} \le c_p' \|D^m(v - q)\|_{L^p(T)}$. Since $D^m q = 0$ we deduce the assertion.   □

The Bramble–Hilbert lemma implies a bound for the interpolation error by a finite element. This bound is the basis for deriving convergence rates.

Corollary 3.1 (Interpolation Stability) *Let $(T, \mathscr{P}_T, \mathscr{K}_T)$ be a finite element with $\mathscr{P}_{m-1} \subset \mathscr{P}_T$, and $|\cdot|_s$ a seminorm on $W^{m,p}(T)$ with $|v|_s \le c_S \|v\|_{W^{m,p}(T)}$ for all $v \in W^{m,p}(T)$. Then we have*

$$|v - \mathscr{I}_T v|_s \le c_{IS} \|D^m v\|_{L^p(T)}$$

for all $v \in W^{m,p}(T)$.

Proof We define $F(v) = |v - \mathscr{I}_T v|_s$ and note that F is sublinear. With the dual basis of Proposition 3.2, we have $\mathscr{I}_T(v) = \sum_{j=0}^{R} \chi_j(v)\varphi_j$. Using $|\chi_j(v)| \leq c_b \|v\|_{W^{m,p}(T)}$ for all $v \in W^{m,p}(T)$ and $j = 0, 1, \ldots, R$, it follows that

$$|F(v)| \leq |v|_s + |\mathscr{I}_T v|_s \leq \left(c_S + Rc_b \max_{j=0,\ldots,R} |\varphi_j|_s\right) \|v\|_{W^{m,p}(T)},$$

i.e., F is bounded. We have $F(q) = 0$ for all $q \in \mathscr{P}_T$, and hence the conditions of the Bramble–Hilbert lemma are satisfied, which implies the estimate.   □

Remark 3.3 The importance of the Bramble–Hilbert lemma lies in the homogeneity and related scaling properties of the seminorm in $W^{m,p}(T)$ with respect to affine transformations.

3.1.3 Affine Transformations

It is most convenient to define a finite element on a reference domain $\widehat{T}$, and then transfer it to general domains T by an affine transformation, cf. Fig. 3.3.

Definition 3.4 The *diameter* h_T and *inner radius* ϱ_T of a set $T \subset \mathbb{R}^d$ are defined by

$$h_T = \mathrm{diam}(T) = \sup\left\{|x - y| : x, y \in T\right\},$$

$$\varrho_T = \sup\{r : r > 0, x \in T, B_r(x) \subset T\}.$$

We say that a simplex with vertices $z_0, z_1, \ldots, z_d \in \mathbb{R}^d$ is nondegenerate, if its vertices are noncollinear.

Proposition 3.3 (Transformation Estimates) *Let* $\widehat{T} = \mathrm{conv}\{0, e_1, \ldots, e_d\}$ *with the canonical basis* $(e_1, e_2, \ldots, e_d)$ *for* $\mathbb{R}^d$*. Let* $T = \mathrm{conv}\{z_0, z_1, \ldots, z_d\} \subset \mathbb{R}^d$ *be a nondegenerate simplex in* $\mathbb{R}^d$*. There exists a unique affine diffeomorphism* $\Phi_T : \widehat{T} \to T$ *with* $\Phi_T(0) = z_0$ *and* $\Phi_T(e_j) = z_j$ *for* $j = 1, 2, \ldots, d$*. For every* $v \in W^{m,p}(T)$ *and* $\hat{v} = v \circ \Phi_T \in W^{m,p}(\widehat{T})$*, we have for all* $0 \leq k \leq m$ *that*

$$|v|_{W^{k,p}(T)} \leq c_{at}\, \varrho_T^{-k} |\det B|^{1/p} |\hat{v}|_{W^{k,p}(\widehat{T})},$$

$$|\hat{v}|_{W^{k,p}(\widehat{T})} \leq c'_{at}\, h_T^{-k} |\det B|^{-1/p} |v|_{W^{k,p}(T)},$$

with constants $c_{at}, c'_{at} > 0$ *that are independent of* v *and* T*, and* $B = D\Phi_T$*.*

Fig. 3.3 Diameter h_T and inner radius ϱ_T of a triangle T (*left*); transformation from a reference element (*right*)

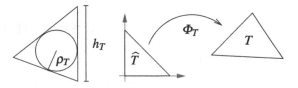

Proof

(i) We define the matrix $B \in \mathbb{R}^{d \times d}$ and the vector $b \in \mathbb{R}^d$ by

$$B = [z_1 - z_0, z_2 - z_0, \ldots, z_d - z_0], \quad b = z_0.$$

Due to the nondegeneracy of T, B is regular. The transformation

$$\Phi_T : \widehat{T} \to \mathbb{R}^d, \quad \hat{x} \mapsto B\hat{x} + b,$$

then satisfies $\Phi_T(0) = z_0$ and $\Phi_T(e_j) = z_j$ for $j = 1, 2, \ldots, d$. Since the image of a convex set under an affine transformation is convex, we have that $\Phi_T(\widehat{T}) = T$. Its inverse is given by

$$\Phi_T^{-1} : T \to \widehat{T}, \quad x \mapsto B^{-1}x - B^{-1}b.$$

Hence Φ_T is a diffeomorphism with $D\Phi_T = B$, which is uniquely defined by the points $z_0, z_1, \ldots, z_d$.

(ii) Given any $z \in \mathbb{R}^d$ with $|z| = \varrho_{\widehat{T}}$, there exist $\xi, \eta \in \widehat{T}$ such that $z = \xi - \eta$. Then since $Bz = \Phi_T(\xi) - \Phi_T(\eta)$, we deduce that

$$\|B\| = \sup_{z \in \mathbb{R}^d, |z| = \varrho_{\widehat{T}}} \varrho_{\widehat{T}}^{-1} |Bz| \leq \varrho_{\widehat{T}}^{-1} \sup_{\xi, \eta \in \widehat{T}} |\Phi_T(\xi) - \Phi_T(\eta)| = \varrho_{\widehat{T}}^{-1} h_T.$$

Analogously, by exchanging the roles of T and $\widehat{T}$, we find that

$$\|B^{-1}\| \leq \varrho_T^{-1} h_{\widehat{T}}.$$

(iii) For a function or vector field $f \in L^1(T; \mathbb{R}^m)$ we set

$$\hat{f} = f \circ \Phi_T.$$

The transformation formula shows that

$$\int_{\widehat{T}} \hat{f} \, d\hat{x} = \int_{\Phi_T(\widehat{T})} \hat{f} \circ \Phi_T^{-1} |\det D\Phi_T^{-1}| \, dx = |\det B|^{-1} \int_T f \, dx.$$

(iv) Let $w \in C^\infty(T)$. We have

$$\nabla w = \nabla(\hat{w} \circ \Phi_T^{-1}) = (D\Phi_T)^{-\top}(\widehat{\nabla w}) \circ \Phi_T^{-1} = (D\Phi)^{-\top} \widehat{\nabla w} \circ \Phi_T^{-1},$$

and therefore

$$\widehat{\nabla w} = (\nabla w) \circ \Phi_T = (D\Phi)^{-\top} \widehat{\nabla} \hat{w}.$$

Similarly, we have

$$\widehat{\nabla}\hat{w} = \widehat{\nabla}(w \circ \Phi_T) = D\Phi_T^\top (\nabla w) \circ \Phi_T = (D\Phi_T)^\top \widehat{\nabla w}.$$

Using that $0 < \varrho_{\widehat{T}} \le h_{\widehat{T}} \le c'$, we thus have that

$$|\widehat{\nabla w}| \le c\varrho_T^{-1}|\widehat{\nabla}\hat{w}|, \quad |\widehat{\nabla}\hat{w}| \le ch_T|\widehat{\nabla w}|.$$

More generally, using that $D\Phi_T$ is constant, it follows for every $\alpha \in \mathbb{N}_0^d$ that

$$|\widehat{\partial^\alpha w}| \le c\varrho_T^{-|\alpha|} \max_{|\beta|=|\alpha|} |\widehat{\partial^\beta}\hat{w}|, \quad |\widehat{\partial^\alpha}\hat{w}| \le ch_T^{|\alpha|} \max_{|\beta|=|\alpha|} |\widehat{\partial^\beta w}|,$$

where the maxima are taken over all $\beta \in \mathbb{N}_0^d$ with $|\beta| = |\alpha|$.

(v) Applying the transformation formula with $f = |\partial^\alpha w|^p$, we find that

$$\int_T |\partial^\alpha w(x)|^p \, dx = |\det B| \int_T |\widehat{\partial^\alpha w}(\hat{x})|^p \, d\hat{x}$$

$$\le c|\det B|\varrho_T^{-p|\alpha|} \max_{|\beta|=|\alpha|} \int_T |\widehat{\partial^\beta}\hat{w}(\hat{x})|^p \, d\hat{x}.$$

Using the transformation formula with $f(x) = |\partial^\beta w(x)|^p$ shows that

$$\int_T |\widehat{\partial^\alpha}\hat{w}(\hat{x})|^p \, d\hat{x} \le ch_T^{p|\alpha|} \max_{|\beta|=|\alpha|} \int_{\widehat{T}} |\widehat{(\partial^\beta w)}(\hat{x})|^p \, d\hat{x}$$

$$= c|\det B|^{-1}h_T^{p|\alpha|} \max_{|\beta|=|\alpha|} \int_T |\partial^\beta w(x)|^p \, dx.$$

A combination of the estimates and a density argument prove the proposition.
□

3.1.4 Interpolation Estimate

In order to determine the dependence of the interpolation error on the diameter of an element, we assume that a finite element is obtained by an affine transformation from a reference element. For a functional $\hat{\chi} : C^\infty(\widehat{T}) \to \mathbb{R}$ and a diffeomorphism $\Phi_T : \widehat{T} \to T$, we define $\hat{\chi} \circ \Phi_T^{-1} : C^\infty(T) \to \mathbb{R}$

$$[\hat{\chi} \circ \Phi_T^{-1}](v) = \chi(v \circ \Phi_T^{-1})$$

for every $v \in C^\infty(T)$.

Theorem 3.2 (Interpolation Error) *Assume that* $(\widehat{T}, \widehat{\mathscr{P}}, \widehat{\mathscr{K}})$ *is a reference finite element with complete polynomial degree* $m - 1 \geq 0$, *and a set* $\widehat{T} \subset \mathbb{R}^d$. *Let* $(T, \mathscr{P}, \mathscr{K})$ *be a finite element that is obtained by an affine transformation from the reference element with the affine diffeomorphism* $\Phi_T : \widehat{T} \to T$, *i.e., we have*

$$T = \Phi_T(\widehat{T}), \quad \mathscr{P} = \{\hat{q} \circ \Phi_T^{-1} : \hat{q} \in \widehat{\mathscr{P}}\}, \quad \mathscr{K} = \{\hat{\chi} \circ \Phi_T^{-1} : \hat{\chi} \in \widehat{\mathscr{K}}\}.$$

Then for every $v \in W^{m,p}(T)$, *the interpolant* $\mathscr{I}_T v \in P_T$ *satisfies*

$$|v - \mathscr{I}_T v|_{W^{k,p}(T)} \leq c_{\mathscr{I}} h_T^m \varrho_T^{-k} |v|_{W^{m,p}(T)}$$

with a constant $c_{\mathscr{I}} = c_{\mathscr{I}}(d, m, \widehat{T})$ *for all* $0 \leq k \leq m$.

Proof For $\hat{v} = v \circ \Phi_T \in W^{k,m}(\widehat{T})$ and its interpolant $\mathscr{I}_{\widehat{T}} \hat{v}$, it follows from Corollary 3.1 that

$$|\hat{v} - \mathscr{I}_{\widehat{T}} \hat{v}|_{W^{k,p}(\widehat{T})} \leq \hat{c}_{IS} |\hat{v}|_{W^{m,p}(\widehat{T})}.$$

Using that $(\mathscr{I}_T v) \circ \Phi_T = \mathscr{I}_{\widehat{T}} \hat{v}$, we verify with the transformation estimates from Proposition 3.3 that

$$|v - \mathscr{I}_T v|_{W^{k,p}(T)} \leq c \varrho_T^{-k} |\det B|^{1/p} |\hat{v} - \mathscr{I}_{\widehat{T}} \hat{v}|_{W^{k,p}(\widehat{T})}$$

$$\leq c \varrho_T^{-k} |\det B|^{1/p} |\hat{v}|_{W^{k,p}(\widehat{T})}$$

$$\leq c h_T^m \varrho_T^{-k} |\det B|^{1/p} |\det B|^{-1/p} |v|_{W^{m,p}(T)}.$$

This proves the estimate.   □

3.1.5   Affine Families

We consider a bounded, polyhedral Lipschitz domain $\Omega \subset \mathbb{R}^d$.

Definition 3.5 A *(conforming) triangulation* $\mathscr{T}_h$ of a domain $\Omega \subset \mathbb{R}^d$ is a set $\mathscr{T}_h = \{T_1, T_2, \ldots, T_L\}$ of closed intervals, triangles, or tetrahedra for $d = 1, 2, 3$, respectively, called *elements*, such that $\overline{\Omega} = \cup_{T \in \mathscr{T}_h} T$ and the intersection of distinct $T_1, T_2 \in \mathscr{T}_h$ is either empty or an entire subsimplex, cf. Fig. 3.4.

We define spaces of piecewise polynomial functions by mapping a reference finite element to the elements of a triangulation.

Fig. 3.4 Uniform conforming triangulation, nonconforming triangulation with a hanging node, and locally refined conforming triangulation (*from left to right*)

Definition 3.6 Given a triangulation $\mathcal{T}_h$ of Ω, an *affine family (of finite elements)* is a family

$$\left(T, \mathcal{P}_T, \mathcal{K}_T\right)_{T \in \mathcal{T}_h},$$

such that each finite element $(T, \mathcal{P}_T, \mathcal{K}_T)$ is obtained by an affine transformation from a common reference finite element $(\widehat{T}, \widehat{\mathcal{P}}, \widehat{\mathcal{K}})$, i.e., there exists a family of affine diffeomorphisms $(\Phi_T)_{T \in \mathcal{T}_h}$ such that

$$T = \Phi_T(\widehat{T}), \quad \mathcal{P} = \{\hat{q} \circ \Phi_T^{-1} : \hat{q} \in \widehat{\mathcal{P}}\}, \quad \mathcal{K} = \{\widehat{\chi} \circ \Phi_T^{-1} : \widehat{\chi} \in \widehat{\mathcal{K}}\}.$$

for every $T \in \mathcal{T}_h$.

We assemble the local interpolants to obtain an approximating function on the domain Ω.

Definition 3.7 Let $\mathcal{T}_h$ be a triangulation of the Lipschitz domain $\Omega \subset \mathbb{R}^d$, and $(T, \mathcal{P}_T, \mathcal{K}_T)_{T \in \mathcal{T}_h}$ an affine family. The *global interpolant* $\mathcal{I}_{\mathcal{T}} : W^{m,p}(\Omega) \to L^\infty(\Omega)$ is defined by

$$(\mathcal{I}_{\mathcal{T}} v)|_T = \mathcal{I}_T(v|_T)$$

for all $T \in \mathcal{T}_h$. The affine family is called a C^r-*element* if $\mathcal{I}_{\mathcal{T}} v \in C^r(\overline{\Omega})$ for every $v \in C^r(\overline{\Omega}) \cap W^{m,p}(\Omega)$.

Remark 3.4 The local degrees of freedom shared by different finite elements in an affine family are called *global degrees of freedom*.

Examples 3.4

(i) The $P1$-interpolant $\mathcal{I}_h u$ of a function $u \in W^{2,2}(\Omega)$ is the uniquely defined continuous function that is affine on every element $T \in \mathcal{T}_h$ and satisfies $\mathcal{I}_h u(z) = u(z)$ for all vertices z of elements in $T \in \mathcal{T}_h$. It is a C^0-element.
(ii) The $P0$-element is not a C^0-element.

We are now in position to state an error estimate for approximating sufficiently regular functions on the domain Ω. We recall that every piecewise polynomial function $v \in C^{m-1}(\overline{\Omega})$ belongs to $W^{m,p}(\Omega)$.

Proposition 3.4 (Global Interpolation Estimate) *Assume that the affine family* $(T, \mathcal{P}_T, \mathcal{K}_T)_{T \in \mathcal{T}_h}$ *is* C^{m-1}-*regular. For every* $v \in W^{m,p}(\Omega)$ *and* $0 \le k \le m$, *we have*

$$|v - \mathscr{I}_\mathscr{T} v|_{W^{k,p}(\Omega)} \le c_\mathscr{I} \max_{T \in \mathcal{T}_h} h_T^m \varrho_T^{-k} |v|_{W^{m,p}(\Omega)}.$$

Proof The estimate follows from Proposition 3.2 by summing over all elements $T \in \mathcal{T}_h$. □

For nondegenerate simplices, we expect that $\varrho_T \ge ch_T$. For a sequence of triangulations, this property has to hold uniformly.

Definition 3.8 A family of (conforming) triangulations $(\mathcal{T}_h)_{h>0}$ is called *(uniformly shape) regular* if there exists a constant $c_{\text{usr}} > 0$ such that

$$\sup_{h>0} \sup_{T \in \mathcal{T}_h} h_T \varrho_T^{-1} \le c_{\text{usr}}.$$

The index h in a family of triangulations $(\mathcal{T}_h)_{h>0}$ typically refers to a characteristic or maximal size of the elements in $\mathcal{T}_h$, e.g., it is assumed that $\max_{T \in \mathcal{T}_h} h_T \le ch$ for all $h > 0$. Nevertheless, for a sequence of locally refined triangulations, we may have $\max_{T \in \mathcal{T}_h} h_T = \max_{T' \in \mathcal{T}_{h'}} h_{T'}$ for two different triangulations $\mathcal{T}_h$ and $\mathcal{T}_{h'}$. In this case h may refer to an average mesh-size, e.g., defined by the number of elements in $\mathcal{T}_h$.

Remark 3.5 For shape regularity, a *minimum angle condition*, requiring that the angles of triangles be uniformly bounded from below by a positive number, is sufficient. A weaker *maximum angle condition* is sufficient for a robust interpolation estimate.

Example 3.5 We consider the $P1$-finite element on the triangulations $\mathcal{T}_1$ and $\mathcal{T}_2$ displayed in Fig. 3.5 and the function $u(x_1, x_2) = 1 - x_1^2$ for $x = (x_1, x_2) \in \mathbb{R}^2$. For $\varepsilon > 0$ and the triangulation $\mathcal{T}_1^\varepsilon = \{T\}$ with

$$T = \text{conv}\{(-1, 0), (1, 0), (0, \varepsilon)\},$$

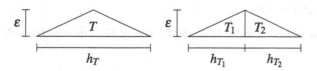

Fig. 3.5 A triangle that violates the minimum angle condition if $\varepsilon/h_T \to 0$ (*left*) and triangles that satisfy the maximum angle condition even for $\varepsilon/h_{T_\ell} \to 0$, $\ell = 1, 2$ (*right*)

we have $\mathscr{I}_1 u(x_1, x_2) = x_2/\varepsilon$. For the triangulation $\mathscr{T}_2^\varepsilon = \{T_1, T_2\}$ with

$$T_1 = \mathrm{conv}\{(-1,0),(0,0),(0,\varepsilon)\}, \quad T_2 = \mathrm{conv}\{(0,0),(1,0),(0,\varepsilon)\},$$

we have $\mathscr{I}_2 u(x_1, x_2) = 1 - |x_1|$. Obviously, only $\mathscr{I}_2 u$ is a useful approximation to u for $0 < \varepsilon \ll 1$. Both triangulations violate a minimum angle condition, but $\mathscr{T}_2$ satisfies a maximum angle condition.

Remark 3.6 Isoparametric families of finite elements are obtained by using polynomial diffeomorphisms $\Phi_T : \widehat{T} \to T$ such that every component of Φ_T belongs to $\widehat{\mathscr{P}}$. This allows for the accurate approximation of curved boundaries.

3.2 P1-Approximation of the Poisson Problem

3.2.1 P1-Finite Element Method

We discuss in this section the approximation of the Poisson problem with a low order finite element method. In particular, we let $u \in H_\mathrm{D}^1(\Omega)$ be the unique weak solution of the Poisson problem

$$-\Delta u = f \text{ in } \Omega, \quad u|_{\Gamma_\mathrm{D}} = 0, \quad \partial_n u|_{\Gamma_\mathrm{N}} = g.$$

Here $\Omega \subset \mathbb{R}^d$ is a bounded Lipschitz domain with polyhedral boundary, $\Gamma_\mathrm{D} \subset \partial\Omega$ is assumed to be closed and of positive surface measure, and we set $\Gamma_\mathrm{N} = \partial\Omega \setminus \Gamma_\mathrm{D}$. We assume that $f \in L^2(\Omega)$ and $g \in L^2(\Gamma_\mathrm{N})$. The function $u \in H_\mathrm{D}^1(\Omega)$ thus satisfies

$$\int_\Omega \nabla u \cdot \nabla v \, \mathrm{d}x = \int_\Omega f v \, \mathrm{d}x + \int_{\Gamma_\mathrm{N}} g v \, \mathrm{d}s$$

for all $v \in H_\mathrm{D}^1(\Omega)$. We recall that we also write $H_0^1(\Omega)$ instead of $H_\mathrm{D}^1(\Omega)$ in case $\Gamma_\mathrm{D} = \partial\Omega$.

Definition 3.9 For a triangulation $\mathscr{T}_h$, we let $\mathscr{N}_h$ denote the set of vertices of elements called *nodes* and $\mathscr{S}_h$ the set of $(d-1)$-dimensional sides of elements in $\mathscr{T}_h$, i.e., endpoints of intervals, edges of triangles, or faces of tetrahedra if $d = 1, 2, 3$, respectively, cf. Fig. 3.6.

We always assume that the Dirichlet boundary Γ_D is matched exactly by sides in $\mathscr{T}_h$, i.e., that

$$\Gamma_\mathrm{D} = \bigcup_{S \in \mathscr{S}_h, S \subset \Gamma_\mathrm{D}} S.$$

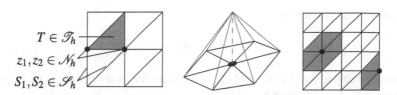

$$T \in \mathscr{T}_h$$
$$z_1, z_2 \in \mathscr{N}_h$$
$$S_1, S_2 \in \mathscr{S}_h$$

Fig. 3.6 Element $T \in \mathscr{T}_h$, nodes $z_1, z_2 \in \mathscr{N}_h$, and sides $S_1, S_2 \in \mathscr{S}_h$ *(left)*, nodal basis functions φ_z *(middle)*, and supports of nodal basis functions φ_z for different nodes $z \in \mathscr{N}_h$ *(right)*

This implies that also the closure of the Neumann boundary is partitioned by sides of elements in $\mathscr{T}_h$. We abbreviate the L^2-norm by

$$\| \cdot \| = \| \cdot \|_{L^2(\Omega)}$$

in what follows and recall that the seminorm

$$v \mapsto \| \nabla v \|$$

defines a norm on $H_D^1(\Omega)$ due to Poincaré's inequality. It is induced by a scalar product so that the space $H_D^1(\Omega)$ equipped with this norm is a Hilbert space.

Definition 3.10 The *P1-finite element space* subordinated to a triangulation $\mathscr{T}_h$ of Ω is the space

$$\mathscr{S}^1(\mathscr{T}_h) = \{ v_h \in C(\overline{\Omega}) : v_h|_T \in \mathscr{P}_1(T) \text{ for all } T \in \mathscr{T}_h \}.$$

The subset of functions in $\mathscr{S}^1(\mathscr{T}_h)$, satisfying homogeneous Dirichlet conditions on a subset $\Gamma_D \subset \partial\Omega$, is defined by

$$\mathscr{S}_D^1(\mathscr{T}_h) = \mathscr{S}^1(\mathscr{T}_h) \cap H_D^1(\Omega).$$

If $\Gamma_D = \partial\Omega$, we also write $\mathscr{S}_0^1(\mathscr{T}_h)$ instead of $\mathscr{S}_D^1(\mathscr{T}_h)$. The *nodal basis* of $\mathscr{S}^1(\mathscr{T}_h)$ is the family $(\varphi_z : z \in \mathscr{N}_h)$ with functions $\varphi_z \in \mathscr{S}^1(\mathscr{T}_h)$ satisfying $\varphi_z(y) = \delta_{zy}$ for all $z, y \in \mathscr{N}_h$.

The *P1 finite element approximations* of the Poisson problem associated with a triangulation $\mathscr{T}_h$ is defined as the unique function $u_h \in \mathscr{S}_D^1(\mathscr{T}_h)$ that satisfies

$$\int_\Omega \nabla u_h \cdot \nabla v_h \, dx = \int_\Omega f v_h \, dx + \int_{\Gamma_N} g v_h \, ds$$

for all $v_h \in \mathscr{S}_D^1(\mathscr{T}_h)$. The existence and uniqueness of u_h are consequences of the Poincaré inequality and Lax–Milgram lemma. For the practical computation of u_h, we use the nodal basis $(\varphi_z : z \in \mathscr{N}_h \setminus \Gamma_D)$ of $\mathscr{S}^1(\mathscr{T}_h)$ and let $U = (u_z : z \in \mathscr{N}_h \setminus \Gamma_D)$ be the coefficient vector of $u_h \in \mathscr{S}_D^1(\mathscr{T}_h)$. The vector is the unique solution of the linear system of equations

$$AU = b,$$

in which the *stiffness matrix* $A = (A_{zy})_{z,y \in \mathcal{N}_h \setminus \Gamma_D}$ and the *load vector* $b = (b_z)_{z \in \mathcal{N}_h \setminus \Gamma_D}$ are for $z, y \in \mathcal{N}_h \setminus \Gamma_D$ given by

$$A_{zy} = \int_\Omega \nabla \varphi_z \cdot \nabla \varphi_y \, dx, \quad b_z = \int_\Omega f \varphi_z \, dx + \int_{\Gamma_N} g \varphi_z \, ds.$$

The coercivity of the involved bilinear form implies that A is positive definite.

Definition 3.11 The *nodal interpolant* of a function $v \in C(\overline{\Omega})$ is defined by

$$\mathscr{I}_h v = \sum_{z \in \mathcal{N}_h} v(z) \varphi_z.$$

We have the following approximation result, which is a consequence of the Bramble–Hilbert lemma.

Theorem 3.3 (Nodal Interpolation Estimates) *For a regular family of triangulations* $(\mathscr{T}_h)_{h>0}$ *such that* $\max_{T \in \mathscr{T}_h} h_T \leq ch$ *and* $v \in H^2(\Omega)$, *we have that* $\mathscr{I}_h v \in \mathscr{S}^1(\mathscr{T}_h)$, *and*

$$h^{-1} \|v - \mathscr{I}_h v\| + \|\nabla(v - \mathscr{I}_h v)\| \leq c_{\mathscr{I}} h \|D^2 v\|.$$

Moreover, if $v|_{\Gamma_D} = 0$, *then* $\mathscr{I}_h v|_{\Gamma_D} = 0$.

Proof The estimates follow from the stability of interpolation and the transformation estimates if $1 \leq p < \infty$. □

3.2.2   Error Estimates

An important property of the Galerkin approximation $u_h \in \mathscr{S}_D^1(\mathscr{T}_h)$ is that the approximation error $u - u_h$ satisfies the *Galerkin orthogonality*

$$\int_\Omega \nabla(u - u_h) \cdot \nabla v_h \, dx = 0$$

for all $v_h \in \mathscr{S}_D^1(\mathscr{T}_h)$. The interpretation of this identity is that $u_h \in \mathscr{S}_D^1(\mathscr{T}_h)$ is the H^1-projection of the exact solution $u \in H_D^1(\Omega)$ onto the subspace $\mathscr{S}_D^1(\mathscr{T}_h)$. In particular, it satisfies a (quasi-) *best-approximation* property or, more generally, the conditions of *Céa's lemma* are satisfied, i.e., we have

$$\|\nabla(u - u_h)\| \leq \inf_{v_h \in \mathscr{S}^1_D(\mathscr{T}_h)} \|\nabla(u - v_h)\|.$$

The density of smooth functions in $H^1_D(\Omega)$ implies convergence $u_h \to u$ in $H^1_D(\Omega)$ as $h \to 0$. If $u \in H^2(\Omega)$, then one obtains a convergence rate.

Corollary 3.2 (Approximation Error) *If the weak solution of the Poisson problem satisfies $u \in H^2(\Omega) \cap H^1_D(\Omega)$, then we have*

$$\|\nabla(u - u_h)\| \leq c_\mathscr{I} h \|D^2 u\|.$$

Proof The error estimate follows from the best-approximation property and the nodal interpolation estimates. □

For the proof of optimal error estimates in $L^2(\Omega)$, a stronger assumption than $u \in H^2(\Omega)$ is required, namely that the Poisson problem be H^2-regular. In this case, the unique weak solution $z \in H^1_D(\Omega)$ of the Poisson problem

$$-\Delta z = e \text{ in } \Omega, \quad z|_{\Gamma_D} = 0, \quad \partial_\nu z = 0 \text{ on } \Gamma_N$$

satisfies $\|D^2 z\| \leq c_2 \|e\|$. If $e = u - u_h$ is the approximation error, then Green's formula and Galerkin orthogonality yield that, for every $z_h \in \mathscr{S}^1_D(\mathscr{T}_h)$, we have

$$\int_\Omega e^2 \, dx = \int_\Omega e(-\Delta z) \, dx = \int_\Omega \nabla e \cdot \nabla z \, dx = \int_\Omega \nabla e \cdot \nabla(z - z_h) \, dx.$$

With Hölder's inequality, the assumed bound for $\|D^2 z\|$ and the choice $z_h = \mathscr{I}_h z$, we find that

$$\|e\|^2 \leq \|\nabla e\| \|\nabla(z - z_h)\| \leq c_\mathscr{I} h \|\nabla e\| \|D^2 z\| \leq c_\mathscr{I} c_2 h \|\nabla e\| \|e\|.$$

Incorporating the estimate $\|\nabla e\| \leq c_\mathscr{I} h \|D^2 u\|$ proves the following result.

Theorem 3.4 (Aubin–Nitsche Lemma) *If the Poisson problem is H^2-regular, then we have*

$$\|u - u_h\| \leq c_\mathscr{I}^2 c_2 h^2 \|D^2 u\|.$$

We note that sufficient for H^2-regularity is that $\Omega \subset \mathbb{R}^2$ be convex and $\Gamma_D = \partial\Omega$.

Remark 3.7 By interpolating Green's function associated with the Poisson problem on $\Omega \subset \mathbb{R}^2$ with $\Gamma_D = \partial\Omega$, one can show that if the Poisson problem is H^2-

regular, if $\mathcal{T}_h$ is quasiuniform, i.e., all elements have a comparable diameter, and if $u \in C^2(\overline{\Omega})$, then we have

$$\|u - u_h\|_{L^\infty(\Omega)} \leq ch^2(1 + |\log h|)\|D^2u\|_{L^\infty(\Omega)}.$$

3.2.3   Discrete Maximum Principle

For the Poisson problem we have a maximum principle, which states that if $f \geq 0$, then also $u \geq 0$. This important physical property is not guaranteed for the discretization by finite elements. In fact, an additional requirement has to be satisfied, which implies that the system matrix is an M-matrix.

Lemma 3.4 (Monotonicity) *Assume that the matrix $A \in \mathbb{R}^{n \times n}$ satisfies $A_{ii} \geq 0$ for $i = 1, 2, \ldots, n$, and $A_{ij} \leq 0$ for $i, j = 1, 2, \ldots, n$, with $i \neq j$. Moreover, assume that A is irreducible and diagonally dominant, i.e., we have*

$$|a_{ii}| \geq \sum_{j=1,\ldots,n, j\neq i} |a_{ij}|$$

for $i = 1, 2, \ldots, n$, with strict inequality for one index, and there is no partition $I, J \subset \{1, 2, \ldots, n\}$ with $a_{ij} = 0$ for all $(i, j) \in I \times J$. Then we have that

$$A^{-1} \geq 0,$$

i.e., the entries $(a_{ij}^{(-1)})_{i,j=1,\ldots,n}$ are nonnegative.

Proof Standard arguments within the convergence analysis of the Jacobi iteration show that the diagonal dominance and irreducibility of A imply that A is regular, that the diagonal part $D \in \mathbb{R}^{n \times n}$ of A is regular, and that the spectral radius of the iteration matrix $M = -D^{-1}(A - D)$ is strictly smaller than 1. Hence the Neumann series with M converges and we have

$$(I - M)^{-1} = \sum_{k=0}^{\infty} M^k.$$

Due to the assumptions of the proposition we have $M \geq 0$, and therefore also the inequality $(I - M)^{-1} \geq 0$. By definition of M, we deduce that $D^{-1}A = I - M$, i.e., $A^{-1} = D(I - M)^{-1}$, which proves the inequality. $\square$

The result leads to a criterion for the validity of a discrete maximum principle.

Fig. 3.7 Interior edge
$S = T_1 \cap T_2$ with opposite
angles α_1 and α_2

Proposition 3.5 (Discrete Maximum Principle) *Assume that* $\Gamma_D = \partial\Omega$, *and the triangulation* $\mathcal{T}_h$ *of* Ω *is such that*

$$A_{zy} = \int_{\Omega} \nabla\varphi_z \cdot \nabla\varphi_y \, dx \le 0$$

for all distinct $z, y \in \mathcal{N}_h \cap \Omega$. *Then whenever* $f \ge 0$, *we have* $u_h \ge 0$.

Proof Exercise.   □

Precise requirements on the triangulation that imply the conditions of the proposition are available if $d = 2$.

Remarks 3.8

(i) If $d = 2$, then the conditions of the proposition are satisfied if and only if $\mathcal{T}_h$ is *weakly acute*, i.e., if every sum of two angles opposite to an interior edge is bounded by π. This follows from the relation

$$\int_{T_1 \cup T_2} \nabla\varphi_z \cdot \nabla\varphi_y \, dx = -\frac{1}{2}(\cot\alpha_1 + \cot\alpha_2) = -\frac{1}{2}\frac{\sin(\alpha_1 + \alpha_2)}{\sin(\alpha_1)\sin(\alpha_2)}$$

for neighboring triangles T_1, T_2 with common edge $S = \text{conv}\{z, y\}$, cf. Fig. 3.7.

(ii) If $d = 3$, then a sufficient condition for the proposition is that every angle between two faces of a tetrahedron be bounded by $\pi/2$.

3.2.4 Quadrature

In general, it is impossible to discretize the bilinear form related to an elliptic partial differential equation exactly. In particular, the integrals that define

$$a(u, v) = \int_{\Omega} \nabla u \cdot \nabla v \, dx, \quad \ell(v) = \int_{\Omega} fv \, dx$$

have to be approximated using appropriate quadrature rules. This introduces approximations a_h and ℓ_h on the employed subspace V_h, e.g., with quadrature rules $(\kappa_m^T, \xi_m^T)_{m=1,\dots,M}$ on the elements $T \in \mathcal{T}_h$,

$$a_h(u_h, v_h) = \sum_{T \in \mathcal{T}_h} \sum_{m=1}^{M} \kappa_m^T [\nabla u_h(\xi_m^T)] \cdot [\nabla v_h(\xi_m^T)]$$

and

$$\ell_h(v_h) = \sum_{T \in \mathscr{T}_h} \sum_{m=1}^{M} \kappa_m^T f(\xi_m^T) v_h(\xi_m^T).$$

These approximations require $V_h \subset C(\overline{\Omega})$, but the approximate bilinear forms cannot be applied to functions in $V = H_D^1(\Omega)$. An abstract framework for an error analysis that respects this aspect is provided by the next result.

Proposition 3.6 (First Strang Lemma) *Assume that associated with a family of subspaces $(V_h)_{h>0}$ of the Hilbert space V, we are given bilinear forms $a_h : V_h \times V_h \to \mathbb{R}$ that are uniformly coercive, i.e., there exists $\alpha > 0$ such that*

$$a_h(v_h, v_h) \geq \alpha \|v_h\|_V^2$$

for all $v_h \in V_h$ and $h > 0$. We then have that there exists $c > 0$ such that

$$\|u - u_h\|_V \leq c_{SL} \inf_{v_h \in V_h} \left(\|u - v_h\|_V + \|a_h(v_h, \cdot) - a(v_h, \cdot)\|_{V_h'} + \|\ell_h - \ell\|_{V_h'} \right),$$

where $\|\mu\|_{V_h'} = \sup_{v_h \in V_h \setminus \{0\}} \mu(v_h) / \|v_h\|_V$.

Proof Let $v_h \in V_h$. For every $r_h \in V_h$ we have

$$a_h(u_h - v_h, r_h) = \ell_h(r_h) - \ell(r_h) + a(u - v_h, r_h) + a(v_h, r_h) - a_h(v_h, r_h)$$

$$\leq \left(\|\ell_h - \ell\|_{V_h'} + k_a \|u - v_h\|_V + \|a(v_h, \cdot) - a_h(v_h, \cdot)\|_{V_h'} \right) \|r_h\|_V.$$

For $r_h = u_h - v_h$ we obtain with the coercivity of a_h that

$$\alpha \|u_h - v_h\|_V \leq \|\ell_h - \ell\|_{V_h'} + k_a \|u - v_h\|_V + \|a(v_h, \cdot) - a_h(v_h, \cdot)\|_{V_h'}.$$

The triangle inequality $\|u - u_h\| \leq \|u - v_h\| + \|u_h - v_h\|$ implies the estimate. $\qquad \square$

We specify the estimate for the case $V_h = \mathscr{S}_D^1(\mathscr{T}_h)$, for which we assume that $a_h = a$. Notice that the elementwise Hessian of every function $v_h \in \mathscr{S}_D^1(\mathscr{T}_h)$ vanishes.

Example 3.6 Setting $x_T = (d+1)^{-1} \sum_{z \in \mathscr{N}_h \cap T} z$, the midpoint rule

$$\int_T \phi \, dx \approx Q_T(\phi) = |T| \phi(x_T)$$

is exact for functions $\phi \in \mathscr{P}_1(T)$. The Bramble–Hilbert lemma and transformation formulas lead to the estimate

$$\left| I_T(\phi) - Q_T(\phi) \right| \leq c h_T^2 \|D^2 \phi\|_{L^2(T)}.$$

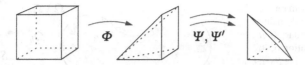

Fig. 3.8 Diffeomorphism Φ that maps the unit cube to a square-based pyramid; and diffeomorphisms Ψ and Ψ' that map halves of the pyramid into a tetrahedron

By summing over the elements $T \in \mathscr{T}_h$, we obtain with the Poincaré inequality the estimate

$$\left|\ell(v_h) - \ell_h(v_h)\right| \leq ch^2 \sum_{T \in \mathscr{T}_h} \|D^2(fv_h)\|_{L^2(T)}$$

$$\leq ch^2 \|f\|_{W^{2,\infty}(\Omega)} \|\nabla v_h\|_{L^2(\Omega)},$$

i.e., the effect of quadrature is of a higher order if f is sufficiently regular.
(ii) Let $Q = (0,1)^d$ be the unit cube in $\mathbb{R}^d$ and let

$$P = \left\{y \in \mathbb{R}^d : 0 \leq y_j \leq y_1 \leq 1, j = 2, \ldots, d\right\}$$

be a pyramid of height 1 with base $\{1\} \times (0,1)^{d-1}$. *Duffy's transformation*

$$\Phi : Q \to P, \quad (\xi_1, \xi_2, \ldots, \xi_d) \to (\xi_1, \xi_1\xi_2, \ldots, \xi_1\xi_d)$$

defines a diffeomorphism Φ with $\det D\Phi = \xi_1^{d-1}$, cf. Fig. 3.8. By the identity

$$\int_P f(y)\, \mathrm{d}y = \int_Q f(\xi_1, \xi_1\xi_2, \ldots, \xi_1\xi_d)\xi_1^{d-1}\, \mathrm{d}\xi$$

quadrature rules can be transferred from Q to P, and then to two copies of a tetrahedron if $d = 3$. Note that if f has a singularity at the origin, then the transformed integrand has a weaker singularity.

3.2.5 Boundary Approximation

The boundary of a Lipschitz domain $\Omega \subset \mathbb{R}^d$ is in general not polyhedral, and triangulations resolve the boundary only approximately. We consider here a two-dimensional situation and the case $\Gamma_\mathrm{D} = \partial\Omega$. Moreover, the boundary is assumed to be (piecewise) C^2-regular, i.e., it can locally be parametrized by a C^2 function. We let $\mathscr{T}_h$ be a triangulation such that all boundary nodes belong to $\partial\Omega$. The

Fig. 3.9 Approximate
triangulation of a domain
with curved boundary (*left*);
local parametrization of $\partial\Omega$
over a boundary side S (*right*)

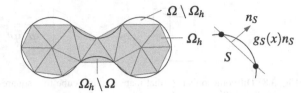

triangulation $\mathcal{T}_h$ defines the approximate domain

$$\Omega_h = \text{int}\left(\bigcup_{T\in\mathcal{T}_h} T\right).$$

If Ω is convex, then $\Omega_h \subset \Omega$, but in general we have neither $\Omega \subset \Omega_h$ nor $\Omega_h \subset \Omega$,
cf. Fig. 3.9. We follow [7].

The condition that boundary nodes belong to the boundary of Ω implies that the
discrete boundary $\partial\Omega_h$ is a piecewise Lagrange interpolant of $\partial\Omega$.

Lemma 3.5 (Complement Area) *For every* $x \in \Omega \setminus \Omega_h$ *and* $y \in \Omega_h \setminus \Omega$ *there exist*
$x' \in \Omega_h$ *and* $y' \in \Omega$ *such that*

$$|x - x'| + |y - y'| \le ch^2.$$

Moreover, we have

$$|\Omega \setminus \Omega_h| + |\Omega_h \setminus \Omega| \le ch^2.$$

Proof For h sufficiently small and every boundary side $S \subset \partial\Omega_h$, there exists a
function $g_S \in C^2(S)$ such that

$$\partial\Omega = \bigcup_{S\in\mathcal{S}_h, S\subset\partial\Omega} \{x + g_S(x)n_S : x \in S\},$$

where n_S is the outer unit normal to $\partial\Omega_h$ on S, cf. Fig. 3.9. Since $g_S(z) = 0$ for the
nodes $z \in \mathcal{N}_h \cap S$, we have that $|g_S(x)| \le ch^2$. The sets $\Omega \setminus \Omega_h$ and $\Omega_h \setminus \Omega$ are
therefore subsets of a strip of width ch^2 around $\partial\Omega$, and hence their measure is of
order $\mathcal{O}(h^2)$.   □

On the difference set $\Omega_h \setminus \Omega$, we need a directional Poincaré inequality. We
assume without loss of generality that $S \subset \mathbb{R}^{d-1}$ and $n_S = e_d$.

Lemma 3.6 (Directional Poincaré Inequality) *Let* $S \subset \mathbb{R}^{d-1}$ *and for a nonnega-*
tive function $g \in C^2(S)$, *define*

$$\omega = \{(x', x_d) \in S \times (0, \infty) : 0 < x_d < g(x')\}.$$

For every function $v \in H^1(\omega)$ with $v|_S = 0$, we have

$$\|v\|_{L^2(\omega)} \leq \|g\|_{L^\infty(S)} \|\partial_d v\|_{L^2(\omega)}.$$

Proof Assume that $v \in C^1(\overline{\omega})$. For every $x' \in S$ and $0 < x_d < g(x')$ we have, using $v(x',0) = 0$, that

$$v(x',x_d) = \int_0^{x_d} \partial_d v(x',t)\, dt.$$

Hölder's inequality implies that

$$|v(x',x_d)| \leq g(x')^{1/2} \left(\int_0^{g(x')} |\partial_d v(x',t)|^2\, dt \right)^{1/2}.$$

Taking squares and integrating over $x_d \in (0, g(x'))$ shows that

$$\int_0^{g(x')} |v(x',x_d)|^2\, dx_d \leq g(x')^2 \int_0^{g(x')} |\partial_d v(x',t)|^2\, dt.$$

Integrating over $x' \in S$ and incorporating a density argument implies the estimate.
□

We show that under certain regularity assumptions the approximate treatment of the domain Ω does not lead to a reduced convergence rate.

Proposition 3.7 (Boundary Approximation) *Let $u \in H_0^1(\Omega)$ be the weak solution of*

$$-\Delta u = f \text{ in } \Omega, u = 0 \text{ on } \partial\Omega,$$

and assume that $u \in H^2(\Omega) \cap W^{1,\infty}(\Omega)$. Let $\tilde{u} \in H^2(\Omega \cup \Omega_h)$ and $\tilde{f} \in L^2(\Omega \cup \Omega_h)$ be extensions of u and f to $\Omega \cup \Omega_h$, respectively. If the Galerkin approximation $u_h \in \mathscr{S}_0^1(\mathscr{T}_h)$, defined by

$$\int_{\Omega_h} \nabla u_h \cdot \nabla v_h\, dx = \int_{\Omega_h} f v_h\, dx$$

for all $v_h \in \mathscr{S}_0^1(\mathscr{T}_h)$, is extended by zero to $\Omega \cup \Omega_h$, then we have

$$\|\nabla(\tilde{u} - u_h)\|_{L^2(\Omega \cup \Omega_h)} \leq ch.$$

Proof We first note that for every $w_h \in \mathscr{S}_0^1(\mathscr{T}_h)$, which is extended by zero to $\Omega \cup \Omega_h$, we have, using that $-\Delta u = f$,

$$\int_{\Omega_h} \nabla \tilde{u} \cdot \nabla w_h \, dx = -\int_{\Omega_h \cap \Omega} \Delta \tilde{u} \, w_h \, dx - \int_{\Omega_h \setminus \Omega} \Delta \tilde{u} \, w_h \, dx$$

$$= \int_{\Omega_h \cap \Omega} f w_h \, dx - \int_{\Omega_h \setminus \Omega} \Delta \tilde{u} \, w_h \, dx.$$

Since we also have

$$\int_{\Omega_h} \nabla u_h \cdot \nabla w_h \, dx = \int_{\Omega_h} \tilde{f} w_h \, dx = \int_{\Omega_h \cap \Omega} f w_h \, dx + \int_{\Omega_h \setminus \Omega} \tilde{f} w_h \, dx,$$

we deduce the perturbed Galerkin orthogonality

$$\int_{\Omega_h} \nabla (\tilde{u} - u_h) \cdot \nabla w_h \, dx = -\int_{\Omega_h \setminus \Omega} (\Delta \tilde{u} + \tilde{f}) w_h \, dx.$$

We thus have, for every $v_h \in \mathscr{S}_0^1(\mathscr{T}_h)$, abbreviating $w_h = u_h - v_h$,

$$\|\nabla(u_h - v_h)\|_{L^2(\Omega_h)}^2 = \int_{\Omega_h} \nabla(u_h - v_h) \cdot \nabla w_h \, dx$$

$$= \int_{\Omega_h} \nabla(\tilde{u} - v_h) \cdot \nabla w_h \, dx + \int_{\Omega_h \setminus \Omega} (\Delta \tilde{u} + \tilde{f}) w_h \, dx$$

$$\leq \|\nabla(\tilde{u} - v_h)\|_{L^2(\Omega_h)} \|\nabla w_h\|_{L^2(\Omega_h)} + \|\Delta \tilde{u} + \tilde{f}\|_{L^2(\Omega \setminus \Omega_h)} \|w_h\|_{L^2(\Omega_h \setminus \Omega)}.$$

Incorporating the directional Poincaré inequality on every boundary side S and dividing by $\|\nabla w_h\|_{L^2(\Omega_h)} = \|\nabla(u_h - v_h)\|_{L^2(\Omega_h)}$ lead to

$$\|\nabla(u_h - v_h)\|_{L^2(\Omega_h)} \leq \|\nabla(\tilde{u} - v_h)\|_{L^2(\Omega_h)} + ch^2 \|\Delta \tilde{u} + \tilde{f}\|_{L^2(\Omega \setminus \Omega_h)},$$

where we used that $\|g_S\|_{L^\infty(S)} \leq ch^2$ for a local parametrization g_S of $\partial \Omega$ over S. The triangle inequality and the choice $v_h = \mathscr{I}_h \tilde{u}$ prove that

$$\|\nabla(\tilde{u} - u_h)\|_{L^2(\Omega_h)} \leq \|\nabla(\tilde{u} - v_h)\|_{L^2(\Omega_h)} + \|\nabla(u_h - v_h)\|_{L^2(\Omega_h)}$$

$$\leq ch \|D^2 \tilde{u}\|_{L^2(\Omega_h)} + ch^2 \|\Delta \tilde{u} + \tilde{f}\|_{L^2(\Omega \setminus \Omega_h)}.$$

Finally, on $\Omega \setminus \Omega_h$ we have $u_h = 0$, and hence

$$\|\nabla(\tilde{u} - u_h)\|_{L^2(\Omega \setminus \Omega_h)}^2 = \|\nabla u\|_{L^2(\Omega \setminus \Omega_h)}^2 \leq \|\nabla u\|_{L^\infty(\Omega \setminus \Omega_h)} |\Omega \setminus \Omega_h|^{1/2} \leq ch.$$

A combination of the estimates proves the result. $\square$

An abstract framework for the analysis of nonconforming discretizations is provided by the second Strang lemma.

Proposition 3.8 (Second Strang Lemma) *Let $W = V + V_h$, and assume that $a :$ $W \times W \to \mathbb{R}$ is bilinear and continuous with respect to a norm $\| \cdot \|_h$, and assume that a_h is coercive on V_h. Let $u_h \in V_h$ satisfy $a_h(u_h, v_h) = \ell_h(v_h)$ for all $v_h \in V_h$, and let $u \in V$ be such that $a(u, v) = \ell(v)$ for all $v \in V$. Then there exists $c > 0$ such that*

$$c^{-1} \|u - u_h\|_h \leq \inf_{v_h \in V_h} \|u - v_h\|_h + \|a_h(u, \cdot) - \ell_h\|_{V_h'}.$$

Proof Exercise.   □

Remark 3.9 In the case of higher order methods, isoparametric families have to be considered to retain the expected convergence rates.

3.2.6 Discrete Inequalities

For $P1$-finite element functions various inequalities are available, that do not hold for Sobolev functions. They exploit the finite dimensionality of the finite element spaces.

Lemma 3.7 (Norm Equivalence) *For every $1 \leq p < \infty$ there exists $c > 0$ such that for all $v_h \in \mathscr{S}^1(\mathscr{T}_h)$, we have*

$$c^{-1} \|v_h\|_{L^p(\Omega)} \leq \left(\sum_{z \in \mathscr{N}_h} h_z^d |v_h(z)|^p \right)^{1/p} \leq c \|v_h\|_{L^p(\Omega)}.$$

Moreover, we have $\|v_h\|_{L^\infty(\Omega)} = \max_{z \in \mathscr{N}_h} |v_h(z)|$ for every $v_h \in \mathscr{S}^1(\mathscr{T}_h)$.

Proof For $T \in \mathscr{T}_h$ the expressions $\|v_h\|_{L^p(T)}$ and $\left(h_T^d \sum_{z \in \mathscr{N}_h \cap T} |v_h(z)|^p \right)^{1/p}$ are norms on the finite-dimensional space $\mathscr{S}^1(\mathscr{T}_h)|_T$. Hence they are equivalent and a transformation argument shows that the constant is independent of h_T and h_z. The asserted estimate follows from a summation over $T \in \mathscr{T}_h$.   □

Further inequalities are available, when the diameters of the elements in a triangulation do not vary too strongly.

Definition 3.12 A family of triangulations $(\mathscr{T}_h)_{h>0}$ is called *quasiuniform* if there exists $c > 0$ such that $c^{-1} h \leq h_T \leq ch$ for all $h > 0$ and all $T \in \mathscr{T}_h$.

Lemma 3.8 (Inverse Estimates) *For $v_h \in \mathscr{S}^1(\mathscr{T}_h)$ and $1 \leq r, p \leq \infty$ we have*

$$\|\nabla v_h\|_{L^p(T)} \leq c h_T^{-1} \|v_h\|_{L^p(T)}$$

and

$$\|v_h\|_{L^p(T)} \le ch_T^{d(r-p)/(pr)}\|v_h\|_{L^r(T)}.$$

If the family $(\mathscr{T}_h)_{h>0}$ is quasiuniform, then we have

$$\|\nabla v_h\|_{L^p(\Omega)} \le ch^{-1}\|v_h\|_{L^p(\Omega)}$$

and

$$\|v_h\|_{L^p(\Omega)} \le ch^{\min\{0,d(r-p)/(pr)\}}\|v_h\|_{L^r(\Omega)}.$$

Proof To prove the first estimate, consider the space $V_T = \mathscr{S}^1(\mathscr{T}_h)|_T/\mathbb{R}$, i.e., functions $v_h \in \mathscr{S}^1(\mathscr{T}_h)$ with $\int_T v_h\,dx = 0$. The expressions $\|\nabla v_h\|_{L^p(T)}$ and $h_T^{-1}\|v_h\|_{L^p(T)}$ are equivalent norms on the finite-dimensional space V_T. Using the elementary estimate $\|v_h - \overline{v}_h\|_{L^p(T)} \le 2\|v_h\|_{L^p(T)}$ for $v_h \in \mathscr{S}_D^1(\mathscr{T}_h)$ and $\overline{v}_h = |T|^{-1}\int_T v_h\,dx$, a transformation argument proves the first estimate. A similar argument proves the second estimate. The third estimate follows from a summation of the first estimate over $T \in \mathscr{T}_h$ and $h_T^{-1} \le ch^{-1}$ due to the assumed quasiuniformity of $\mathscr{T}_h$. To prove the last estimate we first note that it follows directly from Hölder's inequality if $r \ge p$. Otherwise, we use $\left(\sum_{j=1}^L |x_j|^r\right)^{1/r} \le \left(\sum_{j=1}^L |x_j|^p\right)^{1/p}$ for every $L \in \mathbb{N}$ and $x \in \mathbb{R}^L$ and deduce that

$$\left(\sum_{T\in\mathscr{T}_h} \|v_h\|_{L^r(T)}^p\right)^{1/p} \le \left(\sum_{T\in\mathscr{T}_h} \|v_h\|_{L^r(T)}^r\right)^{1/r} = \|v_h\|_{L^r(\Omega)}.$$

With the corresponding elementwise estimates, this implies the global estimate for quasiuniform triangulations.   □

Inverse inequalities have important implications.

Example 3.7 The nodal interpolation operator $\mathscr{I}_h$ related to a triangulation $\mathscr{T}_h$ defines a bounded linear operator $\mathscr{I}_h : W_D^{1,\infty}(\Omega) \to W_D^{1,\infty}(\Omega)$. To verify this, let $v \in W_D^{1,\infty}(\Omega)$ and $T \in \mathscr{T}_h$. Lemma 3.8 shows that for every $q \in \mathbb{R}$ we have that

$$\|\nabla(\mathscr{I}_h v - q)\|_{L^\infty(T)} \le ch_T^{-1}\|\mathscr{I}_h v - q\|_{L^\infty(T)}$$

$$\le ch_T^{-1}\|\mathscr{I}_h v - v\|_{L^\infty(T)} + ch_T^{-1}\|v - q\|_{L^\infty(T)}.$$

A nodal interpolation estimate, the choice $q = |T|^{-1}\int_T v\,dx$, and a Poincaré inequality in $L^\infty(T)$ yield that

$$\|\nabla\mathscr{I}_h v\|_{L^\infty(T)} \le c\|\nabla v\|_{L^\infty(T)},$$

which implies that $\|\nabla \mathscr{I}_h v\|_{L^\infty(\Omega)} \le c\|\nabla v\|_{L^\infty(\Omega)}$. Note that $\mathscr{I}_h$ is not a bounded operator on $W_D^{1,2}(\Omega)$ if $d > 1$.

3.3 Implementation of *P*1- and *P*2-Methods

3.3.1 *P*1-Method

For ease of presentation we assume that the right-hand side functions in the considered Poisson problem allow for an exact integration.

Assumption 3.1 (Data Approximation) *We assume that* $u_D = \tilde{u}_{D,h}|_{\Gamma_D}$ *for a function* $\tilde{u}_{D,h} \in \mathscr{S}^1(\mathscr{T}_h)$ *and* $f \in \mathscr{L}^0(\mathscr{T}_h)$ *and* $g \in \mathscr{L}^0(\mathscr{T}_h)|_{\Gamma_N}$ *are piecewise constant.*

We decompose the unknown function as $u_h = \tilde{u}_h + \tilde{u}_{D,h}$, and compute the uniquely defined function $\tilde{u}_h \in \mathscr{S}_D^1(\mathscr{T}_h)$ with

$$\int_\Omega \nabla \tilde{u}_h \cdot \nabla v_h \, dx = \int_\Omega f v_h \, dx + \int_{\Gamma_N} g v_h \, ds - \int_\Omega \nabla \tilde{u}_{D,h} \cdot \nabla v_h \, dx$$

for all $v_h \in \mathscr{S}_D^1(\mathscr{T}_h)$. We let $\widetilde{U} = (\widetilde{U}_y : y \in \mathscr{N}_h \setminus \Gamma_D)$ be the coefficients of $\tilde{u}_h$ with respect to the nodal basis restricted to the *free nodes* in $\mathscr{N}_h \setminus \Gamma_D$. For every $T \in \mathscr{T}_h$ and every $S \in \mathscr{S}_h$, we let x_T and x_S denote their midpoints. The discrete formulation is thus equivalent to the linear system of equations

$$\sum_{y \in \mathscr{N}_h \setminus \Gamma_D} \widetilde{U}_y \int_\Omega \nabla \varphi_y \cdot \nabla \varphi_z \, dx = \sum_{T \in \mathscr{T}_h} f(x_T) \int_T \varphi_z \, dx + \sum_{S \subset \mathscr{S}_h \cap \overline{\Gamma}_N} g(x_S) \int_S \varphi_z \, ds$$

$$- \sum_{y \in \mathscr{N}_h} \tilde{u}_{D,h}(y) \int_\Omega \nabla \varphi_y \cdot \nabla \varphi_z \, dx$$

for all $z \in \mathscr{N}_h \setminus \Gamma_D$, i.e.,

$$\tilde{s}\widetilde{U} = \tilde{b},$$

with a symmetric matrix $\tilde{s} \in \mathbb{R}^{N \times N}$ and $\tilde{b} \in \mathbb{R}^N$ for $N = |\mathscr{N}_h \setminus \Gamma_D|$. The integrals that define the matrix and the vector on the right-hand side are computed by decomposing the integral as a sum over elements, e.g.,

$$\int_\Omega \nabla \varphi_z \cdot \nabla \varphi_y \, dx = \sum_{T \in \mathscr{T}_h : z, y \in T} \int_T \nabla \varphi_z \cdot \nabla \varphi_y \, dx.$$

The triangulation of Ω and the partition of the boundary $\partial\Omega$ are defined by the arrays c4n, n4e, Db, and Nb that specify the coordinates of the nodes, the vertices of the elements, and the vertices of the sides on Γ_D and $\overline{\Gamma}_N$, respectively. In particular, the $n_C \times d$ array c4n defines the coordinates of the nodes and implicitly a *global enumeration* of the nodes. The $n_E \times (d+1)$ array n4e defines the elements by specifying the positions of their vertices via their numbers. This defines a *local enumeration* of the nodes of every element. Similarly, the $n_{Db} \times d$ and $n_{Nb} \times d$ arrays Db and Nb define the vertices of the sides belonging to Γ_D and $\overline{\Gamma}_N$, respectively. The arrays thus define mappings

$$\texttt{c4n}: \mathcal{N}_h \to \mathbb{R}^d, \quad \texttt{n4e}: \mathcal{T}_h \to \mathcal{N}_h^{d+1},$$

and

$$\texttt{Db}: \mathcal{S}_h \cap \Gamma_D \to \mathcal{N}_h^d, \quad \texttt{Nb}: \mathcal{S}_h \cap \overline{\Gamma}_N \to \mathcal{N}_h^d.$$

For a triangulation consisting of two triangles and with four nodes, the arrays are displayed in Fig. 3.10.

In the following assumption we identify an element with an ordered list of nodes, denoted $T \equiv (z_0, z_1, \ldots, z_d)$.

Assumption 3.2 (Orientation) *We assume that the list of elements defines an ordering of the nodes of elements that induces a positive orientation of T, i.e., if $T \equiv (z_0, z_1, \ldots, z_d)$ for $T \in \mathcal{T}_h$ and $z_0, z_1, \ldots, z_d \in \mathcal{N}_h$ such that $T = \mathrm{conv}\{z_0, z_1, \ldots, z_d\}$, then the vectors $\tau_\ell = z_\ell - z_0$, $\ell = 1, 2, \ldots, d$, satisfy*

$$\tau_1 > 0, \quad \tau_2 \cdot \tau_1^\perp > 0, \quad \tau_3 \cdot (\tau_1 \times \tau_2) > 0$$

for $d = 1, 2, 3$, respectively.

To compute the system matrix $\tilde{s}$ and the vector $\tilde{b}$, we note some elementary identities for the nodal basis functions.

$\Omega = (0,1)^2$, $\Gamma_D = \{0,1\} \times \{0\}$, $\Gamma_N = \partial\Omega \setminus \Gamma_D$

```
c4n = [0,0;1,0;1,1;0,1];
n4e = [1,2,3;1,3,4];
Db  = [1,2];
Nb  = [2,3;3,4;4,1];
```

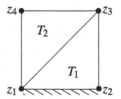

Fig. 3.10 Triangulation of the unit square and corresponding arrays

Lemma 3.9 (Elementwise Gradients) *Let* $T \equiv (z_0, z_1, \ldots, z_d)$ *be a simplex with vertices* $z_0, z_1, \ldots, z_d \in \mathbb{R}^d$ *and define*

$$X_T = \begin{bmatrix} 1 & 1 & \cdots & 1 \\ z_0 & z_1 & \cdots & z_d \end{bmatrix} \in \mathbb{R}^{(d+1) \times (d+1)}.$$

We then have that the volume $|T|$ *is given by* $|T| = (1/d!) \det X_T$, *and with the identity matrix* $I_d \in \mathbb{R}^{d \times d}$ *that*

$$\left[\nabla \varphi_{z_0}|_T, \ldots, \nabla \varphi_{z_d}|_T \right]^{\mathsf{T}} = X_T^{-1} \begin{bmatrix} 0 \\ I_d \end{bmatrix}.$$

Proof The proof follows from noting that the nodal basis function associated with z_j is for $x \in T$ given by

$$\varphi_{z_j}(x) = \frac{1}{d!|T|} \det \begin{bmatrix} 1 & 1 & \cdots & 1 \\ x & z_{j+1} & \cdots & z_{j+d} \end{bmatrix},$$

where subscripts are understood modulo d, together with Laplace's formula and Cramer's rule. □

Some additional identities are required for computing the vector $\tilde{b}$.

Lemma 3.10 (Right-Hand Side) *For a side* $S = \mathrm{conv}\{z_0, z_1, \ldots, z_{d-1}\} \in \mathscr{S}_h$, *the surface area* $|S|$ *is given by*

$$|S| = \begin{cases} 1 & \text{if } d = 1, \\ |z_1 - z_0| & \text{if } d = 2, \\ |(z_2 - z_0) \times (z_1 - z_0)|/2 & \text{if } d = 3. \end{cases}$$

Moreover, for $T \in \mathscr{T}_h$, $S \in \mathscr{S}_h$ *and* $z \in T \cap S$, *we have*

$$\int_T \varphi_z \, \mathrm{d}x = \frac{|T|}{d+1}, \quad \int_S \varphi_z \, \mathrm{d}s = \frac{|S|}{d}.$$

Proof The proof of the formula for $|S|$ follows from elementary geometric identities. The integrals over T and S are computed with the help of an affine transformation to a reference element. □

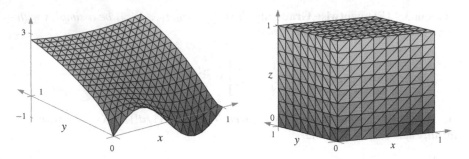

Fig. 3.11 Finite element approximations of two- (*left*) and three-dimensional (*right*) Poisson problems

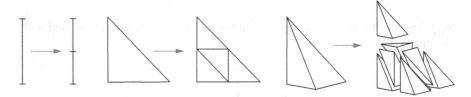

Fig. 3.12 Red-refinement of one-, two-, and three-dimensional simplices to define refined triangulations

The finite element functions visualized in Fig. 3.11 were computed with the refinement method illustrated in Fig. 3.12 and the MATLAB routines shown in Figs. 3.14, 3.15, and 3.16. Figure 3.14 shows a MATLAB implementation of the $P1$-method in which the matrix $\tilde{s}$ corresponds to the array s(fNodes,fNodes). We input the space dimension and the number of refinements of a coarse triangulation. The routine red_refine.m carries out the refinements of the triangulation by dividing every element into 2^d similar subelements. The operation s\b solves a linear system of equations and the command sparse(I,J,X,nC,nC) assembles a sparse matrix $s \in \mathbb{R}^{n_C \times n_C}$ by specifying its entries via lists $I, J, X \in \mathbb{R}^L$ and $s_{ij} = \sum_{\ell \in \{1,\dots,L\}, I_\ell = i, J_\ell = j} X_\ell$. The initial triangulation of the unit cube $(0, 1)^d$ is provided by the routine triang_cube.m shown in Fig. 3.13. Results of two- and three-dimensional numerical experiments carried out with the implementation are shown in Fig. 3.11.

3.3.2 Uniform Refinement

Coarse triangulations can be refined with the MATLAB routine red_refine.m displayed in Fig. 3.15. The refinement procedure subdivides every d-dimensional simplex into 2^d subsimplices by bisecting its one-dimensional subsimplices and appropriately connecting new nodes as illustrated in Fig. 3.12. The routine also

```
function [c4n,n4e,Db,Nb] = triang_cube(d)
if d == 1
    c4n = [0;1]; n4e = [1,2]; Db = 1; Nb = 2;
elseif d == 2
    c4n = [0,0;1,0;1,1;0,1]; n4e = [1,2,3;1,3,4];
    Db = [1,2]; Nb = [2,3;3,4;4,1];
elseif d == 3
    c4n = [0,0,0;1,0,0;1,1,0;0,1,0;0,0,1;1,0,1;1,1,1;0,1,1];
    n4e = [1,2,3,7;1,6,2,7;1,5,6,7;1,8,5,7;1,4,8,7;1,3,4,7];
    Db = [1,2,3;1,4,3];
    Nb = [2,3,7;2,7,6;1,2,6;1,6,5;5,6,7;1,8,5;5,7,8;1,4,8; ...
        4,7,8;3,4,7];
end
```

Fig. 3.13 MATLAB routine that provides a triangulation for the unit cube $(0, 1)^d$

provides *prolongation matrices* that allow for computing the coefficients of a given finite element function on the coarse triangulation with respect to the nodal basis on the refined triangulation by a matrix vector multiplication. For continuous, piecewise affine functions, this is realized with the matrix P1 and for elementwise constant functions with the matrix P0. The realization of the refinement routine is based on appropriate data structures for the edges in a triangulation. These are computed in the subroutine edge_data and provide mappings

$$\text{edges} : \mathscr{E}_h \to \mathscr{N}^2, \quad \text{el2edges} : \mathscr{T}_h \to \mathscr{E}^{i_d},$$

where $i_d = 0, 1, 3, 6$ for $d = 0, 1, 2, 3$, which specify the nodes that are the endpoints of a given edge and the edges of an element, respectively. Furthermore, the arrays

$$\text{Db2edges} : \mathscr{S}_h \cap \Gamma_{\mathrm{D}} \to \mathscr{E}^{i_{d-1}}, \quad \text{Nb2edges} : \mathscr{S}_h \cap \overline{\Gamma}_{\mathrm{N}} \to \mathscr{E}^{i_{d-1}},$$

determine the edges which belong to a boundary side on Γ_{D} or Γ_{N}.

```
function p1_poisson(d,red)
[c4n,n4e,Db,Nb] = triang_cube(d);
for j = 1:red
    [c4n,n4e,Db,Nb,P0,P1] = red_refine(c4n,n4e,Db,Nb);
end
[nC,d] = size(c4n); nE = size(n4e,1); nNb = size(Nb,1);
dNodes = unique(Db); fNodes = setdiff(1:nC,dNodes);
u = zeros(nC,1); tu_D = zeros(nC,1); b = zeros(nC,1);
ctr = 0; ctr_max = (d+1)^2*nE;
I = zeros(ctr_max,1); J = zeros(ctr_max,1); X = zeros(ctr_max,1);
for j = 1:nE
    X_T = [ones(1,d+1);c4n(n4e(j,:),:)'];
    grads_T = X_T\[zeros(1,d);eye(d)];
    vol_T = det(X_T)/factorial(d);
    mp_T = sum(c4n(n4e(j,:),:),1)/(d+1);
    for m = 1:d+1
        b(n4e(j,m)) = b(n4e(j,m))+(1/(d+1))*vol_T*f(mp_T);
        for n = 1:d+1
            ctr = ctr+1; I(ctr) = n4e(j,m); J(ctr) = n4e(j,n);
            X(ctr) = vol_T*grads_T(m,:)*grads_T(n,:)';
        end
    end
end
s = sparse(I,J,X,nC,nC);
for j = 1:nNb
    if d == 1
        vol_S = 1;
    elseif d == 2
        vol_S = norm(c4n(Nb(j,1),:)-c4n(Nb(j,2),:));
    elseif d == 3
        vol_S = norm(cross(c4n(Nb(j,3),:)-c4n(Nb(j,1),:),...
            c4n(Nb(j,2),:)-c4n(Nb(j,1),:)),2)/2;
    end
    mp_S = sum(c4n(Nb(j,:),:),1)/d;
    for k = 1:d
        b(Nb(j,k)) = b(Nb(j,k))+(1/d)*vol_S*g(mp_S);
    end
end
for j = 1:nC
    tu_D(j) = u_D(c4n(j,:));
end
b = b-s*tu_D; u(fNodes) = s(fNodes,fNodes)\b(fNodes); u = u+tu_D;
if d == 1; plot(c4n(n4e),u(n4e));
elseif d == 2; trisurf(n4e,c4n(:,1),c4n(:,2),u);
elseif d == 3; trisurf([Db;Nb],c4n(:,1),c4n(:,2),c4n(:,3),u);
end

function val = f(x); val = 1;
function val = g(x); val = 1;
function val = u_D(x); val = sin(2*pi*x(:,1));
```

Fig. 3.14 MATLAB implementation of the *P*1 finite element method for the Poisson problem

```
function [c4n,n4e,Db,Nb,P0,P1] = red_refine(c4n,n4e,Db,Nb)
nE = size(n4e,1); [nC,d] = size(c4n);
[edges,el2edges,Db2edges,Nb2edges] = edge_data(n4e,Db,Nb);
nS = size(edges,1);
newNodes = nC+(1:nS)';
newIndices = reshape(newNodes(el2edges),size(el2edges));
idxElements = refinement_rule(d);
n4e = [n4e,newIndices];
n4e = n4e(:,idxElements);
n4e = reshape(n4e',d+1,[])';
newCoord = .5*(c4n(edges(:,1),:)+c4n(edges(:,2),:));
c4n = [c4n;newCoord];
idx_Bdy = refinement_rule(d-1);
if size(Db,1) > 0
    newDb = reshape(newNodes(Db2edges),size(Db2edges));
    Db = [Db,newDb]; Db = Db(:,idx_Bdy); Db = reshape(Db',d,[])';
end
if size(Nb,1) > 0
    newNb = reshape(newNodes(Nb2edges),size(Nb2edges));
    Nb = [Nb,newNb]; Nb = Nb(:,idx_Bdy); Nb = reshape(Nb',d,[])';
end
P0 = sparse(1:size(n4e,1),reshape( repmat(1:nE,2^d,1),1,[]),1);
P1 = sparse([1:nC,newNodes',newNodes'],...
    [1:nC,edges(:,1)',edges(:,2)'],...
    [ones(1,nC),.5*ones(1,2*nS)],nC+nS,nC);

function idx = refinement_rule(d)
switch d
 case 0; idx = 1;
 case 1; idx = [1 3,3 2];
 case 2; idx = [1 4 5,5 6 3,4 2 6,6 5 4];
 case 3; idx = [1 5 6 7,5 2 8 9,6 8 3 10,7 9 10 4,...
           5 6 7 9,9 6 8 5,6 7 9 10,10 8 9 6];
end

function [edges,el2edges,Db2edges,Nb2edges] = edge_data(n4e,Db,Nb)
[nE,nV] = size(n4e); d = nV-1;
idx = [0,1,3,6]; Bdy = [Db;Nb]; nEdges = idx(d+1)*nE;
nDb = idx(d)*size(Db,1); nNb = idx(d)*size(Nb,1);
switch d
 case 1; edges = n4e;
 case 2; edges = [reshape(n4e(:,[1 2,1 3,2 3])',2,[])';Bdy];
 case 3; edges = reshape(n4e(:,[1 2,1 3,1 4,2 3,2 4,3 4])',2,[])';
     edges = [edges;reshape(Bdy(:,[1 2,1 3,2 3])',2,[])'];
end
[edges,~,edgeNumbers] = unique(sort(edges,2),'rows','first');
el2edges = reshape(edgeNumbers(1:nEdges),idx(d+1),[])';
Db2edges = reshape(edgeNumbers(nEdges+(1:nDb))',idx(d),[])';
Nb2edges = reshape(edgeNumbers(nEdges+nDb+(1:nNb))',idx(d),[])';
```

Fig. 3.15 MATLAB implementation of a uniform refinement procedure

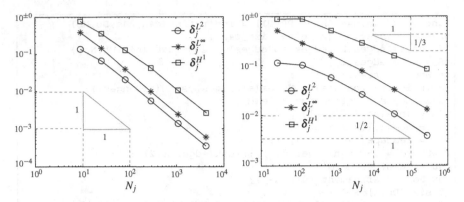

Fig. 3.16 Discrete errors versus degrees of freedom with a logarithmic scaling used for both axes in a two- (*left*) and three-dimensional (*right*) Poisson problem

3.3.3   Experimental Convergence Rates

The rate of convergence of a finite element method for a boundary value problem depends on the discretization used and on the regularity properties of the problem under consideration, which are often difficult to identify. A natural requirement on a numerical scheme is that it be *quasioptimal* in the sense that for smooth solutions the convergence rates are comparable to the interpolation error. If this cannot be proved, then this has to be checked experimentally. For this, one chooses a nontrivial, sufficiently regular function $u \in C^k(\overline{\Omega})$ and defines right-hand sides in the boundary value problem so that this function is the exact solution.

Example 3.8 Let $\Omega = (0, 1)^d$ and define

$$u(x) = \sin(\pi x_1) \sin(\pi x_2) \ldots \sin(\pi x_d)$$

for $x = (x_1, x_2, \ldots, x_d) \in \Omega$. Let $f(x) = -\Delta u(x) = d\pi^2 u(x)$ and $u_D = u|_{\partial\Omega} = 0$. Then $u \in C^\infty(\overline{\Omega})$ is the exact solution of the Poisson problem with homogeneous Dirichlet boundary conditions on $\Gamma_D = \partial\Omega$.

For a sequence of triangulations $(\mathcal{T}_j)_{j=1,2,\ldots}$ of Ω with maximal mesh-sizes $(h_j)_{j=1,2,\ldots}$, one then computes the finite element solutions $(u_j)_{j=1,2,\ldots}$ and determines *discrete approximation errors*

$$\delta_j^{L^p} = \|u_j - \mathscr{I}_j u\|_{L^p(\Omega)}, \quad \delta_j^{H^1} = \|\nabla(u_j - \mathscr{I}_j u)\|_{L^2(\Omega)},$$

where $\mathscr{I}_j u$ is the nodal interpolant associated with the finite element method. The quantities δ_j are equivalent to norms of the errors $e_j = u - u_j$ up to nodal interpolation errors, e.g.,

$$\left| \delta_j^{H^1} - \| \nabla(u - \mathscr{I}_j u) \|_{L^2(\Omega)} \right| \leq \| \nabla e_j \|_{L^2(\Omega)} \leq \delta_j^{H^1} + \| \nabla(u - \mathscr{I}_j u) \|_{L^2(\Omega)}.$$

To identify a relation

$$\delta_j \approx c h_j^\alpha$$

with a constant $c > 0$ and a parameter $\alpha \geq 0$, one computes the logarithm of the quotients δ_j / δ_{j-1}, i.e.,

$$\alpha \approx \alpha_j = \frac{\log(\delta_j / \delta_{j-1})}{\log(h_j / h_{j-1})} = \frac{\log(\delta_j) - \log(\delta_{j-1})}{\log(h_j) - \log(h_{j-1})}.$$

We refer to the quotient on the right-hand side as a *logarithmic slope*. If the values $(\alpha_j)_{j=2,3,...}$ tend to converge to a number $\alpha \in \mathbb{R}$, then α is called an *experimental convergence rate*. Tables 3.1 and 3.2 display the errors $\delta_j^{L^2}$, $\delta_j^{L^\infty}$, and $\delta_j^{H^1}$ for approximating the Poisson problem specified in Example 3.8 on a sequence of uniform triangulations obtained from red-refinements of coarse triangulations $\mathscr{T}_0$ of

Table 3.1 Discrete errors and experimental convergence rates in a two-dimensional Poisson problem. A superconvergence phenomenon occurs for the discrete H^1 error

$d = 2, h_j'$	N_j	$\delta_j^{L^2}$	$\delta_j^{L^\infty}$	$\delta_j^{H^1}$	$\alpha_j^{L^2}$	$\alpha_j^{L^\infty}$	$\alpha_j^{H^1}$
2^{-1}	9	0.1355	0.3831	0.7663	–	–	–
2^{-2}	25	0.0621	0.1375	0.3021	1.1252	1.4781	1.3427
2^{-3}	81	0.0183	0.0375	0.0850	1.7603	1.8758	1.8298
2^{-4}	289	0.0048	0.0096	0.0219	1.9376	1.9693	1.9561
2^{-5}	1089	0.0012	0.0024	0.0055	1.9842	1.9923	1.9889
2^{-6}	4225	0.0003	0.0006	0.0014	1.9960	1.9981	1.9972

Table 3.2 Discrete errors and experimental convergence rates in a three-dimensional Poisson problem

$d = 3, h_j'$	N_j	$\delta_j^{L^2}$	$\delta_j^{L^\infty}$	$\delta_j^{H^1}$	$\alpha_j^{L^2}$	$\alpha_j^{L^\infty}$	$\alpha_j^{H^1}$
2^{-1}	27	0.1133	0.5065	0.8773	–	–	–
2^{-2}	125	0.1016	0.2804	0.8803	0.1572	0.8529	−0.0050
2^{-3}	729	0.0568	0.1610	0.5109	0.8373	0.8010	0.7851
2^{-4}	4913	0.0262	0.0781	0.2860	1.1170	1.0427	0.8370
2^{-5}	35937	0.0107	0.0330	0.1579	1.2913	1.2433	0.8567
2^{-6}	274625	0.0039	0.0132	0.0849	1.4474	1.3250	0.8960

$\Omega = (0, 1)^d$ into $d!$ triangles and tetrahedra for $d = 2$ and $d = 3$, respectively. For simplicity we displayed the quantity $h'_j = h_j/d^{1/2}$. In the two-dimensional situation corresponding to the results in Table 3.1, we observe that all discrete errors decay quadratically, i.e., whenever the mesh-size is decreased by a factor of $1/2$, the errors nearly decrease by a factor of $1/4$. These are the optimal convergence rates in L^2 and L^∞, but this does not coincide with the expected linear convergence in H^1. In fact, we observe the *superconvergence phenomenon* that the differences of numerical solutions u_j and interpolants $\mathscr{I}_j u$ converge faster to zero in H^1 than the error $u - u_j$. This phenomenon is related to symmetry properties of the triangulations and does not occur for $d = 3$, as can be seen from the results displayed in Table 3.2. In particular, the numbers $(\alpha_j)_{j=2,3,\ldots}$ vary more strongly than in the two-dimensional situation, which reflects a larger *preasymptotic range*, i.e., the expected convergence rates and the asymptotic behavior for sufficiently small mesh-sizes are observed later.

The experimental convergence rate can also be obtained as the average slope of a polygonal curve, obtained by linearly interpolating pairs (h_j, δ_j), $j = 1, 2, \ldots$, and plotting this curve with a logarithmic scaling for both axes. Alternatively and more commonly, one interpolates the pairs (N_j, δ_j), $j = 1, 2, \ldots$, with the *numbers of degrees of freedom N_j*, i.e., the dimensions of the finite element spaces related to the triangulations, e.g., $N_j = \dim \mathscr{S}_0^1(\mathscr{T}_j)$. For quasiuniform triangulations, we have that $N_j \sim h_j^{-d}$, i.e., we expect a relation

$$\delta_j \approx c N_j^{-\alpha/d}.$$

Including a gradient triangle in the plot allows for a comparison with the expected convergence rate. In Fig. 3.16 we plotted the numbers from Tables 3.1 and 3.2. We observe a logarithmic slope of approximately -1 for the discrete errors with respect to the degrees of freedom and $d = 2$, corresponding to a quadratic convergence rate with respect to the mesh-size. For $d = 3$, the observed slopes are approximately $-1/2$ and $-1/3$, corresponding to experimental convergence rates $3/2$ and 1 for the discrete errors in L^p and H^1, respectively. The experimental convergence rates for the errors in L^p, $p = 2, \infty$ are below the expected nearly quadratic convergence rates which appear to be related to the larger preasymptotic range in the three-dimensional situation.

Remarks 3.10

(i) Using Galerkin orthogonality, the error $u - u_h$ in $H_0^1(\Omega)$ for the approximation of the Poisson problem can be obtained from the identity

$$\|\nabla(u - u_h)\|_{L^2(\Omega)}^2 = \|\nabla u_h\|_{L^2(\Omega)}^2 - \|\nabla u\|_{L^2(\Omega)}^2,$$

assuming that the right-hand side is approximated exactly. Considering the discrete error $\tilde{e}_h = u_h - \mathscr{I}_h u$ allows for a simple evaluation of norms related to inner products, e.g., if $\widetilde{E}_h$ is the coefficient vector of the finite element function $\tilde{e}_h$, then we have

$$\|\nabla \tilde{e}_h\|^2_{L^2(\Omega)} = a(\tilde{e}_h, \tilde{e}_h) = \widetilde{E}_h^T A \widetilde{E}_h$$

with the stiffness matrix A.

(ii) The occurrence of superconvergence phenomena is closely related to the fact that finite difference methods on rectangular grids are exact for quadratic solutions, and that finite element methods on triangulations with right-angled triangles are up to the treatment of the right-hand sides equivalent to finite difference methods.

3.3.4 Isoparametric P2-Method

We consider a situation in which the domain $\Omega \subset \mathbb{R}^2$ is partitioned into elements that are the images of the reference triangle $\widehat{T}$ under quadratic degree diffeomorphisms, i.e., $\Omega = \cup_{T \in \mathcal{T}_h} T$, where for every $T \in \mathcal{T}_h$, there exist $D_1^T, \ldots, D_6^T \in \mathbb{R}^2$ such that

$$\Psi_T : \widehat{T} \to T, \quad \hat{x} \mapsto \sum_{j=1}^{6} D_j^T \widehat{\varphi}_j(\hat{x})$$

is a diffeomorphism. The functions $\widehat{\varphi}_1, \widehat{\varphi}_2, \ldots, \widehat{\varphi}_6$ are hierarchical basis functions of $\mathscr{P}_2(\widehat{T})$. In particular, the functions

$$\widehat{\varphi}_1(\hat{x}) = 1 - \hat{x}_1 - \hat{x}_2, \quad \widehat{\varphi}_2(\hat{x}) = \hat{x}_1, \quad \widehat{\varphi}_3(\hat{x}) = \hat{x}_2$$

define a basis for $\mathscr{P}_1(T)$, which is complemented by the functions

$$\widehat{\varphi}_4 = 4\widehat{\varphi}_1\widehat{\varphi}_2, \quad \widehat{\varphi}_5 = 4\widehat{\varphi}_2\widehat{\varphi}_3, \quad \widehat{\varphi}_6 = 4\widehat{\varphi}_3\widehat{\varphi}_1.$$

The functions are associated with the nodes $\hat{z}_1 = 0$, $\hat{z}_2 = (1,0)$, $\hat{z}_3 = (0,1)$, $\hat{z}_4 = (1/2, 0)$, $\hat{z}_5 = (1/2, 1/2)$, and $\hat{z}_6 = (0, 1/2)$ on the reference triangle $\widehat{T}$. Two of the basis functions are sketched in Fig. 3.17. Typical transformations Ψ_T are illustrated in Fig. 3.18.

The coefficients D_1^T, D_2^T, D_3^T coincide with the vertices $P_1^T, P_2^T, P_3^T \in \mathbb{R}^2$ of the element T. The position of a node on the edge of T between P_j^T and P_{j+1}^T is either specified by P_{j+3}^T in which case we set $D_{j+3}^T = P_{j+3}^T - (P_j^T + P_{j+1}^T)/2$, or it is unspecified in which case we set $D_{j+3}^T = 0$. The correction in the definition of D_{j+3}^T is necessary due to the use of a hierarchical basis. For a nodal basis of $\mathscr{P}_2(\widehat{T})$, one would set $D_{j+3}^T = P_{j+3}^T$. With these settings we have $\Psi_T(\hat{z}_j) = P_j^T$ for $j = 1, 2, \ldots, 6$.

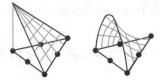

Fig. 3.17 Hierarchical basis functions of quadratic polynomials on the reference triangle $\widehat{T}$

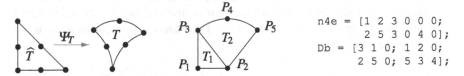

```
n4e = [1 2 3 0 0 0;
       2 5 3 0 4 0];
Db  = [3 1 0; 1 2 0;
       2 5 0; 5 3 4];
```

Fig. 3.18 Nonaffine, quadratic diffeomorphism (*left*); triangulation of a curved domain (*middle*); data structures of the triangulation (*right*)

For every element $T \in \mathcal{T}_h$, we let $K_T \subset \{1, 2, \ldots, 6\}$ be an index set corresponding to the specified nodes of a triangle. We always have $\{1, 2, 3\} \subset K_T$. We then define the set of nodes via

$$\mathcal{N}_h = \{z \in \mathbb{R}^2 : T \in \mathcal{T}_h, k \in K_T, z = P_k^T\}.$$

With every node $z \in \mathcal{N}_h$, we associate a basis function φ_z that is obtained by transformations of basis functions on $\widehat{T}$, i.e.,

$$\varphi_z(x) = \begin{cases} \widehat{\varphi}_k \circ \Psi_T^{-1}(x), & \Psi_T(\hat{z}_k) = z, \\ 0, & \text{otherwise.} \end{cases}$$

We have $\varphi_z \in C(\overline{\Omega})$ for all $z \in \mathcal{N}_h$ and the isoparametric $P2$-finite element space as

$$\mathscr{S}^{2,iso}(\mathcal{T}_h) = \{v_h \in C(\overline{\Omega}) : v_h = \sum_{z \in \mathcal{N}_h} \alpha_z \varphi_z\}.$$

The discretization of the Poisson problem with Dirichlet boundary conditions on the entire boundary then consists in finding $u_h \in \mathscr{S}^{2,iso}(\mathcal{T}_h)$ with $u_h(z) = u_D(z)$ for all $z \in \mathcal{N}_h \cap \partial\Omega$, and

$$\int_\Omega \nabla u_h \cdot \nabla v_h \, dx = \int_\Omega f v_h \, dx$$

for all $v_h \in \mathscr{S}^{2,iso}(\mathcal{T}_h)$ with $v_h|_{\partial\Omega} = 0$. This formulation is equivalent to a linear system of equations with a system matrix that is assembled from the element stiffness matrices

$$M_{k\ell}^T = \int_T ([D\Psi_T]^{-\top} \nabla \widehat{\varphi}_k) \cdot ([D\Psi_T]^{-\top} \nabla \widehat{\varphi}_\ell) |\det D\Psi_T| \, d\hat{x}$$

$$= \sum_{m=1}^M \kappa_m [D\Psi_T(\xi_k)]^{-\top} \nabla \widehat{\varphi}_k(\xi_m) \cdot ([D\Psi_T(\xi_k)]^{-\top} \nabla \widehat{\varphi}_\ell) |\det D\Psi_T(\xi_k)|,$$

for $k, \ell = 1, 2, \ldots, 6$. Analogously, the right-hand side of the linear system of equations results from the element contributions

$$b_k^T = \int_T f \widehat{\varphi}_k \circ \Psi_T^{-1} \, dx = \int_T (f \circ \Psi_T) \widehat{\varphi}_k |\det D\Psi_T| \, d\hat{x}.$$

To realize the boundary condition $u_h(z) = u_D(z)$ for all $z \in \mathscr{N}_h \cap \partial\Omega$, we set

$$\alpha_z = \begin{cases} u_D(z) & \text{if } z \text{ is a vertex,} \\ u_D(z) - (u_D(z_{S,1}) + u_D(z_{S,2}))/2 & \text{if } z \text{ is a midpoint.} \end{cases}$$

An isoparametric finite element approximation of a two-dimensional Poisson problem is shown in Fig. 3.19. The underlying implementation shown in Figs. 3.20 and 3.21 specifies the isoparametric triangulation via the arrays

$$\texttt{c4n} : \mathscr{N}_h \to \mathbb{R}^2, \quad \texttt{n4e} : \mathscr{T}_h \to (\mathscr{N}_h \cup \{0\})^6, \quad \texttt{Db} : \mathscr{S}_h \cap \Gamma_D \to (\mathscr{N}_h \cup \{0\})^3.$$

The array $\texttt{c4n}$ provides the positions of the nodes, which are either vertices of elements or midpoints of sides. The nodes that define an element are specified via the array $\texttt{n4e}$. Here the index 0 represents an unspecified side midpoint, and this implicitly defines the sets K_T. Similarly, $\texttt{Db}$ defines the sides of elements that belong to the Dirichlet boundary. The subroutine $\texttt{quad\_p2\_iso}$ defines a 7-point Gaussian quadrature rule on $\widehat{T}$ that is exact for polynomials in $\mathscr{P}_5(\widehat{T})$.

Fig. 3.19 Isoparametric finite element approximation of a two-dimensional Poisson problem

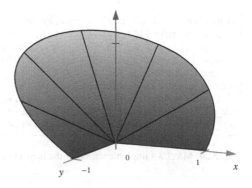

```
function p2_iso_2d
[c4n,n4e,Db] = triang_p2_iso();
nC = size(c4n,1); nE = size(n4e,1);
A = sparse(nC,nC); b = zeros(nC,1); u = zeros(nC,1);
fNodes = setdiff(1:nC,unique(Db));
[phi,d1_phi,d2_phi,kappa] = quad_p2_iso();
N = [1,1,0;0,1,1;1,0,1]/2;
for j = 1:nE
    K_T = find(n4e(j,:));
    P = zeros(6,2);
    P(K_T,:) = c4n(n4e(j,K_T),:);
    P(4:6,:) = P(4:6,:)+((n4e(j,4:6)==0)'*[1,1]).*(N*P(1:3,:));
    D(1:3,:) = P(1:3,:);
    D(4:6,:) = P(4:6,:)-(N*P(1:3,:));
    M = zeros(6,6);
    det_D_Psi = zeros(1,size(kappa,2));
    for m = 1:size(kappa,2)
        D_Psi = [d1_phi(m,:);d2_phi(m,:)]*D;
        D_phi_transp = D_Psi\[d1_phi(m,:);d2_phi(m,:)];
        det_D_Psi(m) = abs(det(D_Psi));
        M = M+kappa(m)*(D_phi_transp'*D_phi_transp)*det_D_Psi(m);
    end
    A(n4e(j,K_T),n4e(j,K_T)) = ...
        A(n4e(j,K_T),n4e(j,K_T))+M(K_T,K_T);
    val_b = kappa.*det_D_Psi.*f(phi*D)'*phi;
    b(n4e(j,K_T)) = b(n4e(j,K_T))+val_b(K_T)';
end
idx = find(Db(:,3));
u(unique(Db(:,1:2))) = u_D(c4n(unique(Db(:,1:2)),:));
u(Db(idx,3))=u_D(c4n(Db(idx,3),:))-(u(Db(idx,1))+u(Db(idx,2)))/2;
b = b-A*u;
u(fNodes) = A(fNodes,fNodes)\b(fNodes);
show_p2_iso(c4n,n4e,u)

function val = f(x)
val = 0*ones(size(x,1),1);

function val = u_D(x)
[phi,r] = cart2pol(x(:,1),x(:,2));
phi = phi+2*pi*(phi<0);
val = r.^(2/3).*sin(2/3*phi);

function [phi,d1_phi,d2_phi,kappa] = quad_p2_iso()
pos = [6-sqrt(15),9+2*sqrt(15),6+sqrt(15),9-2*sqrt(15),7]/21;
r = pos([1,2,1,3,3,4,5])'; s = pos([1,1,2,4,3,3,5])';
wts = [155-sqrt(15),155+sqrt(15),270]/2400;
kappa = wts([1,1,1,2,2,2,3]);
one = ones(size(kappa,2),1);
phi = [1-r-s,r,s,4*r.*(1-r-s),4*r.*s,4*s.*(1-r-s)];
d1_phi = [-one,one,0*one,4*(1-2*r-s),4*s,-4*s];
d2_phi = [-one,0*one,one,-4*r,4*r,4*(1-r-2*s)];
```

Fig. 3.20 MATLAB implementation of the isoparametric *P*2 method in two dimensions

```
function [c4n,n4e,Db] = triang_p2_iso()
phi = (0:pi/8:3*pi/2)';
c4n = [0,0;[cos(phi),sin(phi)]];
n4e = [1 2 4 0 3 0;1 4 6 0 5 0;1 6 8 0 7 0;1 8 10 0 9 0;
    1 10 12 0 11 0;1 12 14 0 13 0];
Db = [1 2 0;2 4 3;4 6 5;6 8 7;8 10 9;10 12 11;12 14 13];
```

```
function show_p2_iso(c4n,n4e,u)
nE = size(n4e,1); fine = 5;
c4n_ref = [0 0;1 0;0 1]; n4e_ref = [1 2 3];
Db_ref = [1 2;2 3;3 1];
for j = 1:fine
    [c4n_ref,n4e_ref,Db_ref] = ...
        red_refine(c4n_ref,n4e_ref,Db_ref,[]);
end
r = c4n_ref(:,1); s = c4n_ref(:,2);
psi = [1-r-s,r,s,4*r.*(1-r-s),4*r.*s,4*s.*(1-r-s)];
N = [1,1,0;0,1,1;1,0,1]/2;
hold off;
for j = 1:nE
    K_T = find(n4e(j,:));
    P = zeros(6,2);
    P(K_T,:) = c4n(n4e(j,K_T),:);
    P(4:6,:) = P(4:6,:)+((n4e(j,4:6)==0)'*[1,1]).*(N*P(1:3,:));
    D(1:3,:) = P(1:3,:);
    D(4:6,:) = P(4:6,:)-(N*P(1:3,:));
    c4n_def_T = psi*D;
    U = zeros(6,1);
    U(K_T) = u(n4e(j,K_T));
    u_def = psi*U;
    trisurf(n4e_ref,c4n_def_T(:,1),c4n_def_T(:,2),u_def);
    shading flat; hold on;
    for k = 1:size(Db_ref,1);
        X = c4n_def_T(Db_ref(k,:),1);
        Y = c4n_def_T(Db_ref(k,:),2);
        Z = u_def(Db_ref(k,:));
        plot3(X,Y,Z,'k');
    end
end
view(22,36); hold off;
```

Fig. 3.21 Specification of an isoparametric *P*2 triangulation for a part of the unit disk (*top*); visualization of an isoparametric *P*2 function (*bottom*)

3.4 $P1$-Approximation of Evolution Equations

3.4.1 Heat Equation

We consider an initial boundary value problem for the heat equation, defined by initial data $u_0 \in H_0^1(\Omega)$ and the right-hand side $f \in C([0, T]; L^2(\Omega))$ on a time interval $[0, T]$, and in a spatial domain $\Omega \subset \mathbb{R}^d$ by

$$\begin{cases} \partial_t u(t, x) - \Delta u(t, x) = f(t, x) & \text{for all } (t, x) \in (0, T] \times \Omega, \\ u(t, x) = 0 & \text{for all } (t, x) \in (0, T] \times \partial\Omega, \\ u(0, x) = u_0(x) & \text{for all } x \in \Omega. \end{cases}$$

To define a weak formulation, we use the notation

$$(v, w) = \int_\Omega v \cdot w \, dx, \quad \|v\| = \|v\|_{L^2(\Omega)},$$

for functions or vector fields $v, w \in L^2(\Omega; \mathbb{R}^\ell)$. We say that $\phi \in C^1([0, T]; X)$ if there exists $\psi \in C^0([0, T]; X)$ such that

$$\lim_{\delta \to 0} \frac{\phi(t + \delta) - \phi(t)}{\delta} = \psi(t)$$

in X for all $t \in (0, T)$ and ψ is continuous on $[0, T]$. In this case we denote $\partial_t \phi = \psi$. Multiplying the heat equation with a function $v \in H_0^1(\Omega)$ and integrating-by-parts leads to the following notion of a weak solution.

Definition 3.13 A *weak solution* for the initial boundary value problem defined by the heat equation is a function $u \in C^1([0, T), H_0^1(\Omega))$, such that $u(0) = u_0$, and

$$\big(\partial_t u(t), v\big) + \big(\nabla u(t), \nabla v\big) = \big(f(t), v\big)$$

for all $v \in H_0^1(\Omega)$ and all $t \in (0, T)$.

The existence of weak solutions can be established via time or space discretizations and subsequent limit passages; their uniqueness is a consequence of the following result.

Proposition 3.9 (Energy Law) *Let $u \in C^1([0, T]; H_0^1(\Omega))$ be a weak solution of the heat equation. We then have*

$$\sup_{t \in [0, T]} \|u(t)\|^2 + \int_0^T \|\nabla u(t)\|^2 \, dt \le 2\|u_0\|^2 + 2c_P^2 \int_0^T \|f(t)\|^2 \, dt$$

with a Poincaré constant $c_P > 0$.

Proof Exercise.   □

We define a family of numerical schemes for approximating weak solutions. Special instances are explicit and implicit Euler schemes and the Crank–Nicolson scheme for the parameters $\theta = 0, 1, 1/2$, respectively. We denote by d_t the backward difference quotient, defined for the step-size $\tau > 0$ and the sequence $(a^k)_{k=0,\dots,K}$ by

$$d_t a^k = \tau^{-1}\left(a^k - a^{k-1}\right)$$

for $k = 1, 2, \dots, K$.

Algorithm 3.1 (θ-Midpoint Scheme) *Given $\theta \in [0, 1]$, a triangulation $\mathscr{T}_h$ of Ω, a step-size $\tau > 0$, and $u_h^0 \in \mathscr{S}^1(\mathscr{T}_h)$, compute for $k = 1, 2, \dots, K$ with $K = \lfloor T/\tau \rfloor$ functions $u_h^k \in \mathscr{S}_0^1(\mathscr{T}_h)$ such that*

$$\left(d_t u_h^k, v_h\right) + \left(\nabla[\theta u_h^k + (1 - \theta)u_h^{k-1}], \nabla v_h\right) = \left(f(t_{k-1+\theta}), v_h\right)$$

for all $v_h \in \mathscr{S}_0^1(\mathscr{T}_h)$ and with $t_{k-1+\theta} = (k - 1 + \theta)\tau$.

The existence and uniqueness of numerical solutions $(u_h^k)_{k=0,\dots,K}$ follows inductively with the Lax–Milgram lemma. We have unconditional stability if $\theta \geq 1/2$ and conditional stability if $\theta < 1/2$ as the following proposition shows. We let $c_{\mathrm{inv}} > 0$ be such that

$$\|\nabla v_h\| \leq c_{\mathrm{inv}} h^{-1}\|v_h\|$$

for all $v_h \in \mathscr{S}_0^1(\mathscr{T}_h)$ if $\mathscr{T}_h$ is quasiuniform.

Proposition 3.10 (Discrete Stability) *Suppose that $(z_h^k)_{k=0,\dots,K} \subset \mathscr{S}_0^1(\mathscr{T}_h)$ and $(b_k)_{k=0,\dots,K} \subset H_0^1(\Omega)'$ satisfy*

$$\left(d_t z_h^k, v_h\right) + \left(\nabla[\theta z_h^k + (1 - \theta)z_h^{k-1}], \nabla v_h\right) = b_k(v_h)$$

for all $v_h \in \mathscr{S}_0^1(\mathscr{T}_h)$. If $\theta \geq 1/2$, we have

$$\max_{k=1,\dots,K} \|z_h^k\|^2 + \tau \sum_{k=1}^{K} \|\nabla[\theta z_h^k + (1 - \theta)z_h^{k-1}]\|^2 \leq 2\|z_h^0\|^2 + 2\tau \sum_{k=1}^{K} \|b_k\|_{H_0^1(\Omega)'}^2.$$

Suppose that $\mathscr{T}_h$ is quasiuniform and $c_{\mathrm{inv}}^2 \tau h^{-2} \leq 1/2$ if $\theta < 1/2$. Then

$$\max_{k=1,\dots,K} \frac{1}{2}\|z_h^k\|^2 + \tau \sum_{k=1}^{K} \|\nabla z_h^k\|^2 = 2\|z_h^0\|^2 + 2\tau \sum_{k=1}^{K} \|b_k\|_{H_0^1(\Omega)'}^2.$$

Proof

(i) Assume that $\theta \geq 1/2$. Noting that

$$z_h^{k,\theta} = \theta z_h^k + (1-\theta)z_h^{k-1} = (z_h^k + z_h^{k-1})/2 + (\theta - 1/2)\tau d_t z_h^k,$$

the choice of $v_h = z_h^{k,\theta}$ in Algorithm 3.1 yields that

$$\frac{d_t}{2}\|z_h^k\|^2 + \left(\theta - \frac{1}{2}\right)\tau\|d_t z_h^k\|^2 + \frac{1}{2}\|\nabla z_h^{k,\theta}\|^2 \leq \frac{1}{2}\|b_k\|_{H_0^1(\Omega)'}^2 + \frac{1}{2}\|\nabla z_h^{k,\theta}\|^2.$$

A summation over $k = 1, 2, \dots, L$, for every $1 \leq L \leq K$ and multiplication by τ, imply the estimate.

(ii) Assume that $\theta < 1/2$. The choice of $v_h = z_h^k$ and the identity

$$z_h^{k,\theta} = \theta z_h^k + (1-\theta)z_h^{k-1} = z_h^k - (1-\theta)\tau d_t z_h^k$$

lead with the binomial formula $(a-b)a = (a-b)^2/2 + (a^2 - b^2)/2$, and hence

$$(d_t z_h^k, z_h^k) = \frac{\tau}{2}\|d_t z_h^k\|^2 + \frac{d_t}{2}\|z_h^k\|^2$$

to

$$\frac{d_t}{2}\|z_h^k\|^2 + \frac{\tau}{2}\|d_t z_h^k\|^2 + \|\nabla z_h^k\|^2 = (1-\theta)\frac{\tau}{2}\left(d_t\|\nabla z_h^k\|^2 + \tau\|\nabla d_t z_h^k\|^2\right) + b_k(z_h^k).$$

Summing over $k = 1, 2, \dots, L$, multiplying by τ, and estimating $(1-\theta) \leq 1$ show that

$$\frac{1}{2}\|z_h^L\|^2 + \frac{\tau^2}{2}\sum_{k=1}^{L}\|d_t z_h^k\|^2 + \frac{\tau}{2}\sum_{k=1}^{L}\|\nabla z_h^k\|^2$$

$$\leq \frac{1}{2}\|z_h^0\|^2 + \frac{\tau}{2}\sum_{k=1}^{L}\|b_k\|_{H_0^1(\Omega)'}^2 + \frac{\tau}{2}\left(\|\nabla z_h^L\|^2 + \tau^2\sum_{k=1}^{L}\|\nabla d_t z_h^k\|^2\right).$$

We incorporate the inverse estimates

$$\|\nabla d_t z_h^k\| \leq c_{\mathrm{inv}}h^{-1}\|d_t z_h^k\|, \quad \|\nabla z_h^L\| \leq c_{\mathrm{inv}}h^{-1}\|z_h^L\|$$

to verify the stability estimate for $\theta < 1/2$. $\qquad\qquad\square$

A maximum principle for the heat equation with $f = 0$ states that

$$u(t, x) \leq \max_{x' \in \Omega} u_0(x')$$

for all $t \in (0, T)$. A discrete version holds for a modified implicit Euler scheme under a structural condition on the triangulation.

Definition 3.14 The *discrete (or lumped) L^2 product* with the nodal interpolation operator $\mathscr{I}_h : C(\overline{\Omega}) \to \mathscr{S}^1(\mathscr{T}_h)$ is defined by

$$(v, w)_h = \int_\Omega \mathscr{I}_h(vw) \, dx = \sum_{z \in \mathscr{N}_h} \beta_z v(z) w(z)$$

for all $v, w \in C(\overline{\Omega})$ and with $\beta_z = \int_\Omega \varphi_z \, dx$ for all $z \in \mathscr{N}_h$.

The interpretation of the discrete L^2 product is a numerical integration in the evaluation of the L^2 product with a trapezoidal rule. It is represented by a diagonal matrix with positive entries on the diagonal. Provided that the matrix representing the discretized Laplace operator is an M-matrix, a discrete maximum principle holds.

Proposition 3.11 (Discrete Maximum Principle) *Let* $(u_h^k)_{k=0,\dots,K} \subset \mathscr{S}_0^1(\mathscr{T}_h)$ *satisfy*

$$(d_t u_h^k, v_h)_h + (\nabla u_h^k, \nabla v_h) = 0$$

for $k = 1, 2, \dots, K$ and all $v_h \in \mathscr{S}_0^1(\mathscr{T}_h)$. If $\mathscr{T}_h$ is such that

$$A_{zy} = \int_\Omega \nabla \varphi_z \cdot \nabla \varphi_y \, dx \le 0$$

for all distinct $z, y \in \mathscr{N}_h$, then we have $u_h^k(z) \le \max_{y \in \mathscr{N}_h} u_h^0(y)$ for all $k = 1, 2, \dots, K$ and all $z \in \mathscr{N}_h$.

Proof Exercise.   □

To obtain optimal consistency estimates, we use the H^1-projection onto the finite element space $\mathscr{S}_0^1(\mathscr{T}_h)$ instead of the nodal interpolation operator.

Proposition 3.12 (H^1-Projection) *For $v \in H^1(\Omega)$ let $Q_h v \in \mathscr{S}_0^1(\mathscr{T}_h)$ be defined by*

$$(\nabla Q_h v, \nabla w_h) = (\nabla v, \nabla w_h)$$

for all $w_h \in \mathscr{S}_0^1(\mathscr{T}_h)$. If the Poisson problem in Ω with homogeneous Dirichlet boundary conditions on $\partial\Omega$ is H^2-regular, then we have

$$h^{-1} \|v - Q_h v\| + \|\nabla(v - Q_h v)\| \le c_Q h \|D^2 v\|$$

for all $v \in H^2(\Omega) \cap H_0^1(\Omega)$.

Proof Exercise.   □

We verify the consistency of the numerical scheme only for the case $\theta = 1$, i.e., the implicit Euler scheme.

Proposition 3.13 (Consistency) *If $u \in C^1([0, T]; H^2(\Omega)) \cap C^2([0, T]; L^2(\Omega))$, $\theta = 1$, and the Poisson problem is H^2-regular, then we have that*

$$(d_t Q_h u(t_k), v_h) + (\nabla Q_h u(t_k), \nabla v_h) = (f(t_k), v_h) + \mathscr{C}_{h,\tau}(t_k; v_h)$$

with functionals $\mathscr{C}_{h,\tau}(t_k) \in H_0^1(\Omega)'$ such that

$$\sum_{k=1}^{k} \|\mathscr{C}_{h,\tau}(t_k)\|_{H_0^1(\Omega)'}^2 \le c(\tau^2 + h^4).$$

Proof For $k = 1, 2, \ldots, K$ we have, using $(\nabla(Q_h u(t_k) - u(t_k)), \nabla v_h) = 0$ and the equation satisfied by u, that

$$(d_t Q_h u(t_k), v_h) + (\nabla Q_h u(t_k), \nabla v_h) = (f(t_k), v_h) + (d_t Q_h u(t_k) - \partial_t u(t_k), v_h)$$
$$= (f(t_k), v_h) + \mathscr{C}_{h,\tau}(t_k; v_h).$$

Incorporating estimates for the H^1-projection, we find that the functional $\mathscr{C}_h(t_k; v)$, for every $v \in H_0^1(\Omega)$ with $\|\nabla v\| \le 1$, satisfies the estimate

$$\mathscr{C}_{h,\tau}(t_k; v) = (d_t Q_h u(t_k) - d_t u(t_k), v) + (d_t u(t_k) - \partial_t u(t_k), v)$$
$$= \frac{1}{\tau} \int_{t_{k-1}}^{t_k} ((Q_h - 1) u_t, v) \, ds + \frac{1}{\tau} \int_{t_{k-1}}^{t_k} (s - t_{k-1})(u_{tt}, v) \, ds$$
$$\le ch^2 \left(\int_{t_{k-1}}^{t_k} \|D^2 u_t\|^2 \, dt \right)^{1/2} + c\tau \left(\int_{t_{k-1}}^{t_k} \|u_{tt}\|^2 \, dt \right)^{1/2}.$$

This implies the asserted bound.   □

Remark 3.11 Note that due to the identity $(\nabla[Q_h u(t_k) - u(t_k)], \nabla v_h) = 0$ the consistency term is simplified. This is crucial to obtain optimal convergence rates and is known as *Wheeler's trick*.

Theorem 3.5 (Error Estimate) *Under the condition of Proposition 3.13, and if $u_h^0 = \mathscr{I}_h u_0$, then we have*

$$\max_{k=1,\ldots,K} \|u(t_k) - u_h^k\|^2 \le c(\tau^2 + h^4),$$

$$\tau \sum_{k=1}^{K} \|\nabla[u(t_k) - u_h^k]\|^2 \le c(\tau^2 + h^2).$$

Proof We decompose the error $u(t_k) - u_h^k$ as

$$u(t_k) - u_h^k = [u(t_k) - Q_h u(t_k)] + [Q_h u(t_k) - u_h^k] = y^k + z_h^k.$$

The approximation errors y^k are estimated with the help of Proposition 3.12. A bound for the error contributions z_h^k follows from the consistency bound of Proposition 3.13, and the stability result of Proposition 3.10. The error estimate then follows with the triangle inequality.   □

Remark 3.12 For the Crank–Nicolson scheme corresponding to $\theta = 1/2$, one can prove the error bound $\max_{k=1,\dots,K} \|u(t_k) - u_h^k\| \le c(\tau^2 + h^2)$ under appropriate regularity conditions. A maximum principle fails in general.

3.4.2   Wave Equation

We consider an initial boundary value problem for the wave equation, defined by initial data $u_0 \in H_0^1(\Omega)$ and $v_0 \in L^2(\Omega)$, and a right-hand side $f \in C([0, T]; L^2(\Omega))$, on a time interval $[0, T]$ and in a spatial domain $\Omega \subset \mathbb{R}^d$ by

$$\begin{cases} \partial_t^2 u(t, x) - \Delta u(t, x) = f(t, x) & \text{for all } (t, x) \in (0, T] \times \Omega, \\ u(t, x) = 0 & \text{for all } (t, x) \in (0, T] \times \partial\Omega, \\ u(0, x) = u_0(x) & \text{for all } x \in \Omega, \\ \partial_t u(0, x) = v_0(x) & \text{for all } x \in \Omega. \end{cases}$$

Multiplying the wave equation with a function $v \in H_0^1(\Omega)$, and integrating-by-parts leads to a notion of a weak solution.

Definition 3.15 A *weak solution* for the initial boundary value problem defined by the wave equation is a function $u \in C^2([0, T), H_0^1(\Omega))$, such that $u(0) = u_0$, $\partial_t u(0) = v_0$, and

$$(\partial_t^2 u(t), v) + (\nabla u(t), \nabla v) = (f(t), v)$$

for all $v \in H_0^1(\Omega)$ and all $t \in (0, T)$.

The existence of weak solutions can be established via time or space discretizations and subsequent passages to a limit; their uniqueness is a consequence of the following result.

Proposition 3.14 (Energy Conservation) *Assume that $f = 0$ and the function $u \in C^2([0, T]; H_0^1(\Omega))$ is a weak solution for the wave equation. Then the total energy*

$$E(t) = \frac{1}{2}\left(\|\partial_t u(t)\|^2 + \|\nabla u(t)\|^2\right)$$

is constant for $t \in [0, T]$ with $E(0) = (\|v_0\|^2 + \|\nabla u_0\|^2)/2$.

Proof Exercise. □

To obtain a discrete energy conservation principle, we use the following midpoint scheme to approximate weak solutions. Due to the second-order time derivative, a time-step involves three time levels. A suitable choice of a function u_h^1 to initialize the iteration is discussed below; a suboptimal choice is given by $u_h^1 = \mathscr{I}_h[u_0 + \tau v_0]$.

Algorithm 3.2 (Midpoint Scheme) *Given a triangulation $\mathscr{T}_h$ of Ω, a step-size $\tau > 0$, and $u_h^0, u_h^1 \in \mathscr{S}_0^1(\mathscr{T}_h)$, compute for $k = 2, 3, \ldots, K$ with $K = \lfloor T/\tau \rfloor$ the functions $u_h^k \in \mathscr{S}_0^1(\mathscr{T}_h)$ such that*

$$\left(d_t^2 u_h^k, v_h\right) + \frac{1}{4}\left(\nabla[u_h^k + 2u_h^{k-1} + u_h^{k-2}], \nabla v_h\right) = \left(f(t_{k-1}), v_h\right)$$

for all $v_h \in \mathscr{S}_0^1(\mathscr{T}_h)$.

Note that $d_t^2 a^k = (a^k - 2a^{k-1} + a^{k-2})/\tau^2$ for a sequence $(a^k)_{k=0,\ldots,K}$. The existence of approximations follows with the Lax–Milgram lemma. If $f = 0$, then we have a discrete conservation principle.

Proposition 3.15 (Discrete Energy Conservation) *Assume that $f = 0$. Then the discrete total energy of the iterates of the midpoint scheme, defined by*

$$E_{h,\tau}^k = \frac{1}{2}\left(\|d_t u_h^k\|^2 + \|\nabla u_h^{k-1/2}\|^2\right),$$

is constant for $k = 1, 2, \ldots, K$, where $u_h^{k-1/2} = (u_h^k + u_h^{k-1})/2$.

Proof We choose

$$v_h = \frac{1}{2}\left(d_t u_h^k + d_t u_h^{k-1}\right)$$

in the numerical scheme. With a binomial formula we find that

$$\left(d_t^2 u_h^k, v_h\right) = \frac{1}{2\tau}\left(d_t u_h^k - d_t u_h^{k-1}, d_t u_h^k + d_t u_h^{k-1}\right) = \frac{1}{2\tau}\left(\|d_t u_h^k\|^2 - \|d_t u_h^{k-1}\|^2\right).$$

Similarly, noting that with the notation $u_h^{k-3/2} = (u_h^{k-1} + u_h^{k-2})/2$, we have

$$\frac{1}{4}\left(u_h^k + 2u_h^{k-1} + u_h^{k-2}\right) = \frac{1}{2}\left(u_h^{k-1/2} + u_h^{k-3/2}\right),$$

$$\frac{1}{2}\left(d_t u_h^k + d_t u_h^{k-1}\right) = \frac{1}{\tau}\left(u_h^{k-1/2} - u_h^{k-3/2}\right),$$

and we deduce that

$$\frac{1}{4}\left(\nabla[u_h^k + 2u_h^{k-1} + u_h^{k-2}], \nabla v_h\right) = \frac{1}{2\tau}\left(\|\nabla u_h^{k-1/2}\|^2 - \|\nabla u_h^{k-3/2}\|^2\right).$$

Combining the identities, and summing over $k = 2, 3, \ldots, K'$, proves the asserted identity.   □

A modification of the energy conservation result leads to a discrete stability estimate.

Proposition 3.16 (Discrete Stability) *Let* $(z_h^k)_{k=0,\ldots,K} \subset \mathscr{S}_0^1(\mathscr{T}_h)$ *and* $(b_k)_{k=0,\ldots,K} \subset H_0^1(\Omega)'$ *satisfy*

$$(d_t^2 z_h^k, v_h) + \frac{1}{4}(\nabla[z_h^k + 2z_h^{k-1} + z_h^{k-2}], \nabla v_h) = b_k(v_h)$$

for all $v_h \in \mathscr{S}_0^1(\mathscr{T}_h)$. *We then have*

$$\|d_t z_h^k\|^2 + \frac{1}{2}\|\nabla z_h^{k-1/2}\|^2 \le \|d_t z_h^1\|^2 + \|\nabla z_h^{1/2}\|^2 + \frac{\tau}{2}\sum_{k=2}^{K}\|b_k\|_{H_0^1(\Omega)'}^2.$$

Proof Exercise.   □

We use the H^1-projection to determine the consistency error of the numerical scheme.

Proposition 3.17 (Consistency) *If the Poisson problem is* H^2-*regular, and if* $u \in C^2([0, T]; H^2(\Omega)) \cap C^4([0, T]; L^2(\Omega))$, *then we have*

$$(d_t^2 Q_h u(t_k), v_h) + (\nabla Q_h \bar{u}(t_{k-1}), \nabla v_h) = (f(t_{k-1}), v_h) + \mathscr{C}_{h,\tau}(t_k; v_h),$$

with $\bar{u}(t_{k-1}) = \big(u(t_k) + 2u(t_{k-1}) + u(t_{k-2})\big)/4$ *and functionals* $\mathscr{C}_{h,\tau}(t_k) \in H_0^1(\Omega)'$ *satisfying*

$$\max_{k=2,\ldots,K}\|\mathscr{C}_{h,\tau}(t_k)\|_{H_0^1(\Omega)'} \le c(\tau^2 + h^2).$$

Proof (Sketched) Using that $(\nabla Q_h \bar{u}(t_{k-1}), \nabla v_h) = (\nabla \bar{u}(t_{k-1}), \nabla v_h)$ and the equation satisfied by u evaluated at t_{k-1}, we find that

$$(d_t^2 Q_h u(t_k), v_h) + (\nabla Q_h \bar{u}(t_{k-1}), \nabla v_h) - (f(t_{k-1}), v_h)$$

$$= (d_t^2[Q_h u(t_k) - u(t_k)], v_h) + (d_t^2 u(t_k) - \partial_t^2 u(t_{k-1}), v_h).$$

The approximation properties of Q_h and Taylor expansions lead to the estimate.   □

The combination of the discrete stability result and the consistency bounds imply the following error estimate.

Theorem 3.6 (Error Estimate) *Under the conditions of Proposition 3.17, and if* $\Delta u_0 \in H^2(\Omega)$, $u_h^0 = \mathscr{I}_h u_0$, *and*

$$u_h^1 = \mathscr{I}_h\left[u_0 + \tau v_0 + \frac{\tau^2}{2}\left(\Delta u_0 + f(0)\right)\right],$$

then we have

$$\max_{k=1,\dots,K} \|\nabla(u(t_{k-1/2}) - u_h^{k-1/2})\| \le c(h + \tau^2).$$

Proof (Sketched) We split the error according to

$$u(t_k) - u_h^k = \left[u(t_k) - Q_h u(t_k)\right] + \left[Q_h u(t_k) - u_h^k\right] = y^k + z_h^k.$$

A bound for the error contributions y^k follows from the approximation properties of the H^1-projection. The stability and the discrete consistency bound lead to estimate

$$\|d_t z_h^k\|^2 + \frac{1}{2}\|\nabla z_h^{k-1/2}\|^2 \le \|d_t z_h^1\|^2 + \|\nabla z_h^{1/2}\|^2 + cT(\tau^4 + h^4).$$

Using the initial conditions, and the equation satisfied by u in a Taylor expansion of u at $t = 0$ shows that

$$u(t_1) = u(0) + \tau \partial_t u(0) + \frac{\tau^2}{2}\partial_t^2 u(0) + \mathcal{O}(\tau^3)$$

$$= u_0 + \tau v_0 + \frac{\tau^2}{2}\left(\Delta u_0 + f(0)\right) + \mathcal{O}(\tau^3).$$

This implies that $\|d_t z_h^1\| \le (\tau^2 + h^2)$ and $\|\nabla z_h^{1/2}\| \le c(\tau^2 + h)$. Combining the estimates proves the asserted bound.   □

Remark 3.13 A refinement of the proof leads to the additional L^∞-L^2 error estimate $\max_{k=0,\dots,K} \|u(t_k) - u_h^k\| \le c(h^2 + \tau^2)$.

3.4.3 Implementations

For a simple implementation, certain assumptions about the approximation of the data functions are made.

Assumption 3.3 (Data Approximation) *We assume that*

$$u_0 = u_{0,h}, \ v_0 = v_{0,h} \in \mathscr{S}^1(\mathscr{T}_h), \quad f = f_h \in C([0,T]; \mathscr{S}^1(\mathscr{T}_h)).$$

The θ-midpoint scheme for approximating the heat equation then computes the sequence $(u_h^k)_{k=0,\ldots,K} \subset \mathscr{S}_0^1(\mathscr{T}_h)$ with

$$\left(d_t u_h^k, v_h\right) + \left(\nabla[\theta u_h^k + (1-\theta)u_h^{k-1}], \nabla v_h\right) = \left(f_h(t_{k,\theta}), v_h\right)$$

for all $v_h \in \mathscr{S}_0^1(\mathscr{T}_h)$. The nontrivial coefficients $U^k = (U_y^k : y \in \mathscr{N}_h \cap \Omega)$ of u_h^k satisfy the equation

$$\frac{1}{\tau} \sum_{y \in \mathscr{N}_h \cap \Omega} U_y^k \left((\varphi_z, \varphi_y) + \theta(\nabla \varphi_y, \nabla \varphi_z)\right)$$

$$= \frac{1}{\tau} \sum_{y \in \mathscr{N}_h \cap \Omega} U_y^{k-1}\left((\varphi_z, \varphi_y) + (1-\theta)(\nabla \varphi_y, \nabla \varphi_z)\right) + \sum_{y \in \mathscr{N}_h} f_h(t_{k,\theta}, y)(\varphi_z, \varphi_y)$$

for every $z \in \mathscr{N}_h \cap \Omega$. The implementation thus requires computing the L^2-inner products of the nodal basis functions. These can be replaced by simplified discrete versions based on numerical integration (or mass lumping) as introduced in Definition 3.14 to guarantee a maximum principle if $\theta = 1$. The midpoint scheme for approximating the wave equation computes the sequence $(u_h^k)_{k=0,\ldots,K} \subset \mathscr{S}_0^1(\mathscr{T}_h)$ with

$$\left(d_t^2 u_h^k, v_h\right) + \left(\nabla[u_h^k + 2u_h^{k-1} + u_h^{k-2}], \nabla v_h\right) = \left(f_h(t_{k-1}), v_h\right)$$

for all $v_h \in \mathscr{S}_0^1(\mathscr{T}_h)$. As in the case of the numerical scheme for the heat equation, the resulting linear system of equations involves matrices defined by quantities (φ_z, φ_y) and $(\nabla \varphi_z, \nabla \varphi_y)$ for $z, y \in \mathscr{N}_h$.

Lemma 3.11 (Mass Matrices) *For $T \in \mathscr{T}_h$ with $T = \mathrm{conv}\{z_0, z_1, \ldots, z_d\}$, we have for $0 \leq m, n \leq d$, that*

$$\int_T \varphi_{z_m} \varphi_{z_n} \, dx = \frac{|T|(1 + \delta_{mn})}{(d+1)(d+2)}, \qquad \int_T \mathscr{I}_h[\varphi_{z_m} \varphi_{z_n}] \, dx = \frac{|T|\delta_{mn}}{d+1}.$$

Proof The identities follow from elementary calculations on a reference element and a transformation to T. □

Figure 3.22 displays MATLAB implementations of the θ-midpoint scheme for the heat equation and the midpoint scheme for the wave equation. The routine fe_matrices.m is shown in Fig. 3.23 and provides the stiffness and mass matrices. The parameter α in the codes determines the time-step size via $\tau = h^\alpha/4$. The visualization is done with the routine show_p1.m, which is displayed in Fig. 3.24.

```
function p1_theta_heat_diri(d,red)
T = 10; theta = 1/2; alpha = 1;
[c4n,n4e,Db,Nb] = triang_cube(d); Db = [Db;Nb]; Nb = [];
for j = 1:red
    [c4n,n4e,Db,Nb] = red_refine(c4n,n4e,Db,Nb);
end
nC = size(c4n,1); h = 2^(-red); tau = h^alpha/4; K = floor(T/tau);
dNodes = unique(Db); fNodes = setdiff(1:nC,dNodes);
u_old = u_0(c4n); u_new = zeros(nC,1);
[s,m,m_lumped] = fe_matrices(c4n,n4e);
for k = 1:K
    t_k_theta = (k-1+theta)*tau;
    b = (1/tau)*m*u_old-(1-theta)*s*u_old+m*f(t_k_theta,c4n);
    X = (1/tau)*m+theta*s;
    u_new(fNodes) = X(fNodes,fNodes)\b(fNodes);
    show_p1(c4n,n4e,Db,Nb,u_new);
    axis([0,1,0,1,0,.25,0,.25]); pause(.1);
    u_old = u_new;
end

function val = f(t,x); val = ones(size(x,1),1);
function val = u_0(x); val = sin(2*pi*x(:,1)).*sin(2*pi*x(:,2));
```

```
function p1_midpoint_wave_diri(d,red)
T = 10; alpha = 1;
[c4n,n4e,Db,Nb] = triang_cube(d); Db = [Db;Nb]; Nb = [];
for j = 1:red
    [c4n,n4e,Db,Nb] = red_refine(c4n,n4e,Db,Nb);
end
nC = size(c4n,1); h = 2^(-red); tau = h^alpha/4; K = floor(T/tau);
dNodes = unique(Db); fNodes = setdiff(1:nC,dNodes);
u_old_1 = u_0(c4n)+tau*v_0(c4n)+(tau^2/2)*(Del_u_0(c4n)+f(0,c4n));
u_old_2 = u_0(c4n); u_new = zeros(nC,1);
[s,m,m_lumped] = fe_matrices(c4n,n4e);
for k = 2:K
    b = (1/tau^2)*m*(2*u_old_1-u_old_2)...
        -(1/4)*s*(2*u_old_1+u_old_2)+m*f((k-1)*tau,c4n);
    X = (1/tau^2)*m+(1/4)*s;
    u_new(fNodes) = X(fNodes,fNodes)\b(fNodes);
    show_p1(c4n,n4e,Db,Nb,u_new);
    axis([0,1,0,1,-1,1,-1,1]); pause(.05);
    u_old_2 = u_old_1; u_old_1 = u_new;
end

function val = f(t,x); val = -ones(size(x,1),1);
function val = u_0(x); val = zeros(size(x,1),1);
function val = v_0(x); val = 4*sin(pi*x(:,1)).*sin(pi*x(:,2));
function val = Del_u_0(x); val = zeros(size(x,1),1);
```

Fig. 3.22 MATLAB implementation of time-stepping finite element schemes for the heat and wave equations

```
function [s,m,m_lumped,vol_T] = fe_matrices(c4n,n4e)
[nC,d] = size(c4n); nE = size(n4e,1);
m_loc = (ones(d+1,d+1)+eye(d+1))/((d+1)*(d+2));
ctr = 0; ctr_max = (d+1)^2*nE;
I = zeros(ctr_max,1); J = zeros(ctr_max,1);
X_s = zeros(ctr_max,1); X_m = zeros(ctr_max,1);
m_lumped_diag = zeros(nC,1); vol_T = zeros(nE,1);
for j = 1:nE
    X_T = [ones(1,d+1);c4n(n4e(j,:),:)'];
    grads_T = X_T\[zeros(1,d);eye(d)];
    vol_T(j) = det(X_T)/factorial(d);
    for m = 1:d+1
        for n = 1:d+1
            ctr = ctr+1; I(ctr) = n4e(j,m); J(ctr) = n4e(j,n);
            X_s(ctr) = vol_T(j)*grads_T(m,:)*grads_T(n,:)';
            X_m(ctr) = vol_T(j)*m_loc(m,n);
        end
        m_lumped_diag(n4e(j,m)) = m_lumped_diag(n4e(j,m))...
            +vol_T(j)/(d+1);
    end
end
s = sparse(I,J,X_s,nC,nC); m = sparse(I,J,X_m,nC,nC);
m_lumped = diag(m_lumped_diag);
```

Fig. 3.23 MATLAB routine that computes the mass matrix, the lumped mass matrix, and the stiffness matrix associated with $P1$-finite elements

```
function show_p1(c4n,n4e,Db,Nb,u)
d = size(c4n,2);
if d == 1
    plot(c4n(n4e),u(n4e));
elseif d == 2
    trisurf(n4e,c4n(:,1),c4n(:,2),u);
elseif d == 3
    trisurf([Db;Nb],c4n(:,1),c4n(:,2),c4n(:,3),u);
end
```

Fig. 3.24 MATLAB routine for displaying a $P1$-finite element function

References

A discussion of the historical development of the finite element method can be found in [9]. The monograph [5] provides a comprehensive overview over finite element methods and their analysis. Constructive existence theories for interpolating finite element functions are stated in [3]; a more practical approach is followed in [2]. An important contribution to the numerical analysis of finite element methods for parabolic equations is the article [22]; further results can be found in the monograph [21]. Classical articles on error estimates for finite element methods are the references [1, 14, 15, 17, 18]; see [6] for quadrature rules on simplices.

Further references on finite element methods for partial differential equations are [4, 7, 8, 10–13, 16, 19, 20].

1. Aubin, J.P.: Behavior of the error of the approximate solutions of boundary value problems for linear elliptic operators by Galerkin's and finite difference methods. Ann. Scuola Norm. Sup. Pisa (3) **21**, 599–637 (1967)
2. Braess, D.: Finite Elements, 3rd edn. Cambridge University Press, Cambridge (2007). URL http://dx.doi.org/10.1017/CBO9780511618635
3. Brenner, S.C., Scott, L.R.: The mathematical theory of finite element methods. Texts in Applied Mathematics, vol. 15, 3rd edn. Springer, New York (2008). URL http://dx.doi.org/10.1007/978-0-387-75934-0
4. Carstensen, C.: Wissenschaftliches Rechnen (1997).   Lecture Notes, University of Kiel, Germany
5. Ciarlet, P.G.: The finite element method for elliptic problems. Classics in Applied Mathematics, vol. 40. Society for Industrial and Applied Mathematics (SIAM), Philadelphia, PA (2002). URL http://dx.doi.org/10.1137/1.9780898719208
6. Duffy, M.G.: Quadrature over a pyramid or cube of integrands with a singularity at a vertex. SIAM J. Numer. Anal. **19**(6), 1260–1262 (1982). URL http://dx.doi.org/10.1137/0719090
7. Dziuk, G.: Theorie und Numerik Partieller Differentialgleichungen. Walter de Gruyter GmbH & Co. KG, Berlin (2010). URL http://dx.doi.org/10.1515/9783110214819
8. Ern, A., Guermond, J.L.: Theory and practice of finite elements. Applied Mathematical Sciences, vol. 159. Springer, New York (2004). URL http://dx.doi.org/10.1007/978-1-4757-4355-5
9. Gander, M.J., Wanner, G.: From Euler, Ritz, and Galerkin to modern computing. SIAM Rev. **54**(4), 627–666 (2012). URL http://dx.doi.org/10.1137/100804036
10. Grossmann, C., Roos, H.G.: Numerical treatment of partial differential equations. Universitext. Springer, Berlin (2007). URL http://dx.doi.org/10.1007/978-3-540-71584-9
11. Hackbusch, W.: Elliptic differential equations. Springer Series in Computational Mathematics, vol. 18. Springer, Berlin (1992). URL http://dx.doi.org/10.1007/978-3-642-11490-8
12. Knabner, P., Angermann, L.: Numerical methods for elliptic and parabolic partial differential equations. Texts in Applied Mathematics, vol. 44. Springer, New York (2003)
13. Larson, M.G., Bengzon, F.: The finite element method: theory, implementation, and applications. Texts in Computational Science and Engineering, vol. 10. Springer, Heidelberg (2013). URL http://dx.doi.org/10.1007/978-3-642-33287-6
14. Nitsche, J.: Ein Kriterium für die Quasi-Optimalität des Ritzschen Verfahrens. Numer. Math. **11**, 346–348 (1968)
15. Rannacher, R.: Zur L^∞-Konvergenz linearer finiter Elemente beim Dirichlet-Problem. Math. Z. **149**(1), 69–77 (1976)
16. Rannacher, R.: Numerische Mathematik 2 (Numerik partieller Differentialgleichungen) (2008).   URL http://numerik.iwr.uni-heidelberg.de/~lehre/notes/.   Lecture Notes, University of Heidelberg, Germany
17. Rannacher, R., Scott, R.: Some optimal error estimates for piecewise linear finite element approximations.   Math. Comp. **38**(158), 437–445 (1982).   URL http://dx.doi.org/10.2307/2007280
18. Schatz, A.H., Wahlbin, L.B.: Maximum norm estimates in the finite element method on plane polygonal domains. I. Math. Comp. **32**(141), 73–109 (1978)
19. Schwab, C.: p- and hp-finite element methods.   Numerical Mathematics and Scientific Computation. The Clarendon Press, Oxford University Press, New York (1998)
20. Strang, G., Fix, G.: An Analysis of the Finite Eelement Method, 2nd edn. Wellesley-Cambridge Press, Wellesley, MA (2008)
21. Thomée, V.: Galerkin finite element methods for parabolic problems. Springer Series in Computational Mathematics, vol. 25, 2nd edn. Springer, Berlin (2006)
22. Wheeler, M.F.: A priori L_2 error estimates for Galerkin approximations to parabolic partial differential equations. SIAM J. Numer. Anal. **10**, 723–759 (1973)

Part II
Local Resolution and Iterative Solution

Chapter 4
Local Resolution Techniques

4.1 Local Resolution of Corner Singularities

4.1.1 Corner Singularities

We recall that a solution of the Poisson problem on the partial disk

$$\Omega = \{r(\cos\phi, \sin\phi) : 0 < r < 1, 0 < \phi < \gamma\},$$

for $\gamma \in (0, 2\pi)$, and with Dirichlet boundary conditions on $\partial\Omega$, is given by the function

$$u(r, \phi) = r^{\pi/\gamma} \sin(\phi\pi/\gamma).$$

The gradient of u has a singularity at the origin if and only if $\gamma \in (\pi, 2\pi)$, i.e., if and only if Ω is nonconvex. The function u belongs to the broken Sobolev space $H^{1+\pi/\gamma}(\Omega)$, and via interpolation, one can prove the error estimate

$$\|\nabla(u - u_h)\| \leq \inf_{v_h \in \mathscr{S}^1(\mathscr{T}_h)} \|\nabla(u - v_h)\| \leq ch^{\min\{1, \pi/\gamma\}},$$

where $\|\cdot\|$ denotes the norm in $L^2(\Omega)$. Numerical experiments show that this estimate is optimal on quasiuniform meshes. Since the solution is smooth away from the origin, it appears natural to try to use meshes with a finer mesh-size near the origin. More generally, solutions of elliptic equations on planar polygonal domains have singularities at re-entrant corners, i.e., corner points with interior angle greater than π. The following theorem specifies this fact.

© Springer International Publishing Switzerland 2016
S. Bartels, *Numerical Approximation of Partial Differential Equations*,
Texts in Applied Mathematics 64, DOI 10.1007/978-3-319-32354-1_4

Fig. 4.1 Polygonal domain
in $\mathbb{R}^2$ with corners
$P_1, P_2, \dots, P_L$; corners P_1
and P_5 are re-entrant

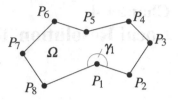

Theorem 4.1 (Additive Decomposition, See [12, Chap. 2]) *Let $\Omega \subset \mathbb{R}^2$ be a bounded polygonal Lipschitz domain, and $P_\ell \in \mathbb{R}^2$, $1 \le \ell \le L$, such that*

$$\partial\Omega = \bigcup_{\ell=1}^{L} \mathrm{conv}\,\{P_\ell, P_{\ell+1}\},$$

where $P_{L+1} = P_1$. For each corner point P_ℓ, $1 \le \ell \le L$, let $\varepsilon_\ell > 0$ and $\phi_\ell, \gamma_\ell \in (0, 2\pi)$ be such that

$$\Omega \cap B_{\varepsilon_\ell}(P_\ell) = \{P_\ell + r(\cos\phi, \sin\phi) : 0 < r < \varepsilon_\ell, \ \phi_\ell < \phi < \phi_\ell + \gamma_\ell\},$$

cf. Fig. 4.1. For $1 \le \ell \le L$ and $k \in \mathbb{N}$, define $\alpha_{\ell k} = k\pi/\gamma_\ell$ and the singularity function

$$S_{\ell k} = r^{\alpha_{\ell k}} \sin(\alpha_{\ell k}\phi),$$

with $r(x) = |x - P_\ell|$ and $\phi(x) = \arg(x - P_\ell) + \phi_\ell$. Let $f \in L^2(\Omega)$ and $u \in H_0^1(\Omega)$ be the weak solution of the Poisson problem $-\Delta u = f$ in Ω with $u|_{\partial\Omega} = 0$. For every $s > 0$ there exist numbers $b_{\ell k} \in \mathbb{R}$, $1 \le \ell \le L$, $k \in \mathbb{N}$, a function $u_{\mathrm{reg}} \in H^{1+s}(\Omega)$, and functions $\chi_\ell \in C_0^\infty(\mathbb{R}^2)$ with $\chi_\ell|_{B_{\varepsilon_\ell}(P_\ell)} = 1$, such that

$$u = u_{\mathrm{reg}} + \sum_{\ell=1}^{L} \sum_{k \in \mathbb{N}} b_{\ell k} \chi_\ell S_{\ell k}.$$

4.1.2 Graded Grids

To improve the performance of the finite element method, we construct triangulations with smaller triangles in the vicinity of corner singularities.

Definition 4.1 A *graded grid* with *graduation strength* $\beta \ge 1$ and maximal width $h = 1/J$ for $J \in \mathbb{N}$ is the triangulation of the reference element

$$T_{\mathrm{ref}} = \mathrm{conv}\{(0, 0), (1, 0), (0, 1)\},$$

Fig. 4.2 Graded grid defined on the reference element T_{ref} and transformation to a macro-triangle $T \subset \mathbb{R}^2$

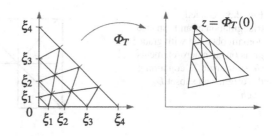

defined by setting $\xi_j = (j/J)^\beta, j = 0, 1, \ldots, J$, and partitioning each strip

$$\mathscr{L}_j = \text{conv}\{(\xi_{j-1}, 0), (0, \xi_{j-1}), (\xi_j, 0), (0, \xi_j)\},$$

$j = 1, 2, \ldots, J$, into $2j - 1$ triangles as shown in Fig. 4.2.

With an affine transformation $\Phi_T : T_{\text{ref}} \to T$, we can map the graded grid from the reference element to a general macro-triangle $T \subset \mathbb{R}^2$, with a refinement towards the vertex $z = \Phi_T(0)$.

The following remarks collect important properties of graded grids.

Remarks 4.1

(i) For a fixed $\beta \geq 1$, the graded grids on T_{ref} with maximal width $h = 1/J$ define a sequence of triangulations $(\mathscr{T}_J^\beta)_{J \in \mathbb{N}}$ that satisfy a minimal angle condition.

(ii) The diameter h_T of each triangle T contained in the strip $\mathscr{L}_j$, using $h = 1/J$ and $(j/J)^{\beta-1} \leq 1$, satisfies the estimate

$$h_T \leq c_\beta \sqrt{2} \frac{\xi_j}{j} = c_\beta \sqrt{2} (j/J)^\beta \frac{1}{j} = c_\beta \sqrt{2} (j/J)^{\beta-1} J^{-1} \leq c_\beta' J^{-1} = c_\beta' h.$$

(iii) If T is a triangle in the strip $\mathscr{L}_j$, then for all $x \in T$ we have

$$|x| \geq \frac{1}{\sqrt{2}} \xi_j = \frac{1}{\sqrt{2}} (j/J)^\beta,$$

and hence, since $h_T \leq c_\beta \sqrt{2} J^{-1} (j/J)^{\beta-1}$,

$$h_T \leq c_\beta' J^{-1} |x|^{(\beta-1)/\beta}.$$

The left plot in Fig. 4.3 displays a graded triangulation of the L-shaped domain $\Omega = (-1, 1)^2 \setminus ([0, 1] \times [0, -1])$, obtained by mapping the graded grid of the reference triangle T_{ref} to the six macro-elements of the coarse triangulation.

Fig. 4.3 Graded
triangulation of an L-shaped
domain obtained with graded
grids on six macro-elements
(*left*); triangle T_0 containing
the origin, and subset D_β
(*right*)

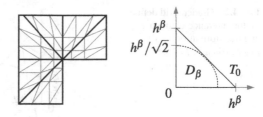

4.1.3 Approximation on Graded Grids

For a number $\alpha > 0$ and a 2π-periodic function $v \in C^2([0, 2\pi])$, we consider the
singularity function

$$u_\alpha(r, \phi) = r^\alpha v(\alpha\phi),$$

in polar coordinates with respect to a corner point, which is assumed to coincide
with the origin. We define $A_{\alpha,v}(\phi) \in \mathbb{R}^2$ by

$$\nabla u_\alpha(x) = \begin{bmatrix} \cos\phi & \sin\phi \\ \sin\phi & -\cos\phi \end{bmatrix} \begin{bmatrix} \alpha r^{\alpha-1} v(\alpha\phi) \\ \alpha r^{\alpha-1} v'(\alpha\phi) \end{bmatrix} = \alpha r^{\alpha-1} A_{\alpha,v}(\phi).$$

With these settings we have the following result about the approximation of the
singularity function near the origin.

Proposition 4.1 (Local Lower Bound) *For* $T_0 = \mathrm{conv}\{(0,0), (h^\beta, 0), (0, h^\beta)\}$ *and*
$\alpha \neq 1$, *we have*

$$\inf_{v_h \in \mathscr{P}_1(\mathbb{R}^2)} \|\nabla(u_\alpha - v_h)\|_{L^2(T_0)} \geq c_{\alpha,v} h^{\alpha\beta},$$

with a constant $c_{\alpha,v} > 0$ *that does not depend on* h.

Proof We consider the subset $D_\beta = B_{h^\beta/\sqrt{2}}(0) \cap T_0$, cf. Fig. 4.3, and note that

$$\inf_{v_h \in \mathscr{P}_1(\mathbb{R}^2)} \|\nabla(u_\alpha - v_h)\|_{L^2(T_0)} \geq \inf_{v_h \in \mathscr{P}_1(\mathbb{R}^2)} \|\nabla(u_\alpha - v_h)\|_{L^2(D_\beta)}.$$

Since every affine function $v_h \in \mathscr{P}_1(\mathbb{R}^2)$ satisfies $v_h(x) = a + b \cdot x$ with $a \in \mathbb{R}$ and
$b \in \mathbb{R}^2$, it follows that

$$\inf_{v_h \in \mathscr{P}_1(\mathbb{R}^2)} \|\nabla(u_\alpha - v_h)\|_{L^2(D_\beta)} = \inf_{b \in \mathbb{R}^2} \|\nabla u_\alpha - b\|_{L^2(D_\beta)}.$$

A straightforward calculation shows that the optimal choice is given by

$$\bar{b} = \frac{1}{|D_\beta|} \int_{D_\beta} \nabla u_\alpha \, dx.$$

This implies that

$$\inf_{v_h \in \mathscr{P}_1(\mathbb{R}^2)} \|\nabla(u_\alpha - v_h)\|^2_{L^2(D_\beta)} = \|\nabla u_\alpha\|^2_{L^2(D_\beta)} - |D_\beta| \, |\bar{b}|^2.$$

Using $\nabla u_\alpha = \alpha r^{\alpha-1} A_{\alpha,v}(\phi)$, we have

$$\int_{D_\beta} |\nabla u_\alpha|^2 \, dx = \int_0^{h^\beta/\sqrt{2}} \int_0^{\pi/2} \alpha^2 r^{2\alpha-1} |A_{\alpha,v}(\phi)|^2 \, dr \, d\phi$$

$$= \frac{\alpha h^{2\alpha\beta}}{2^{\alpha+1}} \|A_{\alpha,v}\|^2_{L^2(0,\pi/2)}.$$

For the average of ∇u_α we have

$$|D_\beta|\bar{b} = \int_0^{h^\beta/\sqrt{2}} \int_0^{\pi/2} \alpha r^\alpha A_{\alpha,v}(\phi) \, d\phi \, dr = \frac{\alpha}{\alpha+1} \frac{h^{\beta(\alpha+1)}}{2^{(\alpha+1)/2}} \frac{\pi}{2} \bar{A}_{\alpha,v},$$

where $\bar{A}_{\alpha,v} = (2/\pi) \int_0^{\pi/2} A_{\alpha,v}(\phi) \, d\phi$. With $|D_\beta| = \pi h^{2\beta}/8$, we deduce that

$$\inf_{v_h \in \mathscr{P}_1(\mathbb{R}^2)} \|\nabla(u_\alpha - v_h)\|^2_{L^2(D_\beta)} = \frac{\alpha h^{2\alpha\beta}}{2^{\alpha+1}} \left(\|A_{\alpha,v}\|^2_{L^2(0,\pi/2)} - \frac{4\alpha}{(\alpha+1)^2} \frac{\pi}{2} |\bar{A}_{\alpha,v}|^2 \right).$$

Noting that $(\pi/2)|\bar{A}_{\alpha,v}|^2 \leq \|A_{\alpha,v}\|^2_{L^2(0,\pi/2)}$, we find that the right-hand side is positive unless $\alpha = 1$. □

The local lower bound implies that the best approximation error of u_α on a graded grid is bounded from below by $h^{\beta\alpha} = J^{-\beta\alpha}$. To obtain the optimal convergence rate

$$\inf_{v_h \in \mathscr{S}^1(\mathscr{T}_h)} \|\nabla(u_\alpha - v_h)\| \leq ch,$$

we thus have to choose the grading strength $\beta \geq 1/\alpha$. The following proposition shows that this is sufficient.

Proposition 4.2 (Global Upper Bound) *Let $\mathscr{T}_h$ be the graded grid on T_{ref} with grading strength $\beta = 1/\alpha$ and maximal width $h = 1/J$. We then have*

$$\|\nabla(u_\alpha - \mathscr{I}_h u_\alpha)\|_{L^2(T_{\mathrm{ref}})} \leq c_{\mathrm{grad}} c_{\alpha,v} h (1 + |\log(h)|)^{1/2},$$

where $\mathscr{I}_h$ is the nodal interpolation associated with $\mathscr{T}_h$.

Proof

(i) We first study the contribution of the triangle T_0 with corners 0, $(h^\beta, 0)$, and $(0, h^\beta)$. Using $\nabla u_\alpha = \alpha r^{\alpha-1} A_{\alpha,v}(\phi)$, and enlarging the domain of integration, we have that

$$\int_{T_0} |\nabla u_\alpha|^2 \, dx \le \alpha^2 \int_0^{\pi/2} \int_0^{h^\beta} r^{2\alpha-1} |A_{\alpha,v}(\phi)|^2 \, dr \, d\phi = \frac{\alpha}{2} \|A_{\alpha,v}\|_{L^2(0,\pi/2)}^2 h^{2\alpha\beta}.$$

The gradient of $\mathscr{I}_h u_\alpha$ on T_0 is given by the difference quotients in the two coordinate directions, i.e.,

$$\partial_1 \mathscr{I}_h u_\alpha = h^{-\beta} \left(u_\alpha(h^\beta, 0) - u_\alpha(0,0) \right) = h^{\alpha\beta-\beta} v(0),$$

$$\partial_2 \mathscr{I}_h u_\alpha = h^{-\beta} \left(u_\alpha(h^\beta, \pi/2) - u_\alpha(0,0) \right) = h^{\alpha\beta-\beta} v(\alpha\pi/2).$$

With $|T_0| = h^{2\beta}/2$ it follows that

$$\int_{T_0} |\nabla \mathscr{I}_h u_\alpha|^2 \, dx \le h^{2\alpha\beta} \|v\|_{L^\infty(0,2\pi)}^2.$$

The triangle inequality leads to

$$\|\nabla(u_\alpha - \mathscr{I}_h u_\alpha)\|_{L^2(T_0)} \le \|\nabla u_\alpha\|_{L^2(T_0)} + \|\nabla \mathscr{I}_h u_\alpha\|_{L^2(T_0)} \le c_{\alpha,v} h^{\alpha\beta}.$$

(ii) To estimate the interpolation error on the union of triangles $T \in \mathscr{T}_h \setminus \{T_0\}$, we use that $u_\alpha|_T \in H^2(T)$, so that standard interpolation estimates lead to

$$\|\nabla(u_\alpha - \mathscr{I}_h u_\alpha)\|_{L^2(T)} \le c_{\mathscr{I}} h_T \|D^2 u_\alpha\|_{L^2(T)}.$$

Direct calculations show that

$$|D^2 u_\alpha(r, \phi)| \le \tilde{c}_{\alpha,v} r^{\alpha-2}.$$

We recall that for triangles $T \in \mathscr{T}_h \setminus \{T_0\}$ and $x \in T$, we have $h_T \le c_\beta J^{-1} |x|^{(\beta-1)/\beta}$. This implies that

$$\int_{T_{\mathrm{ref}} \setminus T_0} |\nabla(u_\alpha - \mathscr{I}_h u_\alpha)|^2 \, dx \le c_{\mathscr{I}}^2 \sum_{T \in \mathscr{T}_h \setminus \{T_0\}} h_T^2 \int_T |D^2 u_\alpha|^2 \, dx$$

$$\le c_{\mathscr{I}}^2 \tilde{c}_{\alpha,v}^2 \sum_{T \in \mathscr{T}_h \setminus \{T_0\}} \int_T h_T^2 r^{2\alpha-4} \, dx$$

$$\le c_{\mathscr{I}}^2 \tilde{c}_{\alpha,v}^2 c_\beta^2 J^{-2} \sum_{T \in \mathscr{T}_h \setminus \{T_0\}} \int_T r^{2(\alpha-1/\beta)-2} \, dx.$$

We enlarge the domain of integration using that

$$\bigcup_{T \in \mathcal{T}_h \setminus \{T_0\}} T \subset \{ r(\cos \phi, \sin \phi) : h^\beta / \sqrt{2} \leq r \leq 1, 0 \leq \phi \leq \pi/2 \},$$

and incorporate the choice $\beta = 1/\alpha$ to verify that

$$J^{-2} \sum_{T \in \mathcal{T}_h \setminus \{T_0\}} \int_T r^{2(\alpha-1/\beta)-2} \, dx \leq J^{-2} \int_{h^\beta/\sqrt{2}}^1 \int_0^{\pi/2} r^{-1} \, d\phi \, dr$$

$$= J^{-2} \frac{\pi}{2} [0 - \log (h^\beta/\sqrt{2})].$$

(iii) Combining the upper bounds implies the estimate. $\qquad\square$

The estimate of the theorem can be applied to bound the approximation error for Poisson problems in twodimensional polygonal domains. The decomposition theorem with $s = 1$ provides functions $u_{\text{reg}} \in H^2(\Omega)$ and $u_{\text{sing}} \in H^1(\Omega)$ such that

$$u = u_{\text{reg}} + u_{\text{sing}}.$$

With Céa's lemma, nodal interpolation estimates, the linearity of $\mathcal{I}_h$, and Proposition 4.2, for appropriately graded triangulations we deduce that

$$\|\nabla(u - u_h)\| \leq \|\nabla(u - \mathcal{I}_h u)\|$$

$$\leq \|\nabla(u_{\text{reg}} - \mathcal{I}_h u_{\text{reg}})\| + \|\nabla(u_{\text{sing}} - \mathcal{I}_h u_{\text{sing}})\|$$

$$\leq c_{\mathcal{I}} h \|D^2 u_{\text{reg}}\| + c_{\text{grad}} c_\Omega h (1 + |\log(h)|).$$

Up to the logarithmic factor, which can be omitted for practically relevant meshsizes, we have thus derived the same convergence rate as in the case of a smooth solution and a sequence of quasiuniform triangulations.

4.1.4 Realization

Figures 4.4 and 4.5 show a sequence of graded triangulations and a finite element solution on a graded triangulation, respectively, obtained with the MATLAB implementation shown in Fig. 4.6. The graded triangulation of the reference element is mapped to the elements in the macro triangulation of the L-shaped domain. Reocurring nodes are eliminated with the help of the MATLAB command unique.

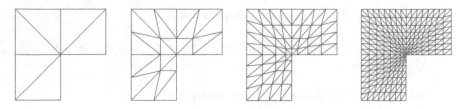

Fig. 4.4 Graded triangulations with grading strength $\beta = 3/2$ and $N = 1, 2, 4, 8$ of an L-shaped domain

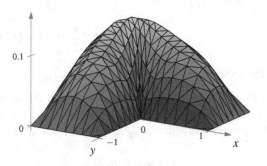

Fig. 4.5 Finite element approximation on a graded triangulation

4.2 Error Control and Adaptivity

4.2.1 Local Inequalities

In order to obtain a flexible error control on locally refined triangulations, we need local Poincaré and trace inequalities. We let $(\mathcal{T}_h)_{h>0}$ be a sequence of uniformly shape-regular triangulations of the bounded Lipschitz domain Ω with Dirichlet boundary $\Gamma_D \subset \partial\Omega$. We recall that $\mathcal{N}_h$ is the set of all vertices of elements and $(\varphi_z : z \in \mathcal{N}_h)$ is the nodal basis of the finite element space $\mathcal{S}^1(\mathcal{T}_h)$.

Definition 4.2 For a node $z \in \mathcal{N}_h$, we define the *node patch* $\omega_z \subset \Omega$ by

$$\omega_z = \operatorname{supp} \varphi_z.$$

We let $h_z = \operatorname{diam}(\omega_z)$ be the diameter of ω_z.

Due to the assumed uniform shape-regularity, the diameters of elements and patches are locally equivalent.

```
function p1_graded(N,beta)
[c4n,n4e,Db] = Lshape_graded(N,beta);
nC = size(c4n,1);
dNodes = unique(Db); fNodes = setdiff(1:nC,dNodes);
[s,m] = fe_matrices(c4n,n4e);
u = zeros(nC,1); b = m*f(c4n);
u(fNodes) = s(fNodes,fNodes)\b(fNodes);
show_p1(c4n,n4e,Db,[],u);

function [c4n,n4e,Db] = Lshape_graded(N,beta)
c4n_macro = [-1 -1;0 -1;-1 0;0 0;1 0;-1 1;0 1;1 1];
n4e_macro = [4 5 8;4 8 7;4 7 6;4 6 3;4 3 1;4 1 2];
nE_macro = size(n4e_macro,1);
[c4n_micro,n4e_micro] = grad_grid_ref(N,beta);
nC_mi = size(c4n_micro,1);
n4e = []; c4n = [];
for j = 1:nE_macro
    phi_0 = (1-c4n_micro(:,1)-c4n_micro(:,2));
    phi_1 = c4n_micro(:,1); phi_2 = c4n_micro(:,2);
    c4n_transf = phi_0*c4n_macro(n4e_macro(j,1),:)...
        +phi_1*c4n_macro(n4e_macro(j,2),:)...
        +phi_2*c4n_macro(n4e_macro(j,3),:);
    n4e = [n4e;n4e_micro+(j-1)*nC_mi];
    c4n = [c4n;c4n_transf];
end
[c4n,~,j] = unique(c4n,'rows');
n4e = j(n4e);
Db = find_bdy_sides(n4e);

function [c4n,n4e] = grad_grid_ref(N,beta)
c4n = [0,0]; n4e = [1 2 3];
for j = 1:N
    xi = (j/N)^beta;
    c4n = [c4n;ones(j+1,1)*[xi,0]+(0:j)'/j*[-xi,xi]];
end
for j = 1:N-1
    n4e = [n4e;j*(j+1)/2+[(1:j)',j+2+(1:j)',1+(1:j)']];
    n4e = [n4e;j*(j+1)/2+[(1:j+1)',j+1+(1:j+1)',j+2+(1:j+1)']];
end

function bdy = find_bdy_sides(n4e)
all_sides = [n4e(:,[1,2]);n4e(:,[2,3]);n4e(:,[3,1])];
[sides,~,j] = unique(sort(all_sides,2),'rows');
valence = accumarray(j(:),1);
bdy = sides(valence==1,:);

function val = f(x); val = ones(size(x,1),1);
```

Fig. 4.6 Approximation of the Poisson problem on an L-shaped domain using a graded triangulation

Remarks 4.2

(i) There exists an h-independent constant $c_{\ell oc} > 0$ such that if $T \in \mathcal{T}_h$ and $z \in \mathcal{N}_h \cap T$, then we have

$$h_T \leq h_z \leq c_{\ell oc} h_T.$$

(ii) For every $z \in \mathcal{N}_h$, we have

$$\overline{\omega}_z = \cup_{T \in \mathcal{T}_h, z \in T} \, T$$

(iii) The patches $(\omega_z : z \in \mathcal{N}_h)$ have a *finite overlap*, i.e., every $x \in \Omega$ belongs at most to $d + 1$ patches $\omega_{z_1}, \omega_{z_2}, \ldots, \omega_{z_{d+1}}$. Conversely, the closure of every element patch $\overline{\omega}_z$ contains at most an h-independent number of elements in $\mathcal{T}_h$.

On the patches ω_z we have a local Poincaré inequality that bounds the error for the approximation of a function by a constant on ω_z.

Lemma 4.1 (Local Poincaré Inequality) *For $v \in H_D^1(\Omega)$ and $z \in \mathcal{N}_h$, set $v_z = 0$ if $z \in \Gamma_D$ and*

$$v_z = |\omega_z|^{-1} \int_{\omega_z} v \, dx$$

otherwise. For every $h > 0$ and $z \in \mathcal{N}_h$, there exists a constant $c_{P,z} > 0$, such that for all $v \in H_D^1(\Omega)$ we have

$$\|v - v_z\|_{L^2(\omega_z)} \leq c_{P,z} h_z \|\nabla v\|_{L^2(\omega_z)}.$$

The constant $c_{P,z} > 0$ does not depend on the diameter of ω_z, but may depend on the shape of ω_z.

Proof

(i) For $z \in \mathcal{N}_h$, define $\widehat{\omega}_z = h_z^{-1}(\omega_z - z)$. Then $\mathrm{diam}(\widehat{\omega}_z) = 1$ and

$$\Phi_z : \widehat{\omega}_z \to \omega_z, \quad \hat{x} \mapsto h_z \hat{x} + z$$

is an affine diffeomorphism with $D\Phi_z = h_z I$, cf. Fig. 4.7.

Fig. 4.7 Node patch ω_z of a node $z \in \mathcal{N}_h$ (*left*). Scaling and translation of a patch ω_z by a diffeomorphism Φ_z (*right*)

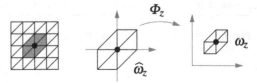

(ii) By the standard Poincaré inequality, we have for every $\hat{w} \in H^1(\widehat{\omega}_z)$

$$\|\hat{w}\|_{L^2(\widehat{\omega}_z)} \leq c_{P,z}\|\nabla \hat{w}\|_{L^2(\widehat{\omega}_z)},$$

provided that

$$\int_{\widehat{\omega}_z} \hat{w}\, d\hat{x} = 0 \quad \text{or} \quad \hat{w}|_{\hat{\gamma}_z} = 0$$

for a closed subset $\hat{\gamma}_z \subset \partial\widehat{\omega}_z$ with positive surface measure.

(iii) For $v \in H_D^1(\Omega)$ and v_z as in the lemma, we let $w = v - v_z \in H^1(\omega_z)$ and set

$$\hat{w} = w \circ \Phi_z.$$

Because of the definition of v_z, the function $\hat{w}$ satisfies the conditions for the standard Poincaré inequality on $\widehat{\omega}_z$.

(iv) With the transformation formula, we obtain

$$\int_{\omega_z} w^2\, dx = \int_{\widehat{\omega}_z} \hat{w}^2 |\det D\Phi_z|\, d\hat{x}$$

$$= h_z^d \|\hat{w}\|_{L^2(\widehat{\omega}_z)}^2 \leq c_{P,z}^2 h_z^d \|\nabla \hat{w}\|_{L^2(\widehat{\omega}_z)}^2$$

$$= c_{P,z}^2 h_z^d \int_{\widehat{\omega}_z} |\nabla(w \circ \Phi_z)|^2\, d\hat{x} = c_{P,z}^2 h_z^d \int_{\widehat{\omega}_z} |D\Phi_z^\top (\nabla w) \circ \Phi_z|^2\, d\hat{x}$$

$$= c_{P,z}^2 h_z^d h_z^2 \int_{\widehat{\omega}_z} |(\nabla w) \circ \Phi_z|^2\, d\hat{x} = c_{P,z}^2 h_z^2 \int_{\omega_z} |\nabla w|^2\, dx,$$

which proves the estimate. $\qquad \square$

To formulate a local trace inequality, we require some notation for the sides of elements.

Definition 4.3 We denote by $\mathscr{S}_h$ the set of all *sides* of elements in $\mathscr{T}_h$, i.e., $(d-1)$-dimensional subsimplices of elements in $\mathscr{T}_h$,

$$\mathscr{S}_h = \{S \subset \mathbb{R}^d : S = \text{conv}\{z_1, z_2, \ldots, z_d\},\ z_1, z_2, \ldots, z_{d+1} \in \mathscr{N}_h,$$

$$T \in \mathscr{T}_h,\ T = \text{conv}\{z_1, z_2, \ldots, z_{d+1}\} \in \mathscr{T}_h\}.$$

For every $S \in \mathscr{S}_h$, we set $h_S = \text{diam}(S)$.

The following lemma provides a bound on the L^2 norm of the trace $v|_S$ for a function $v \in H^1(\Omega)$ and a side $S \in \mathscr{S}_h$.

Lemma 4.2 (Local Trace Inequality) *Let $S \in \mathscr{S}_h$ and $T_S \in \mathscr{T}_h$ be such that $S \subset \partial T_S$. There exists a constant $c_{\mathrm{Tr}} > 0$, such that for every $v \in H^1(\Omega)$ we have*

$$\|v\|_{L^2(S)}^2 \leq c_{\mathrm{Tr}}^2 \left(h_S^{-1} \|v\|_{L^2(T_S)}^2 + h_S \|\nabla v\|_{L^2(T_S)}^2 \right).$$

Proof Exercise.   □

Remark 4.3 Note that if $v \in H^1(\Omega)$ and $S = T_1 \cap T_2$, then we have $(v_{T_1})|_S = (v_{T_2})|_S$, i.e., it does not matter from which side of S the trace is taken.

4.2.2 Quasi-Interpolation

In order to apply the nodal interpolation operator to a Sobolev function, we need to guarantee that the function is continuous. Quasi-interpolation operators are defined by local averages instead of nodal values, and thereby avoid this restriction.

Definition 4.4 For $v \in L^1(\Omega)$ and $z \in \mathscr{N}_h$, let

$$v_z = \begin{cases} |\omega_z|^{-1} \int_{\omega_z} v \, dx & \text{if } z \in \mathscr{N}_h \setminus \Gamma_{\mathrm{D}}, \\ 0 & \text{if } z \in \mathscr{N}_h \cap \Gamma_{\mathrm{D}}, \end{cases}$$

and define the *Clément quasi-interpolant* $\mathscr{J}_h v \in \mathscr{S}_{\mathrm{D}}^1(\mathscr{T}_h)$ of v by

$$\mathscr{J}_h v = \sum_{z \in \mathscr{N}_h} v_z \varphi_z.$$

To provide uniform bounds on the stability and approximation estimates for the quasi-interpolant, we make the following assumption on the shapes of node patches in a sequence of triangulations. It is satisfied if only a finite number of patch shapes occur in the sequence of triangulations.

Assumption 4.1 (Patch Shapes) *There exists a constant $c_P > 0$, such that for all $h > 0$ and all $z \in \mathscr{N}_h$ we have $c_{P,z} \leq c_P$.*

With this assumption we can prove that $\mathscr{J}_h$ defines a uniformly bounded linear operator on $H_{\mathrm{D}}^1(\Omega)$ and that $\mathscr{J}_h v$ approximates a given function $v \in H_{\mathrm{D}}^1(\Omega)$. We note that the case $\Gamma_{\mathrm{D}} = \emptyset$ is admissible in the following theorem.

Theorem 4.2 (Quasi-Interpolation Estimates) *There exists a constant $c_{\mathscr{J}} > 0$, such that for all $v \in H_{\mathrm{D}}^1(\Omega)$ we have*

$$\|\nabla \mathscr{J}_h v\| + \|h_{\mathscr{T}}^{-1}(v - \mathscr{J}_h v)\| + \|h_{\mathscr{S}}^{-1/2}(v - \mathscr{J}_h v)\|_{L^2(\cup \mathscr{S}_h)} \leq c_{\mathscr{J}} \|\nabla v\|,$$

where the mesh-size functions $h_{\mathscr{T}} : \Omega \to \mathbb{R}$ and $h_{\mathscr{S}} : \cup \mathscr{S}_h \to \mathbb{R}$ are defined by $h_{\mathscr{T}}|_T = h_T$ and $h_{\mathscr{S}}|_S = h_S$ for all $T \in \mathscr{T}_h$ and $S \in \mathscr{S}_h$, respectively.

Proof

(i) We note that $\sum_{\varphi \in \mathscr{N}_h} \varphi_z = 1$ in Ω, and therefore $\sum_{z \in \mathscr{N}_h} \nabla \varphi_z = 0$ in Ω. We thus have that

$$\nabla \mathscr{J}_h v = \nabla \sum_{z \in \mathscr{N}_h} v_z \varphi_z = \sum_{z \in \mathscr{N}_h} (v_z - v) \nabla \varphi_z.$$

We use this identity and the fact that $\operatorname{supp} \varphi_z = \omega_z$ to verify that

$$\|\nabla \mathscr{J}_h v\|^2 = \int_\Omega \left(\sum_{z \in \mathscr{N}_h} (v_z - v) \nabla \varphi_z \right) \cdot \nabla \mathscr{J}_h v \, dx$$

$$\leq \sum_{z \in \mathscr{N}_h} \int_{\omega_z} |v_z - v| \, |\nabla \varphi_z| \, |\nabla \mathscr{J}_h v| \, dx.$$

An inverse estimate shows that

$$\|\nabla \varphi_z\|_{L^\infty(\omega_z)} \leq c_{\mathrm{inv}} h_z^{-1} \|\varphi_z\|_{L^\infty(\omega_z)} = c_{\mathrm{inv}} h_z^{-1}$$

for all $z \in \mathscr{N}_h$. With Hölder's inequality, the local Poincaré inequality, and the Cauchy–Schwarz inequality, we deduce that

$$\|\nabla \mathscr{J}_h v\|^2 \leq \sum_{z \in \mathscr{N}_h} \|\nabla \varphi_z\|_{L^\infty(\omega_z)} \|v_z - v\|_{L^2(\omega_z)} \|\nabla \mathscr{J}_h v\|_{L^2(\omega_z)}$$

$$\leq c_{\mathrm{inv}} c_P \sum_{z \in \mathscr{N}_h} \|\nabla v\|_{L^2(\omega_z)} \|\nabla \mathscr{J}_h v\|_{L^2(\omega_z)}$$

$$\leq c_{\mathrm{inv}} c_P \left(\sum_{z \in \mathscr{N}_h} \|\nabla v\|_{L^2(\omega_z)}^2 \right)^{1/2} \left(\sum_{z \in \mathscr{N}_h} \|\nabla \mathscr{J}_h v\|_{L^2(\omega_z)}^2 \right)^{1/2}$$

$$\leq c_{\mathrm{inv}} c_P (d + 1) \|\nabla v\| \|\nabla \mathscr{J}_h v\|.$$

In the last estimate we used the finite overlap of the patches, i.e.,

$$\sum_{z \in \mathscr{N}_h} \|\nabla v\|_{L^2(\omega_z)}^2 \leq (d + 1) \sum_{T \in \mathscr{T}_h} \|\nabla v\|_{L^2(T)}^2 = (d + 1) \|\nabla v\|^2.$$

A division by $\|\nabla \mathscr{J}_h v\|$ proves the first estimate.

(ii) For any $\psi \in L^2(\Omega)$ we have, using again $\sum_{z \in \mathcal{N}_h} \varphi_z = 1$, that

$$\int_\Omega \psi(v - \mathcal{J}_h v)\,dx = \int_\Omega \psi\left(v - \sum_{z \in \mathcal{N}_h} v_z \varphi_z\right) dx$$

$$= \int_\Omega \psi\left(\sum_{z \in \mathcal{N}_h} v \varphi_z - \sum_{z \in \mathcal{N}_h} v_z \varphi_z\right) dx$$

$$= \sum_{z \in \mathcal{N}_h} \int_{\omega_z} \psi(v - v_z)\varphi_z\,dx$$

$$\leq \sum_{z \in \mathcal{N}_h} \int_{\omega_z} |\psi|\,|v - v_z|\,dx.$$

Arguing as in (i) we find that

$$\int_\Omega \psi(v - \mathcal{J}_h v)\,dx \leq c_P \sum_{z \in \mathcal{N}_h} \|\psi\|_{L^2(\omega_z)} h_z \|\nabla v\|_{L^2(\omega_z)}$$

$$\leq c_P c_{\ell oc} \sum_{z \in \mathcal{N}_h} \|\nabla v\|_{L^2(\omega_z)} \|h_{\mathcal{T}}\psi\|_{L^2(\omega_z)}$$

$$\leq c_P c_{\ell oc}(d+1)\|\nabla v\|\|h_{\mathcal{T}}\psi\|.$$

We choose $\psi = h_{\mathcal{T}}^{-2}(v - \mathcal{J}_h v)$ and divide by $\|h_{\mathcal{T}}\psi\| = \|h_{\mathcal{T}}^{-1}(v - \mathcal{J}_h v)\|$ to verify the second estimate.

(iii) For each $S \in \mathcal{S}_h$, let $T_S \in \mathcal{T}_h$ be such that $S \subset \partial T_S$. The local trace inequality, the fact that each element occurs at most $d+1$ times, and the first two estimates of the theorem imply that

$$\|h_{\mathcal{S}}^{-1/2}(v - \mathcal{J}_h v)\|_{L^2(\cup \mathcal{S}_h)}^2 = \sum_{S \in \mathcal{S}_h} h_S^{-1}\|v - \mathcal{J}_h v\|_{L^2(S)}^2$$

$$= c_{\mathrm{Tr}}^2 \sum_{S \in \mathcal{S}_h} \left(h_S^{-2}\|v - \mathcal{J}_h v\|_{L^2(T_S)}^2 + \|\nabla(v - \mathcal{J}_h v)\|_{L^2(T_S)}^2\right)$$

$$\leq d c_{\mathrm{Tr}}^2 \sum_{T \in \mathcal{T}_h} \left(c_{\ell oc}^2 h_T^{-2}\|v - \mathcal{J}_h v\|_{L^2(T)}^2 + \|\nabla(v - \mathcal{J}_h v)\|_{L^2(T)}^2\right)$$

$$= d c_{\mathrm{Tr}}^2 \left(c_{\ell oc}^2 \|h_{\mathcal{T}}^{-1}(v - \mathcal{J}_h v)\|^2 + \|\nabla(v - \mathcal{J}_h v)\|^2\right)$$

$$\leq d c_{\mathrm{Tr}}^2 (c_{\ell oc}^2 c_1^2 + c_2^2)\|\nabla v\|^2.$$

This proves the third estimate and completes the proof.   □

Remarks 4.4

(i) The estimate implies that $\mathscr{J}_h$ defines a bounded linear operator on $H_D^1(\Omega)$. Moreover, if h is the maximal mesh-size of $\mathscr{T}_h$, so that $h_\mathscr{T} \leq h$ in Ω, then we have the approximation property

$$\|v - \mathscr{J}_h v\| \leq c_2 h \|\nabla v\|.$$

This yields that $\mathscr{J}_h v \to v$ in $H^1(\Omega)$ as $h \to 0$ with convergence rate $\mathscr{O}(h)$.

(ii) The operator $\mathscr{J}_h$ is not a projection, i.e., in general we have $\mathscr{J}_h v_h \neq v_h$ for $v_h \in \mathscr{S}^1(\mathscr{T}_h)$.

4.2.3 A Posteriori Error Estimate

With the help of the quasi-interpolant we are able to derive an *a posteriori error* estimate that allows us to control the approximation error with computable quantities. For this, the following definition is needed.

Definition 4.5 Let $u_h \in \mathscr{S}^1(\mathscr{T}_h)$ and $S \in \mathscr{S}_h$ be an interior side, i.e., $S = T_1 \cap T_2$ for two distinct elements $T_1, T_2 \in \mathscr{T}_h$, cf. Fig. 4.8. With the outer unit normals $n_{T_1,S}$ and $n_{T_2,S}$ on S to T_1 and T_2, respectively, we define the *jump* of ∇u_h in normal direction across S by

$$[\![\nabla u_h \cdot n_S]\!] = \nabla u_h|_{T_1} \cdot n_{T_1,S} + \nabla u_h|_{T_2} \cdot n_{T_2,S}.$$

For $S \subset \Gamma_D$ we set $[\![\nabla u_h \cdot n_S]\!] = 0$.

Definition 4.6 For $u_h \in \mathscr{S}^1(\mathscr{T}_h)$ and $T \in \mathscr{T}_h$, we define the *refinement indicator* $\eta_T(u_h)$ by

$$\eta_T^2(u_h) = h_T^2 \|f\|_{L^2(T)}^2 + \sum_{S \in \mathscr{S}_h, S \subset \partial T} h_S \|[\![\nabla u_h \cdot n_S]\!]\|_{L^2(S)}^2$$

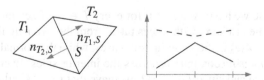

Fig. 4.8 Neighboring elements for a side (*left*); large (*solid line*) and small (*dashed line*) jumps of the derivative of one-dimensional piecewise affine functions (*right*)

and the *error estimator* $\eta_{\mathscr{R}}(u_h)$ by

$$\eta_{\mathscr{R}}^2(u_h) = \sum_{T \in \mathscr{T}_h} \eta_T^2(u_h).$$

We assume for simplicity that $\Gamma_D = \partial\Omega$ in what follows.

Theorem 4.3 (A Posteriori Error Estimate) *Let* $u \in H_0^1(\Omega)$ *be the weak solution of the Poisson problem*

$$-\Delta u = f \text{ in } \Omega, \quad u|_{\partial\Omega} = 0,$$

and $u_h \in \mathscr{S}_0^1(\mathscr{T}_h)$ *its Galerkin approximation. There exists a constant* $c_R > 0$*, such that we have the* reliability *estimate*

$$\|\nabla(u - u_h)\| \le c_R \, \eta_{\mathscr{R}}(u_h).$$

Proof We let $e = u - u_h$ be the approximation error and note that due to Galerkin orthogonality and the equation satisfied by u, we have

$$\|\nabla e\|^2 = \int_\Omega \nabla e \cdot \nabla(e - \mathscr{J}_h e)\,dx$$

$$= \int_\Omega \nabla u \cdot \nabla(e - \mathscr{J}_h e)\,dx - \int_\Omega \nabla u_h \cdot \nabla(e - \mathscr{J}_h e)\,dx$$

$$= \int_\Omega f(e - \mathscr{J}_h e)\,dx - \int_\Omega \nabla u_h \cdot \nabla(e - \mathscr{J}_h e)\,dx.$$

We write the last term as a sum over all $T \in \mathscr{T}_h$ and then integrate-by-parts on each $T \in \mathscr{T}_h$, i.e.,

$$\int_\Omega \nabla u_h \cdot \nabla(e - \mathscr{J}_h e)\,dx = \sum_{T \in \mathscr{T}_h} \int_T \nabla u_h \cdot \nabla(e - \mathscr{J}_h e)\,dx$$

$$= \sum_{T \in \mathscr{T}_h} \left(\int_T (-\Delta u_h)(e - \mathscr{J}_h e)\,dx + \int_{\partial T} (\nabla u_h \cdot n_T)(e - \mathscr{J}_h e)\,ds \right).$$

Since $u_h|_T$ is affine we have $\Delta u_h|_T = 0$ for every $T \in \mathscr{T}_h$. Each interior side occurs exactly twice in the last sum, with traces taken from the two neighboring elements. The trace of $(e - \mathscr{J}_h e)$ is the same in both cases, while the normals are in opposite directions and ∇u_h is discontinuous across the interior sides. Every boundary side occurs exactly once and since $\Gamma_D = \partial\Omega$, we have that $(e - \mathscr{J}_h e)|_{\partial\Omega} = 0$. With the definition of $[\nabla u_h \cdot n_S]$ we thus deduce that

$$\sum_{T \in \mathscr{T}_h} \int_{\partial T} (\nabla u_h \cdot n)(e - \mathscr{J}_h e)\,ds = \sum_{S \in \mathscr{S}_h} \int_S [\nabla u_h \cdot n_S](e - \mathscr{J}_h e)\,ds.$$

On combining the identities, and using Hölder and Schwarz inequalities, we obtain that

$$\|\nabla e\|^2 = \sum_{T \in \mathscr{T}_h} \int_T f(e - \mathscr{I}_h e)\, dx - \sum_{S \in \mathscr{S}_h} \int_S [\nabla u_h \cdot n_S](e - \mathscr{I}_h e)\, ds$$

$$\leq \sum_{T \in \mathscr{T}_h} \|f\|_{L^2(T)} \|e - \mathscr{I}_h e\|_{L^2(T)} + \sum_{S \in \mathscr{S}_h} \|[\nabla u_h \cdot n_S]\|_{L^2(S)} \|e - \mathscr{I}_h e\|_{L^2(S)}$$

$$\leq \left(\sum_{T \in \mathscr{T}_h} h_T^2 \|f\|_{L^2(T)}^2 \right)^{1/2} \left(\sum_{T \in \mathscr{T}_h} h_T^{-2} \|e - \mathscr{I}_h e\|_{L^2(T)}^2 \right)^{1/2}$$

$$+ \left(\sum_{S \in \mathscr{S}_h} h_S \|[\nabla u_h \cdot n_S]\|_{L^2(S)}^2 \right)^{1/2} \left(\sum_{S \in \mathscr{S}_h} h_S^{-1} \|e - \mathscr{I}_h e\|_{L^2(S)}^2 \right)^{1/2}$$

$$\leq \left(\sum_{T \in \mathscr{T}_h} h_T^2 \|f\|_{L^2(T)}^2 \right)^{1/2} \|h_{\mathscr{T}}^{-1}(e - \mathscr{I}_h e)\|$$

$$+ \left(\sum_{S \in \mathscr{S}_h} h_S \|[\nabla u_h \cdot n_S]\|_{L^2(S)}^2 \right)^{1/2} \|h_{\mathscr{S}}^{-1/2}(e - \mathscr{I}_h e)\|_{L^2(\cup \mathscr{S}_h)}.$$

With the estimates for the Clément interpolant and a division by $\|\nabla e\|$, we deduce that

$$\|\nabla e\| \leq c_{\mathscr{I}} \left(\sum_{T \in \mathscr{T}_h} h_T^2 \|f\|_{L^2(T)}^2 \right)^{1/2} + c_{\mathscr{I}} \left(\sum_{S \in \mathscr{S}_h} h_S \|[\nabla u_h \cdot n_S]\|_{L^2(S)}^2 \right)^{1/2}.$$

Using the inequality $a + b \leq \sqrt{2}(a^2 + b^2)^{1/2}$ proves the estimate. □

Remarks 4.5

(i) Note that in general we do not have for every $T \in \mathscr{T}_h$ that

$$\|\nabla(u - u_h)\|_{L^2(T)} \leq c\eta_T(u_h),$$

i.e., the error estimate is a global one.

(ii) If a higher-order finite element method is used, or if Neumann boundary conditions $\partial_n u = g$ on a subset $\Gamma_N \subset \partial\Omega$ are imposed, then one obtains the error indicators

$$\eta_T^2(u_h) = h_T^2 \|f + \Delta u_h\|_{L^2(T)}^2 + \sum_{S \in \mathscr{S}_h, S \subset \partial T} h_S \|[\nabla u_h \cdot n_S]\|_{L^2(S)}^2,$$

where $[\nabla u_h \cdot n_S]$ is for $S \subset \partial T \cap \overline{\Gamma}_N$ defined by

$$[\nabla u_h \cdot n_S] = \nabla u_h \cdot n - g.$$

The terms $f + \Delta u_h$ and $\nabla u_h \cdot n - g$ are the *residuals* of the approximation u_h. The jumps of ∇u_h are a measure of discontinuity of u_h. The estimator $\eta_{\mathscr{R}}$ is therefore called a *residual estimator*.

4.2.4 Efficiency

The error estimator $\eta_{\mathscr{R}}$ defines up to certain constants a reliable computable error bound. To complete its justification as a criterion for local mesh refinement, it is desirable to show that it is optimal in an appropriate sense. This will be done locally for the error indicators. We follow [26].

Definition 4.7 For a side $S \in \mathscr{S}_h$, we define the *side patch* ω_S as the interior of the union of its neighboring elements, i.e.,

$$\omega_S = \mathrm{int}\Big(\bigcup_{T \in \mathscr{T}_h, S \subset \partial T} T \Big).$$

The *element patch* ω_T of an element $T \in \mathscr{T}_h$ is defined by

$$\omega_T = \bigcup_{z \in \mathscr{N}_h, z \in T} \omega_z.$$

We define locally supported functions in $H_{\mathrm{D}}^1(\Omega)$ associated with elements and sides, cf. Fig. 4.9.

Lemma 4.3 (Bubble Functions)

(i) *There exist constants $c_{e,1}, c_{e,2} > 0$, such that for every $h > 0$ and every $T \in \mathscr{T}_h$ with $T = \mathrm{conv}\{z_1, z_2, \ldots, z_{d+1}\}$, the element bubble function $b_T = \varphi_{z_1}\varphi_{z_2} \cdots \varphi_{z_{d+1}} \in H^1(\Omega) \cap C(\overline{\Omega})$ satisfies*

$$\mathrm{supp}\, b_T \subset T, \quad \int_T b_T = c_{e,1}|T|, \quad \|\nabla b_T\|_{L^2(T)} \leq c_{e,2}h_T^{d/2-1}.$$

(ii) *There exist constants $c_{s,1}, c_{s,2} > 0$ such that for every $h > 0$ and every $S \in \mathscr{S}_h$ with $S = \mathrm{conv}\{z_1, z_2, \ldots, z_d\}$ the side bubble function $b_S = \varphi_{z_1}\varphi_{z_2} \cdots \varphi_{z_d} \in H^1(\Omega) \cap C(\overline{\Omega})$ satisfies*

$$\mathrm{supp}\, b_S \subset \omega_S, \quad \int_S b_S = c_{s,1}|S|, \quad \|\nabla b_S\|_{L^2(\omega_S)} \leq c_{s,2}h_S^{d/2-1}.$$

Proof Exercise.   □

Fig. 4.9 Element and side
bubble functions b_T and b_S

With the help of the bubble functions, we show that the error indicators also define a lower bound for the approximation error. For simplicity we restrict to $d = 2$, $\Gamma_D = \partial\Omega$, and a piecewise constant function $f \in L^2(\Omega)$.

Proposition 4.3 (Local Efficiency) *Assume $d = 2$, $\Gamma_N = \emptyset$, and $f \in L^2(\Omega)$ is elementwise constant on $\mathcal{T}_h$. There exists a constant $c_E > 0$, such that for every $T \in \mathcal{T}_h$ we have*

$$c_E\, \eta_T^2(u_h) \le \left\| \nabla(u - u_h) \right\|_{L^2(\omega_T)}^2.$$

Proof

(i) For an element $T \in \mathcal{T}_h$ and the element bubble function $b_T \in H_0^1(\Omega)$, we have

$$\int_\Omega \nabla u \cdot \nabla b_T \, dx = \int_\Omega f b_T \, dx.$$

An integration-by-parts on T in combination with the identities $\Delta u_h|_T = 0$ and $b_T|_{\partial T} = 0$ shows that

$$\int_T \nabla u_h \cdot \nabla b_T \, dx = 0.$$

Therefore we have

$$\int_T f_T b_T \, dx = \int_T \nabla(u - u_h) \cdot \nabla b_T \, dx.$$

Since f is constant on T with value f_T and sign σ_T, we have

$$\sigma_T \int_T f_T b_T \, dx = |f_T| \int_T b_T \, dx = c_{e,1} |f_T| |T| = c_{e,1} |T|^{1/2} \| f \|_{L^2(T)}.$$

Combining the previous two identities, incorporating Hölder's inequality, and using that $\| \nabla b_T \|_{L^2(T)} \le c_{e,2}$ imply that

$$c_{e,1} |T|^{1/2} \| f \|_{L^2(T)} = \sigma_T \int_T \nabla(u - u_h) \cdot \nabla b_T \, dx \le c_{e,2} \| \nabla(u - u_h) \|_{L^2(T)}.$$

With the estimate $c_{loc} h_T \le |T|^{1/2}$, we find that

$$h_T \| f \|_{L^2(T)} \le c_{el} \| \nabla(u - u_h) \|_{L^2(T)}.$$

(ii) For an interior side $S = T_1 \cap T_2$ and the side bubble function $b_S \in H_0^1(\Omega)$, we have

$$\int_{\omega_S} \nabla u \cdot \nabla b_S \, dx = \int_\omega f b_S \, dx.$$

Integration-by-parts on T_1 and T_2 and $\Delta u_h|_{T_j} = 0$ and $b_S|_{\partial T_j \setminus S} = 0$ for $j = 1, 2$, prove that

$$\int_{\omega_S} \nabla u_h \cdot \nabla b_S \, dx = \int_{T_1} \nabla u_h \cdot \nabla b_S \, dx + \int_{T_2} \nabla u_h \cdot \nabla b_S \, dx$$

$$= \int_S (\nabla u_h \cdot n_{T_1}) b_S \, ds + \int_S (\nabla u_h \cdot n_{T_2}) b_S \, ds$$

$$= \int_S [\![\nabla u_h \cdot n_S]\!] b_S \, ds.$$

Subtracting the last two identities leads to

$$\int_{\omega_S} \nabla (u - u_h) \cdot \nabla b_S \, dx = -\int_S [\nabla u_h \cdot n_S] b_S \, ds + \int_{\omega_S} f b_S \, dx.$$

Since $[\nabla u_h \cdot n_S]$ is constant on S with sign σ_S, we have that

$$\sigma_S \int_S [\![\nabla u_h \cdot n_S]\!] b_S \, ds = c_{s,1} |[\![\nabla u_h \cdot n_S]\!]| |S| = c_{s,1} \| [\![\nabla u_h \cdot n_S]\!] \|_{L^2(S)} |S|^{1/2}.$$

Combining the previous identities, incorporating Hölder's inequality, and using that $\|\nabla b_S\|_{L^2(\omega_S)} \le c_{s,2}$ and $\|b_S\|_{L^2(\omega_S)} \le |\omega_S|^{1/2}$ prove that

$$c_{s,1} \left\| [\![\nabla u_h \cdot n_S]\!] \right\|_{L^2(S)} |S|^{1/2} = -\sigma_S \int_{\omega_S} \nabla (u - u_h) \cdot \nabla b_S \, dx + \sigma_S \int_{\omega_S} f b_S \, dx$$

$$\le c_{s,2} \|\nabla (u - u_h)\|_{L^2(\omega_S)} + \|f\|_{L^2(\omega_S)} |\omega_S|^{1/2}.$$

With the estimate $|\omega_S|^{1/2} \le c_{\ell oc} h_S$ and the identity $h_S = |S|$, we find that with a constant $c_{\text{side}} > 0$ we have

$$c_{\text{side}}^{-1} h_S^{1/2} \left\| [\![\nabla u_h \cdot n_S]\!] \right\|_{L^2(S)} \le \|\nabla (u - u_h)\|_{L^2(\omega_S)} + h_S \|f\|_{L^2(\omega_S)}.$$

(iii) Combining the estimates for $h_T\|f\|_{L^2(T)}$ and $h_S^{1/2}\|[\![\nabla u_h \cdot n_S]\!]\|_{L^2(S)}$ implies that

$$\eta_T^2(u_h) = h_T^2\|f\|_{L^2(T)}^2 + \sum_{S\in\mathscr{S}_h, S\subset\partial T} h_S\|[\![\nabla u_h \cdot n_S]\!]\|_{L^2(S)}^2$$

$$\leq h_T^2\|f\|_{L^2(T)}^2 + c_{\text{side}}^2 \sum_{S\in\mathscr{S}_h, S\subset\partial T} \|\nabla(u-u_h)\|_{L^2(\omega_S)}^2 + h_S^2\|f\|_{L^2(\omega_S)}^2$$

$$\leq c_E^{-2}\|\nabla(u-u_h)\|_{L^2(\omega_T)}^2,$$

which proves the estimate.   □

Remark 4.6 Due to the finite overlap of the patches $(\omega_T : T \in \mathscr{T}_h)$, we obtain the global efficiency estimate

$$c_E' \, \eta_{\mathscr{R}}(u_h) \leq \|\nabla(u-u_h)\|.$$

Hence the quantity $\eta_{\mathscr{R}}(u_h)$ is equivalent to the error $\|\nabla(u-u_h)\|$. It is important that an error estimator is reliable and efficient, e.g., the trivial estimator $\eta(u_h) = 1$ is reliable but not efficient, while $\eta(u_h) = 0$ is efficient but not reliable. Both estimators are not useful for error estimation and mesh refinement.

4.2.5 Adaptive Mesh Refinement

The a posteriori error estimate leads to an adaptive mesh refinement strategy, which automatically refines an initial triangulation in regions in which the refinement indicators $\eta_T^2(u_h)$ are relatively large.

Algorithm 4.1 (Solve-Estimate-Mark-Refine)   *Choose a triangulation $\mathscr{T}_0$, parameters $\theta \in (0, 1]$, $\varepsilon_{\text{stop}} > 0$, and set $k = 0$.*

(1) Compute the Galerkin approximation $u_k \in \mathscr{S}_0^1(\mathscr{T}_k)$.
(2) For every $T \in \mathscr{T}_k$ compute the refinement indicator $\eta_T^2(u_k)$.
(3) Stop if $\eta_{\mathscr{R}}(u_k) \leq \varepsilon_{\text{stop}}$.
(4) Choose a set of marked elements $\mathscr{M}_k \subset \mathscr{T}_k$ such that

$$\sum_{T\in\mathscr{M}_k} \eta_T^2(u_k) \geq \theta \sum_{T\in\mathscr{T}_k} \eta_T^2(u_k).$$

(5) Refine every $T \in \mathscr{M}_k$ and further elements to obtain a new, locally refined conforming triangulation $\mathscr{T}_{k+1}$.
(6) Set $k \to k + 1$ and go to (1).

Remark 4.7 It is necessary to refine further elements in Step (5) of the algorithm to avoid *hanging nodes* and nonconforming triangulations, cf. Fig. 4.10.

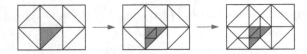

Fig. 4.10 Refinement of a marked element and further elements to avoid hanging nodes

Fig. 4.11 Red-, green-, and blue-refinement of a triangle

To realize the conformity requirement, we use three different refinements of a triangle, cf. Fig. 4.11. For every triangle $T \in \mathcal{T}_k$, we select one of its longest edges and call it the longest edge.

Definition 4.8

 (i) A *red-refinement* of a triangle consists in partitioning it into four subtriangles by connecting the midpoints of its edges.
 (ii) A *green-refinement* of a triangle consists in partitioning it into two subtriangles by connecting the midpoint of the longest edge with the vertex opposite to it.
(iii) A *blue-refinement* of a triangle consists in partitioning it into three subtriangles by first performing a green-refinement and then connecting the new vertex with the midpoint of one of the two unrefined edges.

The refinement procedure in Step (5) of the Algorithm 4.1 is implemented by appropriately marking edges for refinement. To guarantee shape regularity, we ensure that if an element is refined, then the longest edge is refined.

Algorithm 4.2 (Red-Green-Blue Refinement) *Let $\mathcal{M}_k \subset \mathcal{T}_k$ be a set of elements marked for refinement.*

(1) Mark all edges $S \in \mathcal{S}_k$ which belong to a marked element $T \in \mathcal{M}_k$.
(2) Mark further edges in such a way that for every element $T \in \mathcal{T}_k$, for which one of its edges $S \subset \partial T$ is marked, also the longest edge is marked for refinement.
(3) Refine each $T \in \mathcal{T}_k$ which has a marked edge by red-, blue-, or green-refinement if it has three, two, or one marked edge, respectively.

An alternative to red-green-blue refinement is the refinement by *edge bisection*, which is of particular importance in three-dimensional situations. It is based on the definition of a mapping $R : \mathcal{T} \to \mathcal{E}$ that associates with every element a *refinement edge*, e.g., the longest edge of the element or the edge opposite to the newest vertex. In a typical refinement procedure, the reference edges of marked elements are marked for refinement. These and other edges are bisected in such a way that mesh conformity is guaranteed. This can be achieved by repeatedly bisecting edges until no hanging nodes or sides are contained in the triangulation. An approach that avoids nonconforming triangulations at all stages of the refinement procedure

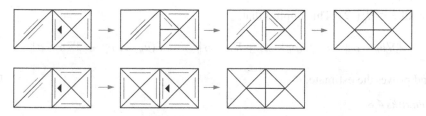

Fig. 4.12   Edges are bisected until no hanging nodes are contained in the triangulation (*upper row*); refinement edges of neighboring elements are marked and only compatible edge patches are bisected (*lower row*)

only bisects a marked edge if it is the refinement edge for all elements it belongs to. This strategy is called a *compatible edge patch refinement*. This is done in a recursive way, so that all marked edges are after a number of additional refinements, the refinement edges of all elements in their patches. Figure 4.12 illustrates the two approaches.

4.2.6   Constant-Free Error Estimation

The error estimator $\eta_{\mathscr{R}}(u_h)$ is of limited use for practical error estimation, since the reliability estimate involves an unknown constant. The following theorem provides a basis for constructing estimators that avoid this deficiency. We follow [7].

Theorem 4.4 (Prager–Synge) *Let $u \in H_0^1(\Omega)$ be the weak solution of the Poisson problem with the right-hand side $f \in L^2(\Omega)$, and let $u_h \in H_0^1(\Omega)$ be an arbitrary approximation. For every $\sigma \in L^2(\Omega; \mathbb{R}^d)$ such that $-\operatorname{div}\sigma = f$ in the weak sense, i.e.,*

$$\int_\Omega \sigma \cdot \nabla v \, dx = \int_\Omega f v \, dx$$

for all $v \in H_0^1(\Omega)$, we have

$$\|\nabla(u - u_h)\| \le \|\nabla u_h - \sigma\|.$$

Proof By definition of the weak solution and the assumption on σ, we have

$$\int_\Omega (\nabla u - \sigma) \cdot \nabla v \, dx = 0$$

for all $v \in H_0^1(\Omega)$. This implies that

$$\|\nabla u_h - \sigma\|^2 = \|\nabla(u_h - u) + \nabla u - \sigma\|^2 = \|\nabla(u_h - u)\|^2 + \|\nabla u - \sigma\|^2$$

and proves the estimate.   □

Remarks 4.8

(i) The function $\sigma \in L^2(\Omega; \mathbb{R}^d)$ can be obtained with the Raviart–Thomas mixed
   finite element method if f is elementwise constant. Alternatively, a continuous
   approximation of ∇u_h can be used.
(ii) Other concepts for deriving error estimates with explicit constants use local
   problems to represent the error or employ specific upper bounds for constants
   involved in the estimates for the quasi-interpolation operator.
(iii) *Asymptotical exactness* of an error estimator η_h refers to the property that
   $\|\nabla(u - u_h)\|/\eta_h \to 1$ as $h \to 0$.

4.2.7 Implementation

A MATLAB realization of the adaptive refinement algorithm is shown in Fig. 4.13
and 4.14, where for simplicity we used the marking criterion

$$\mathscr{M}_k = \{T \in \mathscr{T}_k : \eta_T(u_k) \geq \theta \max_{T' \in \mathscr{T}_k} \eta_{T'}(u_k)\}.$$

The subroutine `comp_estimators` computes the refinement indicators

$$\eta_T^2(u_k) = h_T^2 \|f\|_{L^2(T)}^2 + h_T^{2-d} \sum_{S \subset \partial T} |S| \|[\![\nabla u_k \cdot n_S]\!]\|_{L^2(S)}^2,$$

where we used an equivalent redefinition for a simpler computation, i.e., noting that
$h_T^{2-d}|S| \leq c h_S$. In particular, using the identity

$$\nabla \varphi_z|_T = -n_{S_z}/\varrho_z = -n_S|S|/(d|T|),$$

for the side $S \subset \partial T$ with opposite vertex $z \in T$, we can efficiently compute the
quantities

$$|S|^2 |[\![\nabla u_k \cdot n_S]\!]|^2 = |S| \|[\![\nabla u_k \cdot n_S]\!]\|_{L^2(S)}^2.$$

The routine `sides.m` displayed in Fig. 4.15 provides data structures related to the
sides of elements. It defines mappings

$$\texttt{s4e} : \mathscr{T}_h \to \mathscr{S}_h^{d+1}, \quad \texttt{sign\_s4e} : \mathscr{T}_h \to \{\pm 1\}^{d+1},$$

```
function p1_adaptive(red) % Nb = []; d = 2;
theta = .5; eps_stop = 5e-2; error_bound = 1;
c4n = [-1,-1;0,-1;-1,0;0,0;1,0;-1,1;0,1;1,1];
n4e = [1,2,4;4,3,1;3,4,7;7,6,3;4,5,8;8,7,4];
Db = [1,2;2,4;4,5;5,8;8,7;7,6;6,3;3,1];
for j = 1:red
    marked = ones(size(n4e,1),1);
    [c4n,n4e,Db] = rgb_refine(c4n,n4e,Db,marked);
end
while error_bound > eps_stop
    %%% solve
    fNodes = setdiff(1:size(c4n,1),unique(Db));
    u = zeros(size(c4n,1),1);
    [s,m] = fe_matrices(c4n,n4e);
    b = m*f(c4n);
    u(fNodes) = s(fNodes,fNodes)\b(fNodes);
    show_p1(c4n,n4e,Db,[],u); view(0,90); pause(.05)
    %%% estimate
    eta = comp_estimators(c4n,n4e,Db,u);
    error_bound = sqrt(sum(eta.^2))
    %%% mark
    marked = (eta>theta*max(eta));
    %%% refine
    if error_bound > eps_stop
        [c4n,n4e,Db] = rgb_refine(c4n,n4e,Db,marked);
    end
end

function eta = comp_estimators(c4n,n4e,Db,u)
[s4e,~,n4s,s4Db] = sides(n4e,Db,[]);
[~,d] = size(c4n); nE = size(n4e,1); nS = size(n4s,1);
eta_S = zeros(nS,1); eta_T_sq = zeros(nE,1);
for j = 1:nE
    X_T = [ones(1,d+1);c4n(n4e(j,:),:)'];
    grads_T = X_T\[zeros(1,d);eye(d)];
    vol_T = det(X_T)/factorial(d);
    h_T = vol_T^(1/d);
    mp_T = sum(c4n(n4e(j,:),:),1)/(d+1);
    eta_T_sq(j) = h_T^2*vol_T*(f(mp_T));
    nabla_u_T = grads_T'*u(n4e(j,:));
    normal_times_area = -grads_T*vol_T*d;
    eta_S(s4e(j,:)) = h_T^((2-d)/2)*eta_S(s4e(j,:))...
        +normal_times_area*nabla_u_T;
end
eta_S(s4Db) = 0;
eta_S_T_sq = sum(eta_S(s4e).^2,2);
eta = (eta_T_sq+eta_S_T_sq).^(1/2);

function val = f(x); val = ones(size(x,1),1);
```

Fig. 4.13 Adaptive approximation of the Poisson problem; the routine iterates the steps solve, estimate, mark, and refine until a stopping criterion is reached

```matlab
function [c4n,n4e,Db] = rgb_refine(c4n,n4e,Db,marked)
% edges n4e(:,[1,3]) have to be longest edges
[edges,el2edges,Db2edges] = edge_data_2d(n4e,Db);
nC = size(c4n,1); nEdges = size(edges,1); tmp = 1;
markedEdges = zeros(nEdges,1);
markedEdges(reshape(el2edges(marked==1,[1 2 3]),[],1)) = 1;
while tmp > 0
    tmp = nnz(markedEdges);
    el2markedEdges = markedEdges(el2edges);
    el2markedEdges(el2markedEdges(:,1)+el2markedEdges(:,3)>0,2)=1;
    markedEdges(el2edges(el2markedEdges==1))= 1;
    tmp = nnz(markedEdges)-tmp;
end
newNodes = zeros(nEdges,1);
newNodes(markedEdges==1) = (1:nnz(markedEdges))'+nC;
newInd = newNodes(el2edges);
red   = newInd(:,1) >  0 & newInd(:,2) >  0 & newInd(:,3) >  0;
blue1 = newInd(:,1) >  0 & newInd(:,2) >  0 & newInd(:,3) == 0;
blue3 = newInd(:,1) == 0 & newInd(:,2) >  0 & newInd(:,3) >  0;
green = newInd(:,1) == 0 & newInd(:,2) >  0 & newInd(:,3) == 0;
remain= newInd(:,1) == 0 & newInd(:,2) == 0 & newInd(:,3) == 0;
n4e_red  = [n4e(red,1),newInd(red,[3 2]),...
            newInd(red,[2 1]),n4e(red,3),...
            newInd(red,3),n4e(red,2),newInd(red,1),...
            newInd(red,:)];
n4e_red = reshape(n4e_red',3,[])';
n4e_blue1 = [n4e(blue1,2),newInd(blue1,2),n4e(blue1,1) ...
             n4e(blue1,2),newInd(blue1,[1 2]),...
             newInd(blue1,[2 1]),n4e(blue1,3)];
n4e_blue1 = reshape(n4e_blue1',3,[])';
n4e_blue3 = [n4e(blue3,1),newInd(blue3,3),newInd(blue3,2),...
             newInd(blue3,2),newInd(blue3,3),n4e(blue3,2),...
             n4e(blue3,3),newInd(blue3,2),n4e(blue3,2)];
n4e_blue3 = reshape(n4e_blue3',3,[])';
n4e_green = [n4e(green,2),newInd(green,2),n4e(green,1),...
             n4e(green,3),newInd(green,2),n4e(green,2)];
n4e_green = reshape(n4e_green',3,[])';
n4e = [n4e(remain,:);n4e_red;n4e_blue1;n4e_blue3;n4e_green];
newCoord =.5*(c4n(edges(markedEdges==1,1),:)...
    +c4n(edges(markedEdges==1,2),:));
c4n = [c4n;newCoord];
newDb = newNodes(Db2edges); ref = newDb>0; Db_old = Db(~ref,:);
Db_new = [Db(ref,1),newDb(ref),newDb(ref),Db(ref,2)];
Db_new = reshape(Db_new',2,[])'; Db = [Db_old;Db_new];

function [edges,el2edges,Db2edges] = edge_data_2d(n4e,Db)
nE = size(n4e,1); nEdges = 3*nE; nDb = size(Db,1);
edges = [reshape(n4e(:,[2 3,1 3,1 2])',2,[])';Db];
[edges,~,edgeNumbers] = unique(sort(edges,2),'rows','first');
el2edges = reshape(edgeNumbers(1:nEdges),3,[])';
Db2edges = reshape(edgeNumbers(nEdges+(1:nDb)),1,[])';
```

Fig. 4.14 MATLAB routine for the local refinement of a triangulation

```
function [s4e,sign_s4e,n4s,s4Db,s4Nb,e4s] = sides(n4e,Db,Nb)
nE = size(n4e,1); d = size(n4e,2)-1;
nDb = size(Db,1); nNb = size(Nb,1);
if d == 2
    Tsides = [n4e(:,[2,3]);n4e(:,[3,1]);n4e(:,[1,2])];
else
    Tsides = [n4e(:,[2,4,3]);n4e(:,[1,3,4]);...
        n4e(:,[1,4,2]);n4e(:,[1,2,3])];
end
[n4s,i2,j] = unique(sort(Tsides,2),'rows');
s4e = reshape(j,nE,d+1); nS = size(n4s,1);
sign_s4e = ones((d+1)*nE,1); sign_s4e(i2) = -1;
sign_s4e = reshape(sign_s4e,nE,d+1);
[~,~,j2] = unique(sort([n4s;Db;Nb],2),'rows');
s4Db = j2(nS+(1:nDb)); s4Nb = j2(nS+nDb+(1:nNb));
e4s = zeros(size(n4s,1),2);
e4s(:,1) = mod(i2-1,nE)+1;
i_inner = setdiff(1:(d+1)*nE,i2);
e4s(j(i_inner),2) = mod(i_inner-1,nE)+1;
```

Fig. 4.15 MATLAB routine that provides data structures related to sides of elements

that specify the sides of elements and signs of the elements relative to the sides, according to the local enumeration defined by the opposite nodes. The signs are chosen in such a way that if a side belongs to the boundary, then its associated element is assigned a positive sign. The mappings

$$n4s : \mathscr{S}_h \to \mathscr{N}_h^d, \quad s4Db : \mathscr{S}_h \cap \Gamma_D \to \mathscr{S}_h, \quad s4Nb : \mathscr{S}_h \cap \overline{\Gamma}_N \to \mathscr{S}_h,$$

provide the nodes of a side, and inject the boundary sides on Γ_D and Γ_N into the set of all sides $\mathscr{S}_h$. Moreover, it specifies the mapping

$$e4s : \mathscr{S}_h \to \left(\mathscr{T}_h \cup \{0\}\right)^2$$

that determines the adjacent elements of a side, where the index zero indicates that a side has only one neighboring element. The routine rgb_refine.m carries out the red-green-blue refinement of the triangulation for a given set of marked elements. We use the convention that

$$T \equiv (z_1, z_2, z_3) \implies S = \text{conv}\{z_3, z_1\} \text{ is a longest edge of } T.$$

The routine marks all edges of marked elements and then applies a red-, green-, or blue-refinement according to whether one, two, or all edges are marked, and whether the longest edge is among the marked edges. The subroutine edges.m provides the arrays edges and el2edges that specify the endpoints of edges and the edges of elements, respectively. Figures 4.16 and 4.17 show a sequence of adaptively refined triangulations and a corresponding finite element solution.

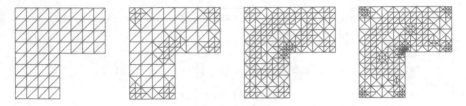

Fig. 4.16   Adaptively refined triangulations of an L-shaped domain

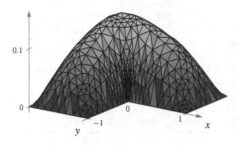

Fig. 4.17   Finite element approximation on an adaptively generated triangulation

4.3   Convergence and Quasioptimality

4.3.1   Strict Error Reduction

The adaptive algorithm generates a sequence of approximations $(u_k)_{k \in \mathbb{N}} \subset H_0^1(\Omega)$ associated with triangulations $(\mathscr{T}_k)_{k \in \mathbb{N}}$ by marking subsets of elements $\mathscr{M}_k \subset \mathscr{T}_k$ for refinement. An important question is whether the sequence $(u_k)_{k \in \mathbb{N}}$ converges to the exact solution of the Poisson problem

$$-\Delta u = f \text{ in } \Omega, \quad u|_{\partial \Omega} = 0.$$

The convergence $u_k \to u$ is not obvious, since parts of Ω may remain unrefined during the adaptive process, cf. Fig. 4.18. Throughout this section we follow [9, 14, 22].

We prove that under certain conditions on the local mesh refinement we have a fixed error reduction which implies convergence $u_k \to u$ in $H^1(\Omega)$ as $k \to \infty$.

Definition 4.9 We say that the triangulation $\mathscr{T}_{k+1}$ is a *refinement* of $\mathscr{T}_k$ if every $T \in \mathscr{T}_{k+1}$ is the union of elements in $\mathscr{T}_k$.

For a refinement $\mathscr{T}_{k+1}$ of $\mathscr{T}_k$, we have the *nestedness* property $\mathscr{S}^1(\mathscr{T}_k) \subset \mathscr{S}^1(\mathscr{T}_{k+1})$.

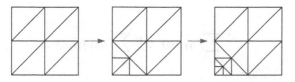

Fig. 4.18   Parts of a domain may remain unrefined in a local mesh refinement strategy

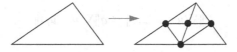

Fig. 4.19   Creation of new inner nodes in the element and on the sides of a marked element

Lemma 4.4 (Discrete Efficiency) *Assume that $f \in L^2(\Omega)$ is elementwise constant on $\mathcal{T}_0$. Let $\mathcal{T}_{k+1}$ be a refinement of $\mathcal{T}_k$, and let u_k and u_{k+1} be the corresponding Galerkin approximations. If $T \in \mathcal{T}_k$ is refined in such a way that new nodes on the interior of sides of T and in the interior of T are created, cf. Fig. 4.19, then we have*

$$c_{dE}^{-1}\, \eta_T(u_k) \leq \|\nabla(u_{k+1} - u_k)\|_{L^2(\omega_T)} \qquad \text{(dEff)}$$

with a k-independent constant $c_{dE} > 0$.

Proof The assumption about the new inner nodes allows us to choose bubble functions $b_{T,h}, b_{S,h} \in \mathcal{S}^1(\mathcal{T}_{k+1})$, and to argue, as in the setting of Proposition 4.3, with u and u_h replaced by u_{k+1} and u_k. $\qquad\square$

Remark 4.9 The assumption on the right-hand side is not restrictive, since replacing f by its elementwise averages leads to an error of order $\mathcal{O}(h_{\max}^2)$ for the maximal mesh-size $h_{\max}$ in $\mathcal{T}_0$.

We combine the reliability estimate

$$\|\nabla(u - u_k)\|^2 \leq c_R^2 \sum_{T \in \mathcal{T}_k} \eta_T^2(u_k) \qquad \text{(Rel)}$$

with the discrete efficiency estimate to prove convergence.

Theorem 4.5 (Strict Error Reduction) *Let f be elementwise constant on $\mathcal{T}_0$, and assume that $\theta \in (0, 1]$ is sufficiently small so that*

$$0 < \frac{c_q \theta^{-1} - 1}{c_q \theta^{-1}} = q^2,$$

where $c_q = K c_{dE}^2 c_R^2$ and K is a uniform upper bound for the overlap of the sets in the families $(\omega_T : T \in \mathcal{T}_k)$. Assume that the marked elements $\mathcal{M}_k \subset \mathcal{T}_k$ are chosen *such that*

$$\sum_{T \in \mathcal{M}_k} \eta_T^2(u_k) \geq \theta \sum_{T \in \mathcal{T}_k} \eta_T^2(u_k), \tag{Mark}$$

and assume that inner nodes on marked elements and their sides are created. Then we have

$$\|\nabla(u - u_{k+1})\| \leq q\|\nabla(u - u_k)\|$$

for all $k \in \mathbb{N}$.

Proof We first note that Galerkin orthogonality of the error $u - u_{k+1}$ to $\mathscr{S}_0^1(\mathcal{T}_{k+1})$ implies that

$$\|\nabla(u - u_k)\|^2 = \|\nabla(u - u_{k+1})\|^2 + \|\nabla(u_{k+1} - u_k)\|^2.$$

In this identity we successively use the reliability estimate (Rel), the marking condition (Mark), the discrete efficiency estimate (dEff), and the bounded overlap K of patches ω_T, to verify that

$$
\begin{aligned}
\|\nabla(u - u_{k+1})\|^2 &= \|\nabla(u - u_k)\|^2 - \|\nabla(u_k - u_{k+1})\|^2 \\
&\leq c_R^2 \sum_{T \in \mathcal{T}_k} \eta_T^2(u_k) - \|\nabla(u_k - u_{k+1})\|^2 \\
&\leq c_R^2 \theta^{-1} \sum_{T \in \mathcal{M}_k} \eta_T^2(u_k) - \|\nabla(u_k - u_{k+1})\|^2 \\
&\leq c_{dE}^2 c_R^2 \theta^{-1} \sum_{T \in \mathcal{M}_k} \|\nabla(u_k - u_{k+1})\|_{L^2(\omega_T)}^2 - \|\nabla(u_k - u_{k+1})\|^2 \\
&\leq K c_{dE}^2 c_R^2 \theta^{-1} \|\nabla(u_k - u_{k+1})\|^2 - \|\nabla(u_k - u_{k+1})\|^2 \\
&= (c_q \theta^{-1} - 1)\|\nabla(u_k - u_{k+1})\|^2 \\
&= (c_q \theta^{-1} - 1)\left(\|\nabla(u - u_k)\|^2 - \|\nabla(u - u_{k+1})\|^2\right).
\end{aligned}
$$

The condition on θ implies the estimate.   □

Remark 4.10 Note that the theorem does not imply an improvement over uniform refinement strategies, e.g., we may choose $\mathcal{M}_k = \mathcal{T}_k$ in every step.

The inner node property is a necessary condition for strict error reduction.

Fig. 4.20 No new nodes are created in the interior of the elements of $\mathcal{T}_0$ and a strict error reduction fails

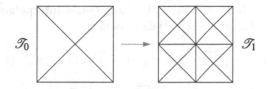

Example 4.1 Let $\mathcal{T}_0$ and $\mathcal{T}_1$ be the triangulations of $\Omega = (0, 1)^2$ shown in Fig. 4.20. If u_0 and u_1 are the corresponding Galerkin approximations of the Poisson problem with $f = 1$, then we have $u_0 = u_1 \neq u$.

4.3.2 Scott–Zhang Quasi-Interpolant

The Clément quasi-interpolant is an essential tool for deriving a posteriori error estimates. It is often desirable to work with a quasi-interpolation operator that is a projection in the sense that it is the identity on $\mathcal{S}^1(\mathcal{T}_h)$. It will be an essential ingredient in the proof of convergence for adaptive methods without an inner node property.

Lemma 4.5 (Dual Basis) *For every $z \in \mathcal{N}_h$, let $S_z \in \mathcal{S}_h$ be such that $z \in S_z$ and with $S_z \subset \Gamma_\mathrm{D}$ if $z \in \Gamma_\mathrm{D}$. Then for every $z \in \mathcal{N}_h$, there exists an affine function $\psi_z \in \mathcal{P}_1(S_z)$ such that*

$$\int_{S_z} \psi_z \varphi_{z'} \, \mathrm{d}s = \delta_{zz'}$$

for all $z' \in \mathcal{N}_h$. Moreover, $\|\psi_z\|_{L^\infty(S_z)} \leq c_{db} h_z^{-(d-1)}$.

Proof Exercise. □

The definition of the Scott–Zhang interpolant uses the dual basis.

Definition 4.10 For every $z \in \mathcal{N}_h$, let $\psi_z \in \mathcal{P}_1(S_z)$ be a dual basis function. For $v \in H^1(\Omega)$, define the *Scott–Zhang quasi-interpolant* $\widetilde{\mathcal{J}}_h : H^1(\Omega) \to \mathcal{S}^1(\mathcal{T}_h)$ by

$$\widetilde{\mathcal{J}}_h v = \sum_{z \in \mathcal{N}_h} v_z \varphi_z, \quad v_z = \int_{S_z} \psi_z v \, \mathrm{d}s.$$

Remark 4.11 Note that $\widetilde{\mathcal{J}}_h$ is well-defined for $v \in H^1(\Omega)$ due to the local trace inequality and $\psi_z \in L^\infty(S_z)$.

The Scott–Zhang quasi-interpolant satisfies similar approximation estimates as the Clément interpolant, but also fulfills a projection property.

Theorem 4.6 (Scott–Zhang Quasi-Interpolant) *There exists $c_{SZ} > 0$, such that for all $v \in H^1(\Omega)$ we have that*

$$\|\nabla \widetilde{\mathcal{J}}_h v\| + \|h_{\mathcal{F}}^{-1}(v - \widetilde{\mathcal{J}}_h v)\| + \|h_{\mathcal{S}}^{-1/2}(v - \widetilde{\mathcal{J}}_h v)\|_{L^2(\bigcup \mathcal{S}_h)} \le c_{SZ}\|\nabla v\|.$$

If $v \in H_D^1(\Omega)$, then $\widetilde{\mathcal{J}}_h v \in \mathscr{S}_D^1(\mathcal{T}_h)$; for every $v_h \in \mathscr{S}^1(\mathcal{T}_h)$, we have $\widetilde{\mathcal{J}}_h v_h = v_h$.

Proof We let $c > 0$ denote a constant that may change in every step.

(i) For every function $v_h \in \mathscr{S}^1(\mathcal{T}_h)$, we have $v_h = \sum_{z' \in \mathcal{N}_h} v_h(z')\varphi_{z'}$, and with the orthogonality properties of the functions ψ_z, $z \in \mathcal{N}_h$, it follows that

$$\widetilde{\mathcal{J}}_h v_h = \sum_{z \in \mathcal{N}_h} \int_{S_z} \psi_z v_h \, ds \, \varphi_z = \sum_{z,z' \in \mathcal{N}_h} v_h(z') \int_{S_z} \psi_z \varphi_{z'} \, ds \, \varphi_z = \sum_{z \in \mathcal{N}_h} v_h(z)\varphi_z,$$

i.e., $\widetilde{\mathcal{J}}_h v_h = v_h$.

(ii) If $v \in H_D^1(\Omega)$, then due to the assumption that $S_z \subset \Gamma_D$ for all $z \in \mathcal{N}_h \cap \Gamma_D$, it follows that $v_z = 0$ for all $z \in \mathcal{N}_h \cap \Gamma_D$ and hence $\widetilde{\mathcal{J}}_h v \in \mathscr{S}_D^1(\mathcal{T}_h)$.

(iii) Using the bound $\|\psi_z\|_{L^\infty(S_z)} \le ch_z^{-(d-1)}$, the estimate $|S_z| \le ch_z^{d-1}$, and the local trace inequality, we infer with Hölder's inequality for $w \in H^1(\Omega)$ that

$$|w_z|^2 \le \|w\|_{L^2(S_z)}^2 \|\psi_z\|_{L^\infty(S_z)}^2 |S_z|$$

$$\le ch_z^{-d+1}\left(h_z^{-1}\|w\|_{L^2(T_z)}^2 + h_z\|\nabla w\|_{L^2(T_z)}^2\right)$$

$$\le ch_z^{-d}\left(\|w\|_{L^2(T_z)}^2 + h_z^2\|\nabla w\|_{L^2(T_z)}^2\right)$$

for $T_z \in \mathcal{T}_h$ with $S_z \subset \partial T_z$. This implies that

$$|w_z|^2 \le ch_z^{-d}\left(\|w\|_{L^2(\omega_{S_z})}^2 + h_{S_z}^2\|\nabla w\|_{L^2(\omega_{S_z})}^2\right).$$

We let ∇^ℓ be the gradient if $\ell = 1$ and the identity if $\ell = 0$. Hence, by an inverse estimate if $\ell = 1$, we have $\|\nabla^\ell \varphi_z^\ell\|_{L^2(T)} \le ch_z^{-\ell+d/2}$, for $\ell = 0, 1$. This allows us to deduce that

$$\|\nabla^\ell \widetilde{\mathcal{J}}_h w\|_{L^2(T)}^2 \le c \sum_{z \in \mathcal{N}_h \cap T} |w_z|^2 \|\nabla^\ell \varphi_z\|_{L^2(\omega_z)}^2$$

$$\le c \sum_{z \in \mathcal{N}_h \cap T} h_z^{d-2\ell} h_z^{-d}\left(\|w\|_{L^2(\omega_{S_z})}^2 + h_{S_z}^2\|\nabla w\|_{L^2(\omega_{S_z})}^2\right)$$

$$\le c \sum_{z \in \mathcal{N}_h \cap T} h_z^{-2\ell}\left(\|w\|_{L^2(\omega_z)}^2 + h_z^2\|\nabla w\|_{L^2(\omega_z)}^2\right).$$

For every $T \in \mathscr{T}_h$ and arbitrary $v_h \in \mathscr{S}^1(\mathscr{T}_h)$ we have, applying the previous estimate to $w = v - v_h$ and using $v_h = \widetilde{\mathscr{J}}_h v_h$, that

$$\|\nabla^\ell(v - \widetilde{\mathscr{J}}_h v)\|_{L^2(T)} = \|\nabla^\ell(v - v_h + v_h - \widetilde{\mathscr{J}}_h v)\|_{L^2(T)}$$

$$\leq \|\nabla^\ell w\|_{L^2(T)} + \|\nabla^\ell \widetilde{\mathscr{J}}_h w\|_{L^2(T)}$$

$$\leq \|\nabla^\ell w\|_{L^2(T)} + c\Big(h_T^{-2\ell} \sum_{z \in \mathscr{N}_h \cap T} \|w\|_{L^2(\omega_z)}^2 + h_T^{2-2\ell}\|\nabla w\|_{L^2(\omega_z)}^2\Big)^{1/2}$$

$$\leq c\big(\|\nabla^\ell(v - v_h)\|_{L^2(\omega_T)} + h_T^{1-\ell}\|\nabla(v - v_h)\|_{L^2(\omega_T)} + h_T^{-\ell}\|v - v_h\|_{L^2(\omega_T)}\big).$$

Choosing $v_h = |\omega_T|^{-1} \int_{\omega_T} v \, dx$, and employing the local Poincaré inequality lead to

$$\|\nabla^\ell(v - \widetilde{\mathscr{J}}_h v)\|_{L^2(T)} \leq c h_T^{1-\ell}\|\nabla v\|_{L^2(\omega_T)}.$$

A summation over $T \in \mathscr{T}_h$ and the finite overlap of $(\omega_T : T \in \mathscr{T}_h)$ imply that

$$\|h_{\mathscr{T}}^{-1}(v - \widetilde{\mathscr{J}}_h v)\| + \|\nabla(v - \widetilde{\mathscr{J}}_h v)\| \leq c\|\nabla v\|.$$

The estimate on the skeleton $\cup \mathscr{S}_h$ follows from the first two bounds, as in the proof of the corresponding estimate for the Clément quasi-interpolant with a trace inequality.   □

Remark 4.12 Since $\widetilde{\mathscr{J}}_h v_h = v_h$ for every $v_h \in \mathscr{S}^1(\mathscr{T}_h)$ and since $\widetilde{\mathscr{J}}_h$ is linear, we have

$$\|\nabla^\ell(v - \widetilde{\mathscr{J}}_h v)\|_{L^2(T)} \leq \|\nabla^\ell(v - v_h)\|_{L^2(T)} + \|\nabla^\ell \widetilde{\mathscr{J}}_h(v - v_h)\|_{L^2(T)}$$

for $\ell = 0, 1$. Choosing $v_h = \mathscr{I}_h v$ if $v \in H^2(\Omega)$, we deduce with nodal interpolation estimates that

$$\|v - \widetilde{\mathscr{J}}_h v\|_{L^2(T)} + h_T\|\nabla(v - \widetilde{\mathscr{J}}_h v)\|_{L^2(T)} \leq c h_T^2 \|D^2 v\|_{L^2(\omega_T)},$$

i.e., the Scott–Zhang quasi-interpolant $\widetilde{\mathscr{J}}_h$ up to an enlarged integration domain on the right-hand side satisfies the same estimates as the nodal interpolation operator $\mathscr{I}_h$.

4.3.3   Discrete Localized Reliability

The Scott–Zhang quasi-interpolant allows us to prove a localized computable upper bound for the difference of two successive Galerkin approximations.

Fig. 4.21 Subset $\mathscr{R}_k$ of
refined elements in the
passage from $\mathscr{T}_k$ to $\mathscr{T}_{k+1}$,
and subdomains Ω_i, $i = 1, 2$,
indicated by different
shadings

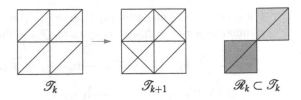

$\mathscr{T}_k \qquad\qquad \mathscr{T}_{k+1} \qquad\qquad \mathscr{R}_k \subset \mathscr{T}_k$

Lemma 4.6 (Discrete Reliability) *Let $\mathscr{T}_{k+1}$ be a refinement of $\mathscr{T}_k$ and let u_{k+1} and u_k be the corresponding Galerkin approximations. Let $\mathscr{R}_k \subset \mathscr{T}_k$ be the subset of refined elements in $\mathscr{T}_k$, i.e.,*

$$\mathscr{R}_k = \mathscr{T}_k \setminus (\mathscr{T}_k \cap \mathscr{T}_{k+1}).$$

Then we have that

$$\|\nabla(u_k - u_{k+1})\|^2 \le c_{dR}^2 \sum_{T \in \mathscr{R}_k} \eta_T^2(u_k). \tag{dRel}$$

Proof Set $\Omega^* = \mathrm{int}\big(\bigcup_{T \in \mathscr{R}_k} T\big)$ and denote by Ω_i, $1 \le i \le I$, the connected components of Ω^*. Let $\mathscr{T}_k^{(i)} = \{T \in \mathscr{T}_k : T \subset \overline{\Omega}_i\}$ be the triangulations of the subdomains, cf. Fig. 4.21. We define $\delta_{k+1} = u_k - u_{k+1} \in \mathscr{S}^1(\mathscr{T}_{k+1})$. and note the important property that

$$\delta_{k+1}|_{\Omega \setminus \Omega^*} \in \mathscr{S}^1(\mathscr{T}_k)|_{\Omega \setminus \Omega^*},$$

i.e., in the unrefined region $\Omega \setminus \Omega^*$ the difference δ_{k+1} belongs to the coarser finite element space. In the remaining parts we apply the Scott–Zhang quasi-interpolant $\widetilde{\mathscr{J}}_k^{(i)}$ related to the subtriangulation $\mathscr{T}_k^{(i)}$ to $\delta_{k+1}|_{\Omega_i}$, $i = 1, 2, \ldots, I$, and set

$$\widetilde{\mathscr{J}}_k \delta_{k+1} = \begin{cases} \widetilde{\mathscr{J}}_k^{(i)} \delta_{k+1}|_{\Omega_i} & \text{in } \Omega_i, \ 1 \le i \le I, \\ \delta_{k+1} & \text{in } \Omega \setminus \Omega^*. \end{cases}$$

Since the Scott–Zhang quasi-interpolant preserves piecewise affine boundary data, we have that $\widetilde{\mathscr{J}}_k \delta_{k+1} \in C(\overline{\Omega})$. In particular, we have $\widetilde{\mathscr{J}}_k \delta_{k+1} \in \mathscr{S}_0^1(\mathscr{T}_k)$. By definition of u_k and u_{k+1}, and since $\mathscr{S}_0^1(\mathscr{T}_k) \subset \mathscr{S}_0^1(\mathscr{T}_{k+1})$ we have

$$\int_\Omega \nabla \delta_{k+1} \cdot \nabla v_k \, dx = \int_\Omega \nabla(u_k - u_{k+1}) \cdot \nabla v_k \, dx = 0,$$

for every $v_k \in \mathscr{S}_0^1(\mathscr{T}_k)$. This implies that

$$\|\nabla \delta_{k+1}\|^2 = \int_\Omega \nabla(u_k - u_{k+1}) \cdot \nabla(\delta_{k+1} - \widetilde{\mathscr{J}}_k \delta_{k+1}) \, dx.$$

The variational formulation satisfied by u_{k+1}, and a $\mathcal{T}_k$-elementwise integration-by-parts lead to

$$\|\nabla\delta_{k+1}\|^2 = \int_\Omega \nabla u_k \cdot \nabla(\delta_{k+1} - \widetilde{\mathcal{J}}_k\delta_{k+1})\,dx - \int_\Omega f(\delta_{k+1} - \widetilde{\mathcal{J}}_k\delta_{k+1})\,dx$$

$$= \sum_{T\in\mathcal{T}_k}\left(\int_{\partial T}(\nabla u_k \cdot n_T)(\delta_{k+1} - \widetilde{\mathcal{J}}_k\delta_{k+1})\,ds\right) - \int_\Omega f(\delta_{k+1} - \widetilde{\mathcal{J}}_k\delta_{k+1})\,dx$$

$$= \sum_{T\in\mathcal{T}_k} -\int_T f(\delta_{k+1} - \widetilde{\mathcal{J}}_k\delta_{k+1})\,dx - \sum_{S\in\mathcal{S}_k}\int_S [\![\nabla u_k \cdot n_S]\!](\delta_{k+1} - \widetilde{\mathcal{J}}_k\delta_{k+1})\,ds.$$

Noting that $(\delta_{k+1} - \widetilde{\mathcal{J}}_k\delta_{k+1}) = 0$ in $\Omega \setminus \Omega^*$ and employing the interpolation estimates for $\mathcal{J}_k$, we argue, as in the proof of the continuous reliability estimate of Proposition 4.3 to verify the bound. $\qquad\square$

Remark 4.13 It is important to note that the upper bound in the proposition is localized to those elements that are refined to obtain $\mathcal{T}_{k+1}$. This is remarkable, since in general we have $u_{k+1} \neq u_k$ in the unrefined region $\Omega \setminus \Omega^*$.

4.3.4 General Convergence

Example 4.1 showed that a strict error reduction cannot be expected for adaptive and even uniform refinements in general. An important observation in that example is that the error estimator decays. To generalize this observation, we redefine the quantity h_T that measures the size of an element by setting

$$h_T = |T|^{1/d}$$

for every element T in a triangulation $\mathcal{T}_k$. We have the equivalence

$$c_{loc}\,\mathrm{diam}(T) \le h_T \le \mathrm{diam}(T)$$

for elements in a regular family of triangulations. In particular, all previous estimates remain valid with this definition of h_T. The important difference is that if T is bisected into subelements $T_1, T_2, \ldots, T_L \subset T$, then this element size strictly decreases, i.e.,

$$h_{T_\ell} \le 2^{-1/2} h_T$$

which is in general not the case for the diameters. We also use the new definition of h_T to define the error indicators. For a triangulation $\mathscr{T}_k$ of Ω, an element $T \in \mathscr{T}_k$ and a function $v \in H^1(\Omega)$ we set

$$\eta_{k,T}^2(v) = h_T^2 \|f\|_{L^2(T)}^2 + h_T \sum_{S \in \mathscr{S}_k, S \subset \partial T} \|[\![\nabla v \cdot n_S]\!]\|_{L^2(S)}^2,$$

where $[\![\nabla v \cdot n_S]\!] = 0$ if $S \subset \Gamma_D$, and

$$\eta_k^2(v) = \sum_{T \in \mathscr{T}_k} \eta_{k,T}^2(v).$$

The new definition of h_T leads to the following reduction property.

Lemma 4.7 (Reduction) *Let $T \in \mathscr{T}_k$ and $T_1, T_2, \ldots, T_L \in \mathscr{T}_{k+1}$ with $L \geq 2$ such that $T_1 \cup T_2 \cup \cdots \cup T_L \subset T$. Then*

$$\sum_{T' \in \{T_1, \ldots, T_L\}} \eta_{k+1,T'}^2(u_k) \leq \frac{1}{\sqrt{2}} \eta_{k,T}^2(u_k).$$

Proof Note that ∇u_k is constant in T and hence does not have a jump across sides within T. We thus have

$$\sum_{T' \in \{T_1, \ldots, T_L\}} \eta_{k+1,T'}^2(u_k)$$

$$= \sum_{T' \in \{T_1, \ldots, T_L\}} \left(h_{T'}^2 \|f\|_{L^2(T')}^2 + h_{T'} \sum_{S' \in \mathscr{S}_{k+1}, S' \subset \partial T'} \|[\![\nabla u_k \cdot n_{S'}]\!]\|_{L^2(S')}^2 \right)$$

$$\leq \frac{1}{\sqrt{2}} \left(h_T^2 \|f\|_{L^2(T)}^2 + h_T \sum_{S \in \mathscr{S}_k, S \subset \partial T} \|[\![\nabla u_k \cdot n_S]\!]\|_{L^2(S)}^2 \right) = \frac{1}{\sqrt{2}} \eta_{k,T}^2(u_k),$$

which proves the estimate.   □

Lemma 4.8 (Perturbation Inequality) *For all $v, w \in \mathscr{S}_0^1(\mathscr{T}_k)$ and $\alpha > 0$, we have*

$$\eta_k^2(v) \leq (1 + \alpha)\, \eta_k^2(w) + (1 + \alpha^{-1}) c_{pi}^2 \|\nabla(v - w)\|^2.$$

Proof If $T, \widetilde{T} \in \mathscr{T}_k$ and $S \in \mathscr{S}_k$ are such that $S = T \cap \widetilde{T}$ we have, using $|a \cdot n_S| \le |a|$ for every $a \in \mathbb{R}^d$,

$$\|[\nabla v \cdot n_S]\|_{L^2(S)} = |S|^{1/2} |(\nabla v|_T - \nabla v|_{\widetilde{T}}) \cdot n_S|$$

$$\le |S|^{1/2} \Big(|(\nabla w|_T - \nabla w|_{\widetilde{T}}) \cdot n_S| + |\nabla(v-w)|_T| + |\nabla(v-w)|_{\widetilde{T}}| \Big)$$

$$= \|[\nabla w \cdot n_S]\|_{L^2(S)} + |S|^{1/2} |T|^{-1/2} \|\nabla(v-w)\|_{L^2(T)}$$

$$\qquad\qquad + |S|^{1/2} |\widetilde{T}|^{-1/2} \|\nabla(v-w)\|_{L^2(\widetilde{T})}$$

$$\le \|[\nabla w \cdot n_S]\|_{L^2(S)} + c_{\ell oc} h_T^{-1/2} \Big(\|\nabla(v-w)\|_{L^2(T)} + \|\nabla(v-w)\|_{L^2(\widetilde{T})} \Big),$$

where we used $|S|^{1/2} \le c_{\ell oc} h_T^{(d-1)/2}$ and $h_T^d \le c_{\ell oc} |T|, |\widetilde{T}|$. This leads to

$$h_T^{1/2} \|[\nabla v \cdot n_S]\|_{L^2(S)} \le h_T^{1/2} \|[\nabla w \cdot n_S]\|_{L^2(S)} + c_{\ell oc} \|\nabla(v-w)\|_{L^2(\omega_S)}.$$

Using Young's inequality in the form $(a+b)^2 \le (1+\alpha)a^2 + (1+\alpha^{-1})b^2$, and the finite overlap K of side patches, we find that

$$\eta_k^2(v) = \sum_{T \in \mathscr{T}_k} \Big(h_T^2 \|f\|_{L^2(T)}^2 + h_T \sum_{S \in \mathscr{S}_k, S \subset \partial T} \|[\nabla v \cdot n_S]\|_{L^2(S)}^2 \Big)$$

$$\le \sum_{T \in \mathscr{T}_k} \Big(h_T^2 \|f\|_{L^2(T)}^2 + h_T \sum_{S \in \mathscr{S}_k, S \subset \partial T} (1+\alpha) \|[\nabla w \cdot n_S]\|_{L^2(S)}^2 \Big)$$

$$\qquad\qquad + \sum_{T \in \mathscr{T}_k} \sum_{S \in \mathscr{S}_k, S \subset \partial T} (1+\alpha^{-1}) c_{\ell oc}^2 \|\nabla(v-w)\|_{L^2(\omega_S)}^2$$

$$\le (1+\alpha)\, \eta_k^2(w) + (1+\alpha^{-1}) c_{\ell oc}^2 K \|\nabla(v-w)\|^2.$$

This proves the asserted estimate. $\qquad\qquad\qquad\qquad\qquad\qquad\qquad\qquad\square$

For the proof of the general convergence result, in addition to the reduction and perturbation properties of the previous two lemmas, we will use the reliability (Rel), the marking condition (Mark), and the Galerkin orthogonality

$$\|\nabla(u - u_{k+1})\|^2 = \|\nabla(u - u_k)\|^2 - \|\nabla(u_k - u_{k+1})\|^2. \tag{GO}$$

Theorem 4.7 (General Contraction) *Let $(\mathscr{T}_k)_{k=1,2,\dots}$ be a sequence of adaptively refined triangulations with corresponding Galerkin approximations $(u_k)_{k=1,2,\dots} \subset H_0^1(\Omega)$. There exist $\gamma > 0$ and $q < 1$ such that*

$$\big(\|\nabla(u - u_{k+1})\|^2 + \gamma\, \eta_{k+1}^2(u_{k+1}) \big)^{1/2} \le q \big(\|\nabla(u - u_k)\|^2 + \gamma\, \eta_k^2(u_k) \big)^{1/2}.$$

Proof We let $\cup \mathcal{M}_k$ denote the subdomain of all elements that are marked for refinement. We then have by the reduction property of Lemma 4.7 and the marking condition (Mark) that

$$
\begin{aligned}
\eta_{k+1}^2(u_k) &= \sum_{T' \in \mathscr{T}_{k+1}, T' \subset \cup \mathcal{M}_k} \eta_{k+1,T'}^2(u_k) + \sum_{T' \in \mathscr{T}_{k+1}, T' \subset \Omega \setminus \cup \mathcal{M}_k} \eta_{k+1,T'}^2(u_k) \\
&\leq 2^{-1/2} \sum_{T \in \mathscr{T}_k, T \subset \cup \mathcal{M}_k} \eta_{k,T}^2(u_k) + \sum_{T \in \mathscr{T}_k, T \not\subset \mathcal{M}_k} \eta_{k,T}^2(u_k) \\
&\leq 2^{-1/2} \sum_{T \in \mathcal{M}_k} \eta_{k,T}^2(u_k) + \sum_{T \in \mathscr{T}_k \setminus \mathcal{M}_k} \eta_{k,T}^2(u_k) \\
&\leq \eta_k^2(u_k) - \left(1 - 2^{-1/2}\right) \sum_{T \in \mathcal{M}_k} \eta_{k,T}^2(u_k) \\
&\leq \left(1 - \theta\left(1 - 2^{-1/2}\right)\right) \eta_k^2(u_k) = \vartheta \, \eta_k^2(u_k).
\end{aligned}
$$

With the perturbation inequality of Lemma 4.8, we deduce that

$$
\eta_{k+1}^2(u_{k+1}) \leq (1 + \alpha) \vartheta \, \eta_k^2(u_k) + (1 + \alpha^{-1}) c_{pi}^2 \| \nabla(u_{k+1} - u_k) \|^2.
$$

For every $\gamma > 0$ with $\gamma(1 + \alpha^{-1}) c_{pi}^2 < 1$, the Galerkin orthogonality (GO) and the reliability estimate (Rel) lead to

$$
\begin{aligned}
\| \nabla(u &- u_{k+1}) \|^2 + \gamma \, \eta_{k+1}^2(u_{k+1}) \\
&\leq \| \nabla(u - u_k) \|^2 - \| \nabla(u_k - u_{k+1}) \|^2 \\
&\quad + \gamma(1 + \alpha) \vartheta \, \eta_k^2(u_k) + \gamma(1 + \alpha^{-1}) c_{pi}^2 \| \nabla(u_{k+1} - u_k) \|^2 \\
&= \| \nabla(u - u_k) \|^2 + \left(\gamma(1 + \alpha^{-1}) c_{pi}^2 - 1\right) \| \nabla(u_{k+1} - u_k) \|^2 + \gamma(1 + \alpha) \vartheta \, \eta_k^2(u_k) \\
&\leq \| \nabla(u - u_k) \|^2 + \gamma(1 + \alpha) \vartheta \, \eta_k^2(u_k) \\
&= \| \nabla(u - u_k) \|^2 - \gamma \beta \, \eta_k^2(u_k) + \gamma\left((1 + \alpha) \vartheta + \beta\right) \eta_k^2(u_k) \\
&\leq \left(1 - \gamma \beta c_R^{-2}\right) \| \nabla(u - u_k) \|^2 + \gamma\left((1 + \alpha) \vartheta + \beta\right) \eta_k^2(u_k).
\end{aligned}
$$

The parameters $\alpha, \beta, \gamma > 0$ are specified as follows:

- choose $\alpha > 0$ such that $(1 + \alpha)\vartheta < 1$,
- choose $\gamma > 0$ such that $\gamma(1 + \alpha^{-1}) c_{pi}^2 < 1$,
- choose $\beta > 0$ such that $(1 + \alpha)\vartheta + \beta < 1$ and $\gamma \beta c_R^{-2} < 1$.

For $q^2 = \max\{1 - \gamma \beta c_R^{-2}, \gamma((1 + \alpha)\vartheta + \beta)\} < 1$, we obtain the asserted estimate.

$\square$

Remarks 4.14

(i) The proof of the general contraction property does not use the efficiency of the estimator. In particular, we do not have to assume that f is elementwise constant.
(ii) Note that if the error does not decrease in an iteration, then the estimator does.

4.3.5 Quasioptimal Meshes

We next investigate whether the triangulations generated by the adaptive algorithm are quasioptimal, i.e., that no overrefinement occurs asymptotically. For this, the set of marked elements has to be chosen in an optimal way. We focus on the convergence analysis and only cite details about precise properties of bisection procedures.

Lemma 4.9 (Closure Control, See [25]) *Let $(\mathscr{T}_k)_{k\in\mathbb{N}}$ be a sequence of regular triangulations obtained by bisection of elements in $\mathscr{M}_k \subset \mathscr{T}_k$ and further elements to avoid hanging nodes from an initial triangulation $\mathscr{T}_0$. For an appropriate choice of reference edges in $\mathscr{T}_0$, there exists a constant $c_{cc} > 0$ such that*

$$|\mathscr{T}_k| - |\mathscr{T}_0| \leq c_{cc} \sum_{j=0}^{k-1} |\mathscr{M}_j|.$$

We assume that the sets of marked elements $\mathscr{M}_k \subset \mathscr{T}_k$ are chosen in an optimal way with respect to cardinality, i.e., that

$$\mathscr{M}_k \subset \mathscr{T}_k \text{ minimal with } \sum_{T\in\mathscr{M}_k} \eta_T^2(u_k) \geq \theta\, \eta_k^2(u_k). \qquad \text{(Mark')}$$

We prove below that in this case, we have the following control on the cardinality of $\mathscr{M}_k$,

$$|\mathscr{M}_k| \leq c_{card}\big(\|\nabla(u - u_k)\|^2 + \gamma\, \eta_k^2(u_k)\big)^{-d/s}, \qquad \text{(Card)}$$

where the parameter $s > 0$ is related to the approximability of the exact solution in the class of regular triangulations obtained by bisections from $\mathscr{T}_0$. This leads to the following theorem.

Theorem 4.8 (Quasioptimality) *Assume that f is elementwise constant on $\mathscr{T}_0$, and $s > 0$ is such that for every $\varepsilon > 0$, there exists a regular triangulation $\mathscr{T}_\varepsilon$ obtained from $\mathscr{T}_0$ by bisection such that*

$$|\mathscr{T}_\varepsilon| - |\mathscr{T}_0| \leq c_u \varepsilon^{-d/s}, \quad \|\nabla(u - u_\varepsilon)\| \leq \varepsilon$$

with the Galerkin approximation $u_\varepsilon \in \mathscr{S}_0^1(\mathscr{T}_\varepsilon)$. *Suppose that* (Mark') *and* (Card) *hold for all* $k \geq 0$ *and that* $\theta \leq (c_{EC}c_{dR})^{-2}$. *Then we have*

$$\|\nabla(u - u_k)\| \leq c_{qo}(|\mathscr{T}_k| - |\mathscr{T}_0|)^{-s/d}.$$

Proof By the contraction property of Theorem 4.7 we have

$$e_k \leq q^{k-j}e_j$$

for some $q < 1$, $k \geq j$, and $e_j^2 = \|\nabla(u - u_j)\|^2 + \gamma\eta_j^2(u_j)$. Therefore, using Lemma 4.9, the optimal marking condition (Mark'), and the cardinality control (Card), we deduce that

$$|\mathscr{T}_k| - |\mathscr{T}_0| \leq c_{cc}\sum_{j=0}^{k-1}|\mathscr{M}_j| \leq c_{cc}\sum_{j=0}^{k-1}e_j^{-d/s} \leq c_{cc}c_{\text{card}}\sum_{j=0}^{k-1}q^{-d(j-k)/s}e_k^{-d/s}$$

$$\leq c_{cc}c_{\text{card}}\,e_k^{-d/s}\sum_{j=0}^{k-1}q^{-d(j-k)/s} \leq \frac{c_{cc}c_{\text{card}}\,e_k^{-d/s}}{1 - q^{d/s}} = c_{qo}e_k^{-d/s},$$

which proves the estimate.   □

Remarks 4.15

(i) Note that $h_k = (|\mathscr{T}_k| - |\mathscr{T}_0|)^{-1/d}$ is the average mesh-size of the locally refined triangulation $\mathscr{T}_k$ so that we have

$$\|\nabla(u - u_k)\| \leq c_{qo}h_k^s.$$

Due to the results on the approximation of the Poisson problem on graded triangulations, we expect $s = 1$. With respect to the numbers of degrees of freedom, the approximation is then optimal.

(ii) If $c_{EC}c_{dR} \gg 1$, then the adaptive algorithm marks only a small number of elements, which may be inefficient.

The coarsest common refinement of regular triangulations $\mathscr{T}_A$ and $\mathscr{T}_B$, obtained from $\mathscr{T}_0$ by bisection, is denoted by

$$\mathscr{T}_{A\oplus B} = \mathscr{T}_A \oplus \mathscr{T}_B,$$

cf. Fig. 4.22. Its cardinality is controlled as follows; we refer to the references for a result of the statement.

Lemma 4.10 (Overlay, See [25]) *For regular triangulations* $\mathscr{T}_A$, $\mathscr{T}_B$, *obtained by bisections from* $\mathscr{T}_0$, *we have*

$$|\mathscr{T}_{A\oplus B}| \leq |\mathscr{T}_A| + |\mathscr{T}_B| - |\mathscr{T}_0|.$$

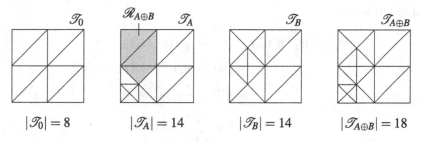

$$|\mathcal{T}_0| = 8 \qquad |\mathcal{T}_A| = 14 \qquad |\mathcal{T}_B| = 14 \qquad |\mathcal{T}_{A\oplus B}| = 18$$

Fig. 4.22 The cardinality of the coarsest common refinement $\mathcal{T}_{A\oplus B}$ of two triangulations $\mathcal{T}_A$ and $\mathcal{T}_B$, obtained from an initial triangulation $\mathcal{T}_0$ by bisection, satisfies $|\mathcal{T}_{A\oplus B}| \le |\mathcal{T}_A| + |\mathcal{T}_B| - |\mathcal{T}_0|$

If $\mathcal{R}_{A\oplus B} = \mathcal{T}_A \setminus (\mathcal{T}_A \cap \mathcal{T}_{A\oplus B}) \subseteq \mathcal{T}_A$ *is the subset of elements that needs to be refined to pass from* $\mathcal{T}_A$ *to* $\mathcal{T}_{A\oplus B}$, *we have*

$$|\mathcal{R}_{A\oplus B}| + |\mathcal{T}_A| \le |\mathcal{T}_{A\oplus B}|.$$

For the proof of (Card), we incorporate the efficiency estimate

$$c_E^{-2}\eta_k^2(u_k) \le \|\nabla(u - u_k)\|^2. \tag{Eff}$$

Applying Lemma 4.6 with $\mathcal{T}_{k+1}$ and u_{k+1} replaced by $\mathcal{T}_*$ and u_* shows that

$$\|\nabla(u_* - u_k)\|^2 \le c_{dR}^2 \sum_{T \in \mathcal{R}_{k\to*}} \eta_{k,T}^2(u_k), \tag{dRel}$$

where $\mathcal{R}_{k\to*} = \mathcal{T}_k \setminus (\mathcal{T}_k \cap \mathcal{T}_*)$.

Lemma 4.11 (Localization) *Suppose that* $\theta \le \theta_* = (c_E c_{dR})^{-2}$. *Let* $\mathcal{T}_*$ *be a refinement of* $\mathcal{T}_k$, *and let* $u_k \in \mathcal{S}_0^1(\mathcal{T}_k)$ *and* $u_* \in \mathcal{S}_0^1(\mathcal{T}_*)$ *be the corresponding Galerkin approximations. Assume that*

$$\|\nabla(u - u_*)\| \le \tilde{q}\|\nabla(u - u_k)\|$$

with $\tilde{q}^2 = (1 - \theta/\theta_*) < 1$. *Then we have*

$$\theta\,\eta_k^2(u_k) \le \sum_{T \in \mathcal{R}_{k\to*}} \eta_{k,T}^2(u_k).$$

Proof By (Eff), Galerkin orthogonality, i.e.,

$$\|\nabla(u - u_k)\|^2 = \|\nabla(u - u_*)\|^2 + \|\nabla(u_* - u_k)\|^2,$$

and (dRel) we have

$$
\begin{aligned}
(1 - \tilde{q}^2) c_E^{-2} \eta_k^2(u_k) &\leq (1 - \tilde{q}^2) \|\nabla(u - u_k)\|^2 \\
&\leq \|\nabla(u - u_k)\|^2 - \|\nabla(u - u_*)\|^2 \\
&= \|\nabla(u_* - u_k)\|^2 \\
&\leq c_{dR}^2 \sum_{T \in \mathscr{R}_{k \to *}} \eta_{k,T}^2(u_k).
\end{aligned}
$$

The choice of $\tilde{q}$ shows that

$$
\frac{1 - \tilde{q}^2}{c_E^2 c_{dR}^2} = \frac{\theta}{\theta_*} \frac{1}{c_E^2 c_{dR}^2} = \theta,
$$

and this proves the estimate.   □

Proposition 4.4 (Optimal Marking) *If* $\theta \leq (c_E c_{dR})^{-2}$ *and* $\mathscr{M}_k \subset \mathscr{T}_k$ *satisfies* (Mark'), *then we have*

$$
|\mathscr{M}_k| \leq c_{\mathrm{card}} e_k^{-d/s},
$$

where $e_k^2 = \|\nabla(u - u_k)\|^2 + \gamma \eta_k^2(u_k)$.

Proof For $\varepsilon > 0$, let $\mathscr{T}_\varepsilon$ be as in Theorem 4.8, and set $\mathscr{T}_* = \mathscr{T}_k \oplus \mathscr{T}_\varepsilon$. Let u_ε and u_* be the corresponding Galerkin approximations. By Lemma 4.10 we have

$$
|\mathscr{R}_{k \to *}| + |\mathscr{T}_k| \leq |\mathscr{T}_*| \leq |\mathscr{T}_k| + |\mathscr{T}_\varepsilon| - |\mathscr{T}_0|,
$$

i.e.,

$$
|\mathscr{R}_{k \to *}| \leq |\mathscr{T}_\varepsilon| - |\mathscr{T}_0| \leq c_u \varepsilon^{-d/s}.
$$

Since $\mathscr{S}_0^1(\mathscr{T}_\varepsilon) \subset \mathscr{S}_0^1(\mathscr{T}_*)$, we have by Céa's lemma that

$$
\|\nabla(u - u_*)\| \leq \|\nabla(u - u_\varepsilon)\| \leq \varepsilon.
$$

With the contraction factor $\tilde{q} < 1$ of Lemma 4.11 and $\varepsilon = \tilde{q} e_k$, we are in the situation of Lemma 4.11 so that

$$
\sum_{T \in \mathscr{R}_{k \to *}} \eta_{k,T}^2(u_k) \geq \theta \, \eta_k^2(u_k).
$$

Thus $\mathscr{R}_{k\to*}$ satisfies the marking criterion and since $\mathscr{M}_k$ is assumed to be optimal among sets with this property, we have

$$|\mathscr{M}_k| \le |\mathscr{R}_{k\to*}|,$$

which by definition of ε implies the estimate. □

4.4 Adaptivity for the Heat Equation

4.4.1 Abstract Error Estimate

We consider the linear heat equation

$$\begin{cases} \partial_t u(t,x) - \Delta u(t,x) = f(t,x) & \text{for all } (t,x) \in (0,T] \times \Omega, \\ u(t,x) = 0 & \text{for all } (t,x) \in (0,T] \times \partial\Omega, \\ u(0,x) = u_0(x) & \text{for all } x \in \Omega. \end{cases}$$

Here, $u_0 \in H_0^1(\Omega)$ and $f \in C(0,T;L^2(\Omega))$ are given functions. A *weak solution* $u : [0,T] \times \Omega \to \mathbb{R}$ is required to satisfy $u(0,\cdot) = u_0$, $u(t,\cdot)|_{\partial\Omega} = 0$ for all $t \in [0,T]$, and

$$\big(\partial_t u(t,\cdot),v\big) + \big(\nabla u(t,\cdot),\nabla v\big) - \big(f,v\big)$$

for all $t \in (0,T)$ and all $v \in H_0^1(\Omega)$, with the L^2-inner product

$$(\phi,\psi) = \int_\Omega \phi \cdot \psi \, \mathrm{d}x$$

for functions or vector fields $\phi, \psi \in L^2(\Omega;\mathbb{R}^m)$. An appropriate function space for the solution is

$$\mathscr{X} = H^1(0,T;L^2(\Omega)) \cap L^\infty(0,T;H_0^1(\Omega)),$$

which consists of all functions $w : (0,T) \times \Omega \to \mathbb{R}$ that are weakly differentiable in time and space, such that $w(t,\cdot)|_{\partial\Omega} = 0$ for almost every $t \in (0,T)$ and

$$\int_0^T \|\partial_t w(t,\cdot)\|^2 \, \mathrm{d}t + \operatorname{ess\,sup}_{t\in[0,T]} \|\nabla w(t,\cdot)\|^2 < \infty.$$

The existence of a unique weak solution can be established by a discretization of the above weak formulation in time and a subsequent passage to a limit. We assume that we are given an approximation $U \in \mathcal{X}$ of the exact solution and aim at controlling the approximation error $u - U$ in appropriate norms.

Definition 4.11 For an arbitrary function $U \in \mathcal{X}$, we define its residual $\mathcal{R}_U$: $[0, T] \to H_0^1(\Omega)'$ by

$$\langle \mathcal{R}_U(t), v \rangle = (\partial_t U(t, \cdot), v) + (\nabla U(t, \cdot), \nabla v) - (f(t), v)$$

for $t \in [0, T]$ and $v \in H_0^1(\Omega)$.

The operator norm of a residual is for $t \in [0, T]$ given by

$$\|\mathcal{R}_U(t)\|_* = \sup_{v \in H_0^1(\Omega) \setminus \{0\}} \frac{\langle \mathcal{R}_U(t), v \rangle}{\|\nabla v\|}.$$

Here we have used the norm $v \mapsto \|\nabla v\|$ on $H_0^1(\Omega)$. By definition, the residual measures the deviation of U in satisfying the weak formulation of the heat equation.

Proposition 4.5 (Error Estimate) *Let u be the weak solution of the heat equation and U an approximation. For the error $e = u - U$, we have that*

$$\sup_{t \in (0,T)} \|e(t)\|^2 + \int_0^T \|\nabla e(t)\|^2 \, dt \leq 2\|u_0 - U(0)\|^2 + 2\int_0^T \|\mathcal{R}_U(t)\|_*^2 \, dt.$$

Proof The definition of $\mathcal{R}_U$ and the weak formulation of the heat equation imply that we have

$$(\partial_t e(t), v) + (\nabla e(t), \nabla v) = -\langle \mathcal{R}_U(t), v \rangle$$

for all $v \in H_0^1(\Omega)$. We choose $v = e(t)$, and note the product rule

$$(\partial_t e(t), e(t)) = \frac{1}{2} \frac{d}{dt} \|e(t)\|^2,$$

to deduce that

$$\frac{1}{2} \frac{d}{dt} \|e(t)\|^2 + \|\nabla e(t)\|^2 \leq \|\mathcal{R}_U(t)\|_* \|\nabla e(t)\| \leq \frac{1}{2} \|\mathcal{R}_U(t)\|_*^2 + \frac{1}{2} \|\nabla e(t)\|^2.$$

Absorbing $\|\nabla e(t)\|^2/2$ and integrating over $(0, T')$ implies the estimate. $\qquad \square$

4.4.2 Residual Bounds

In order to derive computable bounds for the residual of the approximation, we need to specify how it is obtained. We investigate numerical solutions that are computed with an implicit Euler scheme. For this, we assume that we are given a sequence of step-sizes $(\tau_j)_{j=1,\dots,J}$ and time-steps

$$t_j = \sum_{\ell=1}^{j} \tau_\ell$$

for $j = 0, 1, \dots, J$, with the conventions that $t_0 = 0$ and $t_J = T$. Associated with the time-steps $(t_j)_{j=0,\dots,J}$ is a sequence of regular triangulations $(\mathcal{T}^j)_{j=0,\dots,J}$ of Ω. We let $\mathscr{I}_j : C(\overline{\Omega}) \to \mathscr{S}^1(\mathcal{T}^j)$ denote the nodal interpolation operator on $\mathcal{T}^j$ for $j = 0, 1, \dots, J$.

Algorithm 4.3 (Implicit Euler Scheme) *Let $(\tau_j)_{j=1,\dots,J}$ be positive step-sizes and let $(\mathcal{T}^j)_{j=0,\dots,J}$ be regular triangulations of Ω. Let $U^0 \in \mathscr{S}_0^1(\mathcal{T}^0)$ and set $j = 1$.*

(1) Compute the solution $U^j \in \mathscr{S}_0^1(\mathcal{T}^j)$ of

$$\tau_j^{-1}\big(U^j - \mathscr{I}_j U^{j-1}, V\big) + (\nabla U^j, \nabla V) = (f^j, V)$$

for all $V \in \mathscr{S}_0^1(\mathcal{T}^j)$ with $f^j = f(t_j, \cdot)$.
(2) Stop if $t_j = T$; set $j \to j + 1$ and continue with (1) otherwise.

Remarks 4.16

(i) The Lax–Milgram lemma implies the existence of a unique sequence $(U^j)_{j=0,\dots,J}$ that satisfies the equations of the algorithm.
(ii) The incorporation of the nodal interpolation operator $\mathscr{I}_j$ allows for a simple computation of the products $(\mathscr{I}_j U^{j-1}, V)$ when the triangulations $(\mathcal{T}^j)_{j=0,\dots,j}$ are not nested.

With the iterates $(U^j)_{j=0,\dots,J}$, we construct an approximation $U \in \mathscr{X}$.

Definition 4.12 For time-steps $(t_j)_{j=0,\dots,J}$ and functions $(U^j)_{j=0,\dots,J}$, we define the *affine interpolant* $U \in \mathscr{X}$ for $t \in [t_{j-1}, t_j], j = 1, \dots, J$, and $x \in \Omega$ by

$$U(t, x) = \frac{t - t_{j-1}}{\tau_j} U^j(x) + \frac{t_j - t}{\tau_j} U^{j-1}(x).$$

We remark that we have $U(t_j) = U^j$ for $j = 0, 1, \dots, J$ and

$$\partial_t U(t) = \tau_j^{-1}(U^j - U^{j-1})$$

for $t \in (t_{j-1}, t_j)$ and $j = 1, 2, \dots, J$.

Proposition 4.6 (Residual Bound) *Assume that $f \in C^1(0, T; L^2(\Omega))$. We then have*

$$\|\mathscr{R}_U(t)\|_* \leq c_{C\ell}\, \eta^j_{\text{space}} + c_P\, \eta^j_{\text{coarse}} + \eta^j_{\text{temp}}$$

for $t \in (t_{j-1}, t_j)$, where

$$\eta^j_{\text{space}} = \left\| h_{\mathscr{T}^j}\big(\tau_j^{-1}[U^j - \mathscr{I}_j U^{j-1}] - f(t_j)\big)\right\| + \left\| h^{1/2}_{\mathscr{S}^j}[\nabla U^j \cdot n_{\mathscr{S}^j}]\right\|_{L^2(\bigcup \mathscr{S}^j)},$$

$$\eta^j_{\text{coarse}} = \tau_j^{-1}\|U^{j-1} - \mathscr{I}_j U^{j-1}\|,$$

$$\eta^j_{\text{temp}} = \|\nabla(U^j - U^{j-1})\| + c_P \tau_j \sup_{t \in (t_{j-1}, t_j)}\|\partial_t f(t, \cdot)\|.$$

Proof For $t \in (t_{j-1}, t_j)$ and $v \in H^1_0(\Omega)$, we have

$$\langle\mathscr{R}_U(t), v\rangle = \tau_j^{-1}(U^j - U^{j-1}, v) + (\nabla U(t), \nabla v) - (f(t), v)$$

$$= \Big(\tau_j^{-1}(U^j - \mathscr{I}_j U^{j-1}, v) + (\nabla U^j, \nabla v) - (f(t_j), v)\Big)$$

$$\quad + \tau_j^{-1}(\mathscr{I}_j U^{j-1} - U^{j-1}, v) + \Big((\nabla[U(t) - U^j], \nabla v) + (f(t_j) - f(t), v)\Big)$$

$$= I + II + III.$$

By the equation of the implicit Euler scheme, we have, for every $V \in \mathscr{S}^1_0(\mathscr{T}^j)$, that

$$I = \tau_j^{-1}(U^j - \mathscr{I}_j U^{j-1}, v - V) + (\nabla U^j, \nabla(v - V)) - (f(t_j), v - V).$$

Letting $V = \mathscr{J}_h v$ be the Clément quasi-interpolant of v and integrating-by-parts on every element in $\mathscr{T}^j$, we find, as in the proof of a posteriori error estimate for the Poisson problem of Theorem 4.3, that

$$I \leq c_{C\ell}\, \eta^j_{\text{space}}\|\nabla v\|.$$

Hölder and Poincaré inequalities lead to the estimate

$$II \leq \tau_j^{-1}\|U^{j-1} - \mathscr{I}_j U^{j-1}\|\|v\| \leq c_P \tau_j^{-1}\|U^{j-1} - \mathscr{I}_j U^{j-1}\|\|\nabla v\|.$$

Similarly, employing the mean value theorem we find that

$$III \leq \big(\|\nabla[U(t) - U^j]\| + c_P\|f(t_j) - f(t)\|\big)\|\nabla v\|$$

$$\leq \Big(\|\nabla[U^{j-1} - U^j]\| + c_P \tau_j \sup_{t \in (t_{j-1}, t_j)}\|\partial_t f\|\Big)\|\nabla v\|.$$

The combination of the bounds proves the estimate. $\square$

Remark 4.17 The error quantities η^j_{space}, η^j_{coarse}, η^j_{temp} reflect error contributions related to spatial discretization, mesh coarsening, and temporal discretization, respectively.

4.4.3 Adaptive Time-Stepping

For evolution problems like the heat equation, the simple adaptive refinement of triangulations cannot be expected to be efficient since local features of solutions advance in time and space. Therefore, we include a mesh coarsening step in the following adaptive algorithm.

Algorithm 4.4 (Time-Space Adaptive Algorithm) *Let $\mathscr{T}^0$ be a triangulation of Ω, $U^0 \in \mathscr{S}^1_0(\mathscr{T}^0)$ an approximation of u_0 such that $\|U^0 - u_0\| \leq \varepsilon$, and $\tau_1, \varepsilon > 0$.*

(1) Coarsen $\mathscr{T}^j$ as long as $\eta^j_{\text{coarse}} \leq \varepsilon$.
(2) Repeatedly compute $U^j \in \mathscr{S}^1_0(\mathscr{T}^j)$ as the solution of

$$\tau_j^{-1}(U^j - \mathscr{I}_j U^{j-1}, V) + (\nabla U^j, \nabla V) = (f(t_j), V)$$

for all $V \in \mathscr{S}^1_0(\mathscr{T}^j)$, and refine $\mathscr{T}^j$ until $\eta^j_{\text{space}} \leq \varepsilon$.
(3) Stop if $t_j \geq T$.
(4) If $\eta^j_{\text{temp}} \leq \varepsilon$, set $\tau_{j+1} = 2\tau_j$, $t_{j+1} = t_j + \tau_{j+1}$, $\mathscr{T}^{j+1} = \mathscr{T}^j$, and $j \to j+1$, and continue with (1). Otherwise, set $\tau_j = \tau_j/2$, $t_j = t_{j-1} + \tau_j$, and repeat (2).

Remark 4.18 Only few results about adaptivity for time-dependent problems are available. One important aspect is the termination of the algorithm at the final time.

4.4.4 Mesh Coarsening

Removing single nodes from a triangulation is difficult since this might lead to nonconforming triangulations. To avoid this effect, the refinement history of the triangulation has to be stored in an appropriate way. We discuss a strategy that avoids storing the history explicitly and allows for efficient mesh coarsening. For this, we recall that local refinement by bisection can be realized by compatible edge patch bisections. This means that a marked edge is the refinement edge of all elements it belongs to. The refinement of such an edge patch can be reversed without effects on the regularity of the triangulation, cf. Fig. 4.23.

Proposition 4.7 (Coarsening Criterion, See [5]) *A node $z \in \mathcal{N}_f$ in a triangulation $\mathscr{T}_f$ is the result of a compatible edge patch bisection and can thus be coarsened, if and only if it is the newest vertex of all elements it belongs to.*

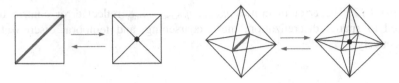

Fig. 4.23 Compatible edge patch bisection and reversion without affecting neighboring elements

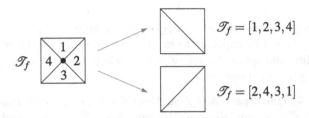

Fig. 4.24 To combine the correct pair of elements in the coarsening procedure, the list of elements determines the left and right son of the father element by the position in the list of elements, i.e., left sons appear before their neighboring right sons

The coarsening criterion of the proposition is easy to check and implement. To obtain the correct predecessors of elements, further information is required. In particular, if $z \in \mathcal{N}_f$ is a node that can be coarsened, and if $T, T' \in \mathcal{T}_f$ are elements that contain the node z and share a side, then we need to determine whether they result from the bisection of a common father element. This is realized by requiring that the left neighbor with respect to the node z and the shared side S appears first in the list of elements. Figure 4.24 illustrates this mechanism to identify the correct history of the grid.

4.4.5 Elliptic Reconstruction

The abstract a posteriori error estimate for the approximation of the heat equation controls the error quantities

$$\max_{t \in [0,T)} \|e(t)\|, \quad \left(\int_0^T \|\nabla e(t)\|^2 \, dt \right)^{1/2}.$$

Assuming that the step-sizes are small compared to mesh-sizes, we expect the first term to be a quadratic- and the second to be a linear order term with respect to the mesh-size h. Since both quantities are controlled by the same estimator, the bound cannot be optimal for the first error quantity. To understand the problem and propose a remedy, we consider a semidiscrete scheme, which consists of a spatial discretization, and specifies the mapping

$$U : [0, T] \rightarrow \mathscr{S}_0^1(\mathscr{T}_h)$$

as the solution of the initial value problem defined by the conditions $U(0) = U_0$ and

$$\big(\partial_t U(t), V\big) + \big(\nabla U(t), \nabla V\big) = \big(f(t), V\big)$$

for all $t \in [0, T]$ and all $V \in \mathscr{S}_0^1(\mathscr{T}_h)$. We assume for simplicity that $f(t) \in \mathscr{S}_0^1(\mathscr{T}_h)$ for all $t \in [0, T]$. We follow [18].

Definition 4.13 The *discrete Laplacian* is the mapping $-\Delta_h : \mathscr{S}_0^1(\mathscr{T}_h) \rightarrow \mathscr{S}_0^1(\mathscr{T}_h)$ defined for $W \in \mathscr{S}_0^1(\mathscr{T}_h)$ by requiring that

$$\big(- \Delta_h W, V\big) = \big(\nabla W, \nabla V\big)$$

be satisfied for all $V \in \mathscr{S}_0^1(\mathscr{T}_h)$.

Remark 4.19 By the Lax–Milgram lemma, the function $-\Delta_h W \in \mathscr{S}_0^1(\mathscr{T}_h)$ is uniquely defined for every $W \in \mathscr{S}_0^1(\mathscr{T}_h)$.

With the discrete Laplace operator the semidiscrete scheme is equivalent to the ordinary differential equations

$$\partial_t U(t) = \Delta_h U(t) + f(t).$$

The residual, interpreted as a functional on $H_0^1(\Omega)$, satisfies

$$\mathscr{R}_U(t) = \partial_t U(t) - \Delta U(t) - f(t) = \Delta_h U(t) - \Delta U(t),$$

and is of order $\mathscr{O}(h)$.

Definition 4.14 For every $t \in (0, T)$, the *elliptic reconstruction* of $U(t)$ is the uniquely defined function $\tilde{u}(t) \in H_0^1(\Omega)$ with

$$\big(\nabla \tilde{u}(t), \nabla v\big) = \big(- \Delta_h U(t), v\big)$$

for all $v \in H_0^1(\Omega)$.

Remark 4.20 We have for every $t \in (0, T)$ that $U(t)$ is the Galerkin approximation of a Poisson problem whose exact solution is $\tilde{u}(t)$.

Because of the identity $-\Delta \tilde{u}(t) = -\Delta_h U(t)$ we have that

$$\partial_t \tilde{u}(t) - \Delta \tilde{u}(t) = f(t) + \partial_t \tilde{u}(t) - \partial_t U(t).$$

It follows that

$$\mathscr{R}_{\tilde{u}}(t) = \partial_t \tilde{u}(t) - \Delta \tilde{u}(t) - f(t) = \partial_t \tilde{u}(t) - \partial_t U(t).$$

Due to L^2-error estimates for approximating the Poisson problem, the right-hand side is of order $\mathscr{O}(h^2)$. To control the approximation error $u - U$, we apply the abstract theorem with the auxiliary function $\tilde{u}$, and then use the triangle inequality, incorporating bounds for $u - \tilde{u}$ from estimates for the approximation of the Poisson problem.

4.4.6 Crank–Nicolson Reconstruction

The a posteriori error analysis carried out for the implicit Euler scheme leads to suboptimal results if the approximations $(U^j)_{j=0,\dots,J}$ are obtained with the Crank–Nicolson scheme, for which we expect quadratic convergence with respect to the temporal discretization parameter. To develop an improvement, we follow [2] and consider a semidiscrete scheme for the heat equation, where only the time-derivative is discretized, i.e., we let $(U^j)_{j=0,\dots,J} \subset H_0^1(\Omega)$ be such that $U^0 = u_0$ and

$$d_t U^j - \Delta U^{j-1/2} = f^{j-1/2}$$

for $j = 1, 2, \dots, J$, with $f^{j-1/2} = f(t_{j-1/2}, \cdot)$ and $t_{j-1/2} = (j - 1/2)\tau$, and with the backward difference quotient d_t and the averages $U^{j-1/2}$ defined by

$$d_t U^j = \frac{U^j - U^{j-1}}{\tau}, \quad U^{j-1/2} = \frac{U^j + U^{j-1}}{2}.$$

For simplicity we assume a uniform step-size $\tau > 0$ and that f is piecewise constant in time. The piecewise affine interpolant, given for $t \in [t_{j-1}, t_j]$ by

$$U(t) = U^{j-1/2} + (t - t_{j-1/2}) d_t U^j,$$

leads to the residual

$$\mathscr{R}_U(t) = \partial_t U(t) - \Delta U(t) - f(t) = -(t - t_{j-1/2}) \Delta d_t U^j$$

which is of order $\mathscr{O}(\tau)$. To define a suitable interpolant $\widehat{U}$, we enforce the identity

$$\partial_t \widehat{U}(t) = \Delta U(t) + f(t),$$

which is achieved by integration, i.e.,

$$\widehat{U}(t) = U^{j-1} + \int_{t_{j-1}}^{t} \partial_t \widehat{U}(s)\, ds$$

$$= U^{j-1} + \int_{t_{j-1}}^{t} \Delta U(s)\, ds + (t - t_{j-1}) f^{j-1/2}.$$

Evaluating the integral by the trapezoidal rule, we obtain

$$\widehat{U}(t) = U^{j-1} + \frac{1}{2}(t - t_{j-1})(\Delta U^{j-1} + \Delta U(t)) + (t - t_{j-1}) f^{j-1/2}.$$

Direct calculations show that $\widehat{U}(t_{j-1}) = U^{j-1}$ and $\widehat{U}(t_j) = U^j$. In particular, for the difference $\widehat{U}(t) - U(t)$ we have

$$\widehat{U}(t) - U(t) = \frac{1}{2}(t - t_{j-1})(t - t_j) d_t \Delta U^j$$

for $t \in [t_{j-1}, t_j]$. By the enforced identity for $\widehat{U}$, we have for the residual that

$$\mathscr{R}_{\widehat{U}}(t) = \partial_t \widehat{U}(t) - \Delta \widehat{U}(t) - f(t) = \Delta \big(U(t) - \widehat{U}(t) \big)$$

$$= \frac{1}{2}(t - t_{j-1})(t - t_j) d_t \Delta^2 U^j,$$

which is of order $\mathscr{O}(\tau^2)$, assuming sufficient spatial regularity. To obtain a posteriori error bound for $u - U$, we apply the abstract error estimate with $\widehat{U}$, which leads to an estimate for $u - \widehat{U}$, and then use the triangle inequality.

References

Precise statements about corner singularities in partial differential equations can be found in [12, 16]. An early contribution to the development of adaptive finite element methods is the reference [3]. Quasi-interpolation operators have been constructed in [8, 11, 23]. The convergence of adaptive finite element methods is the subject of the articles [6, 9, 13, 14, 20, 22, 24, 25]. Aspects of adaptive discretization methods for parabolic equations are addressed in [2, 5, 10, 15, 17, 18]. Textbooks and survey articles on adaptive finite element methods are the references [1, 4, 7, 19, 21, 26].

1. Ainsworth, M., Oden, J.T.: A posteriori error estimation in finite element analysis. Pure and Applied Mathematics (New York). Wiley-Interscience [John Wiley & Sons], New York (2000). URL http://dx.doi.org/10.1002/9781118032824

2. Akrivis, G., Makridakis, C., Nochetto, R.H.: A posteriori error estimates for the Crank-Nicolson method for parabolic equations. Math. Comp. **75**(254), 511–531 (2006). URL http://dx.doi.org/10.1090/S0025-5718-05-01800-4

3. Babuška, I., Rheinboldt, W.C.: Error estimates for adaptive finite element computations. SIAM J. Numer. Anal. **15**(4), 736–754 (1978)

4. Babuška, I., Strouboulis, T.: The finite element method and its reliability. Numerical Mathematics and Scientific Computation. The Clarendon Press, Oxford University Press, New York (2001)

5. Bartels, S., Schreier, P.: Local coarsening of simplicial finite element meshes generated by bisections. BIT **52**(3), 559–569 (2012). URL http://dx.doi.org/10.1007/s10543-012-0378-0

6. Binev, P., Dahmen, W., DeVore, R.: Adaptive finite element methods with convergence rates. Numer. Math. **97**(2), 219–268 (2004). URL http://dx.doi.org/10.1007/s00211-003-0492-7

7. Braess, D.: Finite Elements, 3rd edn. Cambridge University Press, Cambridge (2007). URL http://dx.doi.org/10.1017/CBO9780511618635

8. Carstensen, C.: Quasi-interpolation and a posteriori error analysis in finite element methods. M2AN Math. Model. Numer. Anal. **33**(6), 1187–1202 (1999). URL http://dx.doi.org/10.1051/m2an:1999140

9. Cascon, J.M., Kreuzer, C., Nochetto, R.H., Siebert, K.G.: Quasi-optimal convergence rate for an adaptive finite element method. SIAM J. Numer. Anal. **46**(5), 2524–2550 (2008). URL http://dx.doi.org/10.1137/07069047X

10. Chen, L., Zhang, C.: A coarsening algorithm on adaptive grids by newest vertex bisection and its applications. J. Comput. Math. **28**(6), 767–789 (2010). URL http://dx.doi.org/10.4208/jcm.1004.m3172

11. Clément, P.: Approximation by finite element functions using local regularization. RAIRO Analyse Numérique **9**(R-2), 77–84 (1975)

12. Dauge, M.: Elliptic boundary value problems on corner domains. Lecture Notes in Mathematics, vol. 1341. Springer, Berlin (1988)

13. Diening, L., Kreuzer, C.: Linear convergence of an adaptive finite element method for the p-Laplacian equation. SIAM J. Numer. Anal. **46**(2), 614–638 (2008). URL http://dx.doi.org/10.1137/070681508

14. Dörfler, W.: A convergent adaptive algorithm for Poisson's equation. SIAM J. Numer. Anal. **33**(3), 1106–1124 (1996). URL http://dx.doi.org/10.1137/0733054

15. Eriksson, K., Johnson, C.: Adaptive finite element methods for parabolic problems. I. A linear model problem. SIAM J. Numer. Anal. **28**(1), 43–77 (1991). URL http://dx.doi.org/10.1137/0728003

16. Grisvard, P.: Elliptic problems in nonsmooth domains. Monographs and Studies in Mathematics, vol. 24. Pitman (Advanced Publishing Program), Boston, MA (1985)

17. Kreuzer, C., Möller, C.A., Schmidt, A., Siebert, K.G.: Design and convergence analysis for an adaptive discretization of the heat equation. IMA J. Numer. Anal. **32**(4), 1375–1403 (2012). URL http://dx.doi.org/10.1093/imanum/drr026

18. Makridakis, C., Nochetto, R.H.: Elliptic reconstruction and a posteriori error estimates for parabolic problems. SIAM J. Numer. Anal. **41**(4), 1585–1594 (2003). URL http://dx.doi.org/10.1137/S0036142902406314

19. Mali, O., Neittaanmäki, P., Repin, S.: Accuracy verification methods. Computational Methods in Applied Sciences, vol. 32. Springer, Dordrecht (2014). URL http://dx.doi.org/10.1007/978-94-007-7581-7

20. Morin, P., Nochetto, R.H., Siebert, K.G.: Data oscillation and convergence of adaptive FEM. SIAM J. Numer. Anal. **38**(2), 466–488 (electronic) (2000). URL http://dx.doi.org/10.1137/S0036142999360044

21. Nochetto, R.H., Siebert, K.G., Veeser, A.: Theory of adaptive finite element methods: an introduction. In: Multiscale, Nonlinear and Adaptive Approximation, pp. 409–542. Springer, Berlin (2009). URL http://dx.doi.org/10.1007/978-3-642-03413-8_12

22. Praetorius: Convergence of adaptive finite element methods, personal communication (2008)

23. Scott, L.R., Zhang, S.: Finite element interpolation of nonsmooth functions satisfying boundary conditions. Math. Comp. **54**(190), 483–493 (1990). URL http://dx.doi.org/10.2307/2008497
24. Stevenson, R.: Optimality of a standard adaptive finite element method. Found. Comput. Math. **7**(2), 245–269 (2007). URL http://dx.doi.org/10.1007/s10208-005-0183-0
25. Stevenson, R.: The completion of locally refined simplicial partitions created by bisection. Math. Comp. **77**(261), 227–241 (electronic) (2008). URL http://dx.doi.org/10.1090/S0025-5718-07-01959-X
26. Verfürth, R.: A posteriori error estimation techniques for finite element methods. Numerical Mathematics and Scientific Computation. Oxford University Press, Oxford (2013). URL http://dx.doi.org/10.1093/acprof:oso/9780199679423.001.0001

Chapter 5
Iterative Solution Methods

5.1 Condition Numbers and Multigrid

5.1.1 Conditioning of the Stiffness Matrix

For a triangulation $\mathcal{T}_h$ of a Lipschitz domain $\Omega \subset \mathbb{R}^d$, we let $A_h \in \mathbb{R}^{n \times n}$ denote the finite element stiffness matrix related to the finite element space $\mathcal{S}_0^1(\mathcal{T}_h)$, i.e.,

$$(A_h)_{ij} = \int_\Omega \nabla\varphi_{z_i} \cdot \nabla\varphi_{z_j} \, dx$$

for $i, j = 1, 2, \ldots, n$, and with the nodal basis functions $(\varphi_{z_j} : j = 1, 2, \ldots, n)$ for an enumeration $z_1, z_2, \ldots, z_n \in \mathcal{N}_h \cap \Omega$ of the free nodes. We assume that $\mathcal{T}_h$ is quasiuniform, i.e., that $c_{\mathrm{uni}}^{-1} h \le \min_{T \in \mathcal{T}_h} h_T \le h$, so that we have the *inverse estimate*

$$\|\nabla v_h\|_{L^2(\Omega)} \le c_{\mathrm{inv}} h^{-1} \|v_h\|_{L^2(\Omega)}$$

and the *norm equivalence*

$$c_{\mathrm{eq}}^{-2} \|v_h\|_{L^2(\Omega)}^2 \le h^d \sum_{z \in \mathcal{N}_h} |v_h(z)|^2 \le c_{\mathrm{eq}}^2 \|v_h\|_{L^2(\Omega)}^2$$

for all $v_h \in \mathcal{S}_0^1(\mathcal{T}_h)$, cf. Lemmas 3.8 and 3.7. For the iterative solution of the linear system of equations, e.g., with the conjugate gradient method, the spectral condition number is an important quantity.

Theorem 5.1 (Spectral Conditioning) *We have $\lambda_{\max}(A_h) \le c_{\max} h^{d-2}$ and*

$$\mathrm{cond}_2(A_h) = \frac{\lambda_{\max}(A_h)}{\lambda_{\min}(A_h)} \le c_{\mathrm{cond}} h^{-2}.$$

© Springer International Publishing Switzerland 2016
S. Bartels, *Numerical Approximation of Partial Differential Equations*,
Texts in Applied Mathematics 64, DOI 10.1007/978-3-319-32354-1_5

Proof We use the characterization of $\lambda_{\max}$ and $\lambda_{\min}$ by Rayleigh-quotients, i.e.,

$$\lambda_{\max} = \sup_{y \in \mathbb{R}^n \setminus \{0\}} \frac{A_h y \cdot y}{|y|^2}, \qquad \lambda_{\min} = \inf_{y \in \mathbb{R}^n \setminus \{0\}} \frac{A_h y \cdot y}{|y|^2}.$$

We identify a vector $y \in \mathbb{R}^n$ with a function $v_h \in \mathcal{S}_0^1(\mathcal{T}_h)$ via

$$v_h = \sum_{j=1}^n y_j \varphi_{z_j}.$$

We have

$$\begin{aligned} A_h y \cdot y &= \int_\Omega \nabla v_h \cdot \nabla v_h \, dx \\ &= \|\nabla v_h\|_{L^2(\Omega)}^2 \\ &\leq c_{\mathrm{inv}}^2 h^{-2} \|v_h\|_{L^2(\Omega)}^2 \\ &\leq c_{\mathrm{inv}}^2 c_{\mathrm{eq}}^{-2} h^{d-2} |y|^2, \end{aligned}$$

which proves that $\lambda_{\max}(A_h) \leq ch^{d-2}$. To bound the smallest eigenvalue from below, we use Poincaré's inequality to verify that

$$|y|^2 \leq c_{\mathrm{eq}}^2 h^{-d} \|v_h\|_{L^2(\Omega)}^2 \leq c_{\mathrm{eq}}^2 c_P^2 h^{-d} \|\nabla v_h\|_{L^2(\Omega)}^2.$$

This implies that $\lambda_{\min}(A_h) \geq h^d c_{\mathrm{eq}}^{-2} c_P^{-2}$. $\square$

Remark 5.1 The estimate of the theorem is optimal.

Although the condition number of the stiffness matrix with respect to the Euclidean norm is large for small mesh-sizes, the discrete Poisson problem is not ill-conditioned. In particular, there exist natural norms for which the condition number is optimal.

Proposition 5.1 (Optimal Conditioning) *Let $f, \tilde{f} \in L^2(\Omega)$ and let $u_h, \tilde{u}_h \in \mathcal{S}_0^1(\mathcal{T}_h)$ be the corresponding Galerkin approximations of the Poisson problem. Assume that $u_h \neq 0$. We then have*

$$\frac{\|\nabla(u_h - \tilde{u}_h)\|_{L^2(\Omega)}}{\|\nabla u_h\|_{L^2(\Omega)}} \leq \frac{\|f - \tilde{f}\|_{\mathcal{S}_0^1(\mathcal{T}_h)'}}{\|f\|_{\mathcal{S}_0^1(\mathcal{T}_h)'}},$$

where for $g \in L^2(\Omega)$,

$$\|g\|_{\mathcal{S}_0^1(\mathcal{T}_h)'} = \sup_{v_h \in \mathcal{S}_0^1(\mathcal{T}_h) \setminus \{0\}} \frac{\int_\Omega g v_h \, dx}{\|\nabla v_h\|_{L^2(\Omega)}}.$$

Proof For the difference $u_h - \tilde{u}_h \in \mathscr{S}_0^1(\mathscr{T}_h)$ and every $v_h \in \mathscr{S}_0^1(\mathscr{T}_h)$, we have that

$$\int_\Omega \nabla(u_h - \tilde{u}_h) \cdot \nabla v_h \, dx = \int_\Omega (f - \tilde{f}) v_h \, dx.$$

The choice $v_h = u_h - \tilde{u}_h$ implies that

$$\|\nabla(u_h - \tilde{u}_h)\|_{L^2(\Omega)}^2 = \int_\Omega (f - \tilde{f})(u_h - \tilde{u}_h) \, dx$$

$$\leq \|f - \tilde{f}\|_{\mathscr{S}_0^1(\mathscr{T}_h)'} \|\nabla(u_h - \tilde{u}_h)\|_{L^2(\Omega)}.$$

Combining this with the bound

$$\|f\|_{\mathscr{S}_0^1(\mathscr{T}_h)'} = \sup_{v_h \in \mathscr{S}_0^1(\mathscr{T}_h) \setminus \{0\}} \frac{\int_\Omega f v_h \, dx}{\|\nabla v_h\|_{L^2(\Omega)}} \leq \|\nabla u_h\|_{L^2(\Omega)}$$

implies the estimate. $\qquad\square$

Because of the large condition number $\mathrm{cond}_2(A_h)$ for $0 < h \ll 1$, stationary iteration methods like the Richardson-, the Jacobi-, or the Gauss–Seidel and the conjugate gradient method will typically converge slowly. Figure 5.1 displays the errors $u_h^k - u_h$ in a one-dimensional Poisson problem of the Richardson iteration

$$x^{k+1} = x^k - \omega_h(A_h x^k - b)$$

after $k = 1, 10, 100$ iterations, with $\omega_h = h^{2-d} \sim 1/\lambda_{\max}(A_h)$ and randomly defined initial configuration u_h^0. It is important to observe that rapid oscillations are removed within a few iterations, while the maximal iteration error decays slowly.

For analytical considerations it is helpful to interpret the matrix A_h not as an operator on $\mathbb{R}^n$ but on $\mathscr{S}_0^1(\mathscr{T}_h)$. In particular, we consider the linear operator

$$\widehat{A}_h : \mathscr{S}_0^1(\mathscr{T}_h) \to \mathscr{S}_0^1(\mathscr{T}_h)' \simeq \mathscr{S}_0^1(\mathscr{T}_h)$$

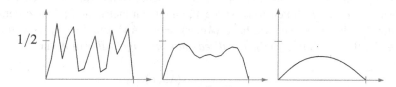

1/2

Fig. 5.1 Iteration error $u_h^k - u_h$ after $k = 1, 10, 100$ Richardson iterations in the discretized Poisson problem $-u'' = 1$ in $(0, 1)$, $u(0) = u(1) = 0$, with $h = 2^{-4}$

defined by $\xi_h = \widehat{A}_h v_h$, such that for all $w_h \in \mathscr{S}_0^1(\mathscr{T}_h)$, we have

$$\int_\Omega \xi_h w_h \, dx = A_h V_h \cdot W_h,$$

where $V_h, W_h \in \mathbb{R}^n$ are the coefficient vectors for $v_h, w_h \in \mathscr{S}_0^1(\mathscr{T}_h)$, respectively. Note that for all $v_h, w_h \in \mathscr{S}_0^1(\mathscr{T}_h)$, we have

$$\int_\Omega (\widehat{A}_h v_h) w_h \, dx = A_h V_h \cdot W_h = \int_\Omega \nabla v_h \cdot \nabla w_h \, dx.$$

The definition of $\widehat{A}_h$ motivates using the L^2-norm and the corresponding inner product on the vector space $\mathscr{S}_0^1(\mathscr{T}_h)$. We then have

$$\lambda_{\max}(\widehat{A}_h) = \sup_{v_h \in \mathscr{S}_0^1(\mathscr{T}_h) \setminus \{0\}} \frac{\int_\Omega (\widehat{A}_h v_h) v_h \, dx}{\|v_h\|_{L^2(\Omega)}^2} \le c_{\mathrm{inv}}^2 h^{-2}.$$

and

$$\lambda_{\min}(\widehat{A}_h) = \inf_{v_h \in \mathscr{S}_0^1(\mathscr{T}_h) \setminus \{0\}} \frac{\int_\Omega (\widehat{A}_h v_h) v_h \, dx}{\|v_h\|_{L^2(\Omega)}^2} \ge c_P^2.$$

The operator $\widehat{A}_h : \mathscr{S}_0^1(\mathscr{T}_h) \to \mathscr{S}_0^1(\mathscr{T}_h)$ may be regarded as a discrete version of the negative Laplace operator, i.e., often one denotes $\widehat{A}_h = -\Delta_h$. The representing matrix differs from A_h by the inverse of the mass matrix.

5.1.2 Two-Grid Iteration

The observation that the error $e_h^k = u_h^k - u_h$ in the Richardson iteration is smooth after a few iterations implies that the function e_h^k can be resolved accurately on a coarser grid with mesh-size $2h$, on which the iteration is less expensive. The following lemma quantifies the smoothing property of the Richardson iteration. Given a scalar product $\langle \cdot, \cdot \rangle$ on an n-dimensional space V, and a symmetric and positive definite operator $\widehat{A} : V \to V$, i.e., we have $\langle \widehat{A}v, w \rangle = \langle v, \widehat{A}w \rangle$ and $\langle \widehat{A}v, v \rangle > 0$ for all $v, w \in V \setminus \{0\}$, we define the *(generalized) maximal eigenvalue* of $\widehat{A}$ via

$$\lambda_{\max} = \sup_{v \in V \setminus \{0\}} \frac{\langle \widehat{A}v, v \rangle}{\langle v, v \rangle}.$$

Throughout what follows we adopt arguments from [14].

Lemma 5.1 (Smoothing Property) *Let* $\widehat{A} : V \to V$ *be a positive definite and symmetric operator. With the Richardson iteration operator*

$$S = I - \lambda_{\max}^{-1}\widehat{A},$$

we have for every $k \in \mathbb{N}$ *that*

$$\|\widehat{A}S^k\| \le k^{-1}\lambda_{\max},$$

where $\| \cdot \|$ *is the operator norm induced by the scalar product on* V.

Proof We let $(v_1, v_2, \dots, v_n)$ be an orthonormal basis of eigenvectors for $\widehat{A}$, associated with the generalized eigenvalues $0 < \lambda_1 \le \lambda_2 \le \cdots \le \lambda_n = \lambda_{\max}$ of $\widehat{A}$, i.e., $\langle \widehat{A}v_j, w \rangle = \lambda_j \langle v_j, w \rangle$ for $j = 1, 2, \dots, n$ and all $w \in V$. For an arbitrary element

$$v = \sum_{j=1}^{n} \alpha_j v_j,$$

we have

$$\langle \widehat{A}(I - \lambda_n^{-1}\widehat{A})^k v, w \rangle = \sum_{j=1}^{n} \lambda_j (1 - \lambda_j/\lambda_n)^k \alpha_j \langle v_j, w \rangle.$$

This shows that

$$\|\widehat{A}S^k v\| \le \lambda_n \max_{j=1,\dots,n} \frac{\lambda_j}{\lambda_n} \left(1 - \frac{\lambda_j}{\lambda_n}\right)^k \|v\|.$$

An exercise proves that

$$\max_{t \in [0,1]} t(1 - t)^k \le \frac{1}{ek} \le k^{-1},$$

and this implies the estimate. $\qquad\square$

The quantity $\|\widehat{A}_h v_h\|$ is a measure of the smoothness of a finite element function v_h. Hence we see that a few iterations with the Richardson matrix improve the smoothness of the iteration error. Note however that the factor $\lambda_{\max} \sim h^{-2}$ is large in case of the Poisson problem. To quantify the idea of representing the iteration error on a coarser grid, we have to analyze the corresponding approximation errors.

Lemma 5.2 (Approximation Property) *Assume that the Poisson problem is H^2-regular, let $\mathcal{T}_h$ and $\mathcal{T}_H$ be triangulations of Ω with mesh-sizes $h = \gamma H$ with $0 < \gamma < 1$. Let $c_h \in \mathcal{S}_0^1(\mathcal{T}_h)$ and $c_H \in \mathcal{S}_0^1(\mathcal{T}_H)$ be the Galerkin approximations of the*

Poisson problem with $r \in L^2(\Omega)$ on the right-hand side. We then have that

$$\|c_h - c_H\|_{L^2(\Omega)} \le c_{\mathrm{ap}} h^2 \|r\|_{L^2(\Omega)}.$$

Proof If $z \in H_0^1(\Omega) \cap H^2(\Omega)$ is the exact solution of the Poisson problem $-\Delta z = r$, then we have that

$$\|z - c_h\|_{L^2(\Omega)} \le c h^2 \|r\|_{L^2(\Omega)},$$
$$\|z - c_H\|_{L^2(\Omega)} \le c H^2 \|r\|_{L^2(\Omega)}.$$

The triangle inequality and $H^2 = \gamma^{-2} h^2$ imply the estimate.   □

Assume that we have carried out k iterations of the Richardson scheme. This defines an approximation $u_h^k \in \mathscr{S}_0^1(\mathscr{T}_h)$ and a residual $r_h \in \mathscr{S}_0^1(\mathscr{T}_h)$ via

$$r_h = b_h - \widehat{A}_h u_h^k = \widehat{A}_h(u_h - u_h^k).$$

Here $b_h \in \mathscr{S}_0^1(\mathscr{T}_h)$ is defined by

$$\int_\Omega b_h v_h \, \mathrm{d}x = \int_\Omega f v_h \, \mathrm{d}x$$

for all $v_h \in \mathscr{S}_0^1(\mathscr{T}_h)$. To obtain the exact discrete solution u_h from the approximation u_h^k, we have to compute the correction $c_h = u_h - u_h^k \in \mathscr{S}_0^1(\mathscr{T}_h)$ which satisfies

$$\int_\Omega \nabla c_h \cdot \nabla v_h \, \mathrm{d}x = \int_\Omega r_h v_h \, \mathrm{d}x$$

for all $v_h \in \mathscr{S}_0^1(\mathscr{T}_h)$. We approximate the solution c_h on the coarser grid $\mathscr{T}_H$.

Theorem 5.2 (Two-Grid Iteration) *Let $u_h^k \in \mathscr{S}_0^1(\mathscr{T}_h)$ be obtained from $k \ge 1$ Richardson iterations with an initial $u_h^0 \in \mathscr{S}_0^1(\mathscr{T}_h)$. Assume that $\mathscr{S}_0^1(\mathscr{T}_H) \subset \mathscr{S}_0^1(\mathscr{T}_h)$ and let $c_H \in \mathscr{S}_0^1(\mathscr{T}_H)$ solve*

$$\int_\Omega \nabla c_H \cdot \nabla v_H \, \mathrm{d}x = \int_\Omega r_h v_H \, \mathrm{d}x = \int_\Omega f v_H \, \mathrm{d}x - \int_\Omega \nabla u_h^k \cdot \nabla v_H \, \mathrm{d}x$$

for all $v_H \in \mathscr{S}_0^1(\mathscr{T}_H)$. For the function $u_h^{k+\mathrm{corr}} = u_h^k + c_H \in \mathscr{S}_0^1(\mathscr{T}_h)$ and the Galerkin approximation $u_h \in \mathscr{S}_0^1(\mathscr{T}_h)$ of the H^2-regular Poisson problem with $f \in L^2(\Omega)$ on the right-hand side, we have

$$\|u_h - u_h^{k+\mathrm{corr}}\|_{L^2(\Omega)} \le c_{tg} k^{-1} \|u_h - u_h^0\|_{L^2(\Omega)}.$$

Proof

(i) The Richardson iteration computes u_h^k recursively via

$$u_h^k = u_h^{k-1} - \lambda_{\max}^{-1}(\widehat{A}_h u_h^{k-1} - b_h).$$

Noting $b_h = \widehat{A}_h u_h$, with the Richardson operator $S = (I - \lambda_{\max}^{-1}\widehat{A}_h)$ we have that

$$u_h - u_h^k = u_h - u_h^{k-1} - \lambda_{\max}^{-1}\widehat{A}_h(u_h - u_h^{k-1}) = S(u_h - u_h^{k-1}).$$

By induction, we see that $u_h - u_h^k = S^k(u_h - u_h^0)$, and with the smoothing property of Lemma 5.1 we find that

$$\|\widehat{A}_h(u_h - u_h^k)\|_{L^2(\Omega)} = \|\widehat{A}_h S^k(u_h - u_h^0)\|_{L^2(\Omega)}$$
$$\leq k^{-1}\lambda_{\max}\|u_h - u_h^0\|_{L^2(\Omega)}.$$

(ii) Defining $r_h = b_h - \widehat{A}_h u_h^k$, and letting $c_h \in \mathscr{S}_0^1(\mathscr{T}_h)$ be the Galerkin approximation of the Poisson problem with r_h on the right-hand side, we have that

$$u_h = u_h^k + c_h.$$

With $c_H \in \mathscr{S}_0^1(\mathscr{T}_H)$ that defines $u_h^{k+\text{corr}} = u_h^k + c_H$, the approximation property of Lemma 5.2 leads to

$$\|u_h - u_h^{k+\text{corr}}\|_{L^2(\Omega)} = \|c_h - c_H\|_{L^2(\Omega)}$$
$$\leq c_{\text{ap}} h^2 \|r_h\|_{L^2(\Omega)}$$
$$= c_{\text{ap}} h^2 \|\widehat{A}_h u_h^k - b_h\|_{L^2(\Omega)}$$
$$= c_{\text{ap}} h^2 \|\widehat{A}_h(u_h^k - u_h)\|_{L^2(\Omega)}.$$

(iii) Combining the previous estimates and noting $\lambda_{\max}(\widehat{A}_h) \leq ch^{-2}$ prove the estimate. □

Remark 5.2 We have proved the convergence of an iterative scheme for solving the discretized Poisson problem with an h-independent reduction factor. To achieve an iteration error, which is comparable to the discretization error $\mathscr{O}(h^2)$, approximately $k \sim h^{-2}$ iterations are required for the two-level iteration defined by the theorem.

5.1.3 Multigrid Algorithm

Computating the coarse-grid correction c_H in the two-grid iteration requires solving a discrete Poisson problem. This can again be done approximately via Richardson iteration and another coarse grid correction. Repeating this idea with a hierarchy of refined triangulations

$$\mathscr{T}_1 \to \mathscr{T}_2 \to \cdots \to \mathscr{T}_L$$

leads to a *multigrid algorithm*. Assuming that $h_L = \gamma^{-L} h_1$, for a mesh-size reduction factor $0 < \gamma < 1$ and $h_1 = \mathscr{O}(1)$, only a finite number of iterations is necessary to achieve an iteration error that is comparable to the discretization error with a complexity $\mathscr{O}(n)$. To formulate the idea in an implementable way, we specify matrices that realize the grid transfer.

Lemma 5.3 (Prolongation Operator) *Let $\mathscr{T}_h$ be a refinement of the triangulation $\mathscr{T}_H$ in the sense that $\mathscr{S}_0^1(\mathscr{T}_H) \subset \mathscr{S}_0^1(\mathscr{T}_h)$. There exists a uniquely defined matrix $P_h^H \in \mathbb{R}^{n \times N}$ such that for every $v_H \in \mathscr{S}_0^1(\mathscr{T}_H)$ with coefficient vector $V_H \in \mathbb{R}^N$, the vector $V_h = P_h^H V_H \in \mathbb{R}^n$ is the coefficient vector of v_H with respect to the nodal basis of $\mathscr{S}_0^1(\mathscr{T}_h)$.*

Proof Exercise.   □

The matrix P_h^H embeds the nodal values of v_H in a larger vector. Its transpose $(P_h^H)^\top \in \mathbb{R}^{N \times n}$ can be regarded as a mapping from $\mathscr{S}_0^1(\mathscr{T}_h)$ to $\mathscr{S}_0^1(\mathscr{T}_H)$.

Definition 5.1 The *restriction operator* $R_H^h : \mathscr{S}_0^1(\mathscr{T}_h) \to \mathscr{S}_0^1(\mathscr{T}_H)$ is, for $v_h \in \mathscr{S}_0^1(\mathscr{T}_h)$ with coefficient vector $V_h \in \mathbb{R}^n$, defined as the function $v_H \in \mathscr{S}_0^1(\mathscr{T}_H)$ with coefficient vector $V_H = R_H^h V_h \in \mathbb{R}^N$ for $R_H^h = (P_h^H)^\top$.

Note that $R_H^h v_h \neq v_h$ and $R_H^h P_h^H v_H \neq v_H$ for $v_h \in \mathscr{S}_0^1(\mathscr{T}_h)$ and $v_H \in \mathscr{S}_0^1(\mathscr{T}_H)$ in general, cf. Fig 5.2.

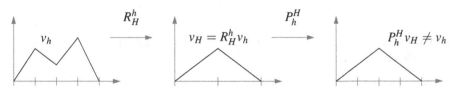

Fig. 5.2 Restriction of a fine-grid function to a coarser grid, and prolongation of a coarse-grid function onto the finer grid

Remarks 5.3

(i) The restriction can alternatively be defined by the L^2-projection onto $\mathscr{S}_0^1(\mathscr{T}_H)$, i.e., $v_H = R_H^h v_h$ is defined by

$$\int_\Omega v_H w_H \, dx = \int_\Omega v_h w_H \, dx$$

for all $w_H \in \mathscr{S}_0^1(\mathscr{T}_H)$. This definition is compatible with Theorem 5.2, but is more expensive to realize.

(ii) An important aspect of the prolongation is that the finite element stiffness matrix only has to be computed on the finest grid, i.e., we have

$$V_H^\top A_H W_H = \int_\Omega \nabla v_H \cdot \nabla w_H \, dx = (P_H^h V_H)^\top A_h (P_H^h W_H)$$

for functions $v_H, w_H \in \mathscr{S}_0^1(\mathscr{T}_H)$ with coefficient vectors $V_H, W_H \in \mathbb{R}^N$.

The multigrid algorithm consists in carrying out a specified number of Richardson iterations, called the *pre-smoothing* procedure, computing the *coarse-grid correction* by restricting the residual to a coarser grid, and carrying out additional Richardson iterations called *post-smoothing*. Since the computation of the coarse-grid correction is also done with this idea, unless we have reached the coarsest level, we obtain a recursive algorithm, which is illustrated in Fig. 5.3.

Algorithm 5.1 (Multigrid) *Let $\mathscr{T}_{\ell_0} < \cdots < \mathscr{T}_L$ be a sequence of refined triangulations, with stiffness matrices A_ℓ, $\ell_0 \le \ell \le L$, let $b_L \in \mathscr{S}_0^1(\mathscr{T}_L)$ be the right-hand side, and let $\nu_{\mathrm{pre}}, \nu_{\mathrm{post}} \in \mathbb{N}$. The approximation $u_L^{\mathrm{mg}} \in \mathscr{S}_0^1(\mathscr{T}_L)$ is defined by*

$$u_L^{\mathrm{mg}} = MG(A_L, b_L, L),$$

with the recursive function $MG : \mathbb{R}^{n_\ell \times n_\ell} \times \mathscr{S}_0^1(\mathscr{T}_\ell) \times \{\ell_0, \dots, L\} \to \mathscr{S}_0^1(\mathscr{T}_\ell)$

$$MG[A_\ell, b_\ell, \ell] = \begin{cases} T_{Ri,\ell}^{\nu_{\mathrm{post}}} MG[P_\ell^\top A_\ell P_\ell, R_\ell(b_\ell - A_\ell T_{Ri,\ell}^{\nu_{\mathrm{pre}}} 0), \ell - 1], & \ell > \ell_0, \\ A_\ell^{-1} b_\ell, & \ell = \ell_0. \end{cases}$$

$$
\begin{array}{ll}
\bullet & \text{pre-smoothing} \\
\circ & \text{post-smoothing} \\
\square & \text{exact solution} \\
\searrow & \text{restriction} \\
\nearrow & \text{prolongation}
\end{array}
$$

Fig. 5.3 Illustration of the recursive multigrid strategy related to a sequence of refined triangulations

Here $T_{Ri,\ell}^\nu w_\ell^0$ denotes the application of ν Richardson iterations on the ℓ-th level with starting value w_ℓ^0, and P_ℓ and R_ℓ are the transfer operators between the levels $\ell - 1$ and ℓ.

Figure 5.4 shows a MATLAB realization of the algorithm. Since the initial triangulation has no interior nodes, we use an exact solution of the linear system of equations on the second level, i.e., we use $\ell_0 = 2$. Figure 5.5 shows the decay of the iteration error for different choices of the smoothing parameters ν_{pre} and ν_{post}.

```
function multigrid(d_tmp,L)
global P d; d = d_tmp;
[c4n,n4e,Db,Nb] = triang_cube(d); Db = [Db;Nb]; Nb = [];
nC = size(c4n,1); fNodes_prev = setdiff(1:nC,unique(Db));
for ell = 1:L
    [c4n,n4e,Db,Nb,~,P1] = red_refine(c4n,n4e,Db,Nb);
    nC = size(c4n,1); fNodes = setdiff(1:nC,unique(Db));
    P{ell} = P1(fNodes,fNodes_prev);
    fNodes_prev = fNodes;
end
u = zeros(nC,1);
[s,m] = fe_matrices(c4n,n4e);
b = m(fNodes,:)*f(c4n);
A = s(fNodes,fNodes);
u(fNodes) = MG(A,b,L);
show_p1(c4n,n4e,Db,Nb,u)

function u = MG(A,b,ell)
global P;
nu_pre = 2; nu_post = 2; ell_0 = 2;
if ell == ell_0
    u = A\b;
else
    u_ini = zeros(size(b,1),1);
    u = richardson(A,b,u_ini,nu_pre,ell);
    A_coarse = P{ell}'*A*P{ell};
    r_coarse = P{ell}'*(b-A*u);
    c = MG(A_coarse,r_coarse,ell-1);
    u = u+P{ell}*c;
    u = richardson(A,b,u,nu_post,ell);
end

function u = richardson(A,b,u,nu,ell)
global d;
h = 2^(-ell); omega = h^(d-2)/10;
for k = 1:nu
    u = u-omega*(A*u-b);
end

function val = f(x)
val = ones(size(x,1),1);
```

Fig. 5.4 MATLAB implementation of the multigrid algorithm

Fig. 5.5 Solution error $\|\nabla(u_h - u_h^{mg,\nu})\|_{L^2(\Omega)}$ for different pre- and postsmoothing strategies in the multigrid solution of a two-dimensional Poisson problem with 3969 degrees of freedom

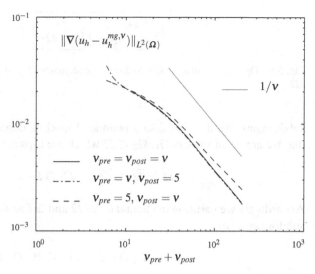

Remark 5.4 On every level we expect that the error be decreased by the factor ν_{pre}^{-1}. Assuming that $L = \log_2(h_L^{-1})$, e.g., $\gamma = 1/2$ and $h_1 = 1/2$, we deduce that we have a total reduction by

$$\nu_{\mathrm{pre}}^{-L} = 2^{\log_2(\nu_{\mathrm{pre}})\log_2(h_L)} = h_L^{\log_2(\nu_{\mathrm{pre}})}.$$

Hence, for $\nu_{\mathrm{pre}} = 4$, we expect an iteration error $\mathcal{O}(h_L^2)$. Noting that $n_\ell = h_\ell^{-d} \sim 2^{\ell d} = n_L 2^{-(\ell-L)d}$, and that the computational complexity is of order $\mathcal{O}(n_\ell)$ on the ℓ-th level, we see that the total computational complexity is of the order

$$n_L \sum_{\ell=1}^{L} 2^{-(\ell-L)d} = \mathcal{O}(n_L).$$

This heuristic argument can be made rigorous, and shows the optimality of the multigrid method, i.e., that it provides an approximation with an iteration error that matches the discretization error with linear computational complexity.

5.2 Domain Decomposition Methods

5.2.1 Transmission Conditions

To avoid large matrices in solving partial differential equations, it is desirable to decompose a domain into subdomains and solve local problems in those

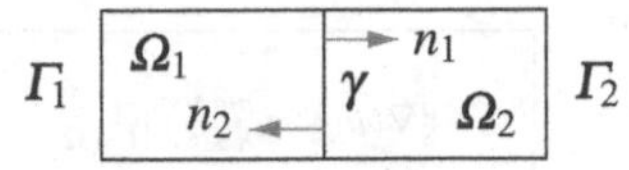

Fig. 5.6 Open partitioning of Ω into disjoint subdomains Ω_1 and Ω_2, and interface $\gamma = \partial\Omega_1 \cap \partial\Omega_2$

subdomains. We thus consider a bounded Lipschitz domain $\Omega \subset \mathbb{R}^d$, and assume that we are given subsets $\Omega_1, \Omega_2 \subset \Omega$ which are Lipschitz domains with

$$\overline{\Omega} = \overline{\Omega}_1 \cup \overline{\Omega}_2, \quad \Omega_1 \cap \Omega_2 = \emptyset.$$

Accordingly we partition the boundary of Ω and define the *interface* γ between the subdomains, i.e.,

$$\gamma = \partial\Omega_1 \cap \partial\Omega_2, \quad \Gamma_j = \partial\Omega \cap \partial\Omega_j$$

for $j = 1, 2$, cf. Fig. 5.6. Throughout this section we follow [5].

To reduce the solution of a partial differential equation to problems on the subdomains Ω_1 and Ω_2, certain weak continuity conditions on the interface have to be satisfied. We assume for simplicity that the Poisson problem in Ω is H^2-regular.

Proposition 5.2 (Transmission Conditions) *Let* $f \in L^2(\Omega)$ *and* $u \in H^2(\Omega) \cap H_0^1(\Omega)$. *We have*

$$-\Delta u = f \text{ in } \Omega, \quad u|_{\partial\Omega} = 0,$$

if and only if the functions $u_j = u|_{\Omega_j}, j = 1, 2$, *solve*

$$-\Delta u_j = f \text{ in } \Omega_j, \quad u_j|_{\Gamma_j} = 0,$$

for $j = 1, 2$, *and satisfy the* transmission conditions

$$u_1 = u_2, \quad \partial_{n_1} u_1 = -\partial_{n_2} u_2,$$

on γ, *where* $\partial_{n_j} u_j = \nabla u_j \cdot n_j$ *on* γ *with the outer unit normal* n_j *to* $\partial\Omega_j$, *i.e.,* $n_2 = -n_1$.

Proof Exercise.   □

Remarks 5.5

(i) If $d = 1$ and $\gamma = \{a\}$, then the conditions require that $u_1(a) = u_2(a)$ and $u_1'(a) = u_2'(a)$. Both conditions are needed, cf. Fig. 5.7.

(ii) Note that at every point on the boundary of a domain we can only impose one condition within a second-order elliptic partial differential equation.

Fig. 5.7 The transmission conditions are not satisfied at the interface $\gamma = \{a\}$ (*left and middle*); both transmission conditions are satisfied (*right*)

5.2.2 Dirichlet–Neumann Method

The transmission condition couples the local problems in the subdomains. We will alternatingly solve the problems with appropriately chosen boundary conditions on the interface to satisfy the transmission conditions.

Algorithm 5.2 (Dirichlet–Neumann Method) *Choose $\lambda^0 \in C(\gamma)$ with $\lambda^0|_{\gamma \cap \Gamma} = 0$, parameters $\theta > 0$, $\varepsilon_{\mathrm{stop}} > 0$, and set $k = 0$.*

(1) Determine $u_1^{k+1} \in H^1(\Omega_1)$ such that

$$-\Delta u_1^{k+1} = f \text{ in } \Omega_1, \quad u_1^{k+1}|_{\Gamma_1} = 0, \quad u_1^{k+1}|_{\gamma} = \lambda^k.$$

(2) Determine $u_2^{k+1} \in H^1(\Omega_2)$ such that

$$-\Delta u_2^{k+1} = f \text{ in } \Omega_2, \quad u_2^{k+1}|_{\Gamma_2} = 0, \quad \partial_{n_2} u_2^{k+1}|_{\gamma} = -\partial_{n_1} u_1^{k+1}|_{\gamma}.$$

(3) Stop if $\|u_1^{k+1} - u_2^{k+1}\|_{L^2(\gamma)} \leq \varepsilon_{\mathrm{stop}}$; otherwise, set

$$\lambda^{k+1} = \theta u_2^{k+1} + (1 - \theta)\lambda^k,$$

set $k \to k + 1$ and continue with (1).

A simple implementation of the Dirichlet–Neumann-method is shown in Fig. 5.8, where we used the stronger stopping criterion $\|u_1^{k+1} - u_2^{k+1}\|_{L^\infty(\gamma)} \leq \varepsilon_{\mathrm{stop}}$. If the iteration becomes stationary, then the functions u_1 and u_2 coincide in the subdomains with the global solution of the Poisson problem. Figure 5.9 shows some iterates of the method in a two-dimensional numerical experiment.

Lemma 5.4 (Consistency) *Every stationary pair (u_1, u_2) for the Dirichlet–Neumann method coincides with the solution $u \in H_0^1(\Omega)$ of the Poisson problem with f on the right-hand side, i.e., $u_j = u|_{\Omega_j}$ for $j = 1, 2$.*

Proof Exercise.   □

The *damping parameter* θ has to be sufficiently small to guarantee convergence.

Example 5.1 If $\Omega = (0, 1)$, $f = 0$, and $\Omega_1 = (0, a)$, $\Omega_2 = (a, 1)$ for $0 < a < 1/2$, then Algorithm 5.2 converges if and only if $\theta < 1$.

```
function dirichlet_neumann(red)
c4n{1} = [0 0;1 0;0 1;1 1];
c4n{2} = [1 0;2 0;1 1;2 1];
n4e{1} = [1 2 4;1 4 3];
n4e{2} = [1 2 4;1 4 3];
Db{1} = [1 2;4 3;3 1];
Db{2} = [1 2;2 4;4 3];
for j = 1:red
    [c4n{1},n4e{1},Db{1}] = red_refine(c4n{1},n4e{1},Db{1},[]);
    [c4n{2},n4e{2},Db{2}] = red_refine(c4n{2},n4e{2},Db{2},[]);
end
for j = 1:2
    nC{j} = size(c4n{j},1);
    dNodes{j} = unique(Db{j});
    [s{j},m{j}] = fe_matrices(c4n{j},n4e{j});
    b{j} = m{j}*f(c4n{j});
end
[~,gamma{1},gamma{2}] = intersect(c4n{1},c4n{2},'rows');
dNodes{1} = union(dNodes{1},gamma{1});
for j = 1:2
    fNodes{j} = setdiff(1:nC{j},dNodes{j});
end
lambda = zeros(length(gamma{1}),1);
theta = 1/4;
eps_stop = 1e-3; diff = 1;
while diff > eps_stop
    %%% Initialize
    u{1} = zeros(nC{1},1); u{2} = zeros(nC{2},1);
    %%% Step (1)
    u{1}(gamma{1}) = lambda;
    b1 = b{1}-s{1}*u{1};
    u{1}(fNodes{1}) = s{1}(fNodes{1},fNodes{1})\b1(fNodes{1});
    %%% Step (2)
    b2 = b{2};
    normal_trans = b{1}-s{1}*u{1};
    b2(gamma{2}) = b2(gamma{2})+normal_trans(gamma{1});
    u{2}(fNodes{2}) = s{2}(fNodes{2},fNodes{2})\b2(fNodes{2});
    %%% Step (3)
    lambda = theta*u{2}(gamma{2})+(1-theta)*lambda;
    diff = max(abs((u{1}(gamma{1})-u{2}(gamma{2}))))
    %%% Visualize
    show_p1(c4n{1},n4e{1},Db{1},[],u{1}); hold on;
    show_p1(c4n{2},n4e{2},Db{2},[],u{2}); hold off;
    pause(1);
end

function val = f(x); val = ones(size(x,1),1);
```

Fig. 5.8 MATLAB implementation of the Dirichlet–Neumann-method

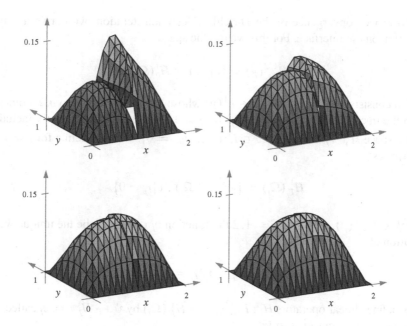

Fig. 5.9 Iterates of the Dirichlet–Neumann method in a two-dimensional Poisson problem

Various other aspects have to be taken into account in the practical realization of the Dirichlet–Neumann-method.

Remarks 5.6

(i) If we replace the boundary condition $\partial_{n_1} u_1^{k+1} = -\partial_{n_2} u_2^{k+1}$ on γ in Step (2) by $\partial_{n_1} u_1^{k+1} = -\partial_{n_2} u_2^k$, then the problems in Steps (1) and (2) can be carried out *in parallel*.

(ii) Notice that we need $\Gamma_2 = \partial \Omega_2 \cap \partial \Omega \neq \emptyset$ in order to have a well-posed problem in Step (2), i.e., Ω_2 should be a *nonfloating domain*.

(iii) The Dirichlet–Neumann-method can be generalized to partitions with more than two nonfloating, nonoverlapping subdomains.

(iv) The normal derivative of u_1^{k+1} on γ does not have to be computed explicitly in Step (2), since for every $v \in H^1(\Omega_2)$ with $v|_{\Gamma_2} = 0$ and an arbitrary extension $\tilde{v} \in H_0^1(\Omega)$, i.e., $\tilde{v}|_{\Omega_2} = v$, we have

$$\int_\gamma \partial_{n_1} u_1^{k+1} v \, ds = \int_{\Omega_1} \nabla u_1^{k+1} \cdot \nabla \tilde{v} \, dx - \int_{\Omega_1} f \tilde{v} \, dx.$$

To prove convergence of the Dirichlet–Neumann iteration, we derive a single equation on the interface. For this we use the space

$$H_{00}^{1/2}(\gamma) = \{v|_\gamma : v \in H_0^1(\Omega)\}$$

which consists of all functions $\psi \in L^1(\gamma)$ whose trivial extensions to $\partial\Omega_j$ coincide with the trace of a function in $H^1(\Omega_j)$ for $j = 1, 2$. In general it is strictly included in the space $H_0^{1/2}(\gamma) = \{v|_\gamma : v \in H^1(\Omega), v|_{\partial\gamma} = 0\}$. We also define for $j = 1, 2$, the spaces

$$H_{\Gamma_j}^1(\Omega_j) = \{v \in H^1(\Omega_j) : v|_{\Gamma_j} = 0\}.$$

For $\psi \in H_{00}^{1/2}(\gamma)$, we let for $j = 1, 2$ the function $v_j \in H_{\Gamma_j}^1(\Omega_j)$ be the unique weak solution of

$$-\Delta v_j = 0 \text{ in } \Omega_j, \quad v_j|_{\Gamma_j} = 0, \quad v_j|_\gamma = \psi.$$

This defines linear operators $H_j : H_{00}^{1/2}(\gamma) \to H_{\Gamma_j}^1(\Omega_j)$ by $\psi \mapsto H_j\psi = v_j$ called the *harmonic extensions of ψ to Ω_j.*

Lemma 5.5 (Norm Equivalence) *The expressions $\|\psi\|_j = \|\nabla H_j\psi\|_{L^2(\Omega_j)}, j = 1, 2$, define equivalent norms on $H_{00}^{1/2}(\gamma)$, i.e., there exists $c_0 \geq 1$, such that for all $\psi \in H_{00}^{1/2}(\gamma)$ we have*

$$c_0^{-1}\|\psi\|_2 \leq \|\psi\|_1 \leq c_0\|\psi\|_2.$$

Proof Exercise.   □

To simplify notation, we define

$$a_j(v, w) = \int_{\Omega_j} \nabla v \cdot \nabla w \, dx, \quad b_j(w) = \int_{\Omega_j} fw \, dx.$$

We then have that $u_1^{k+1} \in H_{\Gamma_1}^1(\Omega_1)$ satisfies

$$a_1(u_1^{k+1}, v_1) = b_1(v_1), \quad u_1^{k+1}|_\gamma = \lambda^k,$$

for all $v_1 \in H_0^1(\Omega_1)$. Moreover, we have that $u_2^{k+1} \in H_{\Gamma_2}^1(\Omega_2)$ satisfies

$$a_2(u_2^{k+1}, v_2) = b_2(v_2) - a_1(u_1^{k+1}, H_1v_2|_\gamma) + b_1(H_1v_2|_\gamma)$$

for all $v_2 \in H_{\Gamma_2}^1(\Omega_2)$, where we used the harmonic extension of $v_2|_\gamma$ to Ω_1 to incorporate the Neumann condition on γ. We let $u \in H_0^1(\Omega)$ be the solution of the

Poisson problem with f on the right-hand side, i.e., u satisfies

$$a_2(u, v_2) = b_1(H_1 v_2|_\gamma) + b_2(v_2) - a_1(u, H_1 v_2|_\gamma)$$

for all $v_2 \in H^1_{\Gamma_2}(\Omega_2)$ with harmonic extension $H_1 v_2|_\gamma \in H^1_{\Gamma_1}(\Omega_1)$ to Ω_1.

Lemma 5.6 (Interface Equation) *Define* $T : H^{1/2}_{00}(\gamma) \to H^{1/2}_{00}(\gamma)$ *by* $T\psi = w_2|_\gamma$, *where* $w_2 \in H^1_{\Gamma_2}(\Omega_2)$ *solves*

$$-\Delta w_2 = 0 \text{ in } \Omega_2, \quad w_2|_{\Gamma_2} = 0, \quad \partial_{n_2} w_2|_\gamma = -\partial_{n_1} H_1 \psi|_\gamma.$$

For the interface error $\delta^k = \lambda^k - u|_\gamma$ *and* $k = 0, 1, \ldots,$ *we then have that*

$$\delta^{k+1} = \theta T \delta^k + (1 - \theta) \delta^k.$$

Proof For $j = 1, 2$ set $e^{k+1}_j = u^{k+1}_j - u|_{\Omega_j}$. We then have that $a_1(e^{k+1}_1, v_1) = 0$ for all $v_1 \in H^1_0(\Omega_1)$, i.e.,

$$e^{k+1}_1 = H_1 \delta^k.$$

Moreover, for all $v_2 \in H^1_{\Gamma_2}(\Omega_2)$ we have that

$$a_2(e^{k+1}_2, v_2) = -a_1(e^{k+1}_1, H_1 v_2|_\gamma).$$

The combination of identities yields

$$a_2(e^{k+1}_2, v_2) = -a_1(H_1 \delta^k, H_1 v_2|_\gamma)$$

for all $v_2 \in H^1_{\Gamma_2}(\Omega_2)$, which is equivalent to

$$T\delta^k = e^{k+1}_2|_\gamma.$$

Together with the error equation on the interface, which follows from subtracting $u|_\gamma$ from the equation for λ^{k+1} in Algorithm 5.2,

$$\delta^{k+1} = \theta e^{k+1}_2|_\gamma + (1 - \theta)\delta^k,$$

we deduce the asserted identity. $\square$

Note that $w_2 \in H^1_{\Gamma_2}(\Omega_2)$ coincides with the harmonic extension of its own boundary data w_2 on γ, which by definition of T is the same as $T\psi$, so that

$$a_2(H_2 T\psi, v_2) = -a_1(H_1 \psi, H_1 v_2|_\gamma)$$

for all $v_2 \in H^1_{\Gamma_2}(\Omega_2)$. For sufficiently small θ, the equation on the interface defines a contraction.

Theorem 5.3 (Contraction) *There exists $\theta^* > 0$ such that, for $0 < \theta < \theta^*$, the linear operator $T_\theta = \theta T + (1 - \theta)\,\mathrm{id} : H_{00}^{1/2}(\gamma) \to H_{00}^{1/2}(\gamma)$ is a contraction.*

Proof From the preceding identity for $a_2(H_2 T\psi, v_2)$ with $v_2 = H_2 T\psi$, we find that

$$a_1(H_1\psi, H_1 T\psi) = -a_2(H_2 T\psi, H_2 T\psi) = -\|\nabla H_2 T\psi\|_{L^2(\Omega_2)}^2 = -\|T\psi\|_2^2,$$

where we used that $H_2 T\psi|_\gamma = T\psi$ and hence $H_1(H_2 T\psi)|_\gamma = H_1 T\psi$. Using the equivalence of $\|\cdot\|_1$ and $\|\cdot\|_2$, we deduce that

$$\|T\psi\|_2^2 = -a_1(H_1\psi, H_1 T\psi) \leq \|\nabla H_1\psi\|_{L^2(\Omega_1)} \|\nabla H_1 T\psi\|_{L^2(\Omega_1)}$$
$$= \|\psi\|_1 \|T\psi\|_1 \leq c_0 \|\psi\|_1 \|T\psi\|_2,$$

which implies that

$$\|T\psi\|_2 \leq c_0 \|\psi\|_1$$

and

$$\|T\psi\|_1 \leq c_0 \|T\psi\|_2 \leq c_0^2 \|\psi\|_1.$$

Similarly, with the identity for $a_2(H_2 T\psi, v_2)$ we find that

$$\|\psi\|_1^2 = a_1(H_1\psi, H_1\psi) = -a_2(H_2 T\psi, H_2\psi)$$
$$\leq \|\nabla H_2 T\psi\|_{L^2(\Omega_2)} \|\nabla H_2\psi\|_{L^2(\Omega_2)}$$
$$= \|T\psi\|_2 \|\psi\|_2 \leq c_0 \|T\psi\|_2 \|\psi\|_1,$$

i.e.,

$$\|T\psi\|_2 \geq c_0^{-1} \|\psi\|_1.$$

We can now estimate the norm of the operator T_θ. We have that

$$\|T_\theta\psi\|_1^2 = \theta^2 \|T\psi\|_1^2 + (1 - \theta)^2 \|\psi\|_1^2 + 2\theta(1 - \theta)a_1(H_1 T\psi, H_1\psi)$$
$$= \theta^2 \|T\psi\|_1^2 + (1 - \theta)^2 \|\psi\|_1^2 - 2\theta(1 - \theta)\|T\psi\|_2^2$$
$$\leq \theta^2 \|T\psi\|_1^2 + (1 - \theta)^2 \|\psi\|_1^2 - 2c_0^{-2}\theta(1 - \theta)\|\psi\|_1^2$$
$$\leq c_0^4 \theta^2 \|\psi\|_1^2 + (1 - \theta)^2 \|\psi\|_1^2 - 2c_0^{-2}\theta(1 - \theta)\|\psi\|_1^2$$
$$= \left(c_0^4 \theta^2 + (1 - \theta)^2 - 2c_0^{-2}\theta(1 - \theta)\right)\|\psi\|_1^2$$
$$= K_\theta^2 \|\psi\|_1^2.$$

A straightforward calculation shows that $K_\theta < 1$ if

$$\theta < \theta^* = \frac{2(1 + c_0^{-2})}{1 + c_0^4 + 2c_0^{-2}},$$

which proves the contraction property. $\qquad\square$

The theorem implies that the functions $u_1^k|_\gamma = \lambda^{k-1}$ converge in $H_{00}^{1/2}(\gamma)$. It remains to show that also the functions u_1^k and u_2^k converge. Then it follows from Lemma 5.4 that the limits u_1 and u_2 coincide with the solution restricted to Ω_1 and Ω_2, respectively.

Lemma 5.7 (Convergence) *Suppose that the sequence $(u_1^k|_\gamma)_{k\in\mathbb{N}}$ converges in $H_{00}^{1/2}(\gamma)$. Then for $j = 1, 2$, the sequence $(u_j^k)_{k\in\mathbb{N}}$ converges in $H^1(\Omega_j)$ to $u|_{\Omega_j}$ as $k \to \infty$.*

Proof Since $u_1^k - u_1^\ell = H_1(u_1^k - u_1^\ell)|_\gamma$ we have that

$$\|\nabla(u_1^k - u_1^\ell)\|_{L^2(\Omega_1)} = \|\nabla H_1(u_1^k - u_1^\ell)|_\gamma\|_{L^2(\Omega_1)} = \||(u_1^k - u_1^\ell)|_\gamma\||_1,$$

and hence, $(u_1^k)_{k\in\mathbb{N}}$ is a Cauchy sequence in $H^1(\Omega_1)$. The identities $u_1^{k+1}|_\gamma = \lambda^k$ and $\lambda^k = \theta u_2^k + (1 - \theta)\lambda^{k-1}$ imply that

$$\theta u_2^k|_\gamma = (u_1^{k+1} - u_1^k)|_\gamma + \theta u_1^k|_\gamma$$

so that the sequence $(u_2^k|_\gamma)_{k\in\mathbb{N}}$ has the same limit as $(u_1^k|_\gamma)_{k\in\mathbb{N}}$. Using that

$$\|\nabla(u_2^k - u_2^\ell)\|_{L^2(\Omega_2)}^2 = -a_1(u_1^k - u_1^\ell, H_1(u_2^k - u_2^\ell)|_\gamma)$$

$$\leq \|\nabla(u_1^k - u_1^\ell)\|_{L^2(\Omega_1)}\||(u_2^k - u_2^\ell)|_\gamma\||_1,$$

we find that $(u_2^k)_{k\in\mathbb{N}}$ is a Cauchy sequence in $H^1(\Omega_2)$. $\qquad\square$

5.2.3 Overlapping Schwarz Method

The overlapping Schwarz method was introduced to establish the existence of solutions for partial differential equations on domains that are representable as unions of simple domains such as disks and rectangles on which the equation can be solved analytically. We assume that

$$\Omega = \Omega_1 \cup \Omega_2$$

such that the *overlap region* $\Omega_{12} = \Omega_1 \cap \Omega_2$ is a Lipschitz domain, and set

$$\gamma_j = \partial\Omega_j \cap \partial\Omega_{12}, \qquad \Gamma_j = \partial\Omega_j \cap \partial\Omega$$

for $j = 1, 2$, cf. Fig. 5.10.

The function $u \in H_0^1(\Omega)$ solves the Poisson problem with $f \in L^2(\Omega)$ on the right-hand side if and only if the functions $u_j = u|_{\Omega_j} \in H^1(\Omega_j)$ solve

$$-\Delta u_j = f \text{ in } \Omega_j, \quad u_j|_{\Gamma_j} = 0$$

for $j = 1, 2$ and satisfy $u_1 = u_2$ in Ω_{12}. In an iterative algorithm, we use the trace of the solution in Ω_2 on γ_1 to define the Dirichlet data for problems in Ω_1 and vice versa.

Algorithm 5.3 (Alternating Schwarz Method) *Choose $u^0 \in H_0^1(\Omega)$, define $u_j^0 = u^0|_{\Omega_j}$ for $j = 1, 2$, let $\varepsilon_{\text{stop}} > 0$, and set $k = 0$.*

(1) Determine $u_1^{k+1} \in H^1(\Omega_1)$ with

$$-\Delta u_1^{k+1} = f \text{ in } \Omega_1, \quad u_1^{k+1}|_{\Gamma_1} = 0, \quad u_1^{k+1}|_{\gamma_1} = u_2^k|_{\gamma_1}.$$

(2) Determine $u_2^{k+1} \in H^1(\Omega_2)$ with

$$-\Delta u_2^{k+1} = f \text{ in } \Omega_2, \quad u_2^{k+1}|_{\Gamma_2} = 0, \quad u_2^{k+1}|_{\gamma_2} = u_1^{k+1}|_{\gamma_2}.$$

(3) Stop if $\|\nabla(u_1^{k+1} - u_2^{k+1})\|_{L^2(\Omega_{12})} \leq \varepsilon_{\text{stop}}$; set $k = k+1$ and continue with Step (1).

No numerical parameter is needed in the algorithm but the speed of convergence depends on the radius of the inner circle of the overlap region, cf. Fig. 5.11.

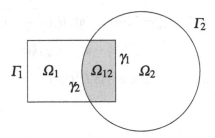

Fig. 5.10 Overlapping partition of Ω with overlap region Ω_{12}

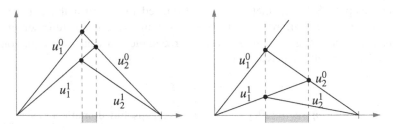

Fig. 5.11 The speed of convergence depends on the diameter of the overlap region Ω_{12}

Remark 5.7 If Steps (1) and (2) are decoupled by using the boundary condition $u_2^{k+1}|_{\gamma_2} = u_1^k|_{\gamma_2}$ in Step (2), then the algorithm can be parallelized.

To analyze the algorithm, we transform the boundary conditions and rewrite the problems in weak form. For this we use the spaces

$$V = H_0^1(\Omega), \quad V_j = \{v \in V : v = 0 \text{ on } \Omega \setminus \Omega_j\},$$

i.e., the functions in $H_0^1(\Omega_j)$ that are extended trivially to the whole domain Ω for $j = 1, 2$. We can then consider the bilinear form a and the linear form b defined on $V \times V$ and V by

$$a(v, w) = \int_\Omega \nabla v \cdot \nabla w \, dx, \quad b(v) = \int_\Omega f v \, dx.$$

Instead of the functions u_1 and u_2, we compute w_1 and w_2 with homogeneous boundary data, i.e., given some $u^0 \in V$, we set $k = 0$ and compute $w_1^k \in V_1$ such that

$$a(w_1^k, v_1) = b(v_1) - a(u^k, v_1)$$

for all $v_1 \in V_1$. We then set $u^{k+1/2} = u^k + w_1^k$ and determine $w_2^k \in V_2$ such that

$$a(w_2^k, v_2) = b(v_2) - a(u^{k+1/2}, v_2)$$

for all $v_2 \in V_2$. The new iterate is $u^{k+1} = u^{k+1/2} + w_2^k$. These two steps are iterated until we have $\|\nabla(u^k - u^{k+1})\|_{L^2(\Omega)} \le \varepsilon_{\text{stop}}$.

Lemma 5.8 (Iterated Projection) *For $j = 1, 2$, let $\mathscr{P}_j : V \to V_j$ denote the orthogonal projection onto V_j with respect to the scalar product $a(\cdot, \cdot)$. We then have that*

$$u - u^{k+1} = (I - \mathscr{P}_2)(I - \mathscr{P}_1)(u - u^k).$$

Proof Using that $a(u, v_1) = b(v_1)$ for all $v_1 \in V_1$, we have

$$a(u^{k+1/2} - u^k, v_1) = a(w_1^k, v_1) = a(u - u^k, v_1),$$

which is equivalent to $u^{k+1/2} - u^k = \mathscr{P}_1(u - u^k)$. Similarly, the identity

$$a(u^{k+1} - u^{k+1/2}, v_2) = a(w_2^k, v_2) = a(u - u^{k+1/2}, v_2)$$

shows that $u^{k+1} - u^{k+1/2} = \mathscr{P}_2(u - u^{k+1/2})$. We thus have that

$$u - u^{k+1/2} = u - u^k + u^k - u^{k+1/2} = (I - \mathscr{P}_1)(u - u^k)$$

and

$$u - u^{k+1} = u - u^{k+1/2} + u^{k+1/2} - u^{k+1} = (I - \mathscr{P}_2)(u - u^{k+1/2}).$$

The combination of the two identities proves the result.   □

Remark 5.8 The lemma explains the terminology *multiplicative iteration*. For the decoupled version of the algorithm one can prove that

$$u - u^{k+1} = (I - \mathscr{P}_2 - \mathscr{P}_1)(u - u^k),$$

which motivates the terminology *additive iteration*.

To prove convergence of the alternating Schwarz method, we use the following result.

Lemma 5.9 *There exists $c_1 \geq 1$ such that for all $v \in V$, we have*

$$\|\nabla v\| \leq c_1 \big(\|\nabla \mathscr{P}_1 v\|^2 + \|\nabla \mathscr{P}_2 v\|^2 \big)^{1/2}.$$

Proof For $v \in V$ let $H_{12}v|_{\gamma_2}$ denote the harmonic extension of $v|_{\gamma_2}$ to Ω_{12}, such that $H_{12}v$ vanishes on γ_1. Let $v_1 \in V_1$ be defined by $v_1 = v$ in $\Omega_1 \setminus \Omega_{12}$ and $v_1 = H_{12}v|_{\gamma_2}$ in Ω_{12}. Let $v_2 \in V_2$ be defined by $v_2 = v$ in $\Omega_2 \setminus \Omega_{12}$ and $v_2 = v - H_{12}v|_{\gamma_2}$ in Ω_{12}. Then $v = v_1 + v_2$ and

$$\|\nabla v_1\|_{L^2(\Omega_1)} + \|\nabla v_2\|_{L^2(\Omega_2)} \leq c_1 \|\nabla v\|_{L^2(\Omega)}.$$

The Cauchy–Schwarz inequality implies that

$$\begin{aligned}
\|\nabla v\|_{L^2(\Omega)}^2 &= a(v_1, v) + a(v_2, v) \\
&= a(v_1, \mathscr{P}_1 v) + a(v_2, \mathscr{P}_2 v) \\
&\leq \|\nabla v_1\|_{L^2(\Omega_1)} \|\nabla \mathscr{P}_1 v\|_{L^2(\Omega)} + \|\nabla v_2\|_{L^2(\Omega_2)} \|\nabla \mathscr{P}_2 v\|_{L^2(\Omega_2)} \\
&\leq \big(\|\nabla v_1\|_{L^2(\Omega_1)}^2 + \|\nabla v_2\|_{L^2(\Omega_2)}^2 \big)^{1/2} \big(\|\nabla \mathscr{P}_1 v\|_{L^2(\Omega_1)}^2 + \|\nabla \mathscr{P}_2 v\|_{L^2(\Omega_2)}^2 \big)^{1/2} \\
&\leq c_1 \|\nabla v\|_{L^2(\Omega)} \big(\|\nabla \mathscr{P}_1 v\|_{L^2(\Omega_1)}^2 + \|\nabla \mathscr{P}_2 v\|_{L^2(\Omega_2)}^2 \big)^{1/2}
\end{aligned}$$

and this proves the estimate.   □

We are now in position to verify convergence of the alternating Schwarz method.

Theorem 5.4 (Convergence) *The operator $(I - \mathscr{P}_2)(I - \mathscr{P}_1)$ is a contraction, i.e., the alternating Schwarz method converges.*

Proof Since $\mathscr{P}_1(I - \mathscr{P}_1)w = 0$ for all $w \in V$, from the previous lemma for $v = (I - \mathscr{P}_1)w$ we deduce that

$$\|\nabla(I - \mathscr{P}_1)w\|_{L^2(\Omega_1)} \leq c_1 \|\nabla\mathscr{P}_2(I - \mathscr{P}_1)w\|_{L^2(\Omega_2)}.$$

Since $\mathscr{P}_2 v$ and $(I - \mathscr{P}_2)v$ are orthogonal with respect to $a(\cdot, \cdot)$, we have that

$$\|\nabla(I - \mathscr{P}_1)w\|^2_{L^2(\Omega_1)} = \|\nabla(I - \mathscr{P}_2)(I - \mathscr{P}_1)w\|^2_{L^2(\Omega_2)} + \|\nabla\mathscr{P}_2(I - \mathscr{P}_1)w\|^2_{L^2(\Omega_2)}$$

$$\geq \|\nabla(I - \mathscr{P}_2)(I - \mathscr{P}_1)w\|^2_{L^2(\Omega_2)} + c_1^{-2}\|\nabla(I - \mathscr{P}_1)w\|^2_{L^2(\Omega_1)},$$

i.e.,

$$\|\nabla(I - \mathscr{P}_2)(I - \mathscr{P}_1)w\|^2_{L^2(\Omega_2)} \leq (1 - c_1^{-2})\|\nabla(I - \mathscr{P}_1)w\|^2_{L^2(\Omega_1)}$$

$$\leq (1 - c_1^{-2})\|\nabla w\|^2_{L^2(\Omega)}.$$

Noting $c_1 \geq 1$, this proves the contraction property. $\qquad\qquad\square$

5.3 Preconditioning

5.3.1 Preconditioned CG Algorithm

The solution of a linear system $Ax = b$ with the conjugate gradient method leads to the convergence result

$$\|x^k - x\|_A \leq 2\left(\frac{\kappa^{1/2} - 1}{\kappa^{1/2} + 1}\right)^k \|x^0 - x\|_A$$

with $\kappa = \text{cond}_2(A)$, provided that A is symmetric and positive definite. For the finite element stiffness matrix of the Poisson problem, we have $\kappa \sim h^{-2}$, and a large number of iterations is required to guarantee $\|x^j - x\|_A \leq \varepsilon$ for some given tolerance $\varepsilon > 0$. If we could construct an invertible matrix C such that $\text{cond}_2(CA) \ll \text{cond}_2(A)$, then we could consider the equivalent linear system $CAx = Cb$. The best possible choice is $C = A^{-1}$, but then the multiplication by C would be as expensive as the direct solution of the original system. Thus a good compromise between reducing the condition number and the cost of matrix-vector multiplications has to be achieved.

Definition 5.2 A regular matrix $C \in \mathbb{R}^{n \times n}$ is a *preconditioner* for the regular matrix $A \in \mathbb{R}^{n \times n}$ if the condition number of CA is asymptotically smaller than the condition number of A, and the computation of matrix-vector products $r \mapsto Cr$ is cheaper than the direct solution of $Ax = b$.

We apply the conjugate gradient algorithm to the equation $CAx = Cb$.

Algorithm 5.4 (PCG Algorithm) *Let $A, C \in \mathbb{R}^{n \times n}$ and $b \in \mathbb{R}^n$. Let $x^0 \in \mathbb{R}^n$, $\varepsilon_{stop} > 0$, and set $r^0 = b - Ax^0$, $d^0 = z^0 = Cr^0$, $k = 0$.*

(1) Set $x^{k+1} = x^k + \alpha_k d^k$, $r^{k+1} = r^k - \alpha_k A d^k$, $z^{k+1} = Cr^{k+1}$, and $d^{k+1} = z^{k+1} + \beta_k d^k$, where

$$\alpha_k = \frac{r^k \cdot z^k}{A d^k \cdot d^k}, \quad \beta_k = \frac{r^{k+1} \cdot z^{k+1}}{r^k \cdot z^k}.$$

(2) Stop if $|r^{k+1}| \leq \varepsilon_{stop}$; set $k \to k+1$ and continue with (1) otherwise.

The standard convergence result requires CA to be positive definite and symmetric, which is often difficult to guarantee. Instead, one ensures a factorization $C = KK^\top$, e.g., a Cholesky factorization, and considers the equivalent linear system

$$(K^\top A K) K^{-1} x = K^\top b.$$

The factorization $C = KK^\top$ is irrelevant for formulating the preconditioned CG algorithm, but could be of interest for an efficient computation of matrix-vector products.

Examples 5.2

(i) Let $D \in \mathbb{R}^{n \times n}$ denote the diagonal part of A and assume that D is invertible. Then $C = D^{-1}$ is called the *diagonal* or *Jacobi preconditioner*.

(ii) Define the diagonal matrix $D \in \mathbb{R}^{n \times n}$ by $d_{ii} = \sum_{j=1}^n |a_{ij}|$ for $i = 1, 2, \ldots, n$. Then $\text{cond}_\infty(D^{-1}A) \leq \text{cond}_\infty(A)$ and $C = D^{-1}$ is called an *equilibration preconditioner*.

(iii) If $x^{k+1} = x^k - M(Ax^k - b)$ is a convergent iteration, then often $C = M$ defines a preconditioner, e.g., if M is the inverse of the lower triangular part of A, then M is called the *Gauss–Seidel preconditioner*. A symmetric factorization is achieved by considering

$$C = \left[(D + L) D^{-1} (D + L)^\top \right]^{-1},$$

where D and L denote the diagonal and lower triangular parts of A, and the inversion is understood in the sense of successive elimination.

(iv) Let $A = LL^\top$ be the Cholesky factorization of A and define $\widetilde{L} \in \mathbb{R}^{n \times n}$ such that the population pattern of A is preserved, i.e.,

$$\widetilde{L}_{ij} = \begin{cases} L_{ij} & \text{if } A_{ij} \neq 0, \\ 0 & \text{else.} \end{cases}$$

Then $A = \widetilde{L}\widetilde{L}^\top + E$ with an error term $E \in \mathbb{R}^{n \times n}$ and $C = (\widetilde{L}\widetilde{L}^\top)^{-1}$ is called the *incomplete Cholesky preconditioner*.

These *blackbox preconditioners* often lead to an improvement in the performance of the conjugate gradient algorithm. Whether they are preconditioners in the sense of Definition 5.2 depends on the specific properties of the problem under consideration.

Remark 5.9 The construction of preconditioners is closely related to certain norm equivalences, i.e., if $c_1 \|x\|_{C^{-1}} \le \|x\|_A \le c_2 \|x\|_{C^{-1}}$ for all $x \in \mathbb{R}^n$, where $\|x\|_B^2 = Bx \cdot x$, then we have $\mathrm{cond}_2(CA) \le c_2/c_1$.

5.3.2 Abstract Multilevel Preconditioner

We follow [5] and assume that we are given a sequence of nested finite-dimensional spaces

$$V_0 \subset V_1 \subset \cdots \subset V_L = V$$

with increasing dimensions $n_\ell = \dim(V_\ell)$ and $n = \dim(V)$. We assume that for $\ell = 0, 1, \ldots, L$, we are given injective linear operators

$$P_\ell : V_\ell \to V.$$

Moreover, we assume that for each space V_ℓ we are given a basis, and accordingly identify operators $T : V_\ell \to V_j$ with matrices $T \in \mathbb{R}^{n_j \times n_\ell}$.

Definition 5.3 For symmetric, positive definite matrices $B_\ell \in \mathbb{R}^{n_\ell \times n_\ell}$, $\ell = 0, 1, \ldots, L$, a *multilevel preconditioner* $C \in \mathbb{R}^{n \times n}$ is defined by

$$C = \sum_{\ell=0}^{L} P_\ell B_\ell^{-1} P_\ell^\top.$$

To justify C as a preconditioner, we first show that it is symmetric and positive definite.

Lemma 5.10 (Symmetry and Definiteness) *The multilevel preconditioner is symmetric and positive definite.*

Proof The symmetry of C is a direct consequence of its definition. The definiteness of B_ℓ^{-1}, $\ell = 0, \ldots, L$, implies that for every $v \in V$, we have

$$v \cdot Cv = \sum_{\ell=0}^{L} (P_\ell^\top v) \cdot (B_\ell^{-1} P_\ell^\top v) \ge 0,$$

i.e., C is positive semidefinite. Since P_L is invertible, we find that C is positive definite. □

The inverse of C can be represented in terms of the matrices B_ℓ, $\ell = 0, 1, \ldots, L$.

Lemma 5.11 (Inverse Matrix) *For every $v \in V$ we have*

$$C^{-1}v \cdot v = \min_{P_\ell v_\ell = v} \sum_{\ell=0}^{L} B_\ell v_\ell \cdot v_\ell,$$

where the minimum is over all tuples $(v_\ell) = (v_0, v_1, \ldots, v_L) \in \prod_{\ell=0}^{L} V_\ell$ with $v = \sum_{\ell=0}^{L} P_\ell v_\ell$.

Proof Since B_ℓ^{-1} is symmetric and positive definite, the Cauchy–Schwarz inequality implies that for $y_\ell, z_\ell \in V_\ell$, we have

$$y_\ell \cdot B_\ell^{-1} z_\ell \leq \left(y_\ell \cdot B_\ell^{-1} y_\ell\right)^{1/2} \left(z_\ell \cdot B_\ell^{-1} z_\ell\right)^{1/2}.$$

For $v = \sum_{\ell=0}^{L} P_\ell v_\ell$ with $v_\ell \in V_\ell$, we thus have that

$$C^{-1}v \cdot v = C^{-1}v \cdot \sum_{\ell=0}^{L} P_\ell v_\ell = \sum_{\ell=0}^{L} P_\ell^\top C^{-1}v \cdot \left(B_\ell^{-1} B_\ell v_\ell\right)$$

$$\leq \sum_{\ell=0}^{L} \left[\left(P_\ell^\top C^{-1}v\right) \cdot B_\ell^{-1}\left(P_\ell^\top C^{-1}v\right)\right]^{1/2} \left[B_\ell v_\ell \cdot B_\ell^{-1} B_\ell v_\ell\right]^{1/2}.$$

The Cauchy–Schwarz inequality in $\mathbb{R}^{L+1}$ and the definition of C lead to

$$C^{-1}v \cdot v \leq \left[\sum_{\ell=0}^{L} \left(P_\ell^\top C^{-1}v\right) \cdot B_\ell^{-1}\left(P_\ell^\top C^{-1}v\right)\right]^{1/2} \left[\sum_{\ell=0}^{L} B_\ell v_\ell \cdot v_\ell\right]^{1/2}$$

$$= \left[C^{-1}v \cdot \left(\sum_{\ell=0}^{L} P_\ell B_\ell^{-1} P_\ell^\top C^{-1}v\right)\right]^{1/2} \left[\sum_{\ell=0}^{L} B_\ell v_\ell \cdot v_\ell\right]^{1/2}$$

$$= \left(C^{-1}v \cdot v\right)^{1/2} \left[\sum_{\ell=0}^{L} B_\ell v_\ell \cdot v_\ell\right]^{1/2}$$

so that

$$C^{-1}v \cdot v \leq \sum_{\ell=0}^{L} B_\ell v_\ell \cdot v_\ell.$$

To prove the assertion, we have to show that the minimum on the right-hand side equals the value on the left-hand side. For this we define for $\ell = 0, 1, \ldots, L$,

$$v_\ell = B_\ell^{-1} P_\ell^\top C^{-1}v.$$

By definition of C, we have $v = \sum_{\ell=0}^{L} P_\ell v_\ell$. Moreover, by using the definitions of v_ℓ and C, we verify that

$$\sum_{\ell=0}^{L} B_\ell v_\ell \cdot v_\ell = \sum_{\ell=0}^{L} C^{-1} v \cdot \left(P_\ell B_\ell^{-1} P_\ell^\top C^{-1} v \right) = C^{-1} v \cdot v,$$

which concludes the proof. □

Together with the lemmas, we obtain useful characterizations of the extremal eigenvalues of a matrix product CA.

Theorem 5.5 (Eigenvalues of CA) *Let $A \in \mathbb{R}^{n \times n}$ be symmetric and positive definite. The minimal and maximal eigenvalue of CA are the extrema of the mapping*

$$v \mapsto \frac{Av \cdot v}{\min_{P_\ell v_\ell = v} \sum_{\ell=0}^{L} B_\ell v_\ell \cdot v_\ell}$$

in the set of all $v \in V \setminus \{0\}$.

Proof The bilinear form $\langle v, w \rangle_{C^{-1}} = C^{-1} v \cdot w$ defines a scalar product on $\mathbb{R}^n$, and the matrix CA is symmetric and positive definite with respect to this scalar product, i.e., $\langle CAv, w \rangle_{C^{-1}} = \langle v, CAw \rangle_{C^{-1}}$ and $\langle CAv, v \rangle_{C^{-1}} > 0$ for all $v, w \in V \setminus \{0\}$. Hence, the eigenvalues of CA are given as the extrema of the Rayleigh quotient

$$R(v) = \frac{\langle CAv, v \rangle_{C^{-1}}}{\langle v, v \rangle_{C^{-1}}}$$

in the set $v \in V \setminus \{0\}$. The characterization of $\langle v, v \rangle_{C^{-1}} = C^{-1} v \cdot v$ in Lemma 5.11 implies the assertion. □

Remark 5.10 Instead of assuming that the operators $P_\ell : V_\ell \to V$ are injective, it is sufficient to assume that for every $v \in V$, there exists $(v_\ell)_{\ell=0,\dots,L} \in \prod_{\ell=0}^{L} V_\ell$ with $v = \sum_{\ell=0}^{L} P_\ell v_\ell$.

5.3.3 BPX Preconditioner

We apply the abstract framework to a sequence of refined, quasiuniform triangulations $\mathcal{T}_0, \mathcal{T}_1, \dots, \mathcal{T}_L$ and $\mathcal{T}_h = \mathcal{T}_L$ with associated finite element spaces

$$\mathcal{S}_0^1(\mathcal{T}_0) \subset \mathcal{S}_0^1(\mathcal{T}_1) \subset \cdots \subset \mathcal{S}_0^1(\mathcal{T}_L) = \mathcal{S}_0^1(\mathcal{T}_h),$$

with maximal mesh-sizes $h_\ell = \gamma^\ell h_0$, $\ell = 0, 1, \dots, L$, for a mesh-size reduction factor $0 < \gamma < 1$, cf. Fig. 5.12. We follow [17].

Fig. 5.12 Sequence of
refined, quasiuniform
triangulations; the associated
finite element spaces are
nested

It is our goal to efficiently solve the discrete Poisson problem on the finest
triangulation $\mathscr{T}_h$, i.e., to approximately determine $u_h \in \mathscr{S}_0^1(\mathscr{T}_h)$ with

$$\int_\Omega \nabla u_h \cdot \nabla v_h \, dx = \int_\Omega f v_h \, dx$$

for all $v_h \in \mathscr{S}_0^1(\mathscr{T}_h)$. This is equivalent to a regular linear system of equations
$Ax = b$ with a positive definite and symmetric matrix $A \in \mathbb{R}^{n \times n}$. We assume that the
Poisson problem in Ω is H^2-regular.

Lemma 5.12 (Stable Decomposition) *For every $v_h \in \mathscr{S}_0^1(\mathscr{T}_h)$, there exists $w_\ell \in$
$\mathscr{S}_0^1(\mathscr{T}_\ell)$, $\ell = 0, 1, \ldots, L$, such that $v_h = \sum_{\ell=0}^L w_\ell$ and*

$$\sum_{\ell=0}^L h_\ell^{-2} \|w_\ell\|_{L^2(\Omega)}^2 \le c_{\mathrm{dec}} \|\nabla v_h\|_{L^2(\Omega)}^2.$$

Proof

(i) By the Lax–Milgram lemma there exists for every $\ell = 0, 1, \ldots, L$, a uniquely
    defined function $v_\ell = Q_\ell v_h \in \mathscr{S}_0^1(\mathscr{T}_\ell)$ such that

$$\int_\Omega \nabla v_\ell \cdot \nabla r_\ell \, dx = \int_\Omega \nabla v_h \cdot \nabla r_\ell \, dx$$

for all $r_\ell \in \mathscr{S}_0^1(\mathscr{T}_\ell)$. Letting $z \in H^2(\Omega) \cap H_0^1(\Omega)$ be the weak solution of the
Poisson problem

$$-\Delta z = v_\ell - v_h,$$

we find that

$$\|v_\ell - v_h\|_{L^2(\Omega)}^2 = \int_\Omega \nabla(z - \mathscr{I}_\ell z) \cdot \nabla(v_\ell - v_h) \, dx$$

$$\le \|\nabla(z - \mathscr{I}_\ell z)\|_{L^2(\Omega)} \|\nabla(v_\ell - v_h)\|_{L^2(\Omega)},$$

where $\mathscr{I}_\ell : H^2(\Omega) \cap H_0^1(\Omega) \to \mathscr{S}_0^1(\mathscr{T}_\ell)$ is the nodal interpolation operator on
$\mathscr{T}_\ell$. Using interpolation estimates and the assumed H^2-regularity

$$\|D^2 z\|_{L^2(\Omega)} \le c_2 \|v_\ell - v_h\|_{L^2(\Omega)}$$

yield that

$$\|(Q_\ell - \mathrm{id})v_h\|_{L^2(\Omega)} = \|v_\ell - v_h\|_{L^2(\Omega)} \leq c_{\mathscr{I}} c_2 h_\ell \|\nabla(v_\ell - v_h)\|_{L^2(\Omega)}.$$

(ii) We formally define $Q_{-1} = 0$, and note that $Q_L = \mathrm{id}$ on $\mathscr{S}_0^1(\mathscr{T}_L)$. Therefore, by a telescope argument, we have

$$v_h = \sum_{\ell=0}^{L} (Q_\ell - Q_{\ell-1})v_h,$$

and we define $w_\ell = (Q_\ell - Q_{\ell-1})v_h \in \mathscr{S}_0^1(\mathscr{T}_\ell)$. Noting that $Q_\ell \circ Q_j = Q_\ell$ for $\ell \leq j$, we have the operator identity

$$Q_\ell - Q_{\ell-1} = \big(\mathrm{id} - Q_{\ell-1}\big)(Q_\ell - Q_{\ell-1}),$$

which allows us to verify that

$$\sum_{\ell=0}^{L} h_\ell^{-2}\|(Q_\ell - Q_{\ell-1})v_h\|_{L^2(\Omega)}^2 \leq c_{\mathscr{I}}^2 c_2^2 \gamma^{-2} \sum_{\ell=0}^{L} \|\nabla(Q_\ell - Q_{\ell-1})v_h\|_{L^2(\Omega)}^2.$$

For $0 \leq \ell, m \leq L$ with $\ell \neq m$, we have that

$$\int_\Omega \big[\nabla(Q_\ell - Q_{\ell-1})v_h\big] \cdot \big[\nabla(Q_m - Q_{m-1})v_h\big]\, \mathrm{d}x = 0,$$

which implies that

$$\Big\|\nabla \sum_{\ell=0}^{L} (Q_\ell - Q_{\ell-1})v_h\Big\|_{L^2(\Omega)}^2 = \sum_{\ell=0}^{L} \|\nabla(Q_\ell - Q_{\ell-1})v_h\|_{L^2(\Omega)}^2.$$

A combination of the estimates proves the lemma. □

The second important ingredient for the analysis of multilevel preconditioners is a variant of the Cauchy–Schwarz inequality.

Lemma 5.13 (Strengthened Cauchy–Schwarz Inequality) *Let $v_\ell \in V_\ell$ and $w_m \in V_m$ for $\ell \geq m$ and assume that $h_\ell \leq \gamma^{\ell-m}h_m$. We then have*

$$\int_\Omega \nabla v_\ell \cdot \nabla w_m\, \mathrm{d}x \leq c_{\mathrm{scs}}\gamma^{(\ell-m)/2}h_\ell^{-1}\|v_\ell\|_{L^2(\Omega)}\|\nabla w_m\|_{L^2(\Omega)}.$$

Proof Let $T \in \mathscr{T}_m$ and note that $\Delta w_m|_T = 0$. An integration-by-parts yields that

$$\int_T \nabla v_\ell \cdot \nabla w_m\, \mathrm{d}x = \int_{\partial T} v_\ell(\nabla w_m \cdot n)\, \mathrm{d}s \leq \|v_\ell\|_{L^2(\partial T)}\|\nabla w_m\|_{L^2(\partial T)}.$$

Since $\nabla w_m|_T$ is constant, we have

$$\|\nabla w_m\|_{L^2(\partial T)} \le c h_m^{-1/2} \|\nabla w_m\|_{L^2(T)}.$$

There exist sides S_i, $i = 1, 2, \dots, I$, in the triangulation $\mathcal{T}_\ell$, such that $S_1 \cup \cdots \cup S_I = \partial T$. With elements $T_i \in \mathcal{T}_\ell$ such that $S_i \subset \partial T_i$ for $i = 1, 2, \dots, I$, cf. Fig. 5.13, and the trace inequality, combined with an inverse estimate,

$$\|v_\ell\|_{L^2(S_i)} \le c\big(h_\ell^{-1/2}\|v_\ell\|_{L^2(T_i)} + h_\ell^{1/2}\|\nabla v_\ell\|_{L^2(T_i)}\big) \le c h_\ell^{-1/2}\|v_\ell\|_{L^2(T_i)},$$

we find that

$$\|v_\ell\|_{L^2(\partial T)} \le c h_\ell^{-1/2}\|v_\ell\|_{L^2(T)}.$$

A combination of the estimates yields that

$$\int_T \nabla v_\ell \cdot \nabla w_m \, \mathrm{d}x \le c(h_\ell/h_m)^{1/2}\|\nabla w_m\|_{L^2(T)} h_\ell^{-1}\|v_\ell\|_{L^2(T)}.$$

Summing this estimate over all $T \in \mathcal{T}_m$, using the Cauchy–Schwarz inequality, and incorporating $h_\ell = \gamma^{\ell-m} h_m$ prove the estimate. $\qquad\square$

The following multilevel preconditioner realizes a simultaneous step of a Richardson iteration on all levels.

Definition 5.4 The *Bramble–Pasciak–Xu* or *BPX preconditioner* is defined by

$$Cv = \sum_{\ell=0}^{L} P_\ell B_\ell^{-1} P_\ell^\top v,$$

where $P_\ell : \mathscr{S}_0^1(\mathcal{T}_\ell) \to \mathscr{S}_0^1(\mathcal{T}_h)$ is the canonical embedding operator, i.e., $P_\ell v_\ell = v_\ell$, and $B_\ell = h_\ell^{d-2}\,\mathrm{id}_\ell$ with the identity map $\mathrm{id}_\ell : \mathscr{S}_0^1(\mathcal{T}_\ell) \to \mathscr{S}_0^1(\mathcal{T}_\ell)$.

The BPX preconditioner is optimal in the sense of the following theorem.

Fig. 5.13 Macro-element $T \in \mathcal{T}_m$ and subelements $T' \in \mathcal{T}_\ell$ for $\ell \ge m$; the boundary of T is resolved by sides of (*shaded*) elements in $\mathcal{T}_\ell$

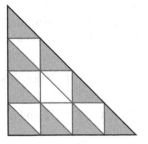

Theorem 5.6 (Optimality of the BPX Preconditioner) *For the BPX precondi-tioner C of the finite element stiffness matrix $A \in \mathbb{R}^{n \times n}$, we have that*

$$\mathrm{cond}_2(CA) \leq c_{\mathrm{bpx}}.$$

Proof

(i) Let $v_h \in \mathscr{S}_0^1(\mathscr{T}_h)$ and let $v_h = \sum_{\ell=0}^{L} w_\ell$ be its decomposition with $w_\ell \in \mathscr{S}_0^1(\mathscr{T}_\ell)$ according to Lemma 5.12. A norm equivalence implies that for $\ell = 0, 1, \ldots, L$, we have

$$c_{\mathrm{eq}}^{-2} \|w_\ell\|_{L^2(\Omega)}^2 \leq \sum_{z \in \mathscr{N}_\ell} h_\ell^d |w_\ell(z)|^2 \leq c_{\mathrm{eq}}^2 \|w_\ell\|_{L^2(\Omega)}^2.$$

In combination with the estimate of Lemma 5.12, we find that

$$\sum_{\ell=0}^{L} B_\ell w_\ell \cdot w_\ell = \sum_{\ell=0}^{L} h_\ell^{d-2} \sum_{z \in \mathscr{N}_\ell} |w_\ell(z)|^2 \leq c \|\nabla v_h\|_{L^2(\Omega)}^2.$$

With the characterization of $\lambda_{\min}(CA)$ of Theorem 5.5, and the identity $Av_h \cdot v_h = \|\nabla v_h\|_{L^2(\Omega)}^2$, we have

$$\lambda_{\min}(CA) = \min_{v_h \in \mathscr{S}_0^1(\mathscr{T}_h) \setminus \{0\}} \frac{Av_h \cdot v_h}{\min_{P_\ell v_\ell = v_h} \sum_{\ell=0}^{L} B_\ell v_\ell \cdot v_\ell} \geq c.$$

(ii) To derive an upper bound for the maximal eigenvalue $\lambda_{\max}(CA)$, let $v_h \in \mathscr{S}_0^1(\mathscr{T}_h)$ and let $v_h = \sum_{\ell=0}^{L} v_\ell$ be an arbitrary decomposition with $v_\ell \in \mathscr{S}_0^1(\mathscr{T}_\ell)$ for $\ell = 0, 1, \ldots, L$. With the strengthened Cauchy–Schwarz inequality, we have that

$$Av_h \cdot v_h \leq 2 \sum_{m=0}^{L} \sum_{\ell=m}^{L} \int_\Omega \nabla v_\ell \cdot \nabla v_m \, dx$$

$$\leq 2 c_{\mathrm{scs}} \sum_{m=0}^{L} \sum_{\ell=m}^{L} \gamma^{(\ell-m)/2} h_\ell^{-1} \|v_\ell\|_{L^2(\Omega)} \|\nabla v_m\|_{L^2(\Omega)}$$

$$\leq 2 c_{\mathrm{scs}} \sum_{\ell,m=0}^{L} \gamma^{|\ell-m|/2} h_\ell^{-1} \|v_\ell\|_{L^2(\Omega)} \|\nabla v_m\|_{L^2(\Omega)}.$$

Defining $\Gamma \in \mathbb{R}^{(L+1) \times (L+1)}$ by $\Gamma_{\ell m} = \gamma^{|\ell-m|/2}$ for $\ell, m = 0, 1, \ldots, L$, we have for vectors $\alpha, \beta \in \mathbb{R}^{L+1}$, that

$$\sum_{\ell,m=0}^{L} \gamma^{|\ell-m|} \alpha_\ell \beta_m = (\Gamma \alpha) \cdot \beta \leq \varrho(\Gamma) |\alpha| |\beta| \leq \frac{1}{1 - \gamma^{1/2}} |\alpha| |\beta|,$$

where $\varrho(\Gamma)$ is the spectral norm of Γ. This implies that

$$Av_h \cdot v_h \leq \frac{2c_{\text{scs}}}{1-\gamma^{1/2}}\Big(\sum_{\ell=0}^{L} h_\ell^{-2}\|v_\ell\|_{L^2(\Omega)}^2\Big)^{1/2}\Big(\sum_{m=0}^{L}\|\nabla v_m\|_{L^2(\Omega)}^2\Big)^{1/2}.$$

Incorporating the inverse estimates $\|\nabla v_m\|_{L^2(\Omega)} \leq ch_m^{-1}\|v_m\|_{L^2(\Omega)}$, $m = 0, 1, \ldots, L$, we deduce that

$$Av_h \cdot v_h \leq c \sum_{\ell=0}^{L} h_\ell^{-2}\|v_\ell\|_{L^2(\Omega)}^2$$

We use the norm equivalence to verify that

$$Av_h \cdot v_h \leq c \sum_{\ell=0}^{L} h_\ell^{d-2} \sum_{z\in\mathcal{N}_\ell} |v_\ell(z)|^2 = c \sum_{\ell=0}^{L} B_\ell v_\ell \cdot v_\ell.$$

The characterization of $\lambda_{\max}(CA)$ of Theorem 5.5 leads to

$$\lambda_{\max}(CA) = \max_{v_h\in\mathscr{S}_0^1(\mathscr{T}_h)\setminus\{0\}} \frac{Av_h \cdot v_h}{\min_{P_\ell v_\ell = v_h} \sum_{\ell=0}^{L} h_\ell^{d-2} v_\ell \cdot v_\ell} \leq c,$$

which implies the estimate for the condition number.   □

The injection operators $P_\ell : \mathscr{S}_0^1(\mathscr{T}_\ell) \to \mathscr{S}_0^1(\mathscr{T}_h)$ are in terms of the nodal bases given by the mapping

$$\big(v_\ell(z)\big)_{z\in\mathcal{N}_\ell} \mapsto \big(v_\ell(z')\big)_{z'\in\mathcal{N}_h}.$$

The representing matrices for the transfer between successive triangulations are provided by the MATLAB routine `red_refine.m`. Hence the operators P_ℓ can be obtained by appropriate matrix products. Figure 5.14 displays a MATLAB implementation of the preconditioned CG algorithm with BPX preconditioner. In Tables 5.1 and 5.2 we displayed corresponding iteration numbers and compared them to those of the CG algorithm without preconditioning. We see that the BPX preconditioning leads to a termination within finitely many iterations for a sequence of uniformly refined triangulations and a fixed stopping criterion. Although the iteration numbers in the unpreconditioned case grow rapidly, the total CPU times of the two solution methods are comparable for all meshes. To benefit fully from the good properties of the BPX preconditioner, a more efficient implementation than the one shown in Fig. 5.14 is required. An alternative is using hierarchical preconditioners which avoid certain redundancies.

```
function bpx_precond_cg(d_tmp,L_tmp)
global h d L P_full; d = d_tmp; L = L_tmp;
[c4n,n4e,Db,Nb] = triang_cube(d); Db = [Db;Nb]; Nb = [];
nC = size(c4n,1); fNodes_prev = setdiff(1:nC,unique(Db));
h = zeros(L,1);
for ell = 1:L
    [c4n,n4e,Db,Nb,~,P1] = red_refine(c4n,n4e,Db,Nb);
    nC = size(c4n,1); fNodes = setdiff(1:nC,unique(Db));
    P{ell} = P1(fNodes,fNodes_prev);
    fNodes_prev = fNodes; h(ell) = 2^(-ell);
end
nfNodes = size(fNodes,2);
P_full{L} = speye(nfNodes);
for ell = L-1:-1:1
    P_full{ell} = P_full{ell+1}*P{ell+1};
end
[s,m] = fe_matrices(c4n,n4e);
A = s(fNodes,fNodes);
b = m(fNodes,:)*f(c4n);
u = zeros(nC,1);
u(fNodes) = cg_precond(u(fNodes),A,b);
show_p1(c4n,n4e,Db,Nb,u)

function x = cg_precond(x,A,b)
r = b-A*x; z = apply_bpx(r); d = z;
rz_old = r'*z; eps = 1e-4;
while sqrt(r'*r) > eps
    alpha = rz_old/(d'*A*d);
    x = x+alpha*d;
    r = r-alpha*A*d;
    z = apply_bpx(r);
    rz_new = r'*z;
    beta = rz_new/rz_old;
    d = z+beta*d;
    rz_old = rz_new;
end

function Cr = apply_bpx(r)
global h d L P_full;
Cr = zeros(size(r));
for ell = 1:L
    Cr = Cr+h(ell)^(2-d)*P_full{ell}*(P_full{ell}'*r);
end

function val = f(x); val = ones(size(x,1),1);
```

Fig. 5.14 MATLAB implementation of the preconditioned conjugate gradient method with BPX preconditioner

Table 5.1 Iteration numbers and total CPU times in seconds for the conjugate gradient algorithm (CG) and its preconditioned version with the BPX preconditioner (BPX) in a two-dimensional Poisson problem

$d = 2, \#\mathcal{N}_h$	9	49	225	961	3969	16129	65025	261121
CG, N_{iter}	3	8	17	34	67	129	251	480
BPX, N_{iter}	4	9	11	12	12	13	13	13
CG, T_{cpu} [s]	0.1	0.1	0.1	0.2	0.8	3.0	12.6	55.2
BPX, T_{cpu} [s]	0.1	0.1	0.1	0.2	0.7	3.1	12.2	49.5

Table 5.2 Iteration numbers and total CPU times in seconds for the conjugate gradient algorithm (CG) and its preconditioned version with the BPX preconditioner (BPX) in a three-dimensional Poisson problem

$d = 3, \#\mathcal{N}_h$	27	343	3375	26791	250047
CG, N_{iter}	5	12	28	62	137
BPX, N_{iter}	5	10	14	18	23
CG, T_{cpu} [s]	0.1	0.4	2.9	23.8	192.3
BPX, T_{cpu} [s]	0.1	0.4	2.9	29.1	188.0

Fig. 5.15 Coarse triangulation $\mathcal{T}_H$, fine triangulation $\mathcal{T}_h$, and compatible overlapping partition of Ω

5.3.4 Two-Level Preconditioning

Let $\mathcal{T}_h$ be a triangulation of Ω and let $(\Omega_j)_{j=1,\dots,J}$ be an overlapping partition of Ω into Lipschitz domains $\Omega_j \subset \Omega$ whose boundaries $\partial\Omega_j$ are matched by edges in $\mathcal{T}_h$. Let $\mathcal{T}_H$ be another triangulation of Ω such that $\mathcal{T}_h$ is a uniform refinement of $\mathcal{T}_H$, cf. Fig. 5.15.

For $j = 1, \dots, J$, let $\mathcal{T}_h^j$ be the induced triangulation $\mathcal{T}_h|_{\Omega_j}$ and let $\widetilde{\mathscr{S}}_0^1(\mathcal{T}_h^j)$ be the space of finite element functions in $\mathscr{S}_0^1(\mathcal{T}_h^j)$ that are extended by zero to Ω. We let

$$P_j : \widetilde{\mathscr{S}}_0^1(\mathcal{T}_h^j) \to \mathscr{S}_0^1(\mathcal{T}_h),$$

$j = 1, \dots, J$, and let

$$P_H : \mathscr{S}_0^1(\mathcal{T}_H) \to \mathscr{S}_0^1(\mathcal{T}_h)$$

denote the embedding of functions into $\mathscr{S}_0^1(\mathscr{T}_h)$, i.e., the identity operator. Letting A_h, A_H, and A_j, $j = 1, \ldots, J$, denote the finite element stiffness matrices related to the spaces $\mathscr{S}_0^1(\mathscr{T}_h)$, $\mathscr{S}_0^1(\mathscr{T}_H)$, and $\mathscr{S}_0^1(\mathscr{T}_h^j)$, respectively, we define the *two-level additive Schwarz preconditioner* $C_{2\ell} \in \mathbb{R}^{n \times n}$ by

$$C_{2\ell} = P_H A_H^{-1} P_H^\top + \sum_{j=1}^J P_j A_j^{-1} P_j^\top.$$

For this preconditioner one can show $\mathrm{cond}_2(C_{2\ell} A_h) \leq c_{2\ell}(1 + H/\delta)$, where δ is the minimal overlap diameter. In the situation depicted in Fig. 5.15 we have $\delta \sim h$.

References

Fundamental contributions to the development of multigrid methods, preconditioning of finite element matrices, and domain decomposition methods are the articles [2–4, 6, 9, 12, 18]. Specialized textbooks on the subjects are the references [7, 10, 11, 13, 15, 16]. The historical development of domain decomposition methods is recapitulated in [8], and the survey article [17] discusses various aspects of preconditioning techniques. Chapters on iterative solution methods are contained in the textbooks [1, 5, 14].

1. Braess, D.: Finite Elements, 3rd edn. Cambridge University Press, Cambridge (2007). URL http://dx.doi.org/10.1017/CBO9780511618635
2. Braess, D., Hackbusch, W.: A new convergence proof for the multigrid method including the V-cycle. SIAM J. Numer. Anal. 20(5), 967–975 (1983). URL http://dx.doi.org/10.1137/0720066
3. Bramble, J.H., Pasciak, J.E., Schatz, A.H.: The construction of preconditioners for elliptic problems by substructuring. I. Math. Comp. 47(175), 103–134 (1986). URL http://dx.doi.org/10.2307/2008084
4. Bramble, J.H., Pasciak, J.E., Xu, J.: Parallel multilevel preconditioners. Math. Comp. 55(191), 1–22 (1990). URL http://dx.doi.org/10.2307/2008789
5. Brenner, S.C., Scott, L.R.: The mathematical theory of finite element methods. Texts in Applied Mathematics, vol. 15, 3rd edn. Springer, New York (2008). URL http://dx.doi.org/10.1007/978-0-387-75934-0
6. Dahmen, W., Kunoth, A.: Multilevel preconditioning. Numer. Math. 63(3), 315–344 (1992). URL http://dx.doi.org/10.1007/BF01385864
7. Elman, H.C., Silvester, D.J., Wathen, A.J.: Finite elements and fast iterative solvers: with applications in incompressible fluid dynamics. Numerical Mathematics and Scientific Computation, 2nd edn. Oxford University Press, Oxford (2014). URL http://dx.doi.org/10.1093/acprof:oso/9780199678792.001.0001
8. Gander, M.J.: Schwarz methods over the course of time. Electron. Trans. Numer. Anal. 31, 228–255 (2008)
9. Griebel, M., Oswald, P.: On the abstract theory of additive and multiplicative Schwarz algorithms. Numer. Math. 70(2), 163–180 (1995). URL http://dx.doi.org/10.1007/s002110050115
10. Hackbusch, W.: Multigrid methods and applications. Springer Series in Computational Mathematics, vol. 4. Springer, Berlin (1985). URL http://dx.doi.org/10.1007/978-3-662-02427-0

11. Hackbusch, W.: Iterative solution of large sparse systems of equations. Applied Mathematical Sciences, vol. 95. Springer, New York (1994). URL http://dx.doi.org/10.1007/978-1-4612-4288-8

12. Lions, P.L.: On the Schwarz alternating method. I. In: First International Symposium on Domain Decomposition Methods for Partial Differential Equations (Paris, 1987), pp. 1–42. SIAM, Philadelphia, PA (1988)

13. Quarteroni, A., Valli, A.: Domain decomposition methods for partial differential equations. Numerical Mathematics and Scientific Computation. The Clarendon Press, Oxford University Press, New York (1999)

14. Rannacher, R.: Numerische Mathematik 2 (Numerik partieller Differentialgleichungen) (2008). URL http://numerik.iwr.uni-heidelberg.de/~lehre/notes/. Lecture Notes, University of Heidelberg, Germany

15. Saad, Y.: Iterative methods for sparse linear systems, second edn. Society for Industrial and Applied Mathematics, Philadelphia, PA (2003). URL http://dx.doi.org/10.1137/1.9780898718003

16. Toselli, A., Widlund, O.: Domain decomposition methods—algorithms and theory. Springer Series in Computational Mathematics, vol. 34. Springer, Berlin (2005)

17. Xu, J., Chen, L., Nochetto, R.H.: Optimal multilevel methods for H(grad), H(curl), and H(div) systems on graded and unstructured grids. In: Multiscale, Nonlinear and Adaptive Approximation, pp. 599–659. Springer, Berlin (2009). URL http://dx.doi.org/10.1007/978-3-642-03413-8_14

18. Yserentant, H.: On the multilevel splitting of finite element spaces. Numer. Math. **49**(4), 379–412 (1986). URL http://dx.doi.org/10.1007/BF01389538

Part III
Constrained and Singularly Perturbed Problems

Chapter 6
Saddle-Point Problems

6.1 Discrete Saddle-Point Problems

6.1.1 Limitations of the Lax–Milgram Framework

Suppose that a fluid is flowing through the fixed domain $\Omega \subseteq \mathbb{R}^d, d = 2, 3$. At each point $x \in \Omega$, we let $u(x) \in \mathbb{R}^d$ be the velocity of the fluid and $p(x) \in \mathbb{R}$ its pressure, cf. Fig. 6.1.

If the process is stationary in the sense that the velocity is the same at all times, and if the fluid is viscous and incompressible, then a simplification of the *Navier–Stokes equations* leads to the *Stokes system*, which specifies u and p as the solution of the partial differential equations

$$-\Delta u + \nabla p = f \quad \text{in } \Omega,$$

$$\operatorname{div} u = 0 \quad \text{in } \Omega,$$

subject to $u|_{\Gamma_D} = 0$, where we assume for simplicity that $\Gamma_D = \partial\Omega$. Here $\Delta u = [\Delta u_1, \dots, \Delta u_d]^\top$. To obtain a variational formulation, we multiply the first equation by $v \in H^1_D(\Omega; \mathbb{R}^d)$, the second one by $q \in L^2(\Omega)$, integrate over Ω, and use integration-by-parts to see that u and p satisfy

$$\int_\Omega \nabla u : \nabla v \, dx - \int_\Omega p \operatorname{div} v \, dx = \int_\Omega f \cdot v \, dx,$$

$$\int_\Omega q \operatorname{div} u \, dx = 0,$$

for all $v \in H^1_D(\Omega; \mathbb{R}^d)$ and $q \in L^2(\Omega)$. Note that ∇u and ∇v are square matrices, and $\nabla u : \nabla v$ is the scalar product defined by the sum $\sum_{j=1}^d \nabla u_j \cdot \nabla v_j$. Abbreviating

© Springer International Publishing Switzerland 2016

S. Bartels, *Numerical Approximation of Partial Differential Equations*,
Texts in Applied Mathematics 64, DOI 10.1007/978-3-319-32354-1_6

Fig. 6.1 Stokes flow
through a domain $\Omega \subset \mathbb{R}^d$
described by a velocity field
$u : \Omega \to \mathbb{R}^d$

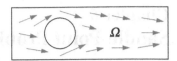

$V = H_{\mathrm{D}}^1(\Omega; \mathbb{R}^d)$ and $Q = L^2(\Omega)$, and introducing the bilinear forms

$$a(u, v) = \int_\Omega \nabla u : \nabla v \, dx, \quad b(v, p) = - \int_\Omega p \, \mathrm{div}\, v \, dx,$$

and the linear functional

$$\ell_1(v) = \int_\Omega f \cdot v \, dx,$$

the weak formulation is equivalent to finding $(u, p) \in V \times Q$ such that

$$(S) \quad \begin{cases} a(u, v) + b(v, p) = \ell_1(v), \\ b(u, q) \qquad\qquad = 0 \end{cases}$$

for all $(v, q) \in V \times Q$. To relate this formulation to the Lax–Milgram lemma, we consider the product space $X = V \times Q$ and define for $x = (u, p)$ and $y = (v, q)$ the bilinear form

$$\Gamma(x, y) = a(u, v) + b(v, p) + b(u, q)$$

and the linear form

$$\ell(y) = \ell_1(v).$$

The formulation (S) is then equivalent to finding $x \in X$ such that

$$\Gamma(x, y) = \ell(y)$$

for all $y \in X$. Unfortunately, the bilinear form is not coercive on X. To see this, we choose $x = (0, p)$ and verify that

$$\Gamma(x, x) = 0.$$

Therefore, we cannot apply the Lax–Milgram theorem to establish existence and uniqueness of a solution (u, p) for (S).

Remark 6.1 By restricting to the subspace $V_0 = \{v \in H_{\mathrm{D}}^1(\Omega; \mathbb{R}^d) : \mathrm{div}\, v = 0\}$, one can prove the existence of a unique velocity field $u \in H_{\mathrm{D}}^1(\Omega; \mathbb{R}^d)$ with the Lax–Milgram lemma via the formulation

$$a(u, v) = \ell_1(v)$$

for all $v \in V_0$. The construction of discrete subspaces of V_0, i.e., of divergence-free finite element functions, is however a nontrivial task. Moreover, the pressure is a relevant quantity in many applications.

6.1.2 Variational Condition Number Estimate

Throughout what follows we adopt arguments from [7] and recall that the *condition number* of a regular matrix $M \in \mathbb{R}^{n \times n}$ is with norms $\| \cdot \|_\ell$ and $\| \cdot \|_r$ defined as the smallest constant $\mathrm{cond}_{\ell r}(M) \geq 0$ such that

$$\frac{\|x - \tilde{x}\|_\ell}{\|x\|_\ell} \leq \mathrm{cond}_{\ell r}(M) \frac{\|b - \tilde{b}\|_r}{\|b\|_r}$$

for all $x, \tilde{x} \in \mathbb{R}^n$ and $b, \tilde{b} \in \mathbb{R}^n$ such that $x, b \neq 0$ and

$$Mx = b, \quad M\tilde{x} = \tilde{b}.$$

The condition number measures the amplification of the relative errors for perturbations of the right-hand sides in solving the linear system $Mx = b$. Defining the induced operator norms

$$\|M\|_{\ell r} = \sup_{y \in \mathbb{R}^d \setminus \{0\}} \frac{\|My\|_r}{\|y\|_\ell}, \quad \|M^{-1}\|_{r\ell} = \sup_{z \in \mathbb{R}^n \setminus \{0\}} \frac{\|M^{-1}z\|_\ell}{\|z\|_r},$$

we have

$$\mathrm{cond}_{\ell r}(M) = \|M\|_{\ell r} \|M^{-1}\|_{r\ell} \geq 1.$$

If $\| \cdot \|_\ell$ and $\| \cdot \|_r$ coincide with the Euclidean norm on $\mathbb{R}^n$, and if M is symmetric and positive definite, then we have $\mathrm{cond}_{\ell r}(M) = \lambda_{\max}/\lambda_{\min}$ with the maximal and minimal eigenvalues $\lambda_{\max}$ and $\lambda_{\min}$ of M. It is often straightforward to construct a norm $\| \cdot \|_\ell$ such that there exists $k \geq 0$ with

$$y^\top M x \leq k \|x\|_\ell \|y\|_\ell$$

for all $x, y \in \mathbb{R}^n$ independently of $n \in \mathbb{N}$. This induces a natural choice for $\| \cdot \|_r$ that leads to a uniform bound for $\|M\|_{\ell r}$. The *dual norm* of $\| \cdot \|_\ell$ is defined by

$$\|z\|_{\ell'} = \sup_{y \in \mathbb{R}^n \setminus \{0\}} \frac{y^\top z}{\|y\|_\ell}.$$

This norm leads to a useful characterization of the operator norm.

Lemma 6.1 (Operator Norm $\|M\|_{\ell r}$) *If $\|\cdot\|_r = \|\cdot\|_{\ell'}$, we have*

$$\|M\|_{\ell r} = \sup_{x,y \in \mathbb{R}^n \setminus \{0\}} \frac{y^\top M x}{\|y\|_\ell \|x\|_\ell}.$$

Proof Due to the assumption and the definition of $\|\cdot\|_{\ell'}$, we have

$$\|M\|_{\ell r} = \sup_{x \in \mathbb{R}^n \setminus \{0\}} \frac{\|Mx\|_r}{\|x\|_\ell}$$

$$= \sup_{x \in \mathbb{R}^n \setminus \{0\}} \frac{\|Mx\|_{\ell'}}{\|x\|_\ell}$$

$$= \sup_{x \in \mathbb{R}^n \setminus \{0\}} \sup_{y \in \mathbb{R}^n \setminus \{0\}} \frac{y^\top M x}{\|x\|_\ell \|y\|_\ell}$$

$$= \sup_{x,y \in \mathbb{R}^n \setminus \{0\}} \frac{y^\top M x}{\|y\|_\ell \|x\|_\ell},$$

which proves the identity.   □

To specify a uniform upper bound for the condition numbers, it remains to bound $\|M^{-1}\|_{r\ell}$. It is desirable to avoid the explicit use of M^{-1} for this.

Lemma 6.2 (Operator Norm $\|M^{-1}\|_{r\ell}$) *If $\|\cdot\|_r = \|\cdot\|_{\ell'}$, then*

$$(\|M^{-1}\|_{r\ell})^{-1} = \inf_{x \in \mathbb{R}^n \setminus \{0\}} \sup_{y \in \mathbb{R}^n \setminus \{0\}} \frac{y^\top M x}{\|y\|_\ell \|x\|_\ell}.$$

Proof With the transformation $z = Mx$, we find that

$$(\|M^{-1}\|_{r\ell})^{-1} = \left(\sup_{z \in \mathbb{R}^n \setminus \{0\}} \frac{\|M^{-1}z\|_\ell}{\|z\|_r} \right)^{-1}$$

$$= \inf_{z \in \mathbb{R}^n \setminus \{0\}} \frac{\|z\|_r}{\|M^{-1}z\|_\ell}$$

$$= \inf_{x \in \mathbb{R}^n \setminus \{0\}} \frac{\|Mx\|_r}{\|x\|_\ell}$$

$$= \inf_{x \in \mathbb{R}^n \setminus \{0\}} \frac{\|Mx\|_{\ell'}}{\|x\|_\ell}$$

$$= \inf_{x \in \mathbb{R}^n \setminus \{0\}} \sup_{y \in \mathbb{R}^n \setminus \{0\}} \frac{y^\top M x}{\|x\|_\ell \|y\|_\ell},$$

which proves the identity.   □

We summarize the observations in the following proposition.

Proposition 6.1 (Condition Number Bound) *Let $M \in \mathbb{R}^{n \times n}$ be regular, $\| \cdot \|_\ell$ a norm on $\mathbb{R}^n$, and $\| \cdot \|_r = \| \cdot \|_{\ell'}$. If $k \geq 0$ and $\gamma > 0$ are such that*

$$\|M\|_{\ell r} = \sup_{x,y \in \mathbb{R}^n \setminus \{0\}} \frac{y^T M x}{\|y\|_\ell \|x\|_\ell} \leq k,$$

$$\|M^{-1}\|_{r\ell}^{-1} = \inf_{x \in \mathbb{R}^n \setminus \{0\}} \sup_{y \in \mathbb{R}^n \setminus \{0\}} \frac{y^T M x}{\|y\|_\ell \|x\|_\ell} \geq \gamma,$$

then we have

$$\mathrm{cond}_{\ell r}(M) \leq k/\gamma.$$

The characterization of the operator norms is particularly useful when the matrix M results from the discretization of the weak formulation of a partial differential equation, and when uniform bounds for operator norms are needed.

Remarks 6.2

(i) Note that the second condition holds if and only if for all $x \in \mathbb{R}^n$, we have

$$\sup_{y \in \mathbb{R}^n \setminus \{0\}} \frac{y^T M x}{\|y\|_\ell \|x\|_\ell} \geq \gamma$$

which is equivalent to $\|Mx\|_{\ell'} \geq \gamma \|x\|_\ell$ for all $x \in \mathbb{R}^n$ and to $\|M^{-1}z\|_\ell \leq \gamma^{-1}\|z\|_r$ for all $z \in \mathbb{R}^n$, i.e., γ^{-1} is an upper bound for the operator norm of M^{-1}.

(ii) The matrix M may result from the discretization of a bilinear form $a : V \times V \to \mathbb{R}$ so that

$$y^T M x = a(u_h, v_h)$$

with respect to an appropriate basis. The bound for the condition number can then be determined in terms of a.

We illustrate the application of the condition number bound for the discretization of an elliptic partial differential equation.

Example 6.1 Assume that $a : V \times V \to \mathbb{R}$ is a bounded and coercive bilinear form with constants $\beta, \alpha > 0$, $V_h = \mathrm{span}\{v_1, v_2, \ldots, v_n\}$ is a finite-dimensional subspace, and $M_{jk} = a(v_j, v_k)$ for $j, k = 1, 2, \ldots, n$. Defining $\|x\|_\ell = \|u_h\|_V$ if $x \in \mathbb{R}^n$ is the coefficient vector of $u_h \in V_h$ with respect to the basis $(v_1, v_2, \ldots, v_n)$, we have

$$\frac{y^T M x}{\|x\|_\ell \|y\|_\ell} = \frac{a(u_h, v_h)}{\|u_h\|_V \|v_h\|_V}$$

and the boundedness and coercivity of a lead to $\text{cond}_{\ell_r}(M) \leq \beta/\alpha$. This upper bound is also the constant that arises in Céa's lemma. For the Poisson problem we thus obtain a uniform upper bound of the condition number with respect to the norm of $H^1(\Omega)$. For the Euclidean norm on $\mathbb{R}^n$, the condition number is of order $\mathcal{O}(h^{-2})$ for typical finite element methods.

6.1.3   Well-Posedness

Motivated by the weak formulation (S) and following [7], we analyze linear systems of equations $Mx = b$, for which $M \in \mathbb{R}^{n \times n}$ and $b \in \mathbb{R}^n$ admit the block structures

$$M = \begin{bmatrix} A & B^\mathsf{T} \\ B & 0 \end{bmatrix}, \quad b = \begin{bmatrix} f \\ g \end{bmatrix}.$$

Here $A \in \mathbb{R}^{n_A \times n_A}$, $B \in \mathbb{R}^{n_B \times n_A}$, and $f \in \mathbb{R}^{n_A}$, $g \in \mathbb{R}^{n_B}$ with $n_A, n_B \in \mathbb{N}$ such that $n = n_A + n_B$. Accordingly, we partition the solution vector $x \in \mathbb{R}^n$ as

$$x = \begin{bmatrix} y \\ z \end{bmatrix}$$

with $y \in \mathbb{R}^{n_A}$ and $z \in \mathbb{R}^{n_B}$. The linear system of equations $Mx = b$ is thus equivalent to the vectorial equations

$$Ay + B^\mathsf{T} z = f, \quad By = g.$$

Provided that B is a surjection, the second equation determines y up to an element from the kernel of B, i.e., in

$$\ker B = \{v \in \mathbb{R}^{n_A} : Bv = 0\}.$$

We note that according to elementary results from linear algebra, we have

$$\mathbb{R}^{n_A} = \ker B \oplus \operatorname{Im} B^\mathsf{T}.$$

In particular, every $v \in \mathbb{R}^{n_A}$ can be written as $v = v_K + v_I$ with $v_K \in \ker B$ and $v_I \in \operatorname{Im} B^\mathsf{T}$ such that $v_K \cdot v_I = 0$. By an appropriate choice of basis, we may assume that with $r = \dim \ker B$, we have

$$
v_K = \begin{bmatrix} v_1 \\ \vdots \\ v_r \\ 0 \\ \vdots \\ 0 \end{bmatrix}, \quad
v_I = \begin{bmatrix} 0 \\ \vdots \\ 0 \\ v_{r+1} \\ \vdots \\ v_{n_A} \end{bmatrix},
$$

and we can identify $v_K = [v_1, \ldots, v_r]^\top$ and $v_I = [v_{r+1}, \ldots, v_{n_A}]^\top$. We then have

$$
Av = \begin{bmatrix} A_{KK} & A_{KI} \\ A_{IK} & A_{II} \end{bmatrix} \begin{bmatrix} v_K \\ v_I \end{bmatrix} = \begin{bmatrix} A_{KK} v_K + A_{KI} v_I \\ A_{IK} v_K + A_{II} v_I \end{bmatrix}
$$

and analogously

$$
Bv = \begin{bmatrix} B_K & B_I \end{bmatrix} \begin{bmatrix} v_K \\ v_I \end{bmatrix} = B_K v_K + B_I v_I = B_I v_I.
$$

In particular, we have $B_I v_I = 0$ if and only if $v_I = 0$. Moreover, up to appropriate identification of vectors, we have

$$
B^\top s = B_I^\top s \in \operatorname{Im} B^\top
$$

for all $s \in \mathbb{R}^{n_B}$, which follows from noting $(B^\top s) \cdot v = s \cdot (Bv) = s \cdot (B_I v_I) = (B_I^\top s) \cdot v_I$. With the decomposition of vectors in $\mathbb{R}^n$, the system $Mx = b$ reads as

$$
\begin{aligned}
A_{KK} y_K + A_{KI} y_I & = f_K, \\
A_{IK} y_K + A_{II} y_I + B_I^\top z & = f_I, \\
B_I y_I & = g.
\end{aligned}
$$

The conditions for unique solvability are:

- equation $A_{KK} y_K = \tilde{f}_K$ is uniquely solvable for every $\tilde{f}_K \in \ker B$;
- equation $B_I y_I = g$ is uniquely solvable for every $g \in \mathbb{R}^{n_B}$;
- equation $B_I^\top z = \tilde{f}_I$ is uniquely solvable for every $\tilde{f}_I \in \operatorname{Im} B^\top$.

To ensure the first condition, we need that $B_I : \operatorname{Im} B^\top \mapsto \mathbb{R}^{n_B}$ be invertible. Since B_I is by construction injective, this means that B has to be surjective, i.e., $\dim \operatorname{Im} B^\top = n_B$ and B_I is a regular square matrix. This automatically yields that also $B_I^\top$ is regular, so that the third condition is satisfied. The second condition holds if $A_{KK} : \ker B \mapsto \ker B$ is invertible.

Proposition 6.2 (Solvability) *The linear system of equations*

$$
\begin{bmatrix} A & B^{\mathsf{T}} \\ B & 0 \end{bmatrix} \begin{bmatrix} y \\ z \end{bmatrix} = \begin{bmatrix} f \\ g \end{bmatrix}
$$

is uniquely solvable for every right-hand side $b = [f, g]^{\mathsf{T}}$, *if and only* $B : \mathbb{R}^{n_A} \to \mathbb{R}^{n_B}$ *is surjective, and the restriction of* A *to* $\ker B$ *defines a bijection.*

Example 6.2 The matrix

$$
\begin{bmatrix} 1 & 0 & 1 \\ 0 & 0 & -1 \\ 1 & -1 & 0 \end{bmatrix}
$$

is invertible. This follows from the proposition by letting A be the upper left 2×2 submatrix and $B = [1, -1]$.

We next investigate conditioning of the linear system. For this, we assume that there exist norms $\| \cdot \|_V$ and $\| \cdot \|_Q$ on $\mathbb{R}^{n_A}$ and $\mathbb{R}^{n_B}$, respectively, such that

$$
v^{\mathsf{T}} A y \leq k_A \|v\|_V \|y\|_V, \qquad v^{\mathsf{T}} B^{\mathsf{T}} z \leq k_B \|v\|_V \|z\|_Q
$$

for all $v, y \in \mathbb{R}^{n_A}$ and all $z \in \mathbb{R}^{n_B}$ with uniformly bounded constants $k_A, k_B \geq 0$. On the product space $\mathbb{R}^{n_A} \times \mathbb{R}^{n_B}$, we employ the norm

$$
\|(y, z)\|_\ell = \|y\|_V + \|z\|_Q,
$$

note that the associated dual norm is given by

$$
\|(f, g)\|_{\ell'} = \max \{ \|f\|_{V'}, \|g\|_{Q'} \},
$$

and we use $\| \cdot \|_r = \| \cdot \|_{\ell'}$. We then have

$$
\|M\|_{\ell r} \leq k_A + 2 k_B.
$$

To provide an upper bound for the condition number, it remains to estimate $\|M^{-1}\|_{r\ell}$, i.e., to specify a constant $c \geq 0$ such that $\|M^{-1} b\|_\ell \leq c \|b\|_r$. In constructing $x = [y, z]^{\mathsf{T}} = M^{-1}[f, g]^{\mathsf{T}}$, we have that

$$
\|y_I\|_V \leq \|B_I^{-1}\|_{Q'V} \|g\|_{Q'},
$$

$$
\|y_K\|_V \leq \|A_{KK}^{-1}\|_{V'V} (\|f_K\|_{V'} + \|A_{KI} y_I\|_{V'})
$$

$$
\leq \|A_{KK}^{-1}\|_{V'V} (\|f_K\|_{V'} + k_A \|y_I\|_V),
$$

$$
\|z\|_Q \leq \|(B_I^{\mathsf{T}})^{-1}\|_{V'Q} (\|f_I\|_{V'} + \|A_{II} y_I\|_{V'} + \|A_{IK} y_K\|_{V'})
$$

$$
\leq \|(B_I^{\mathsf{T}})^{-1}\|_{V'Q} (\|f_I\|_{V'} + k_A \|y_I\|_V + k_A \|y_K\|_V).
$$

Assuming that $\|B_I^{-1}\|_{Q'V}, \|(B_I^\top)^{-1}\|_{V'Q} \le \beta^{-1}$ and $\|A_{KK}^{-1}\|_{V'V} \le \alpha^{-1}$, we obtain

$$\|(y,z)\|_\ell \le c(\alpha^{-1}, \beta^{-1}, k_A)\|(f,g)\|_r,$$

and the constant $c(\alpha^{-1}, \beta^{-1}, k_A)$ provides an upper bound for $\|M^{-1}\|_{r\ell}$. Note that $(B_I^\top)^{-1} = (B_I^{-1})^\top$, and due to the choice of norms

$$\|B_I^{-1}\|_{Q'V} = \|(B_I^\top)^{-1}\|_{V'Q}.$$

It therefore suffices to bound $\|(B_I^\top)^{-1}\|_{V'Q}$ and $\|A_{KK}^{-1}\|_{V'V}$. With the transformation $f_I = B_I^\top q = B^\top q$, we have

$$\|(B_I^\top)^{-1}\|_{V'Q}^{-1} = \inf_{f_I \in \mathbb{R}^{n_A}\setminus\{0\}} \frac{\|f_I\|_{V'}}{\|(B_I^\top)^{-1}f_I\|_Q}$$

$$= \inf_{q \in \mathbb{R}^{n_B}\setminus\{0\}} \frac{\|B_I^\top q\|_{V'}}{\|q\|_Q}$$

$$= \inf_{q \in \mathbb{R}^{n_B}\setminus\{0\}} \frac{\|B^\top q\|_{V'}}{\|q\|_Q}$$

$$= \inf_{q \in \mathbb{R}^{n_B}\setminus\{0\}} \sup_{v \in \mathbb{R}^{n_A}\setminus\{0\}} \frac{v^\top B^\top q}{\|q\|_Q\|v\|_V}$$

$$= \inf_{q \in \mathbb{R}^{n_B}\setminus\{0\}} \sup_{v \in \mathbb{R}^{n_A}\setminus\{0\}} \frac{q^\top Bv}{\|q\|_Q\|v\|_V}.$$

Similarly, with the transformation $v = A_{KK}u$, we verify that

$$\|A_{KK}^{-1}\|_{V'V}^{-1} = \inf_{v \in K\setminus\{0\}} \frac{\|v\|_{V'}}{\|A_{KK}^{-1}v\|_V}$$

$$= \inf_{u \in K\setminus\{0\}} \frac{\|A_{KK}u\|_{V'}}{\|u\|_V}$$

$$= \inf_{u \in K\setminus\{0\}} \sup_{w \in K\setminus\{0\}} \frac{w^\top A_{KK}u}{\|u\|_V\|w\|_V}$$

$$= \inf_{u \in K\setminus\{0\}} \sup_{w \in K\setminus\{0\}} \frac{w^\top Au}{\|u\|_V\|w\|_V}.$$

With these estimates we obtain a bound on the condition number.

Theorem 6.1 (Conditioning) *Assume that we are given the linear systems of equations $Mx = b$ with*

$$M = \begin{bmatrix} A & B^\mathsf{T} \\ B & 0 \end{bmatrix}, \quad b = \begin{bmatrix} f \\ g \end{bmatrix},$$

and assume that there exist constants $k_A, k_B \geq 0$ such that

$$v^\mathsf{T} Ay \leq k_A \|v\|_V \|y\|_V, \quad v^\mathsf{T} Bz \leq k_B \|v\|_V \|z\|_Q,$$

for all $y, v \in \mathbb{R}^{n_A}$ and $z \in \mathbb{R}^{n_B}$. Assume that $\alpha, \beta > 0$ are such that with $K = \ker B$, we have the inf-sup conditions

$$\inf_{v \in K \setminus \{0\}} \sup_{w \in K \setminus \{0\}} \frac{v^\mathsf{T} Aw}{\|v\|_V \|w\|_V} \geq \alpha,$$

$$\inf_{q \in \mathbb{R}^{n_B} \setminus \{0\}} \sup_{v \in \mathbb{R}^{n_A} \setminus \{0\}} \frac{q^\mathsf{T} Bv}{\|q\|_Q \|v\|_V} \geq \beta.$$

Then the condition number of M with respect to $\|(v, s)\|_\ell = \|v\|_V + \|s\|_Q$ and the associated dual norm $\|\cdot\|_r = \|\cdot\|_{\ell'}$ satisfy

$$\mathrm{cond}_{\ell r}(M) \leq c(k_A, k_B, \alpha^{-1}, \beta^{-1}).$$

Remarks 6.3

(i) The second inf-sup condition is equivalent to

$$\|B^\mathsf{T} q\|_{V'} = \sup_{v \in \mathbb{R}^{n_A} \setminus \{0\}} \frac{q^\mathsf{T} Bv}{\|v\|_V} \geq \beta \|q\|_Q$$

for all $q \in \mathbb{R}^{n_B}$. It implies that B^T is injective and B is surjective. In particular, B^T and B have left and right inverses $(B^\mathsf{T})^{-\ell} : \mathrm{Im}\, B^\mathsf{T} = (\ker B)^\perp \to \mathbb{R}^{n_B}$ and $B^{-r} : \mathbb{R}^{n_B} \to \mathbb{R}^{n_A}$, respectively, whose operator norms are bounded by β^{-1}.

(ii) The inf-sup conditions imply the solvability of the linear system of equations.

6.1.4 Constrained Quadratic Minimization

Quadratic minimization problems with a linear constraint consist in finding a solution for the problem

$$\min_{y \in \mathbb{R}^{n_A}} \frac{1}{2} y^\mathsf{T} Ay - f^\mathsf{T} y \quad \text{subject to } By = g.$$

Here $A \in \mathbb{R}^{n_A \times n_A}$ is a symmetric and positive semidefinite and $B \in \mathbb{R}^{n_B \times n_A}$ is a surjective matrix. Noting that

$$By = g \quad \Longleftrightarrow \quad \max_{z \in \mathbb{R}^{n_B}} z^\top (By - g) < \infty,$$

we may equivalently consider solving the *saddle-point* or *min-max problem*

$$\min_{y \in \mathbb{R}^{n_A}} \max_{z \in \mathbb{R}^{n_B}} \frac{1}{2} y^\top A y - f^\top y + z^\top (By - g).$$

Letting $L(y, z) = (1/2) y^\top A y - f^\top y + z^\top (By - g)$ be the *Lagrange functional* associated with the constrained minimization problem, a solution (y, z) of the saddle-point problem is characterized via the inequalities

$$L(y, s) \leq L(y, z) \leq L(r, z)$$

for all $(r, s) \in \mathbb{R}^{n_A} \times \mathbb{R}^{n_B}$. The optimality conditions for this problem require an optimal pair (y, z) to satisfy

$$\begin{bmatrix} A & B^\top \\ B & 0 \end{bmatrix} \begin{bmatrix} y \\ z \end{bmatrix} = \begin{bmatrix} f \\ g \end{bmatrix}.$$

The variable z is also called the *Lagrange multiplier* subject to the constraint $By = g$.

Example 6.3 The Poisson problem $-\Delta u = f$ with Neumann boundary conditions $\partial_n u = g$ on the entire boundary $\Gamma_N = \partial\Omega$ admits a unique solution, if

$$\int_\Omega f \, dx + \int_{\Gamma_N} g \, ds = 0$$

and if one imposes the constraint $\int_\Omega u \, dx = 0$. The solution can then be characterized as the unique minimizer of the functional

$$I(u) = \frac{1}{2} \int_\Omega |\nabla u|^2 \, dx - \int_\Omega f u \, dx - \int_{\Gamma_N} g u \, ds$$

subject to the constraint. Its discretization leads to a constrained quadratic minimization problem.

6.1.5 Iterative Solution

We consider a generalized saddle-point problem of the form

$$
\begin{bmatrix} A & B^\mathsf{T} \\ B & -C \end{bmatrix} \begin{bmatrix} y \\ z \end{bmatrix} = \begin{bmatrix} f \\ g \end{bmatrix}
$$

and assume that

- $A \in \mathbb{R}^{n_A \times n_A}$ is symmetric and positive definite,
- $B \in \mathbb{R}^{n_B \times n_A}$ satisfies $\mathrm{rank}(B) = m \le n$,
- $C \in \mathbb{R}^{n_B \times n_B}$ is symmetric and positive semidefinite.

The system is equivalent to the decoupled system

$$
(BA^{-1}B^\mathsf{T} + C)z = BA^{-1}f - g,
$$

$$
y = A^{-1}(f - B^\mathsf{T} z).
$$

The first equation is called the *Schur complement (equation)*, with the symmetric and positive definite matrix $BA^{-1}B^\mathsf{T} + C$, and allows for the computation of z. Solving the second equation then determines y. The preconditioned *Uzawa algorithm* uses a preconditioner D for the Schur complement and iterates the following steps to approximate z and y.

Algorithm 6.1 (Uzawa Algorithm) *Choose* $z^0 \in \mathbb{R}^{n_B}$, $\tau, \varepsilon_{\mathrm{stop}} > 0$, *and set* $k = 0$.

(1) Solve $Ay^{k+1} = f - B^\mathsf{T} z^k$.
(2) Solve $D(z^{k+1} - z^k) = \tau(By^{k+1} - Cz^k - g)$.
(3) Stop if $\|z^{k+1} - z^k\|_Q \le \varepsilon_{\mathrm{stop}}$; *otherwise set* $k \to k + 1$ *and continue with (1)*.

Since we have

$$
\frac{1}{\tau}D(z^{k+1} - z^k) + (BA^{-1}B^\mathsf{T} + C)z^k = BA^{-1}f - g,
$$

we see that the algorithm is simply a Richardson iteration for the Schur complement. Convergence holds for τ sufficiently small. The preconditioned *Arrow–Hurwicz algorithm* employs preconditioners G and D for A and $BA^{-1}B^\mathsf{T} + C$, respectively.

Algorithm 6.2 (Arrow–Hurwicz Algorithm) *Choose* $y^0 \in \mathbb{R}^n$, $z^0 \in \mathbb{R}^m$ *and parameters* $\omega, \tau, \varepsilon_{\mathrm{stop}} > 0$, *and set* $k = 0$.

(1) Solve $G(y^{k+1} - y^k) = \omega(f - Ay^k - B^\mathsf{T} z^k)$.
(2) Solve $-D(z^{k+1} - z^k) = \tau(g - By^{k+1} + Cz^k)$.
(3) Stop if $\|z^{k+1} - z^k\|_Q \le \varepsilon_{\mathrm{stop}}$; *otherwise set* $k \to k + 1$ *and continue with (1)*.

Remark 6.4 Augmented Lagrangian methods introduce a quadratic term $r|By - g|^2$ with a parameter $r > 0$ in the saddle-point problem. The additional term vanishes for the solution of the saddle point problem, but often improves the performance of the iterative schemes since the matrix $A + rB^\top B$ is positive definite, provided that A is positive definite on the kernel of B.

6.2 Continuous Saddle-Point Problems

6.2.1 Closed Range Theorem

Let X and Z be Banach spaces whose duals are denoted by X' and Z', respectively. Given any $\varphi \in X'$ and $\psi \in Z'$, we write the application of φ to $x \in X$ and of ψ to $z \in Z$ as

$$\langle \varphi, x \rangle = \varphi(x), \quad \langle \psi, z \rangle = \psi(z).$$

Let $L : X \to Z$ be a bounded linear operator. For every fixed $\psi \in Z'$, the mapping

$$x \mapsto \langle \psi, Lx \rangle,$$

specifies an element in X'. This operation defines the *adjoint operator*

$$L' : Z' \to X', \quad \langle L'\psi, x \rangle = \langle \psi, Lx \rangle$$

for all $\psi \in Z'$ and all $x \in X$; it generalizes the transposition of matrices. Throughout this section we follow [6].

Definition 6.1 For a subset $W \subset Z'$, we define its *polar set* $W^\circ \subset Z$ by

$$W^\circ = \{z \in Z : \langle \psi, z \rangle = 0 \text{ for all } \psi \in W\}.$$

If Z is a Hilbert space that is identified with its dual, then W° coincides with the orthogonal complement $W^\perp$ of W. The closed range theorem generalizes the identity

$$\operatorname{Im} L = (\ker L^\top)^\perp,$$

from finite-dimensional to infinite-dimensional situations. Its proof uses the Hahn–Banach theorem, i.e., the existence of a separating hyperplane for two disjoint convex sets.

Theorem 6.2 (Closed Range Theorem) *Let $L : X \to Z$ be a bounded and linear operator. Then $\operatorname{Im} L$ is closed in Z if and only if $\operatorname{Im} L = (\ker L')^\circ$.*

Proof By the definition of $(\ker L')^\circ$, we have

$$(\ker L')^\circ = \{z \in Z : \langle \psi, z \rangle = 0 \text{ for all } \psi \in \ker L'\} = \bigcap_{\psi \in \ker L'} \ker \psi.$$

Since for every $\psi \in \ker L'$ and $x \in X$ we have

$$\langle \psi, Lx \rangle = \langle L'\psi, x \rangle = 0,$$

we see that $\operatorname{Im} L \subset (\ker L')^\circ$.

(i) Assume that there exists $z_0 \in (\ker L')^\circ \setminus \operatorname{Im} L$. Since $\operatorname{Im} L$ is closed, there exists $\varepsilon > 0$ such that $B_\varepsilon(z_0) \cap \operatorname{Im} L = \emptyset$. Applying the separation theorem, cf. Remarks 2.5, we deduce that there exist $\psi \in Z'$ and $m \in \mathbb{R}$ such that

$$\langle \psi, z_0 \rangle > m \geq \langle \psi, Lx \rangle$$

for all $x \in X$. The second inequality can only be true if $m \geq 0$ and $\langle \psi, Lx \rangle = 0$ for all $x \in X$. But this implies that $L'\psi = 0$, i.e., $\psi \in \ker L'$. Since $z_0 \in (\ker L')^\circ$, we obtain the contradiction $\langle \psi, z_0 \rangle = 0$.

(ii) The kernel of every $\psi \in Z'$ is a closed set, and as the intersection of closed sets, $(\ker L')^\circ$ is closed. Hence, also $\operatorname{Im} L$ is closed.   □

Remark 6.5 Continuity of L alone is not sufficient to guarantee that $\operatorname{Im} L$ is closed.

6.2.2   Inf-Sup Condition

The inf-sup condition characterizes the boundedness of left inverse operators as in the discrete situation. We let X, Y, Z be Banach spaces in what follows.

Definition 6.2 A bijective linear operator $L : X \to W$ for $W \subset Z$ is an *isomorphism* if it is bounded and its inverse $L^{-1} : W \to X$ is bounded.

The existence and boundedness of a left inverse of a bounded operator can be expressed by an inf-sup condition.

Lemma 6.3 (Inf-Sup Condition) *Let* $L : X \to Y'$ *be a bounded and linear operator. Then* $L : X \to \operatorname{Im} L$ *is an isomorphism if and only if there exists* $\beta > 0$ *such that*

$$\inf_{x \in X \setminus \{0\}} \sup_{y \in Y \setminus \{0\}} \frac{\langle Lx, y \rangle}{\|x\|_X \|y\|_Y} \geq \beta.$$

Proof

(i) If $L : X \to \operatorname{Im} L$ is an isomorphism, then for every $x \in X$, we have the estimate $\|x\|_X \le c_L^{-1} \|Lx\|_{Y'}$ which is equivalent to

$$c_L \|x\|_X \le \sup_{y \in Y \setminus \{0\}} \frac{\langle Lx, y \rangle}{\|y\|_Y},$$

i.e., to the asserted estimate with $\beta = c_L$.

(ii) Conversely, the inf-sup condition is equivalent to $\|Lx\|_{Y'} \ge \beta \|x\|_X$ for all $x \in X$, which implies that L is injective. Therefore, $L : X \to \operatorname{Im} L$ is a bijection. In particular, for every $\psi \in \operatorname{Im} L$, there exists a unique $x \in X$ with $Lx = \psi$ and

$$\|x\|_X \le \frac{1}{\beta} \|Lx\|_{Y'} = \frac{1}{\beta} \|\psi\|_{Y'},$$

i.e., the left inverse $L^{-\ell} : \operatorname{Im} L \to X$ is bounded.   □

Remarks 6.6

(i) The inf-sup condition is equivalent to

$$\|Lx\|_{Y'} = \sup_{y \in Y \setminus \{0\}} \frac{\langle Lx, y \rangle}{\|y\|_Y} \ge \beta \|x\|_X$$

for some $\beta > 0$ and all $x \subset X$. Injectivity and closedness of the image of L are direct consequences of this bound.

(ii) Equivalent to the statements of the lemma is that $L : X \to Y'$ is injective with bounded left inverse $L^{-\ell}$ with $\|L^{-\ell}\| \le \beta^{-1}$.

If L is an isomorphism, then its image is closed. The converse implication is known as the inverse operator theorem.

Lemma 6.4 (Closedness) *If $L : X \to W$ for $W \subset Z$ is an isomorphism, then $W = \operatorname{Im} L$ is closed.*

Proof We have to show that for every Cauchy sequence $(v_j)_{j \in \mathbb{N}} \subset \operatorname{Im} L$ with $v_j = Lx_j$ for a sequence $(x_j)_{j \in \mathbb{N}} \subset X$, its limit v also belongs to $\operatorname{Im} L$. Since the left inverse is bounded, we have for all $j, k \in \mathbb{N}$ that

$$\|x_j - x_k\|_X \le c_L^{-1} \|v_j - v_k\|_Z,$$

i.e., $(x_j)_{j \in \mathbb{N}}$ is a Cauchy sequence with limit $x \in X$. By the continuity of L, we have $Lx = \lim_{j \to \infty} Lx_j = v$ which proves $v \in \operatorname{Im} L$.   □

Remark 6.7 As a consequence of the previous lemma, we have $\operatorname{Im} L = (\ker L')^\circ$ whenever a bounded linear operator L satisfies the inf-sup condition.

6.2.3   Generalized Lax–Milgram Lemma

For Banach spaces X and Y, and a continuous bilinear form

$$\Gamma : X \times Y \to \mathbb{R},$$

we define the bounded and linear operator $L : X \to Y'$ via

$$Lx = \Gamma(x, \cdot)$$

for all $x \in X$. The problem of finding $x \in X$ such that

$$\Gamma(x, y) = \ell(y)$$

for all $y \in Y$ for a given functional $\ell \in Y'$ is thus equivalent to the operator equation

$$Lx = \ell.$$

Theorem 6.3 (Generalized Lax–Milgram Lemma) *Assume that X and Y are reflexive Banach spaces. The linear operator $L : X \to Y'$ is an isomorphism if and only if the associated bilinear form*

$$\Gamma(x, y) = \langle Lx, y \rangle$$

is bounded, satisfies an inf-sup condition, and is nondegenerate, i.e.,

(a) there exists $c_\Gamma \geq 0$ such that

$$|\Gamma(x, y)| \leq c_\Gamma \|x\|_X \|y\|_Y;$$

(b) there exists $\gamma > 0$ such that

$$\inf_{x \in X \setminus \{0\}} \sup_{y \in Y \setminus \{0\}} \frac{\Gamma(x, y)}{\|x\|_X \|y\|_Y} \geq \gamma;$$

(c) for all $y \in Y \setminus \{0\}$ there exists $x \in X$ with $\Gamma(x, y) \neq 0$.

Proof We note that boundedness of L is equivalent to boundedness of Γ.

(i) Assume that the conditions on Γ are satisfied. Lemma 6.3 then implies that $L : X \to \operatorname{Im} L$ is an isomorphism, and Lemma 6.4 shows that $\operatorname{Im} L$ is closed. The closed range theorem yields that

$$\operatorname{Im} L = (\ker L')^\circ.$$

Note that $L' : Y'' \to X'$ is by identifying $Y'' \simeq Y$ regarded as an operator on Y, i.e., $L' : Y \to X'$, and with $X'' \simeq X$ we have $(L')' \simeq L$. The nondegeneracy condition implies that

$$\ker L' = \{0\},$$

since $L'y = 0$ means that $\Gamma(x, y) = \langle Lx, y \rangle = \langle x, L'y \rangle = 0$ for all $x \in X$. In particular, $(\ker L')^\circ = Y'$ and $\operatorname{Im} L = Y'$. Hence $L : X \to Y'$ is an isomorphism.

(ii) Conversely, if L is an isomorphism, then Lemma 6.3 implies the inf-sup condition. Since $\operatorname{Im} L = Y'$ is closed, we have $Y' = \operatorname{Im} L = (\ker L')^\circ$, which can only be the case if $\ker L' = \{0\}$. In particular, there is no $y \in Y \setminus \{0\}$ with $L'y = 0$, and hence no $y \in Y$, such that for all $x \in X$ we have

$$0 = \langle x, L'y \rangle = \langle Lx, y \rangle = \Gamma(x, y).$$

Therefore the nondegeneracy condition is also satisfied.   □

Remarks 6.8

(i) If we omit the nondegeneracy condition, then the mapping $L : X \to \operatorname{Im} L = (\ker L')^\circ \subseteq Y'$ is an isomorphism.
(ii) If $X = Y$ and Γ is bounded and coercive, then the conditions of the theorem are satisfied.
(iii) The inf-sup condition implies that we have $\| L^{-1} \| \leq \gamma^{-1}$.
(iv) If Γ is symmetric, i.e., $\Gamma(x, y) = \Gamma(y, x)$ for all $x, y \in X$, then the inf-sup condition implies nondegeneracy.

6.2.4   Saddle-Point Problems

We return to the continuous saddle-point formulation which consists in finding $(u, p) \in V \times Q$ such that

$$(S) \quad \begin{cases} a(u, v) + b(v, p) = \ell_1(v), \\ b(u, q) = \ell_2(q), \end{cases}$$

for all $(v, q) \in V \times Q$. To formulate necessary and sufficient conditions for the unique solvability of (S), we analyze properties of the bilinear form b.

Lemma 6.5 (Properties of b) *Let V and Q be Hilbert spaces and let $b : V \times Q \to \mathbb{R}$ be bounded and bilinear. Let*

$$B : V \to Q', \ v \mapsto b(v, \cdot).$$

Then the following statements are equivalent:

(i) *The operator* $B' : Q \to V'$ *satisfies an inf-sup condition, i.e., there exists* $\beta > 0$ *such that*

$$\inf_{q \in Q \setminus \{0\}} \sup_{v \in V \setminus \{0\}} \frac{b(v,q)}{\|v\|_V \|q\|_Q} \geq \beta$$

(ii) *The operator* $B : (\ker B)^\perp \to Q'$ *is an isomorphism and there exists* $\beta > 0$, *such that for all* $v \in (\ker B)^\perp$ *we have*

$$\|Bv\|_{Q'} \geq \beta \|v\|_V.$$

(iii) *The operator* $B' : Q \to (\ker B)^\circ \subseteq V'$ *is an isomorphism and there exists* $\beta > 0$, *such that for all* $q \in Q$ *we have*

$$\|B'q\|_{V'} \geq \beta \|q\|_Q.$$

Proof (i)$\Longleftrightarrow$(iii). The equivalence is the statement of the generalized Lax–Milgram lemma with $X = Q$, $Y = V$, $L = B'$, $L' = B$, $\Gamma(q,v) = b(v,q)$, when the nondegeneracy condition is omitted, cf. Remark 6.8.

(iii) $\Longrightarrow$ (ii). We show that $B : (\ker B)^\perp \to Q'$ satisfies the conditions of the generalized Lax–Milgram lemma with $L = B$, $X = (\ker B)^\perp$, and $Y = Q$. To prove the inf-sup condition, let $v \in (\ker B)^\perp \subset V$. With v we associate the functional $g_v \in (\ker B)^\circ \subset V'$ by defining $\langle g_v, w \rangle = (v, w)_V$ for all $w \in V$. Then, $\|g_v\|_{V'} = \|v\|_V$. Due to the statement (iii) there exists $p \in Q$, such that $B'p = g_v$ and $\|p\|_Q \leq \beta^{-1} \|g_v\|_{V'}$. Using $b(v,p) = \langle B'p, v \rangle = \langle g_v, v \rangle = (v,v)_V$ implies that

$$\sup_{q \in Q \setminus \{0\}} \frac{b(v,q)}{\|q\|_Q} \geq \frac{b(v,p)}{\|p\|_Q} \geq \frac{b(v,p)}{\beta^{-1} \|g_v\|_{V'}} = \beta \|v\|_V,$$

which is the inf-sup condition for B restricted to $(\ker B)^\perp$. To verify the nondegeneracy condition, assume that there exists $q \in Q \setminus \{0\}$, such that $b(v,q) = 0$ for all $v \in (\ker B)^\perp$, which then holds for all $v \in V$. This implies $B'q = 0$ which contradicts $\|B'q\|_{V'} > 0$. Hence, the conditions of the generalized Lax–Milgram lemma are satisfied and B is an isomorphism. The estimate is equivalent to the inf-sup condition.

(ii) $\implies$ (i) For every $q \in Q$, using the bijectivity of B and the estimate $\|Bv\|_{Q'} \geq \beta \|v\|_V$ for all $v \in (\ker B)^\perp$, we have that

$$\|q\|_Q = \sup_{g \in Q' \setminus \{0\}} \frac{\langle g, q \rangle}{\|g\|_{Q'}}$$

$$= \sup_{v \in (\ker B)^\perp \setminus \{0\}} \frac{\langle Bv, q \rangle}{\|Bv\|_{Q'}}$$

$$= \sup_{v \in (\ker B)^\perp \setminus \{0\}} \frac{b(v, q)}{\|Bv\|_{Q'}}$$

$$\leq \sup_{v \in (\ker B)^\perp \setminus \{0\}} \frac{b(v, q)}{\beta \|v\|_V}$$

$$\leq \beta^{-1} \sup_{v \in V \setminus \{0\}} \frac{b(v, q)}{\|v\|_V},$$

which is (i). □

Remark 6.9 The second statement of the lemma is equivalent to the inf-sup condition

$$\inf_{v \in (\ker B)^\perp \setminus \{0\}} \sup_{q \in Q \setminus \{0\}} \frac{b(v, q)}{\|v\|_V \|q\|_Q} \geq \beta > 0,$$

in which the roles of v and q are exchanged and V is replaced by $(\ker B)^\perp$. This inf-sup condition corresponds to the operator $Bv = b(v, \cdot)$.

With the help of the lemma we can specify conditions for the unique solvability of the saddle-point problem (S).

Theorem 6.4 (Brezzi's Splitting Theorem) *Assume that V and Q are Hilbert spaces, $a : V \times V \to \mathbb{R}$ is a symmetric, bounded, and positive semidefinite bilinear form, and $b : V \times Q \to \mathbb{R}$ is a bounded and bilinear from. The operator*

$$L : V \times Q \to V' \times Q', \quad (u, p) \mapsto \big(a(u, \cdot) + b(\cdot, p), b(u, \cdot) \big)$$

is an isomorphism, i.e., (S) is uniquely solvable, if and only if a is coercive on $\ker B$, *i.e., there exists $\alpha > 0$ such that*

$$a(v, v) \geq \alpha \|v\|_V^2$$

for all $v \in \ker B$, and b satisfies an inf-sup condition, i.e., there exists $\beta > 0$ such that

$$\inf_{q \in Q \setminus \{0\}} \sup_{v \in V \setminus \{0\}} \frac{b(v, q)}{\|v\|_V \|q\|_Q} \geq \beta.$$

Proof

(i) Assume that the conditions on a and b are satisfied, in particular, the equivalent statements of Lemma 6.5 are valid. Given $(\ell_1, \ell_2) \in V' \times Q'$, we first let $u_0 \in (\ker B)^{\perp}$ be such that $Bu_0 = \ell_2$ and

$$\|u_0\|_V \le \beta^{-1} \|\ell_2\|_{Q'}.$$

We thus have that

$$b(u_0, q) = \ell_2(q)$$

for all $q \in Q$. We next try to find $u_1 \in \ker B$, such that for some $p \in Q$ we have

$$a(u_0 + u_1, v) + b(v, p) = \ell_1(v)$$

for all $v \in V$. By restricting to $v \in \ker B$, we have $b(v, p) = 0$, and the identity simplifies to the requirement that $u_1 \in \ker B$ be such that

$$a(u_1, v) = \ell_1(v) - a(u_0, v)$$

for all $v \in \ker B$. By the assumed coercivity of a on $\ker B$, the Lax–Milgram lemma implies the existence of a unique solution $u_1 \in \ker B$ with

$$\alpha \|u_1\|_V^2 \le a(u_1, u_1) \le \|\ell_1\|_{V'} \|u_1\|_V + k_a \|u_0\|_V \|u_1\|_V.$$

It remains to determine $p \in Q$ such that

$$b(v, p) = \ell_1(v) - a(u_0 + u_1, v)$$

for all $v \in V$. The right-hand side defines a functional $\varphi \in V'$ and by the construction of u_1, we have $\varphi(v) = 0$ for all $v \in \ker B$. Hence $\varphi \in (\ker B)^{\circ}$. Since $B' : Q \to (\ker B)^{\circ}$ is an isomorphism, there exists a unique $p \in Q$ with $B'p = \varphi$, which is equivalent to the above equation. Moreover, we have

$$\|p\|_Q \le \beta^{-1} \|\varphi\|_{V'}.$$

The pair (u, p) with $u = u_0 + u_1$ satisfies (S), and by construction, we have

$$\|(u, p)\|_{V \times Q} \le c(k_a, \alpha^{-1}, \beta^{-1}) \|(\ell_1, \ell_2)\|_{V' \times Q'}.$$

To prove uniqueness of (u, p), we note that for $(\ell_1, \ell_2) = 0$, the estimate implies $(u, p) = 0$. Together with the linearity of L, we conclude that L is an isomorphism.

(ii) Assume that L is an isomorphism and let $u \in \ker B$. We define $\ell_1 \in V'$ by setting

$$\langle \ell_1, v \rangle = a(u, v)$$

for all $v \in V$. Then, there exists $p \in Q$ such that $(u, p) = L^{-1}(\ell_1, 0)$, i.e., we have

$$a(u, v) + b(v, p) = \langle \ell_1, v \rangle,$$
$$b(u, q) \qquad\qquad = 0,$$

for all $(v, q) \in V \times Q$. Using the Cauchy–Schwarz inequality for the symmetric and positive semidefinite bilinear form a, we find that

$$\|\ell_1\|_{V'} = \sup_{v \in V \setminus \{0\}} \frac{a(u, v)}{\|v\|_V} \le \sup_{v \in V \setminus \{0\}} \frac{a(u, u)^{1/2} a(v, v)^{1/2}}{\|v\|_V}.$$

Since L^{-1} is bounded, we have

$$\|u\|_V \le \|(u, p)\|_{V \times Q} \le \|L^{-1}\| \|(\ell_1, 0)\|_{V' \times Q'} = \|L^{-1}\| \|\ell_1\|_{V'}.$$

The combination of the last two inequalities and the boundedness of a show that

$$\|u\|_V \le \|L^{-1}\| \sup_{v \in \ker B \setminus \{0\}} \frac{a(u, u)^{1/2} a(v, v)^{1/2}}{\|v\|_V} \le \|L^{-1}\| k_a^{1/2} a(u, u)^{1/2}.$$

Since this holds for all $u \in \ker B$, we find that a is coercive on $\ker B$. To prove the inf-sup condition for b, let $\ell_2 \in Q'$ and $(u, p) = L^{-1}(0, \ell_2)$. Then $\|u\|_V \le \|L^{-1}\| \|\ell_2\|_{Q'}$, and for the orthogonal projection $u^\perp$ of u onto $(\ker B)^\perp$, we have

$$\|u^\perp\|_V \le \|u\|_V.$$

Therefore the mapping $\ell_2 \mapsto u \mapsto u^\perp$ is bounded such that $Bu^\perp = \ell_2$, i.e., $B : (\ker B)^\perp \to Q'$ is an isomorphism and Lemma 6.5 implies the inf-sup condition. $\qquad\square$

We illustrate the application of the theorem with a saddle-point formulation of the Poisson problem.

Example 6.4 We consider the Poisson problem $-\Delta p = g$ in $\Omega \subset \mathbb{R}^d$ with Dirichlet boundary condition $p|_{\partial\Omega} = 0$. By introducing $u = \nabla p$, this is equivalent to finding $(u, p) \in L^2(\Omega; \mathbb{R}^d) \times H_0^1(\Omega)$ such that

$$\int_\Omega u \cdot v \, dx - \int_\Omega v \cdot \nabla p \, dx = 0,$$

$$-\int_\Omega u \cdot \nabla q \, dx \qquad\qquad = -\int_\Omega g q \, dx,$$

for all $(v, q) \in L^2(\Omega; \mathbb{R}^d) \times H_0^1(\Omega)$. This is called the *primal mixed form of the Poisson problem*. The bilinear form

$$a(u, v) = \int_\Omega u \cdot v \, dx$$

coincides with the inner product on $L^2(\Omega; \mathbb{R}^d)$ and is thus elliptic on $L^2(\Omega; \mathbb{R}^d)$. The bilinear form

$$b(v, q) = -\int_\Omega v \cdot \nabla q \, dx$$

satisfies an inf-sup condition, since by choosing $v = -\nabla q$, we have

$$\sup_{v \in L^2(\Omega; \mathbb{R}^d) \setminus \{0\}} \frac{b(v, q)}{\|v\|_{L^2(\Omega)}} \geq \frac{b(-\nabla q, q)}{\|\nabla q\|_{L^2(\Omega)}} = \|\nabla q\|_{L^2(\Omega)}$$

for all $q \in H_0^1(\Omega)$.

6.2.5 Perturbed Saddle-Point Problems

In some applications perturbed saddle-point problems occur. These consist in finding $(u, p) \in V \times Q$ such that

$$(S_t) \quad \begin{cases} a(u, v) + b(v, p) = \ell_1(v), \\ b(u, q) - t^2 c(p, q) = \ell_2(q), \end{cases}$$

for all $(v, q) \in V \times Q$. Here $c : Q \times Q \to \mathbb{R}$ is a positive semidefinite symmetric bilinear form and $t \geq 0$ is a small parameter. We follow [6].

Theorem 6.5 (Perturbed Formulation) *Assume that the bilinear forms a and b satisfy the conditions of Brezzi's splitting theorem, and assume that c is bounded and positive semidefinite. Then (S_t) has a unique solution for every $t \in [0, 1]$ and the solution operator $L^{-1} : (\ell_1, \ell_2) \to (u, p)$ is bounded t-independently.*

The proof is based on the following lemma.

Lemma 6.6 (Inf-Sup Condition for Γ_t) *Suppose that the assumptions of the theorem are satisfied, set $|q|_c = c(q,q)^{1/2}$, and assume that for some $\alpha > 0$ and all $u \in V \setminus \{0\}$, we have*

$$\frac{a(u,u)}{\|u\|_V} + \sup_{q \in Q \setminus \{0\}} \frac{b(u,q)}{\|q\|_Q + t|q|_c} \geq \alpha \|u\|_V.$$

Then the bilinear form associated with (S_t), i.e.,

$$\Gamma_t(u,p;v,q) = a(u,v) + b(v,p) + b(u,q) - t^2 c(p,q),$$

satisfies an inf-sup condition with a t-independent constant $\gamma' > 0$, and with the norm

$$\|(v,q)\| = \|v\|_V + \|q\|_Q + t|q|_c$$

on $V \times Q$.

Proof Let $(u,p) \in V \times Q$. We distinguish three cases with a parameter $0 < \delta \leq 1$ to be determined later but which will not depend on (u,p) and t.

Case 1:   Assume that $\|u\|_V + \|p\|_Q \leq \delta^{-1} t|p|_c$. We then have $\Gamma_t(u,p;u,-p) = a(u,u) + t^2 c(p,p) \geq (\delta^2/4)\|(u,p)\|^2$ and hence

$$\frac{\delta^2}{4}\|(u,p)\| \leq \frac{\Gamma_t(u,p;u,-p)}{\|(u,-p)\|} \leq \sup_{(v,q) \neq 0} \frac{\Gamma_t(u,p;v,q)}{\|(v,q)\|},$$

i.e., an inf-sup condition for Γ_t.

Case 2a: Assume that $\|u\|_V + \|p\|_Q > \delta^{-1} t|p|_c$ and $\|u\|_V \leq (\beta/(2k_a))\|p\|_Q$. Since b satisfies the inf-sup condition, we have

$$\beta\|p\|_Q \leq \sup_{v \neq 0} \frac{b(v,p)}{\|v\|_V} \leq \sup_{v \neq 0} \frac{\Gamma_t(u,p;v,0)}{\|v\|_V} + \sup_{v \neq 0} \frac{a(u,v)}{\|v\|_V}$$

$$\leq \sup_{(v,q) \neq 0} \frac{\Gamma_t(u,p;v,q)}{\|(v,q)\|} + k_a\|u\|_V.$$

The assumption that $\|u\|_V \leq (\beta/(2k_a))\|p\|_Q$ leads to

$$k_a\|u\|_V \leq \sup_{(v,q) \neq 0} \frac{\Gamma_t(u,p;v,q)}{\|(v,q)\|}.$$

Incorporating the assumed bound for $t|p|_c$, we deduce the inf-sup condition for Γ_t.

Case 2b: Assume that $\|u\|_V + \|p\|_Q > \delta^{-1}t|p|_c$ and $\|u\|_V > (\beta/(2k_a))\|p\|_Q$. The bounds imply that $\|\|(u,p)\|\| \le (2 + 4k_a/\beta)\|u\|_V$, and for $\delta \le 1/(2 + 4k_a/\beta)$ we have $\delta\|\|(u,p)\|\| \le \|u\|_V$. The assumption of the lemma shows that

$$\alpha\|u\|_V \le \frac{a(u,u)}{\|u\|_V} + \sup_{q\neq 0} \frac{b(u,q)}{\|q\|_Q + t|q|_c}$$

$$\le \frac{a(u,u) + t^2c(p,p)}{\|u\|_V} + \sup_{q\neq 0} \frac{\Gamma_t(u,p;0,q) + t^2c(p,q)}{\|q\|_Q + t|q|_c}.$$

With $c(p,q) \le |p|_c|q|_c$, we deduce that

$$\alpha\|u\|_V \le \frac{a(u,u) + t^2c(p,p)}{\delta\|\|(u,p)\|\|} + \sup_{q\neq 0} \frac{\Gamma_t(u,p;0,q)}{\|q\|_Q + t|q|_c} + t|p|_c.$$

For δ small enough we may assume that $t|p|_c \le \delta(\|u\|_V + \|p\|_Q) \le (\alpha/2)\|u\|_V$. Incorporating the bound $\delta\|\|(u,p)\|\| \le \|u\|_V$ then leads to

$$\frac{\alpha\delta}{2}\|\|(u,p)\|\| \le \frac{a(u,u) + t^2c(p,p)}{\delta\|\|(u,p)\|\|} + \sup_{q\neq 0} \frac{\Gamma_t(u,p;0,q)}{\|q\|_Q + t|q|_c}$$

$$= \frac{\Gamma_t(u,p;u,-p)}{\delta\|\|(u,p)\|\|} + \sup_{q\neq 0} \frac{\Gamma_t(u,p;0,q)}{\|q\|_Q + t|q|_c}$$

$$\le (\delta^{-1} + 1) \sup_{(v,q)\neq 0} \frac{\Gamma_t(u,p;v,q)}{\|\|(v,q)\|\|}.$$

This completes the proof of the inf-sup condition for Γ_t. $\square$

We are now in position to prove Theorem 6.5.

Proof (of Theorem 6.5) The requirement that the unperturbed saddle-point problem with $t = 0$ satisfy the conditions of Brezzi's splitting theorem implies that the bilinear form Γ_0 satisfies an inf-sup condition with a constant $\gamma > 0$. For the pair $(u,0) \in V \times Q$, we therefore have that

$$\gamma\|u\|_V \le \sup_{(v,q)\neq 0} \frac{\Gamma_0(u,0;v,q)}{\|v\|_V + \|q\|_Q} \le \sup_{v\neq 0} \frac{a(u,v)}{\|v\|_V} + \sup_{q\neq 0} \frac{b(u,q)}{\|q\|_Q}.$$

Using $\|q\|_Q + t|q|_c \le (1 + k_c^{1/2})\|q\|_Q$ and the Cauchy–Schwarz inequality for a, we deduce that

$$\gamma\|u\|_V \le k_a^{1/2}a(u,u)^{1/2} + (1 + k_c^{1/2}) \sup_{q\neq 0} \frac{b(u,q)}{\|q\|_Q + t|q|_c}.$$

We use that for nonnegative real numbers x,y,z with $x > 0$ and $x \le y + z$ it follows that $x \le y^2/x + 2z$ (which follows from a case distinction $y \le x$ and $y > x$ with a binomial formula) to estimate

$$\gamma \|u\|_V \le \frac{k_a a(u,u)}{\gamma \|u\|_V} + 2(1 + k_c^{1/2}) \sup_{q \ne 0} \frac{b(u,q)}{\|q\|_Q + t|q|_c}.$$

This is the condition of Lemma 6.6 which therefore implies an inf-sup condition for Γ_t with a positive constant that does not depend on t. Noting that symmetry and the inf-sup condition imply nondegeneracy and verifying continuity, the result follows from the generalized Lax–Milgram lemma. □

Remark 6.10 The theorem motivates treating the optimality conditions related to minimizing

$$u \mapsto \frac{1}{2} \int_\Omega |\nabla u|^2 \, dx + \frac{t^{-2}}{2} \int_\Omega |\operatorname{div} u|^2 \, dx - \int_\Omega f \cdot u \, dx$$

in the set of $u \in H_D^1(\Omega; \mathbb{R}^d)$ for $0 < t \ll 1$ as a perturbed saddle-point problem. By introducing the variable $p = t^{-2} \operatorname{div} u$, we find that every minimizer $u \in H_D^1(\Omega; \mathbb{R}^d)$ satisfies

$$\int_\Omega \nabla u : \nabla v \, dx + \int_\Omega p \operatorname{div} v \, dx = \int_\Omega f \cdot v \, dx,$$

$$\int_\Omega q \operatorname{div} u \, dx - t^2 \int_\Omega pq \, dx = 0,$$

for all $(v,q) \in H_D^1(\Omega; \mathbb{R}^d) \times L^2(\Omega)$. This approach avoids a critical dependence of stability bounds on the parameter t.

6.3 Approximation of Saddle-Point Problems

6.3.1 Generalized Céa Lemma

To approximate a general variational formulation defined by the bilinear form Γ : $X \times Y \to \mathbb{R}$ numerically, we choose finite-dimensional subspaces $X_h \subset X$ and $Y_h \subset Y$. The generalized Lax–Milgram lemma guarantees the existence of a unique solution $x_h \in X_h$ such that

$$\Gamma(x_h, y_h) = \ell(y_h)$$

for all $y_h \in Y_h$, provided that Γ is bounded, and satisfies an inf-sup condition, and is nondegenerate with respect to the spaces X_h and Y_h. In this case we have the following generalization of Céa's lemma.

Theorem 6.6 (Generalized Céa Lemma) *Assume that* $\Gamma : X \times Y \to \mathbb{R}$ *is a bounded bilinear form, such that for every* $\ell \in Y'$, *there exists a unique* $x \in X$ *with*

$$\Gamma(x, y) = \ell(y)$$

for all $y \in Y$, *and such that* $\|x\|_X \leq \gamma^{-1} \|\ell\|_{Y'}$. *Assume that* $(X_h)_{h>0}$ *and* $(Y_h)_{h>0}$ *are families of subspaces of* X *and* Y, *respectively, with the property that for every* $h > 0$, *there exists* $\gamma_h > 0$ *with*

$$\inf_{x_h \in X_h \setminus \{0\}} \sup_{y_h \in Y_h \setminus \{0\}} \frac{\Gamma(x_h, y_h)}{\|x_h\|_X \|y_h\|_Y} \geq \gamma_h,$$

and for every $y_h \in Y \setminus \{0\}$, *there exists* $x_h \in X_h$ *so that* $\Gamma(x_h, y_h) \neq 0$. *Suppose that there exists* $\underline{\gamma} > 0$ *such that* $\gamma_h \geq \underline{\gamma}$ *for all* $h > 0$. *Then for every* $\ell \in Y'$, *there exists a unique* $x_h \in X_h$ *with*

$$\Gamma(x_h, y_h) = \ell(y_h)$$

for all $y_h \in Y_h$, *and such that*

$$\|x - x_h\|_X \leq (1 + k_\Gamma/\underline{\gamma}) \inf_{w_h \in X_h} \|x - w_h\|_X.$$

Proof The generalized Lax–Milgram lemma implies the existence of a uniquely defined solution $x_h \in X_h$. By subtracting the variational formulations, we obtain the Galerkin orthogonality

$$\Gamma(x - x_h, y_h) = 0$$

for all $y_h \in Y_h$. Let $w_h \in X_h$ be arbitrary. The discrete inf-sup condition implies that

$$\gamma_h \|x_h - w_h\|_X \leq \sup_{y_h \in Y_h \setminus \{0\}} \frac{\Gamma(x_h - w_h, y_h)}{\|y_h\|_Y}.$$

Using Galerkin orthogonality and boundedness of Γ, we infer that

$$\gamma_h \|x_h - w_h\|_X \leq \sup_{y_h \in Y_h \setminus \{0\}} \frac{\Gamma(x_h - w_h, y_h)}{\|y_h\|_Y}$$

$$= \sup_{y_h \in Y_h \setminus \{0\}} \frac{\Gamma(x - w_h, y_h)}{\|y_h\|_Y}$$

$$\leq \sup_{y_h \in Y_h \setminus \{0\}} \frac{k_\Gamma \|x - w_h\|_X \|y_h\|_Y}{\|y_h\|_Y}$$

$$= k_\Gamma \|x - w_h\|_X.$$

With the triangle inequality, we deduce that

$$\|x - x_h\|_X \le \|x - w_h\|_X + \|w_h - x_h\|_X$$
$$= (1 + \gamma_h^{-1} k_\Gamma)\|x - w_h\|.$$

Using that $w_h \in X_h$ is arbitrary and that $\gamma_h \ge \underline{\gamma}$ for all $h > 0$ proves the estimate. $\square$

Remarks 6.11

(i) The constant in the estimate of the theorem can be improved. For every $v \in X$ there exists a unique $v_h = P_h v$ with $\Gamma(v_h, y_h) = \Gamma(v, y_h)$ for all $y_h \in Y_h$, and we have $\|P_h\| \le k_\Gamma/\gamma_h$. The operator P_h defines a projection in the sense that $P_h^2 = P_h$ which leads to the nontrivial identity $\|P_h\| = \|I - P_h\|$. With this we deduce that

$$\|x - x_h\|_X = \|(I - P_h)(x - x_h)\|_X = \|(I - P_h)(x - w_h)\|_X \le \|P_h\| \|x - w_h\|_X.$$

(ii) In contrast to boundedness, inf-sup condition and nondegeneracy are in general not inherited by subspaces.

6.3.2 Saddle-Point Problems

As in the continuous setting, it is desirable to formulate conditions on the bilinear forms involved in a saddle-point problem to determine their solvability. We consider again the problem of finding $(u, p) \in V \times Q$ with

$$a(u, v) + b(v, p) = \ell_1(v),$$
$$b(u, q) \qquad\quad = \ell_2(q),$$

for all $(v, q) \in V \times Q$. Here, $\ell_1 \in V'$ and $\ell_2 \in Q'$ are given functionals.

Theorem 6.7 (Babuška–Brezzi Conditions) *Assume that the conditions of Brezzi's splitting theorem are satisfied, and $(V_h)_{h>0}$ and $(Q_h)_{h>0}$ are families of subspaces of the Hilbert spaces V and Q, respectively, such that for all $h > 0$ we have:*

(1) the bilinear form a is elliptic on the discrete kernel of b, i.e., there exists $\alpha_h > 0$ such that for all

$$v_h \in K_h = \{v_h \in V_h : b(v_h, q_h) = 0 \text{ for all } q_h \in Q_h\},$$

we have

$$a(v_h, v_h) \ge \alpha_h \|v_h\|_V^2;$$

(2) *the bilinear form b satisfies the inf-sup condition with respect to V_h and Q_h, i.e., there exists $\beta_h > 0$ such that*

$$\inf_{q_h \in Q_h} \sup_{v_h \in V_h} \frac{b(v_h, q_h)}{\|v_h\|_V \|q_h\|_Q} \geq \beta_h.$$

Suppose that there exist $\underline{\alpha}, \underline{\beta} > 0$ such that $\alpha_h \geq \underline{\alpha}$ and $\beta_h \geq \underline{\beta}$ for all $h > 0$. Then for every $h > 0$, there exists a pair $(u_h, p_h) \in V_h \times Q_h$ such that

$$a(u_h, v_h) + b(v_h, p_h) = \ell_1(v_h),$$
$$b(u_h, q_h) \qquad\qquad = \ell_2(q_h),$$

for all $(v_h, q_h) \in V_h \times Q_h$. Moreover, there exists $c > 0$ such that

$$\|u - u_h\|_V + \|p - p_h\|_Q \leq c \inf_{(v_h, q_h)} \left(\|u - v_h\|_V + \|p - q_h\|_Q \right).$$

Proof The conditions on the subspaces V_h and Q_h imply that the discrete problems for every $h > 0$ are uniquely solvable, i.e., the induced bilinear form Γ on $X_h = Y_h = V_h \times Q_h$ satisfies an inf-sup condition uniformly in h and is nondegenerate for all $h > 0$. The application of the generalized Céa lemma implies the estimate. $\square$

The nondegeneracy and the inf-sup condition are in general not inherited by the subspaces. Also the discrete coercivity has to be verified for every $h > 0$, since in general we have

$$K_h \not\subset \ker B = \{v \in V : b(v, q) = 0 \text{ for all } q \in Q\},$$

i.e., significantly more conditions are involved in the definition of $\ker B$ than in the definition of K_h. In fact, a sharper error estimate can be proved if $K_h \subset \ker B$ for all $h > 0$.

Remark 6.12 The spaces V_h and Q_h have to be carefully chosen so that the operator

$$B_h : K_h^\perp \to Q_h' \simeq Q_h, \quad v_h \mapsto b(v_h, \cdot)$$

is an isomorphism, i.e., $\dim V_h - \dim K_h = \dim Q_h$ is a necessary requirement. Moreover, K_h has to be sufficiently small so that a is coercive on K_h.

We verify the conditions of Theorem 6.7 in a simple model problem.

Example 6.5 Consider the primal mixed formulation of the Poisson problem, and for a sequence of regular triangulations $(\mathcal{T}_h)_{h>0}$, the subspaces

$$V_h = \mathcal{L}^0(\mathcal{T}_h)^d = \{v_h \in L^\infty(\Omega; \mathbb{R}^d) : v_h|_T \in \mathcal{P}_0(T) \text{ for all } T \in \mathcal{T}_h\},$$
$$Q_h = \mathcal{S}_0^1(\mathcal{T}_h) = \{q_h \in C(\overline{\Omega}) : q_h|_T \in \mathcal{P}_1(T) \text{ for all } T \in \mathcal{T}_h\}.$$

The bilinear form

$$a(u, v) = \int_\Omega u \cdot v \, dx$$

is coercive on $V = L^2(\Omega; \mathbb{R}^d)$ and hence uniformly on the subspaces V_h. For the bilinear form $b : V \times Q \to \mathbb{R}$ with $Q = H_0^1(\Omega)$, we find with the choice $v_h = -\nabla q_h$ that

$$\sup_{v_h \in V_h \setminus \{0\}} \frac{b(v_h, q_h)}{\|v_h\|_{L^2(\Omega)}} = \sup_{v_h \in V_h \setminus \{0\}} \frac{-\int_\Omega v_h \cdot \nabla q_h \, dx}{\|v_h\|_{L^2(\Omega)}} \geq \|\nabla q_h\|_{L^2(\Omega)},$$

so that the discrete inf-sup condition for b holds uniformly in h.

The error estimate resulting from Theorem 6.7 in fact coincides up to the involved constants with the estimate one obtains with Céa's lemma for the direct treatment of the Poisson equation. In the case of the Stokes problem, the subspaces have to be chosen more carefully.

Example 6.6 The Stokes problem leads to the bilinear form $b : V \times Q \to \mathbb{R}$ defined by

$$b(v, q) = \int_\Omega q \operatorname{div} v \, dx$$

for $v \in V = H_0^1(\Omega; \mathbb{R}^d)$ and $q \in L_0^2(\Omega) = \{q \in L^2(\Omega) : \int_\Omega q \, dx = 0\}$. The choice of

$$V_h = \mathscr{S}_0^1(\mathscr{T}_h)^d, \quad Q_h = \mathscr{L}^0(\mathscr{T}_h) \cap L_0^2(\Omega)$$

in general does not lead to the inf-sup condition for b. For a triangulation $\mathscr{T}_h$ of $\Omega = (0, 1)^2$ consisting of halved squares with two triangles $T_1, T_2 \in \mathscr{T}_h$ that have all vertices on $\partial\Omega$, cf. Fig. 6.2, we have $\operatorname{div} v_h|_{T_j} = 0$ for $j = 1, 2$ and every $v_h \in V_h$. Hence $\operatorname{div} : V_h \to Q_h$ is not surjective. More generally, for a triangulation $\mathscr{T}_h$ of $\Omega = (0, 1)^2$ with nodes $\mathscr{N}_h$ and consisting of halved squares, we have

$$\dim Q_h = |\mathscr{T}_h| - 1, \quad \dim V_h = 2|\mathscr{N}_h \cap \Omega|$$

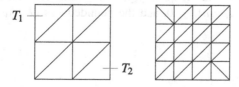

Fig. 6.2 Triangles T_1, T_2 in a triangulation for which all vertices belong to $\partial\Omega$ (*left*); triangulations consisting of halved squares with $|\mathscr{T}_h| = 2|\mathscr{N}_h \cap \Omega| + |\mathscr{N}_h \cap \partial\Omega| - 2$ (*left and right*)

and $|\mathscr{T}_h| = 2|\mathscr{N}_h \cap \Omega| + |\mathscr{N}_h \cap \partial\Omega| - 2$, cf. Fig. 6.2. Since $|\mathscr{N}_h \cap \partial\Omega| \geq 4$, this implies that

$$\dim V_h = \dim Q_h - |\mathscr{N}_h \cap \partial\Omega| + 3 < \dim Q_h,$$

so that div : $V_h \to Q_h$ cannot be surjective.

6.3.3   Fortin Interpolation

In general, it is difficult to verify the Babuška–Brezzi conditions. The following lemma provides a useful equivalent characterization of the discrete inf-sup condition.

Lemma 6.7 (Fortin Criterion) *Suppose that V, Q are Hilbert spaces and there exists $\beta > 0$ such that the bounded bilinear form $b : V \times Q \to \mathbb{R}$ satisfies*

$$\sup_{v \in V \setminus \{0\}} \frac{b(v, q)}{\|v\|_V} \geq \beta \|q\|_Q$$

for all $q \in Q$. Let V_h and Q_h be subspaces of V and Q, respectively, such that there exist $v \in V$ and $q_h \in Q_h$ with $b(v, q_h) \neq 0$. Then the discrete inf-sup condition

$$\sup_{v_h \in V_h \setminus \{0\}} \frac{b(v_h, q_h)}{\|v_h\|_V} \geq c_F' \beta \|q_h\|_Q$$

is satisfied with an h-independent constant $c_F' > 0$ if and only if there exists a uniformly bounded, linear, and nontrivial operator $I_F : V \to V_h$ called a Fortin interpolant, *such that for all $v \in V$, we have*

$$b(v - I_F v, q_h) = 0$$

for all $q_h \in Q_h$.

Proof

(i) Assume that a bounded linear operator $I_F : V \to V_h$ with the specified properties exists, and let $q_h \in Q_h$. The continuous inf-sup condition, the property $b(v - I_F v, q) = 0$, and the boundedness $\|I_F v\|_V \leq c_F \|v\|_V$ imply that

$$\beta \|q_h\|_Q \leq \sup_{v \in V \setminus \{0\}} \frac{b(v, q_h)}{\|v\|_V}$$

$$= \sup_{v \in V \setminus \{0\}} \frac{b(I_F v, q_h)}{\|v\|_V}$$

$$\leq \sup_{v \in V \setminus \{0\}} \frac{b(I_F v, q_h)}{c_F^{-1} \|I_F v\|_V}$$

$$= c_F \sup_{v_h \in \mathrm{Im}\, I_F \setminus \{0\}} \frac{b(v_h, q_h)}{\|v_h\|_V}$$

$$\leq c_F \sup_{v_h \in V_h \setminus \{0\}} \frac{b(v_h, q_h)}{\|v_h\|_V},$$

where c_F denotes the operator norm of I_F. This proves the discrete inf-sup condition.

(ii) Assume that the discrete inf-sup condition is satisfied, and let $v \in V$. With the inner product $(\cdot, \cdot)_V$ of V, we consider the discrete saddle point problem that consists in finding $(v_h, p_h) \in V_h \times Q_h$ with

$$(v_h, w_h)_V + b(w_h, p_h) = 0,$$

$$b(v_h, q_h) = b(v, q_h),$$

for all $(w_h, q_h) \in V_h \times Q_h$. Since the Babuška–Brezzi conditions are satisfied, there exists a unique solution $(v_h, p_h) \in V_h \times Q_h$ such that

$$\|v_h\|_V \leq \|(v_h, p_h)\|_{V \times Q} \leq c_L \|b(v, \cdot)\|_{Q'} \leq c_L k_b \|v\|_V.$$

Setting $I_F v = v_h$ defines a Fortin interpolant. □

6.3.4 Locking and Softening

Saddle-point formulations are often used to avoid the *locking effect* of a standard numerical method. To illustrate this, we consider Hilbert spaces V and Q and the problem of finding $u \in V$ such that

$$a(u, v) + t^{-2}(Bu, Bv)_Q = \ell(v)$$

for all $v \in V$. Here, $a : V \times V \to \mathbb{R}$ is a symmetric, continuous, and coercive bilinear form, $B : V \to Q$ a bounded linear operator, $\ell \in V'$ a functional, $(\cdot, \cdot)_Q$ the inner product on Q, and $0 < t \ll 1$ a small parameter. We follow [6].

Proposition 6.3 (Locking) *Assume that $(V_h)_{h>0}$ is a family of finite-dimensional subspaces of V, and for every $h > 0$, let $u_h \in V_h$ be the Galerkin approximation of u, i.e.,*

$$a(u_h, v_h) + t^{-2}(Bu_h, Bv_h)_Q = \ell(v_h)$$

for all $v_h \in V_h$. Suppose that there exists $\overline{u} \in V$ with $B\overline{u} = 0$, and $\ell(\overline{u}) > 0$. Assume that

$$V_h \cap \ker B = \{0\},$$

in particular, and assume that there exist $\sigma > 0$ and $c_b > 0$ such that $\|Bv_h\|_Q \geq c_b h^\sigma \|v_h\|_V$ for all $v_h \in V_h$. Then for all $h > 0$, we have

$$\|u - u_h\|_V \geq c_1 - c_2 t^2 h^{-2\sigma}$$

with constants $c_1, c_2 > 0$ that do not depend on t and h.

Proof The Lax–Milgram lemma implies the existence of uniquely defined solutions $u \in V$ and $u_h \in V_h$ for every $h > 0$. By replacing $\overline{u}$ by $s\overline{u}$ with a sufficiently small number $s > 0$, we may assume that

$$a(\overline{u}, \overline{u}) \leq \ell(\overline{u}).$$

Since $u \in V$ is the unique minimizer of the functional

$$I_t(v) = \frac{1}{2}a(v, v) + \frac{t^{-2}}{2}(Bv, Bv)_Q - \ell(v),$$

we have that

$$-\ell(u) \leq I_t(u) \leq I_t(\overline{u}) = \frac{1}{2}a(\overline{u}, \overline{u}) - \ell(\overline{u}) \leq -\frac{1}{2}\ell(\overline{u}).$$

This yields that

$$\|\ell\|_{V'} \geq \frac{\ell(u)}{\|u\|_V} \geq \frac{\ell(\overline{u})}{2\|u\|_V},$$

i.e., $\|u\|_V \geq c_1 = \ell(\overline{u})/(2\|\ell\|_{V'})$. On the other hand, we have for the discrete solution $u_h \in V_h$ that

$$t^{-2}c_b^2 h^{2\sigma}\|u_h\|_V^2 \leq a(u_h, u_h) + t^{-2}(Bu_h, Bu_h)_Q = \ell(u_h) \leq \|\ell\|_{V'}\|u_h\|_V,$$

i.e., $\|u_h\|_V \le c_2 t^2 h^{-2\sigma}$. The reverse triangle inequality

$$\|u - u_h\|_V \ge \|u\|_V - \|u_h\|_V$$

implies the estimate. □

Remark 6.13 The proposition states that unless h^σ is small compared to t, the approximation error is large, called a *locking effect* of the numerical method, which occurs when the kernel of B is not sufficiently resolved.

Example 6.7 If $\mathscr{T}_h$ is a triangulation of $\Omega = (0,1)^2$ consisting of halved squares with diagonals parallel to the vector $(1,1)$, then for every $v_h \in \mathscr{S}_0^1(\mathscr{T}_h)^2$ with $\operatorname{div} v_h = 0$, we have $v_h = 0$. Because of the approximation properties of linear finite elements we expect $\sigma = 1$.

A way to avoid the locking effect is to introduce the additional variable $p = t^{-2} B u$ and to consider the perturbed saddle-point formulation

$$a(u,v) \; + (p, Bv)_Q = \ell(v),$$
$$(q, Bu)_Q - t^2 (p,q)_Q = 0.$$

If the Babuška–Brezzi conditions are satisfied, then this formulation can be approximated robustly. This is related to a *softening effect* of the saddle-point formulation.

Proposition 6.4 (Softening) *Assume that the families of subspaces $(V_h)_{h>0}$ and $(Q_h)_{h>0}$ satisfy the Babuša–Brezzi conditions, and let $(u_h, p_h) \in V_h \times Q_h$ be for every $h > 0$ the solution of*

$$a(u_h, v_h) \; + (p_h, Bv_h)_Q = \ell(v_h),$$
$$(q_h, Bu_h)_Q - t^2(p_h, q_h)_Q = 0,$$

for all $(v_h, q_h) \in V_h \times Q_h$. Let $\Pi_h : Q \to Q_h$ denote the orthogonal projection onto Q_h, i.e., for every $q \in Q$, the element $\Pi_h q \in Q_h$ is defined by

$$\left(\Pi_h q, r_h\right)_Q = (q, r_h)_Q$$

for all $r_h \in Q_h$. Then the function $u_h \in V_h$ satisfies

$$a(u_h, v_h) + t^{-2}\left(\Pi_h Bu_h, \Pi_h Bv_h\right)_Q = \ell(v_h)$$

for all $v_h \in V_h$.

Proof The second identity in the saddle-point formulation implies that

$$p_h = t^{-2} \Pi_h B u_h.$$

Since for every $v_h \in V_h$ we have

$$(r_h, B v_h)_Q = (r_h, \Pi_h B v_h)_Q$$

for all $r_h \in Q_h$, the choice $r_h = t^{-2} \Pi_h B u_h$ implies the result. $\square$

Remark 6.14 The interpretation of the proposition is that the variational formulation involves the operator $B_h = \Pi_h \circ B$ instead of B, which increases the discrete kernel, and thereby softens the formulation and avoids a locking effect.

References

A version of the generalized Lax–Milgram lemma can be found in [12]. The formulation of the inf-sup condition and its relevance for the well-posedness of saddle-point problems is due to [3, 7]. Further important contributions to the understanding and approximation of saddle-point problems are the references [1, 2, 13]. A derivation of the inf-sup condition in a finite-dimensional setting is given in [8]. Chapters on saddle-point problems and their efficient numerical solution are contained in [4–6, 9–11].

1. Arnold, D.N.: Discretization by finite elements of a model parameter dependent problem. Numer. Math. **37**(3), 405–421 (1981). URL http://dx.doi.org/10.1007/BF01400318
2. Arnold, D.N., Brezzi, F.: Mixed and nonconforming finite element methods: implementation, postprocessing and error estimates. RAIRO Modél. Math. Anal. Numér. **19**(1), 7–32 (1985)
3. Babuška, I.: Error-bounds for finite element method. Numer. Math. **16**, 322–333 (1970/1971)
4. Benzi, M., Golub, G.H., Liesen, J.: Numerical solution of saddle point problems. Acta Numer. **14**, 1–137 (2005). URL http://dx.doi.org/10.1017/S0962492904000212
5. Boffi, D., Brezzi, F., Fortin, M.: Mixed finite element methods and applications. Springer Series in Computational Mathematics, vol. 44. Springer, Heidelberg (2013). URL http://dx.doi.org/10.1007/978-3-642-36519-5
6. Braess, D.: Finite Elements, 3rd edn. Cambridge University Press, Cambridge (2007). URL http://dx.doi.org/10.1017/CBO9780511618635
7. Brezzi, F.: On the existence, uniqueness and approximation of saddle-point problems arising from Lagrangian multipliers. Rev. Française Automat. Informat. Recherche Opérationnelle Sér. Rouge **8**(R-2), 129–151 (1974)
8. Brezzi, F., Bathe, K.J.: A discourse on the stability conditions for mixed finite element formulations. Comput. Methods Appl. Mech. Eng. **82**(1-3), 27–57 (1990). URL http://dx.doi.org/10.1016/0045-7825(90)90157-H
9. Ern, A., Guermond, J.L.: Theory and practice of finite elements. Applied Mathematical Sciences, vol. 159. Springer, New York (2004). URL http://dx.doi.org/10.1007/978-1-4757-4355-5

10. Fortin, M., Glowinski, R.: Augmented Lagrangian methods. Studies in Mathematics and Its Applications, vol. 15. North-Holland Publishing, Amsterdam (1983)
11. Girault, V., Raviart, P.A.: Finite element methods for Navier-Stokes equations. Springer Series in Computational Mathematics, vol. 5. Springer, Berlin (1986). URL http://dx.doi.org/10.1007/978-3-642-61623-5
12. Nečas, J.: Les méthodes directes en théorie des équations elliptiques. Masson et Cie, Éditeurs, Paris; Academia, Éditeurs, Prague (1967)
13. Xu, J., Zikatanov, L.: Some observations on Babuška and Brezzi theories. Numer. Math. **94**(1), 195–202 (2003). URL http://dx.doi.org/10.1007/s002110100308

Chapter 7
Mixed and Nonstandard Methods

7.1 Mixed Method for the Poisson Problem

7.1.1 Dual Mixed Formulation

For a bounded Lipschitz domain $\Omega \subset \mathbb{R}^d$ with Dirichlet and Neumann boundaries $\Gamma_D \subset \partial\Omega$ and $\Gamma_N = \partial\Omega \setminus \Gamma_D$, functions $g \in L^2(\Omega), p_D \in H^1(\Omega)$, and $\sigma \in L^2(\Gamma_N)$, we consider the Poisson problem

$$-\Delta p = g \text{ in } \Omega, \quad p|_{\Gamma_D} = p_D|_{\Gamma_D}, \quad \partial_n p|_{\Gamma_N} = \sigma.$$

We introduce the variable $u = \nabla p$, and by multiplying the equations $-\operatorname{div} u = g$ and $u = \nabla p$ by functions v and q, and integrating over Ω, obtain the equations

$$\int_\Omega u \cdot v \, dx + \int_\Omega p \operatorname{div} v \, dx = \int_{\Gamma_D} p_D v \cdot n \, ds,$$

$$\int_\Omega q \operatorname{div} u \, dx = -\int_\Omega gq \, dx,$$

provided that $v \cdot n = 0$ on Γ_N. A natural choice of a function space for the function p is $L^2(\Omega)$. Then it is necessary to guarantee that $\operatorname{div} u \in L^2(\Omega)$.

Definition 7.1 The space $H(\operatorname{div}; \Omega)$ consists of all vector fields $u \in L^2(\Omega; \mathbb{R}^d)$ such that there exists $f \in L^2(\Omega)$ with

$$\int_\Omega u \cdot \nabla\phi \, dx = -\int_\Omega f\phi \, dx.$$

© Springer International Publishing Switzerland 2016
S. Bartels, *Numerical Approximation of Partial Differential Equations*,
Texts in Applied Mathematics 64, DOI 10.1007/978-3-319-32354-1_7

for all $\phi \in C_0^\infty(\Omega)$. In this case we denote div $u = f$. The space is equipped with the inner product

$$(u, v)_{H(\mathrm{div};\Omega)} = \int_\Omega u \cdot v \, dx + \int_\Omega \mathrm{div}\, u \, \mathrm{div}\, v \, dx,$$

which induces the norm $\|u\|_{H(\mathrm{div};\Omega)} = (\|u\|_{L^2(\Omega)}^2 + \|\,\mathrm{div}\, u\|_{L^2(\Omega)}^2)^{1/2}$.

The vector fields in $H(\mathrm{div};\Omega)$ have a weak divergence; the space has the following properties.

Remarks 7.1

(i) The space $H(\mathrm{div};\Omega)$ is a Hilbert space.
(ii) The space $C^\infty(\overline{\Omega};\mathbb{R}^d)$ is dense in $H(\mathrm{div};\Omega)$.
(iii) If $u \in L^2(\Omega;\mathbb{R}^d)$ satisfies $u|_{\Omega_i} \in C^1(\overline{\Omega}_i;\mathbb{R}^d)$, $i = 1, 2, \ldots, I$, for a partition $(\Omega_i)_{i=1,\ldots,I}$ of Ω, then we have $u \in H(\mathrm{div};\Omega)$ if and only if $u|_{\Omega_i} \cdot n_i = -u|_{\Omega_j} \cdot n_j$ on every interface $\Gamma_{ij} = \partial\Omega_i \cap \partial\Omega_j$ with the outer unit normals n_i and n_j to Ω_i and Ω_j, respectively, cf. Fig. 7.1.
(iv) For $d \geq 2$ we have that $H^1(\Omega;\mathbb{R}^d)$ is a proper subspace of $H(\mathrm{div};\Omega)$.

In reformulating the Poisson equation, the traces $u \cdot n$ and $v \cdot n$ occurred on parts of $\partial\Omega$. The meaning of these objects is defined via Green's formula.

Proposition 7.1 (Normal Trace) *Let $u \in H(\mathrm{div};\Omega)$. The mapping*

$$q \mapsto \langle u \cdot n, q \rangle_{\partial\Omega} = \int_\Omega u \cdot \nabla q \, dx + \int_\Omega q \, \mathrm{div}\, u \, dx$$

defines a bounded linear operator on $H^1(\Omega)$ denoted $u \cdot n \in H^1(\Omega)'$. Moreover, the mapping $H(\mathrm{div};\Omega) \to H^1(\Omega)'$, $u \mapsto u \cdot n$, is bounded and linear. For $u \in C^\infty(\overline{\Omega};\mathbb{R}^d)$, we have that $u \cdot n$ coincides with $u|_{\partial\Omega} \cdot n$ in the sense that for all $q \in H^1(\Omega)$, we have

$$\langle u \cdot n, q \rangle_{\partial\Omega} = \int_{\partial\Omega} u \cdot n q \, ds.$$

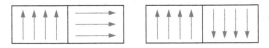

Fig. 7.1 Piecewise smooth, discontinuous vector fields $u : \Omega \to \mathbb{R}^d$ on $\overline{\Omega} = \overline{\Omega}_1 \cup \overline{\Omega}_2$ with $u \notin H(\mathrm{div};\Omega)$ *(left)* and $u \in H(\mathrm{div};\Omega)$ *(right)*

Proof Exercise.   □

The result of the proposition allows us to define a subspace of $H(\text{div}; \Omega)$ with vanishing normal components on a part Γ_N of $\partial\Omega$. For $\sigma \in L^2(\Gamma_N)$ we write $u \cdot n|_{\Gamma_N} = \sigma$ if

$$\langle u \cdot n, q \rangle_{\partial\Omega} = \int_{\Gamma_N} \sigma q \, ds$$

for all $q \in H_D^1(\Omega)$.

Definition 7.2 For a subset $\Gamma_N = \partial\Omega \setminus \Gamma_D$, we let $H_N(\text{div}; \Omega)$ be the space of all $u \in H(\text{div}; \Omega)$ such that $u \cdot n = 0$ on Γ_N.

Remark 7.2 We have that $C_{\Gamma_N}^\infty(\overline{\Omega}; \mathbb{R}^d) = \{v \in C^\infty(\overline{\Omega}; \mathbb{R}^d) : v|_{\Gamma_N} = 0\}$ is dense in $H_N(\text{div}; \Omega)$.

We have constructed the framework to define a saddle-point formulation of the Poisson problem. Using a decomposition of u and modifying g, we may assume that $\sigma = 0$.

Definition 7.3 The *dual mixed formulation* of the Poisson problem consists in finding a pair $(u, p) \in H_N(\text{div}; \Omega) \times L^2(\Omega)$ such that

$$\int_\Omega u \cdot v \, dx + \int_\Omega p \, \text{div} \, v \, dx \; = \langle v \cdot n, p_D \rangle_{\partial\Omega},$$

$$\int_\Omega q \, \text{div} \, u \, dx \qquad\qquad = -\int_\Omega gq \, dx,$$

for all $(v, q) \in H_N(\text{div}; \Omega) \times L^2(\Omega)$.

We apply the abstract saddle-point theory to establish the existence of a unique solution. Alternatively, this follows from the existence of a unique solution $p \in H^1(\Omega)$ for the Poisson problem.

Theorem 7.1 (Well-Posedness) *There exists a unique solution* $(u, p) \in H_N(\text{div}; \Omega) \times L^2(\Omega)$ *for the dual mixed formulation of the Poisson problem.*

Proof We define $V = H_N(\text{div}; \Omega)$ and $Q = L^2(\Omega)$. The right-hand sides in the dual mixed formulation define bounded linear functionals $\ell_1 \in V'$ and $\ell_2 \in Q'$, respectively. Moreover, the formulation defines the bounded bilinear forms

$$a(u, v) = \int_\Omega u \cdot v \, dx, \quad b(v, q) = \int_\Omega q \, \text{div} \, v \, dx$$

on $V \times V$ and $V \times Q$, respectively. To verify the conditions of Brezzi's splitting theorem, we determine the kernel of b with respect to the first argument. We have

$$\ker B = \{v \in V : b(v, q) = 0 \text{ for all } q \in Q\}$$

$$= \left\{v \in H_N(\text{div}; \Omega) : \int_\Omega q \, \text{div} \, v \, dx = 0 \text{ for all } q \in L^2(\Omega)\right\}$$

$$= \{v \in H_N(\text{div}; \Omega) : \text{div} \, v = 0\}.$$

To show that a is elliptic on $\ker B$, we note that $\text{div} \, v = 0$ for $v \in \ker B$ implies that

$$a(v, v) = \int_\Omega v \cdot v \, dx = \|v\|_{L^2(\Omega)}^2 + \|\text{div} \, v\|_{L^2(\Omega)}^2 = \|v\|_{H(\text{div};\Omega)}^2 = \|v\|_V^2.$$

It remains to show that there exists $\beta > 0$, such that for all $q \in L^2(\Omega)$

$$\sup_{v \in V \setminus \{0\}} \frac{b(v, q)}{\|v\|_V} \geq \beta \|q\|_Q.$$

Given $q \in Q = L^2(\Omega)$, let $\phi \in H_D^1(\Omega)$ be the uniquely defined weak solution of

$$-\Delta \phi = q, \quad \phi|_{\Gamma_D} = 0, \quad \partial_n \phi|_{\Gamma_N} = 0,$$

and set $\overline{v} = -\nabla \phi$. Then $\overline{v} \in V = H_N(\text{div}; \Omega)$ with $\text{div} \, \overline{v} = q$ and $\overline{v} \cdot n = \nabla \phi \cdot n = 0$ on Γ_N. With the Poincaré inequality, we find that

$$\|\overline{v}\|_{L^2(\Omega)} = \|\nabla \phi\|_{L^2(\Omega)} \leq c_P \|q\|_{L^2(\Omega)}.$$

Taking squares and adding $\|q\|_{L^2(\Omega)}^2$ on both sides, we obtain the estimate

$$\|\overline{v}\|_{H(\text{div};\Omega)} \leq (1 + c_P^2)^{1/2} \|q\|_{L^2(\Omega)}.$$

Since $\overline{v} \in V$, it follows that

$$\sup_{v \in V \setminus \{0\}} \frac{b(v, q)}{\|v\|_V} \geq \frac{b(\overline{v}, q)}{\|\overline{v}\|_V} = \frac{\int_\Omega q \, \text{div} \, \overline{v} \, dx}{\|\overline{v}\|_{H(\text{div};\Omega)}}$$

$$= \frac{\int_\Omega q^2 \, dx}{\|\overline{v}\|_{H(\text{div};\Omega)}} \geq \frac{\|q\|_{L^2(\Omega)}^2}{(1 + c_P^2) \|q\|_{L^2(\Omega)}} = \frac{1}{(1 + c_P^2)} \|q\|_Q.$$

Hence all conditions of Brezzi's splitting theorem are satisfied, and this implies the assertion. □

Remark 7.3 We have seen in Example 6.6 that the natural choice of finite element spaces $V_h = \mathscr{S}_0^1(\mathscr{T}_h)^d$ and $Q_h = \mathscr{L}^0(\mathscr{T}_h)$ cannot be expected to lead to a stable numerical method. The space V_h has to be enlarged appropriately.

7.1.2 Raviart–Thomas Method

To construct a space V_h that is compatible with the space $Q_h = \mathscr{L}^0(\mathscr{T}_h)$, we omit the continuity property of the space $\mathscr{S}^1(\mathscr{T}_h)^d$.

Definition 7.4 The *Raviart–Thomas finite element space* $\mathscr{R}T^0(\mathscr{T}_h)$ consists of all vector fields $u_h \in H(\mathrm{div}; \Omega)$, such that for every $T \in \mathscr{T}_h$, there exist $a_T \in \mathbb{R}^d$ and $d_T \in \mathbb{R}$ with

$$u_h(x) = a_T + d_T x$$

for all $x \in T$. We let $\mathscr{R}T_N^0(\mathscr{T}_h)$ be the subspace of elements $u_h \in \mathscr{R}T^0(\mathscr{T}_h)$ with $u_h \cdot n = 0$ on Γ_N.

Note that the constants a_T and d_T cannot be chosen arbitrarily, since continuity of the normal component of u_h across the sides of elements has to be guaranteed. We recall that the set $\mathscr{S}_h$ contains the sides of elements in a triangulation $\mathscr{T}_h$.

Lemma 7.1 (Normal Components) *Let $T \in \mathscr{T}_h$ and $u_h(x) = a_T + d_T x$ for $x \in T$. The coefficients $a_T \in \mathbb{R}^d$ and $d_T \in \mathbb{R}$ are uniquely and linearly defined by the $d+1$ numbers*

$$\alpha_j = |S_j|^{-1} \int_{S_j} u_h \cdot n_{S_j}\, ds,$$

for $j = 1, 2, \ldots, d+1$, and the sides $S_1, S_2, \ldots, S_{d+1} \in \mathscr{S}_h$ with $S_j \subset \partial T$.

Proof We first note that the function $x \mapsto u_h(x) \cdot n_S$ is constant on S for every side S of T. To verify this, let $x, y \in S$, so that $x - y$ is a tangent vector to S, i.e., $(x - y) \cdot n_S = 0$, and hence

$$\big(u_h(x) - u_h(y)\big) \cdot n_S = d_T(x - y) \cdot n_S = 0.$$

Due to this observation we have

$$|S|^{-1} \int_S u_h \cdot n_S\, ds = u_h(x) \cdot n_S$$

for an arbitrary point $x \in S$. We let $z \in \mathbb{R}^d$ be the vertex of T at which the sides $S_1, S_2, \ldots, S_d$ intersect, and let $z' \in S_{d+1}$ be an arbitrary point on S_{d+1}. We want to show that for numbers $\alpha_1, \alpha_2, \ldots, \alpha_{d+1}$ there exist uniquely defined coefficients

$a_T \in \mathbb{R}^d$ and $d_T \in \mathbb{R}$ such that

$$a_T \cdot n_{S_j} + d_T z \cdot n_{S_j} = u_h(z) \cdot n_{S_j} = \alpha_j$$

for $j = 1, 2, \ldots, d$, and

$$a_T \cdot n_{S_{d+1}} + d_T z' \cdot n_{S_{d+1}} = u_h(z') \cdot n_{S_{d+1}} = \alpha_{d+1}.$$

These equations are equivalent to the linear system

$$
\begin{bmatrix}
n_{S_1}^\top & z \cdot n_{S_1} \\
\vdots & \vdots \\
n_{S_d}^\top & z \cdot n_{S_d} \\
n_{S_{d+1}}^\top & z' \cdot n_{S_{d+1}}
\end{bmatrix}
\begin{bmatrix}
a_T \\
d_T
\end{bmatrix}
=
\begin{bmatrix}
\alpha_1 \\
\vdots \\
\alpha_d \\
\alpha_{d+1}
\end{bmatrix}.
$$

We denote the matrix on the left-hand side by $A \in \mathbb{R}^{(d+1)\times(d+1)}$. To show that A is regular, let $y \in \mathbb{R}^{d+1}$ with $A^\top y = 0$. This means that

$$y_1 n_{S_1} + \cdots + y_d n_{S_d} = -y_{d+1} n_{S_{d+1}}$$

and

$$y_1 z \cdot n_{S_1} + \cdots + y_d z \cdot n_{S_d} = -y_{d+1} z' \cdot n_{S_{d+1}}.$$

If $y_{d+1} = 0$, then the linear independence of $n_{S_1}, n_{S_2}, \ldots, n_{S_d}$ implies that $y_j = 0$, $j = 1, 2, \ldots, d$. Otherwise, combining the equations we find that

$$z \cdot n_{S_{d+1}} = z' \cdot n_{S_{d+1}},$$

which means that z and z' belong in the same hyperplane with normal $n_{S_{d+1}}$. This cannot be the case, since z is the vertex of T opposite S_{d+1}. Hence, $y = 0$. An alternative proof assumes without loss of generality $z_{d+1} = 0$, so that $u_h(0) \cdot n_{S_j} = 0$ for $j = 1, 2, \ldots, d$, implies that $a_T = 0$, if n_{S_j} is the normal of the side S_j which is opposite z_j for $j = 1, 2, \ldots, d + 1$. But then $u_h(x) \cdot n_{S_{d+1}} = 0$ for every $x \in S_{d+1}$ implies that $z_{d+1} = 0 \in S_{d+1}$ which contradicts the nondegeneracy of T. $\square$

The degrees of freedom of the finite element space $\mathscr{R}T^0(\mathscr{T}_h)$ are thus the normal components on the sides of elements.

Fig. 7.2 Elements $T_-, T_+ \in \mathcal{T}_h$ associated with an interior side $S \in \mathcal{S}_h$ (*left*); basis function ψ_S (*right*)

Definition 7.5

(i) If $S \in \mathcal{S}_h \cap \partial\Omega$ is a boundary side, let $T_- \in \mathcal{T}_h$ be such that $S \subseteq T_- \cap \partial\Omega$, set $T_+ = \emptyset$, and let n_S be the unit normal to S which coincides with the outer unit normal n on $\partial\Omega$.

(ii) If $S \in \mathcal{S}_h$ is an inner side, let $T_+, T_- \in \mathcal{T}_h$ be such that $S = T_+ \cap T_-$, and let n_S be the unit normal to S which points from T_- into T_+.

For $S \in \mathcal{S}_h$ and $T \in \mathcal{T}_h$, we write

$$\delta_{S,T} = \frac{|S|}{(d!)|T|}.$$

Moreover, we let $z_\pm \in T_\pm \cap \mathcal{N}_h$ be the vertex of $T_\pm$ opposite S, cf. Fig. 7.2.

Proposition 7.2 (Raviart–Thomas Basis) *For $S \in \mathcal{S}_h$ and $x \in T_+ \cup T_-$, define*

$$\psi_S(x) = \begin{cases} +\delta_{S,T_+}(z_+ - x) & \text{for } x \in T_+, \\ -\delta_{S,T_-}(z_- - x) & \text{for } x \in T_-, \\ 0 & \text{for } x \in \Omega \setminus (T_+ \cup T_-). \end{cases}$$

The family $(\psi_S : S \in \mathcal{S}_h)$ defines a basis for $\mathcal{R}T^0(\mathcal{T}_h)$ with the property $\psi_S(x) \cdot n_R = \delta_{SR}$ for all $S, R \in \mathcal{S}_h$ and $x \in R$.

Proof

(i) Let $S, R \in \mathcal{S}_h$ and $x \in R$, and let $T_\pm \in \mathcal{T}_h$ be the neighboring elements to S. If $x \notin T_+ \cup T_-$, then $\psi_S(x) \cdot n_R = 0$. If $x \in T_+ \cup T_-$ but $x \notin S$, then the vector $z_\pm - x$ is a tangent vector to R and we have $\psi_S(x) \cdot n_R = 0$. If $x \in S$, then $\pm(z_\pm - x) \cdot n_S$ is the height of the simplex $T_\pm$ with respect to the side S, and we have

$$d!\,|T| = \pm(z_\pm - x) \cdot n_S |S|$$

which proves $\psi_S(x) \cdot n_S = 1$. The arguments also show that the normal components of ψ_S are continuous across the interelement sides, and hence that $\psi_S \in H(\text{div}; \Omega)$. Moreover, from its definition, we see that $\psi_S \in \mathcal{R}T^0(\mathcal{T}_h)$.

Fig. 7.3 Schematical
description of the
Raviart–Thomas finite
element

(ii) Let $u_h \in \mathscr{R}T^0(\mathscr{T}_h)$, define

$$\alpha_S = |S|^{-1} \int_S u_h \cdot n_S \, \mathrm{d}s$$

for all $S \in \mathscr{S}_h$, and set $\tilde{u}_h = \sum_{S \in \mathscr{S}_h} \alpha_S \psi_S$. Then, with $\psi_S(x) \cdot n_R = \delta_{SR}$ for $x \in R$, we find that

$$\tilde{u}_h(x) \cdot n_R = \alpha_R = u_h(x) \cdot n_R$$

for $x \in R$, and with Lemma 7.1 it follows that $\tilde{u}_h = u_h$.   □

Remarks 7.4

(i) Note that for a triangulation of $\Omega = (0,1)^2$ into halved squares of side length $1/n$, we have $|\mathscr{S}_h| = 2n(n+1) + n^2$ and $|\mathscr{T}_h| = 2n^2$, which implies that $\dim \mathscr{R}T^0(\mathscr{T}_h) > \dim \mathscr{L}^0(\mathscr{T}_h)$. In particular, the dimension of $\mathscr{R}T^0(\mathscr{T}_h)$ is larger than $\dim \mathscr{S}^1(\mathscr{T}_h)^2 = 2(n+1)^2$.

(ii) The Raviart–Thomas element is a finite element defined by the triple $(T, \mathscr{P}, \mathscr{K})$ with $\mathscr{P} = \mathscr{P}_0(T)^d + \mathscr{P}_0(T)x$, and $\mathscr{K} = \{\chi_j : j = 1, 2, \ldots, d+1\}$, where χ_j associates with $\phi \in C^\infty(T; \mathbb{R}^d)$ the integral mean of the normal component on $S_j, j = 1, 2, \ldots, d+1$, cf. Fig. 7.3. To give the Raviart–Thomas finite element space the meaning of an affine family, the *Piola transformation* has to be used.

7.1.3   Approximation

To prove the inf-sup condition for the pair of spaces $Q_h = \mathscr{L}^0(\mathscr{T}_h)$ and $V_h = \mathscr{R}T^0_{\mathrm{N}}(\mathscr{T}_h)$, we try to construct a Fortin interpolant.

Proposition 7.3 (Almost-Fortin Interpolant) *For* $u \in H^1(\Omega; \mathbb{R}^d) \cap H_{\mathrm{N}}(\mathrm{div}; \Omega)$, *let* $\mathscr{I}_{\mathscr{R}T} u \in \mathscr{R}T^0_{\mathrm{N}}(\mathscr{T}_h)$ *be defined by requiring*

$$\int_S \mathscr{I}_{\mathscr{R}T} u \cdot n_S \, \mathrm{d}s = \int_S u \cdot n_S \, \mathrm{d}s$$

for all $S \in \mathcal{S}$. Then we have

$$b(u - \mathscr{I}_{\mathscr{R}T}u, q_h) = \int_{\Omega} q_h \operatorname{div}(u - \mathscr{I}_{\mathscr{R}T}u) \, dx = 0$$

for all $q_h \in \mathcal{L}^0(\mathscr{T}_h)$ and

$$\|\mathscr{I}_{\mathscr{R}T}u\|_{H(\operatorname{div};\Omega)} \leq c_F \big(\|u\|_{H(\operatorname{div};\Omega)} + h\|\nabla u\|_{L^2(\Omega)} \big).$$

Proof By Lemma 7.1 the function $\mathscr{I}_{\mathscr{R}T}u \in \mathscr{R}T_N^0(\mathscr{T}_h)$ is well defined. For every $T \in \mathscr{T}_h$, we have by construction of $\mathscr{I}_{\mathscr{R}T}u$ that

$$\int_T \operatorname{div}(u - \mathscr{I}_{\mathscr{R}T}u) \, dx = \int_{\partial T} (u - \mathscr{I}_{\mathscr{R}T}u) \cdot n \, ds = 0.$$

This implies that for every $q_h \in \mathcal{L}^0(\mathscr{T}_h)$, we have

$$\int_{\Omega} q_h \operatorname{div}(u - \mathscr{I}_{\mathscr{R}T}u) \, dx = \sum_{T \in \mathscr{T}_h} q_h|_T \int_T \operatorname{div}(u - \mathscr{I}_{\mathscr{R}T}u) \, dx = 0.$$

Moreover, we have that

$$(\operatorname{div} \mathscr{I}_{\mathscr{R}T}u)|_T = |T|^{-1} \int_T \operatorname{div} u \, dx$$

which implies $\| \operatorname{div} \mathscr{I}_{\mathscr{R}T}u \|_{L^2(T)} \leq \| \operatorname{div} u \|_{L^2(T)}$ for every $T \in \mathscr{T}_h$. By a summation over $T \in \mathscr{T}_h$, we deduce that

$$\| \operatorname{div} \mathscr{I}_{\mathscr{R}T}u \|_{L^2(\Omega)} \leq \| \operatorname{div} u \|_{L^2(\Omega)}.$$

To bound the L^2-norm of $\mathscr{I}_{\mathscr{R}T}u$, we note that with $\alpha_S = |S|^{-1} \int_S u \cdot n_S \, ds$ and the basis $(\psi_S : S \in \mathscr{S}_h)$ of $\mathscr{R}T_N^0(\mathscr{T}_h)$, we have

$$\mathscr{I}_{\mathscr{R}T}u = \sum_{S \in \mathscr{S}_h} \alpha_S \psi_S.$$

For every $T \in \mathscr{T}_h$ it follows that

$$\|\mathscr{I}_{\mathscr{R}T}u\|_{L^2(T)} \leq \sum_{S \in \mathscr{S}_h \cap \partial T} |\alpha_S| \|\psi_S\|_{L^2(T)} \leq \sum_{S \in \mathscr{S}_h \cap \partial T} |\alpha_S| \|\psi_S\|_{L^\infty(\Omega)} |T|^{1/2}.$$

We use the trace inequality

$$\|u\|_{L^2(S)} \leq c \big(h_T^{-1/2} \|u\|_{L^2(T)} + h_T^{1/2} \|\nabla u\|_{L^2(T)} \big)$$

for $T \in \mathscr{T}_h$ and $S \subset \partial T$ to deduce that

$$|S||\alpha_S| \leq \int_S |u \cdot n_S| \, ds \leq \|u\|_{L^2(S)} |S|^{1/2}$$
$$\leq c|S|^{1/2} \big(h_T^{-1/2} \|u\|_{L^2(T)} + h_T^{1/2} \|\nabla u\|_{L^2(T)} \big).$$

With $|S| \sim h_T^{d-1}$, $|T| \sim h_T^d$, and $\|\psi_S\|_{L^\infty(\Omega)} \leq c$, this implies that

$$\|\mathscr{I}_{\mathscr{R}T} u\|_{L^2(T)} \leq c \big(\|u\|_{L^2(T)} + h_T \|\nabla u\|_{L^2(T)} \big).$$

A summation over $T \in \mathscr{T}_h$ leads to the estimate. $\square$

Remark 7.5 The estimate of the proposition does not allow for an extension of $\mathscr{I}_{\mathscr{R}T}$ to an operator on $H(\mathrm{div}; \Omega)$. Although for every $u \in H(\mathrm{div}; \Omega)$, $T \in \mathscr{T}_h$, and $q \in H^1(T)$ we have that the expression

$$\int_{\partial T} u \cdot nq \, ds = \int_T q \, \mathrm{div} \, u \, dx + \int_T u \cdot \nabla q \, dx$$

is well-defined, we cannot restrict the integration on the left-hand side to a subset of ∂T, since characteristic functions of subsets of ∂T in general do not have an extension to a function in $H^1(T)$.

Although the almost-Fortin interpolant of Proposition 7.3 is not defined on $H(\mathrm{div}; \Omega)$, it allows us to prove the discrete inf-sup condition via an approximation argument.

Proposition 7.4 (Discrete Inf-Sup Condition) *There exist h-independent constants $h_0, \beta' > 0$, such that for $0 < h < h_0$ and every $q_h \in \mathscr{L}^0(\mathscr{T}_h)$ we have*

$$\sup_{v_h \in \mathscr{R}T_N^0(\mathscr{T}_h) \setminus \{0\}} \frac{b(v_h, q_h)}{\|v_h\|_{H(\mathrm{div}; \Omega)}} \geq \beta' \|q_h\|_{L^2(\Omega)}.$$

Proof Due to the continuous inf-sup condition, there exists $u \in H_N(\mathrm{div}; \Omega)$ with $\|u\|_{H(\mathrm{div}; \Omega)} = 1$, and

$$\frac{\beta}{2} \|q_h\|_{L^2(\Omega)} \leq b(u, q_h).$$

For every $\varepsilon > 0$ we obtain by regularization a function $u_\varepsilon \in C^\infty(\overline{\Omega}; \mathbb{R}^d)$ with $u_\varepsilon \cdot n = 0$ on Γ_N and

$$\|u - u_\varepsilon\|_{H(\mathrm{div}; \Omega)} \leq c\varepsilon, \quad \|\nabla u_\varepsilon\|_{L^2(\Omega)} \leq c\varepsilon^{-1},$$

where $c_\varepsilon \to 0$ as $\varepsilon \to 0$. We thus have

$$b(u, q_h) = b(u_\varepsilon, q_h) + b(u - u_\varepsilon, q_h) \le b(u_\varepsilon, q_h) + k_b c_\varepsilon \|q_h\|_{L^2(\Omega)}$$

and

$$\|u_\varepsilon\|_{H(\mathrm{div};\Omega)} \le 1 + c_\varepsilon.$$

Incorporating the almost-Fortin interpolant applied to u_ε, this leads to

$$b(u, q_h) \le b(u_\varepsilon, q_h) + k_b c_\varepsilon \|q_h\|_{L^2(\Omega)} = b(\mathscr{I}_{\mathscr{R}T} u_\varepsilon, q_h) + k_b c_\varepsilon \|q_h\|_{L^2(\Omega)}.$$

Moreover, we have that

$$\|\mathscr{I}_{\mathscr{R}T} u_\varepsilon\|_{H(\mathrm{div};\Omega)} \le c_F \big(\|u_\varepsilon\|_{H(\mathrm{div};\Omega)} + h\|\nabla u_\varepsilon\|_{L^2(\Omega)}\big) \le c(1 + h\varepsilon^{-1}).$$

Combining the estimates, we see that

$$\frac{b(\mathscr{I}_{\mathscr{R}T} u_\varepsilon, q_h)}{\|\mathscr{I}_{\mathscr{R}T} u_\varepsilon\|_{H(\mathrm{div};\Omega)}} \ge c(1 + h\varepsilon^{-1})^{-1} \big(b(u, q_h) - k_b c_\varepsilon \|q_h\|_{L^2(\Omega)}\big)$$

$$\ge c(1 + h\varepsilon^{-1})^{-1} (\beta/2 - k_b c_\varepsilon) \|q_h\|_{L^2(\Omega)}.$$

Choosing $\varepsilon = h$ proves the estimate for h sufficiently small. $\qquad\square$

Remark 7.6 An alternative proof of the discrete inf-sup condition uses $v_h = \mathscr{I}_{\mathscr{R}T}(\nabla \phi)$ for $\phi \in H_{\mathrm{D}}^1(\Omega)$, such that $-\Delta \phi = q_h$ in Ω and $\partial_n \phi = 0$ on Γ_{N}, assuming that this Poisson problem is H^2-regular.

To establish the well-posedness of the discretization of the dual mixed formulation of the Poisson problem, it remains to prove coercivity of the bilinear form a on the discrete kernel. Since

$$K_h = \Big\{ v_h \in \mathscr{R}T_{\mathrm{N}}^0(\mathscr{T}_h) : \int_\Omega q_h \operatorname{div} v_h \, \mathrm{d}x \text{ for all } q_h \in \mathscr{L}^0(\mathscr{T}_h)\Big\}$$

$$= \big\{ v_h \in \mathscr{R}T_{\mathrm{N}}^0(\mathscr{T}_h) : \operatorname{div} v_h = 0\big\},$$

we have uniform coercivity, i.e.,

$$a(v_h, v_h) = \int_\Omega v_h \cdot v_h \, \mathrm{d}x = \|v_h\|_{H(\mathrm{div};\Omega)}^2$$

for all $v_h \in K_h$. The Babuška–Brezzi conditions are thus satisfied and we obtain a quasi-best-approximation result. To deduce a convergence rate, we need to construct good approximations.

Lemma 7.2 (Approximation) *There exists a constant $c_F' > 0$, such that for the operator $\mathscr{I}_{\mathscr{R}T}$ from Proposition 7.3 and every $u \in H^1(\Omega; \mathbb{R}^d) \cap H_N(\mathrm{div}; \Omega)$, we have*

$$\|u - \mathscr{I}_{\mathscr{R}T}u\|_{L^2(\Omega)} \le c_F' h \|\nabla u\|_{L^2(\Omega)}.$$

The constant c_F' depends on the shapes of the elements in $\mathscr{T}_h$.

Proof Let $T \in \mathscr{T}_h$. The mapping

$$F : H^1(T) \to \mathbb{R}, \quad u \mapsto \|u - \mathscr{I}_{\mathscr{R}T}u\|_{L^2(T)}$$

is sublinear, bounded on $H^1(T)$, and vanishes on $\mathscr{P}_0(T)^d$. The Bramble–Hilbert lemma thus implies that

$$F(u) = \|u - \mathscr{I}_{\mathscr{R}T}u\|_{L^2(T)} \le c \|\nabla u\|_{L^2(T)}$$

for all $u \in H^1(T; \mathbb{R}^d)$. The transformation formulas show that

$$\|u - \mathscr{I}_{\mathscr{R}T}u\|_{L^2(T)} \le c h_T \|\nabla u\|_{L^2(T)}.$$

A summation over all $T \in \mathscr{T}_h$ proves the estimate.   □

We now derive an error estimate, for which we use the local Poincaré inequality

$$\|\phi - \overline{\phi}\|_{L^2(T)} \le c_P h_T \|\nabla \phi\|_{L^2(T)},$$

which holds for every $T \in \mathscr{T}_h$ and all $\phi \in H^1(T)$ with $\overline{\phi} = |T|^{-1} \int_T \phi \, \mathrm{d}x$.

Theorem 7.2 (Convergence) *Let $(u, p) \in H_N(\mathrm{div}; \Omega) \times L^2(\Omega)$ be the unique solution of the dual mixed formulation of the Poisson problem. There exists a unique pair $(u_h, p_h) \in \mathscr{R}T_N^0(\mathscr{T}_h) \times \mathscr{L}^0(\mathscr{T}_h)$, such that*

$$\int_\Omega u_h \cdot v_h \, \mathrm{d}x + \int_\Omega p_h \, \mathrm{div}\, v_h \, \mathrm{d}x \;\; = \langle v_h \cdot n, p_D \rangle_{\partial\Omega},$$

$$\int_\Omega q_h \, \mathrm{div}\, u_h \, \mathrm{d}x \qquad\qquad\quad = -\int_\Omega g q_h \, \mathrm{d}x$$

for all $(v_h, q_h) \in \mathscr{R}T_N^0(\mathscr{T}_h) \times \mathscr{L}^0(\mathscr{T}_h)$. If $g \in H^1(\Omega)$, $p \in H^1(\Omega)$, and $u \in H^1(\Omega; \mathbb{R}^d)$, then we have

$$\|u - u_h\|_{H(\mathrm{div};\Omega)} + \|p - p_h\|_{L^2(\Omega)} \le c h \big(\|u\|_{H^1(\Omega)} + \|p\|_{H^1(\Omega)} + \|g\|_{H^1(\Omega)} \big),$$

with a constant $c > 0$ that depends on the shapes of the elements in $\mathscr{T}_h$.

Proof Because of the uniform discrete coercivity and the uniform discrete inf-sup condition, the Babuška–Brezzi theorem implies unique solvability of the discrete problem, and the estimate

$$\|u - u_h\|_{H(\text{div};\Omega)} + \|p - p_h\|_{L^2(\Omega)} \leq c\big(\|u - v_h\|_{H(\text{div};\Omega)} + \|p - q_h\|_{L^2(\Omega)}\big)$$

for every $v_h \in \mathscr{R}T_N^0(\mathscr{T}_h)$ and $q_h \in \mathscr{L}^0(\mathscr{T}_h)$. We let $v_h \in \mathscr{R}T_N^0(\mathscr{T}_h)$ be the Fortin interpolant of u. Noting

$$(\text{div }\mathscr{I}_{\mathscr{R}T}u)|_T = |T|^{-1}\int_T \text{div } u \, dx = -|T|^{-1}\int_T g \, dx = -g_T,$$

and with a local Poincaré inequality we find that

$$\|\text{div}(u - \mathscr{I}_{\mathscr{R}T}u)\|_{L^2(\Omega)} = \|g - g_h\|_{L^2(\Omega)} \leq ch\|\nabla g\|_{L^2(\Omega)},$$

where $g_h \in \mathscr{L}^0(\mathscr{T}_h)$ is defined by $g_h|_T = g_T$ for every $T \in \mathscr{T}_h$. With Lemma 7.2 we find that

$$\|u - \mathscr{I}_{\mathscr{R}T}u\|_{L^2(\Omega)} \leq ch\|\nabla u\|_{L^2(\Omega)}.$$

Finally, we choose $q_h \in \mathscr{L}^0(\mathscr{T}_h)$ as the elementwise average of p, and note that

$$\|p - q_h\|_{L^2(\Omega)} \leq ch\|\nabla p\|_{L^2(\Omega)}.$$

This proves the estimate.   □

Remarks 7.7

(i) The regularity assumptions are satisfied, if the Poisson problem is H^2-regular, e.g., if Ω is convex, $\Gamma_D = \partial\Omega$, and $p_D = 0$.

(ii) Compared to the approximation of the Poisson problem in the space $\mathscr{S}^1(\mathscr{T}_h)$, we obtain a linearly convergent approximation of ∇p in the stronger norm of $H(\text{div};\Omega)$ instead of $L^2(\Omega;\mathbb{R}^d)$.

(iii) Higher-order versions of the Raviart–Thomas finite element space lead to higher-order convergence rates, provided that the exact solution is sufficiently regular.

(iv) The dependence of the constant on the shapes of the elements can be omitted if the Piola transformation is used in the proof of Lemma 7.2.

7.1.4 Implementation

A Raviart–Thomas finite element approximation of the dual mixed formulation of the Poisson problem is shown in Fig. 7.4. The implementation uses the bases

$$(\psi_S : S \in \mathscr{S}_h), \quad (\chi_T : T \in \mathscr{T}_h)$$

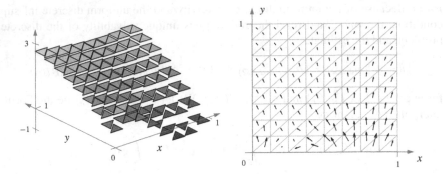

Fig. 7.4 Raviart–Thomas finite element approximation of the dual mixed formulation of a two-dimensional Poisson problem

for the spaces $\mathscr{R}T^0(\mathscr{T}_h)$ and $\mathscr{L}^0(\mathscr{T}_h)$, respectively. The representation of the involved bilinear form b with respect to these bases uses the identity

$$\int_\Omega \operatorname{div} \psi_S \, \chi_T \, \mathrm{d}x = \mp |S|/(d-1)!$$

for $S \in \mathscr{S}_h$, $T \in \mathscr{T}_h$ with $S \subset \partial T$, and $T = T_\pm$. For the discretization of the bilinear form a, we note that

$$\sum_{k=1}^{d+1} \varphi_{z_k}(x) z_k = x, \quad \sum_{k=1}^{d+1} \varphi_{z_k}(x) = 1$$

for all $x \in T = \operatorname{conv}\{z_1, \dots, z_{d+1}\}$. We choose the local enumeration of the sides of an element T that is induced by the local enumeration of the nodes of the element. For an element $T \in \mathscr{T}_h$, we thus have

$$\psi_{S_j}|_T(x) = \sigma_{j,T} \delta_{S_j,T}(z_j - x)$$

for $j = 1, 2, \dots, d + 1$, and with an appropriate sign σ_j. For $1 \le m, n \le d$, we then have

$$\int_T \psi_{S_m} \cdot \psi_{S_n} \, \mathrm{d}x = \sigma_{m,T} \sigma_{n,T} \frac{|S_m||S_n|}{(d!)^2 |T|^2} \int_T (z_m - x) \cdot (z_n - x) \, \mathrm{d}x$$

$$= \sigma_{m,T} \sigma_{n,T} \frac{|S_m||S_n|}{(d!)^2 |T|^2} \sum_{o,p=1}^{d+1} \int_T \varphi_{z_o} \varphi_{z_p} (z_m - z_o) \cdot (z_n - z_p) \, \mathrm{d}x$$

$$= \sigma_{m,T} \sigma_{n,T} \frac{|S_m||S_n|}{(d!)^2 |T|} \sum_{o,p=1}^{d+1} (z_m - z_o) \cdot (z_n - z_p) \frac{1 + \delta_{op}}{(d+1)(d+2)}.$$

The Neumann boundary condition $u_h \cdot n_S = g(x_S)$ for sides $S \in \mathscr{S}_h \cap \overline{\Gamma}_N$ is included via a matrix D and an augmentation of the linear system of equations. An essential component of the MATLAB realization shown in Fig. 7.5 are data structures related to the sides of elements. The routine `sides.m` displayed in Fig. 7.6 provides arrays that define functions

$$\texttt{s4e}: \mathscr{T}_h \to \mathscr{S}_h^{d+1}, \quad \texttt{sign\_s4e}: \mathscr{T}_h \to \{\pm 1\}^{d+1},$$

that specify the sides of elements and define a sign of the element relative to the side, according to the local enumeration defined by the opposite nodes. The mappings

$$\texttt{n4s}: \mathscr{S}_h \to \mathscr{N}_h^d, \quad \texttt{s4Db}: \mathscr{S}_h \cap \Gamma_D \to \mathscr{S}_h, \quad \texttt{s4Nb}: \mathscr{S}_h \cap \overline{\Gamma}_N \to \mathscr{S}_h,$$

provide the nodes of a side, and inject the boundary sides on Γ_D and Γ_N into the set of all sides $\mathscr{S}_h$. The routine `show_rt.m` shown in Fig. 7.6 visualizes a Raviart–Thomas vector field and an elementwise constant quantity.

7.2 Approximation of the Stokes System

7.2.1 Stokes System

The Stokes equations occur as a simplification of the Navier–Stokes equations and model the stationary, viscous, incompressible flow of a fluid through a domain $\Omega \subset \mathbb{R}^d$, $d = 2, 3$. At each point $x \in \Omega$, $u(x) \in \mathbb{R}^d$ is the velocity and $p(x) \in \mathbb{R}$ is the pressure. An outer body force, such as gravity, is modeled by the function $f(x) \in \mathbb{R}^d$. Together with boundary conditions, the Stokes equations state that

$$\begin{cases} -\mu \Delta u + \nabla p = f & \text{in } \Omega, \\ \operatorname{div} u = 0 & \text{in } \Omega, \\ u = u_D & \text{on } \Gamma_D \subset \partial\Omega, \\ (\mu \nabla u - pI)n = g & \text{on } \Gamma_N = \partial\Omega \setminus \Gamma_D. \end{cases}$$

Here the Laplace operator is applied component-wise, I is the $d \times d$ identity matrix, and $\mu > 0$ is the viscosity of the fluid. Note that the first equation can be written as $-\operatorname{div}(\mu \nabla u - pI) = f$. To obtain a weak formulation, we multiply the first equation by $v \in H^1(\Omega; \mathbb{R}^d)$ and integrate-by-parts, and multiply the second equation by $q \in L^2(\Omega)$ and integrate. We assume without loss of generality that $u_D = 0$.

```
function poisson_rt(red)
[c4n,n4e,Db,Nb] = triang_cube(2);
for j = 1:red
    [c4n,n4e,Db,Nb] = red_refine(c4n,n4e,Db,Nb);
end
[s4e,sign_s4e,n4s,s4Db,s4Nb] = sides(n4e,Db,Nb);
nS = size(n4s,1); nE = size(n4e,1);
nNb = size(Nb,1); nDb = size(Db,1);
m_loc = [2,1,1;1,2,1;1,1,2]/12;
A = sparse(nS,nS); B = sparse(nS,nE); D = sparse(nS,nNb);
b = zeros(nS+nE+nNb,1);
shift1 = [2,3,1]; shift2 = [3,1,2];
for j = 1:nE
    area_T = det([1,1,1;c4n(n4e(j,:),:)'])/2;
    mp_T = sum(c4n(n4e(j,:),:))/3;
    for m = 1:3
        length_m = norm(c4n(n4e(j,shift1(m)),:)...
            -c4n(n4e(j,shift2(m)),:));
        for n = 1:3
            length_n = norm(c4n(n4e(j,shift1(n)),:)...
                -c4n(n4e(j,shift2(n)),:));
            for o = 1:3
                for p = 1:3
                    A(s4e(j,m),s4e(j,n)) = A(s4e(j,m),s4e(j,n))...
                        +length_m*length_n/(4*area_T)...
                        *sign_s4e(j,m)*sign_s4e(j,n)*m_loc(o,p)...
                        *(c4n(n4e(j,m),:)-c4n(n4e(j,o),:))...
                        *(c4n(n4e(j,n),:)-c4n(n4e(j,p),:))';
                end
            end
        end
        B(s4e(j,m),j) = -length_m*sign_s4e(j,m);
        b(nS+j) = -area_T*g(mp_T);
    end
end
for j = 1:nDb
    length_S = norm(c4n(Db(j,2),:)-c4n(Db(j,1),:))';
    mp_S = (c4n(Db(j,1),:)+c4n(Db(j,2),:))/2;
    b(s4Db(j)) = p_D(mp_S)*length_S;
end
for j = 1:nNb
    mp_S = (c4n(Nb(j,1),:)+c4n(Nb(j,2),:))/2;
    D(s4Nb(j),j) = 1; b(nS+nE+j) = sigma(mp_S);
end
O1 = sparse(nE,nE+nNb); O2 = sparse(nNb,nE+nNb);
G = [A,B,D;B',O1;D',O2];
x = G\b; u = x(1:nS); p = x(nS+(1:nE));
show_rt(c4n,n4e,u,p,s4e,sign_s4e);

function val = sigma(x); val = ones(size(x,1),1);
function val = p_D(x); val = sin(2*pi*x(:,1));
function val = g(x); val = ones(size(x,1),1);
```

Fig. 7.5 MATLAB implementation of the Raviart–Thomas finite element method

```
function [s4e,sign_s4e,n4s,s4Db,s4Nb,e4s] = sides(n4e,Db,Nb)
nE = size(n4e,1); d = size(n4e,2)-1;
nDb = size(Db,1); nNb = size(Nb,1);
if d == 2
    Tsides = [n4e(:,[2,3]);n4e(:,[3,1]);n4e(:,[1,2])];
else
    Tsides = [n4e(:,[2,4,3]);n4e(:,[1,3,4]);...
        n4e(:,[1,4,2]);n4e(:,[1,2,3])];
end
[n4s,i2,j] = unique(sort(Tsides,2),'rows');
s4e = reshape(j,nE,d+1); nS = size(n4s,1);
sign_s4e = ones((d+1)*nE,1); sign_s4e(i2) = -1;
sign_s4e = reshape(sign_s4e,nE,d+1);
[~,~,j2] = unique(sort([n4s;Db;Nb],2),'rows');
s4Db = j2(nS+(1:nDb)); s4Nb = j2(nS+nDb+(1:nNb));
e4s = zeros(size(n4s,1),2);
e4s(:,1) = mod(i2-1,nE)+1;
i_inner = setdiff(1:(d+1)*nE,i2);
e4s(j(i_inner),2) = mod(i_inner-1,nE)+1;
```

```
function show_rt(c4n,n4e,u,p,s4e,sign_s4e)
nE = size(n4e,1);
u_T = zeros(nE,2); mp_T = zeros(nE,2);
shift1 = [2,3,1]; shift2 = [3,1,2];
for j = 1:nE
    area_T = det([1,1,1;c4n(n4e(j,:),:)'])/2;
    mp_T(j,:) = sum(c4n(n4e(j,:),:))/3;
    for m = 1:3
        side_m = s4e(j,m);
        length_m = ...
            norm(c4n(n4e(j,shift1(m)),:)-c4n(n4e(j,shift2(m)),:));
        signum_m = sign_s4e(j,m);
        u_T(j,:) = u_T(j,:)+u(side_m)*signum_m*length_m...
            *(c4n(n4e(j,m),:)-mp_T(j,:))/(2*area_T);
    end
end
E = reshape(1:3*nE,3,nE)'; n4e_t = n4e';
X = c4n(n4e_t(:),1); Y = c4n(n4e_t(:),2); P = repmat(p,1,3)';
figure(1); clf; subplot(1,2,1); trisurf(E,X,Y,P(:));
subplot(1,2,2); triplot(n4e,c4n(:,1),c4n(:,2),':k'); hold on;
quiver(mp_T(:,1),mp_T(:,2),u_T(:,1),u_T(:,2));
```

Fig. 7.6 MATLAB routine that provides data structures related to sides of elements (*top*); MATLAB routine that visualizes a pair $(u_h, p_h) \in \mathcal{R}T^0(\mathcal{T}_h) \times \mathcal{L}^0(\mathcal{T}_h)$

Definition 7.6 The *weak formulation of the Stokes system* consists in finding $(u, p) \in H_D^1(\Omega; \mathbb{R}^d) \times L^2(\Omega)$ such that

$$\mu \int_\Omega \nabla u : \nabla v \, dx - \int_\Omega p \operatorname{div} v \, dx = \int_\Omega f \cdot v \, dx + \int_{\Gamma_N} g \cdot v \, ds,$$

$$- \int_\Omega q \operatorname{div} u \, dx = 0,$$

for all $(v, q) \in H_D^1(\Omega; \mathbb{R}^d) \times L^2(\Omega)$.

Remark 7.8 If $\Gamma_D = \partial \Omega$, i.e., $\Gamma_N = \emptyset$, then p is only defined up to a constant, and we replace $L^2(\Omega)$ by

$$L_0^2(\Omega) = \{ q \in L^2(\Omega) : \int_\Omega q \, dx = 0 \}.$$

To establish the existence of a solution, we note that the right-hand side defines a continuous functional on $H_D^1(\Omega; \mathbb{R}^d)$. The bilinear form

$$a : H_D^1(\Omega; \mathbb{R}^d) \times H_D^1(\Omega; \mathbb{R}^d) \to \mathbb{R}, \quad a(u, v) = \mu \int_\Omega \nabla u : \nabla v \, dx$$

is elliptic on the entire space $H_D^1(\Omega; \mathbb{R}^d)$. For the bilinear form

$$b : H_D^1(\Omega; \mathbb{R}^d) \times L^2(\Omega) \to \mathbb{R}, \quad b(v, q) = - \int_\Omega q \operatorname{div} v \, dx$$

we have

$$\ker B = \{ v \in H_D^1(\Omega; \mathbb{R}^d) : b(v, q) = 0 \text{ for all } q \in L^2(\Omega) \}$$

$$= \{ v \in H_D^1(\Omega; \mathbb{R}^d) : \operatorname{div} v = 0 \}.$$

It remains to establish the inf-sup condition for b, i.e., that

$$\inf_{q \in L^2(\Omega) \setminus \{0\}} \sup_{v \in H_D^1(\Omega; \mathbb{R}^d) \setminus \{0\}} \frac{- \int_\Omega q \operatorname{div} v \, dx}{\|\nabla v\|_{L^2(\Omega)} \|q\|_{L^2(\Omega)}} \geq \beta > 0.$$

This condition determines mapping properties of the operator

$$\operatorname{div} : H_D^1(\Omega; \mathbb{R}^d) \to L^2(\Omega)$$

and its formal adjoint $\operatorname{div}' : L^2(\Omega) \to (H_D^1(\Omega; \mathbb{R}^d))'$, which coincides with the distributional gradient if $\Gamma_D = \partial \Omega$. In particular, since $\ker \operatorname{div}' = \{0\}$, hence $(\ker \operatorname{div}')^\circ = L^2(\Omega)$, the inf-sup condition is equivalent to the closedness of the image of the divergence as an operator from $H_D^1(\Omega; \mathbb{R}^d)$ into $L^2(\Omega)$, cf. Theorem 6.2. This is distinct from the mapping properties of div : $H(\text{div}; \Omega) \to L^2(\Omega)$, required for the dual mixed formulation of the Poisson problem.

Theorem 7.3 (Divergence Operator, See [11], Chap. II) *Suppose that $\Omega \subset \mathbb{R}^d$ is simply connected and $\Gamma_D = \partial\Omega$. For every $q \in L_0^2(\Omega)$ there exists $v \in H_0^1(\Omega; \mathbb{R}^d)$ such that $\text{div } v = q$ and $\|\nabla v\|_{L^2(\Omega)} \leq \beta^{-1}\|q\|_{L^2(\Omega)}$.*

The result implies the inf-sup condition and hence the well-posedness of the Stokes system if $\Gamma_D = \partial\Omega$; its proof requires deeper analytical arguments.

Assumption 7.1 (Well-Posedness) *We assume that $\Omega \subset \mathbb{R}^d$ and $\Gamma_D \subset \partial\Omega$ are such that the inf-sup condition is valid and hence the weak formulation of the Stokes system is well-posed.*

With this assumption we focus on constructing suitable finite element spaces, in particular, on verifying discrete inf-sup conditions.

Remarks 7.9

(i) In the artificial setting that $\Gamma_D = \emptyset$ and if Ω is convex, we may argue as in the case of the dual mixed formulation of the Poisson problem, and let $\phi \in H_0^1(\Omega)$ for given $q \in L^2(\Omega)$ be such that

$$-\Delta\phi = q \text{ in } \Omega, \quad \phi = 0 \text{ on } \partial\Omega.$$

If this problem is H^2-regular, i.e., if $\|D^2\phi\|_{L^2(\Omega)} \leq c\|q\|_{L^2(\Omega)}$, we may choose $v = \nabla\phi$ to deduce the inf-sup condition from

$$b(v, q) = -\int_\Omega q \, \text{div } v \, dx = -\int_\Omega q\Delta\phi \, dx = \int_\Omega q^2 \, dx$$

and $\|v\|_{H^1(\Omega)} \leq c\|q\|_{L^2(\Omega)}$.

(ii) The inf-sup condition for the Stokes problem is often formulated in form of the equivalent estimate $\|p\|_{L^2(\Omega)} \leq \beta^{-1}\|\nabla p\|_{H_D^1(\Omega)'}$ for all $p \in L^2(\Omega)$, cf. Lemma 6.5.

7.2.2 Unstable Stokes Elements

We show that natural choices of finite element spaces for the discretization of the Stokes system lead to unstable numerical methods, cf. Fig. 7.7. We follow [6, 10].

Example 7.1 The P_1-P_0 element is defined by the spaces

$$V_h = \mathscr{S}_D^1(\mathscr{T}_h)^d, \quad Q_h = \mathscr{L}^0(\mathscr{T}_h).$$

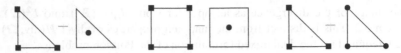

Fig. 7.7 Schematic description of unstable Stokes elements: P_1-P_0 element (*left*), Q_1-Q_0 element (*middle*), and P_1-P_1 element (*right*); boxes indicate the evaluation of a vector field, dots the evaluation of a scalar function

Due to the dimensionality of the spaces, the divergence operator div : $V_h \to Q_h$ is in general not surjective, and hence, a discrete inf-sup condition cannot be satisfied in general, cf. Example 6.6.

On parallelograms a similar effect occurs.

Example 7.2 The Q_1-Q_0 element is for a triangulation $\mathcal{T}_h^q$ consisting of quadrilaterals, defined by the spaces

$$V_h = \{v_h \in H_0^1(\Omega; \mathbb{R}^d) : v_h|_T \in \mathcal{Q}_1(T)^d \text{ for all } T \in \mathcal{T}_h^q\},$$

$$Q_h = \{q_h \in L_0^2(\Omega) : q_h|_T \in \mathcal{Q}_0(T) \text{ for all } T \in \mathcal{T}_h^q\},$$

where $\mathcal{Q}_k(T)$ denotes the polynomials of partial degree k on T, e.g., $\mathcal{Q}_1(T) = \text{span}\{1, x, y, xy\}$. For the triangulation of $\Omega = (0, 1)^2$ consisting of squares with side-lengths h, and the function $\bar{q}_h$ with alternating values ± 1, as indicated in Fig. 7.8, we have that

$$\int_\Omega \bar{q}_h \, \text{div} \, v_h \, dx = 0$$

for all $v_h \in V_h$. Hence a discrete inf-sup condition cannot hold. To verify the identity, let $z \in \mathcal{N}_h \cap \Omega$ and φ_z be the associated nodal basis function, i.e., $\varphi_z(y) = \delta_{zy}$ for all nodes $z, y \in \mathcal{N}_h$. Let $S_1, S_2, \ldots, S_4$ be those edges which have z as an endpoint. With a canonical basis vector $e_j \in \mathbb{R}^2$, we consider the vector field $\varphi_z^j = \varphi_z e_j$ and verify with an elementwise integration-by-parts that

$$\int_\Omega \bar{q}_h \, \text{div} \, \varphi_z^j \, dx = \sum_{T \in \mathcal{T}} \int_{\partial T} \bar{q}_h \varphi_z^j \cdot n_T \, ds = \sum_{k=1}^4 \int_{S_k} \varphi_z^j [\bar{q}_h n] \, ds,$$

where $[\bar{q}_h n]|_{S_k} = \bar{q}_h|_{T_+} n_{T_+} + \bar{q}_h|_{T_-} n_{T_-}$ is the *jump* of $\bar{q}_h n$ across a side $S_k = T_+ \cap T_-$. For every side S_k, we let n_{S_k} be that unit normal to S_k, which points from the neighboring square on which $\bar{q}_h = -1$ into the square with $\bar{q}_h = 1$, cf. Fig 7.8. Noting that $[\bar{q}_h n]|_{S_k} = -2n_{S_k}$ for $k = 1, 2, \ldots, 4$, we find that

$$\int_\Omega \bar{q}_h \, \text{div} \, \varphi_z^j \, dx = -\sum_{k=1}^4 \int_{S_k} 2\varphi_z e_j \cdot n_{S_k} \, ds = -h \sum_{k=1}^4 e_j \cdot n_{S_k} = 0.$$

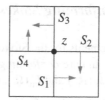

−1	+1	−1	+1
+1	−1	+1	−1
−1	+1	−1	+1
+1	−1	+1	−1

Fig. 7.8 Checkerboard mode leading to the failure of a discrete inf-sup condition for the Q_1-Q_0-element

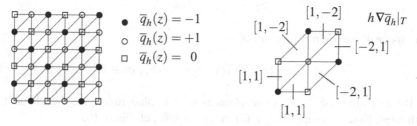

Fig. 7.9 P_1 pressure variable $\bar{q}_h$ that is L^2-orthogonal to the divergence of every P_1 vector field

Since $(\varphi_z e_j : z \in \mathscr{N}_h \cap \Omega, j = 1, 2, \ldots, d)$ is a basis for V_h, we have that $\bar{q}_h$ is orthogonal to the image of V_h under divergence, i.e., the divergence is not surjective.

The P_1-P_1 element is another example of an unstable element.

Example 7.3 The P_1-P_1-element is for a triangulation $\mathscr{T}_h$ of Ω into triangles defined by

$$V_h = \mathscr{S}_0^1(\mathscr{T}_h)^2, \quad Q_h = \mathscr{S}^1(\mathscr{T}_h) \cap L_0^2(\Omega).$$

We assume that $\Omega = (0, 1)^2$, and $\mathscr{T}_h$ is the triangulation of Ω into halved squares of side-lengths h and $\bar{q}_h \in \mathscr{S}^1(\mathscr{T}_h)$ as indicated in Fig. 7.9. Then for $z \in \mathscr{N}_h \cap \Omega$ and a canonical basis vector $e_j, j = 1, 2$, we have

$$\int_\Omega \bar{q}_h \operatorname{div}(\varphi_z e_j) \, dx = -\int_\Omega \nabla \bar{q}_h \cdot (\varphi_z e_j) \, dx = \frac{-h^2}{6} \sum_{T \in \mathscr{T}_h, z \in T} \nabla \bar{q}_h|_T \cdot e_j = 0.$$

This implies that

$$\int_\Omega \bar{q}_h \operatorname{div} v_h \, dx = 0$$

for all $v_h \in \mathscr{S}_0^1(\mathscr{T}_h)^2$. Since the same is true after subtracting a constant from $\bar{q}_h$ to guarantee $\bar{q}_h \in L_0^2(\Omega)$, this proves the failure of the discrete inf-sup condition for the P_1-P_1 element.

7.2.3   Stable Conforming Stokes Elements

We briefly discuss elements for the Stokes system which are stable, i.e., they satisfy an inf-sup condition, and are *conforming* in the sense that the finite element spaces are subspaces of $H_0^1(\Omega; \mathbb{R}^d)$ and $L_0^2(\Omega)$, respectively.

Definition 7.7 The *bubble function* $b_T \in H^1(\Omega)$ associated with an element $T \in \mathcal{T}_h$ with vertices $z_1, z_2, \ldots, z_{d+1} \in \mathcal{N}_h$ is defined by

$$b_T = \varphi_{z_1} \varphi_{z_2} \cdots \varphi_{z_{d+1}}.$$

The *space of bubble functions* on $\mathcal{T}_h$ is

$$\mathcal{B}(\mathcal{T}_h) = \{v \in C(\overline{\Omega}) : v|_T = c_T b_T, \ c_T \in \mathbb{R}\}.$$

By enriching the P_1-P_1 finite element with bubble functions, the element is stabilized. Note that $\operatorname{supp} b_T \subset T$ for every $T \in \mathcal{T}_h$, cf. Fig. 7.10.

Example 7.4 The *MINI* element is defined by

$$V_h = \mathcal{S}_0^1(\mathcal{T}_h)^d \oplus \mathcal{B}(\mathcal{T}_h)^d, \quad Q_h = \mathcal{S}^1(\mathcal{T}_h) \cap L_0^2(\Omega).$$

An exercise shows that if the Poisson problem on Ω is H^2-regular, and if $\mathcal{T}_h$ is quasiuniform, then the MINI-element satisfies the inf-sup condition. A more careful argument proves its general stability. It is called *MINI-element* since it is the conforming element of lowest degree. In fact, the degrees of freedom related to the space $\mathcal{B}(\mathcal{T}_h)^d$ can be eliminated. This is due to orthogonality, i.e.,

$$\int_\Omega \nabla v_h \cdot \nabla b_h \, dx = \sum_{T \in \mathcal{T}_h} \left(-\int_T \Delta v_h b_h \, dx + \int_{\partial T} (\nabla v_h \cdot n) b_h \, ds \right) = 0$$

for all $v_h \in \mathcal{S}_0^1(\mathcal{T}_h)$ and $b_h \in \mathcal{B}(\mathcal{T}_h)$, which follows from $\Delta v_h|_T = 0$ and $b_h|_{\partial T} = 0$ for all $T \in \mathcal{T}_h$. The matrix related to the bilinear form a is therefore block-diagonal, and the block related to $\mathcal{B}(\mathcal{T}_h)^d$ is diagonal.

Fig. 7.10   Element bubble function b_T (*left*), schematic description of the MINI element (*middle*), and schematic description of the quadratic Taylor–Hood element (*right*)

Example 7.5 A k-th order *Taylor–Hood element* is defined by

$$V_h = \mathscr{S}_0^k(\mathscr{T}_h)^d = \{v_h \in H_0^1(\Omega; \mathbb{R}^d) : v_h|_T \in \mathscr{P}_k(T)^d \text{ for all } T \in \mathscr{T}_h\},$$

$$Q_h = \mathscr{S}^{k-1}(\mathscr{T}_h) \cap L_0^2(\Omega) = \{q_h \in C(\overline{\Omega}) : q_h|_T \in \mathscr{P}_{k-1}(T) \text{ for all } T \in \mathscr{T}_h\}.$$

It satisfies the inf-sup condition if $k \geq 2$.

The stability of the Taylor–Hood element is on quasiuniform triangulations established by explicitly constructing for every $q_h \in Q_h$ a vector field $v_h \in V_h$ such that v_h is nonorthogonal to ∇q_h in $L^2(\Omega; \mathbb{R}^d)$, and such that $\|v_h\|_{L^2(\Omega)} \leq c_{TH}\|\nabla q_h\|_{L^2(\Omega)}$. A more careful argument implies its general stability.

7.2.4   Pressure Stabilization

For a triangulation $\mathscr{T}_h$ of Ω, we consider the spaces

$$V_h = \mathscr{S}_0^1(\mathscr{T}_h)^d, \quad Q_h = \mathscr{S}^1(\mathscr{T}_h) \cap L_0^2(\Omega)$$

which fail to satisfy a discrete inf-sup condition. The following lemma shows that this failure can be quantified. We follow [4].

Lemma 7.3 (Perturbed Inf-Sup Condition) *For all $p_h \in Q_h$ we have*

$$\sup_{v_h \in V_h \setminus \{0\}} \frac{\int_\Omega p_h \operatorname{div} v_h \, dx}{\|\nabla v_h\|_{L^2(\Omega)}} \geq \beta' \|p_h\|_{L^2(\Omega)} - c_p h \|\nabla p_h\|_{L^2(\Omega)}.$$

Proof Since $Q_h \subset L_0^2(\Omega)$, the continuous inf-sup condition guarantees that

$$\sup_{v \in H_0^1(\Omega; \mathbb{R}^d) \setminus \{0\}} \frac{\int_\Omega p_h \operatorname{div} v \, dx}{\|\nabla v\|_{L^2(\Omega)}} \geq \beta \|p_h\|_{L^2(\Omega)},$$

i.e., there exists $w \in H_0^1(\Omega; \mathbb{R}^d)$, such that

$$\int_\Omega p_h \operatorname{div} w \, dx \geq \frac{\beta}{2} \|p_h\|_{L^2(\Omega)} \|\nabla w\|_{L^2(\Omega)}.$$

We let $w_h \in V_h$ be the Clément quasi-interpolant of w, i.e., we define

$$w_h = \sum_{z \in \mathscr{N}_h} w_z \varphi_z, \quad w_z = |\omega_z|^{-1} \int_{\omega_z} w \, dx.$$

We then have that, cf. Theorem 4.2,

$$h^{-1}\|w - w_h\|_{L^2(\Omega)} + \|\nabla w_h\|_{L^2(\Omega)} \le c_{C\ell}\|\nabla w\|_{L^2(\Omega)}.$$

This implies that

$$
\begin{aligned}
\int_\Omega p_h \operatorname{div} w_h \, dx &= \int_\Omega p_h \operatorname{div} w \, dx + \int_\Omega p_h \operatorname{div}(w_h - w) \, dx \\
&= \int_\Omega p_h \operatorname{div} w \, dx - \int_\Omega \nabla p_h \cdot (w_h - w) \, dx \\
&\ge (\beta/2)\|p_h\|_{L^2(\Omega)}\|\nabla w\|_{L^2(\Omega)} - \|\nabla p_h\|_{L^2(\Omega)}\|w - w_h\|_{L^2(\Omega)} \\
&\ge (\beta/2)\|p_h\|_{L^2(\Omega)}\|\nabla w\|_{L^2(\Omega)} - c_{C\ell}h\|\nabla p_h\|_{L^2(\Omega)}\|\nabla w\|_{L^2(\Omega)}.
\end{aligned}
$$

Dividing by $\|\nabla w\|_{L^2(\Omega)}$ and using $\|\nabla w_h\|_{L^2(\Omega)} \le c_{C\ell}\|\nabla w\|_{L^2(\Omega)}$ prove the estimate.
□

The lemma motivates introducing a stabilizing term to compensate for the failure of the inf-sup condition. We thus consider the *stabilized discrete Stokes problem*, which consists in finding $(u_h, p_h) \in V_h \times Q_h$ such that

$$
\begin{aligned}
a(u_h, v_h) + b(v_h, p_h) &= \ell(v_h), \\
b(u_h, q_h) - c_h(p_h, q_h) &= 0,
\end{aligned}
$$

for all $(v_h, q_h) \in V_h \times Q_h$, with an appropriately chosen bilinear form $c_h : Q_h \times Q_h \to \mathbb{R}$. With the stabilized saddle-point problem, we associate the augmented bilinear form $\widehat{\Gamma}_h$ on $(V_h \times Q_h)^2$ defined by

$$\widehat{\Gamma}_h((u_h, p_h), (v_h, q_h)) = a(u_h, v_h) + b(v_h, p_h) + b(u_h, q_h) - c_h(p_h. q_h).$$

Provided that c_h is bounded and controls the negative term in the perturbed inf-sup condition, then the bilinear form $\widehat{\Gamma}_h$ defines an isomorphism.

Proposition 7.5 (Well-Posedness) *Assume that $c_h : Q_h \times Q_h \to \mathbb{R}$ is a uniformly bounded symmetric bilinear form which satisfies*

$$|q_h|_{c_h}^2 = c_h(q_h, q_h) \ge c_{stab}h^2\|\nabla q_h\|_{L^2(\Omega)}^2$$

for all $q_h \in Q_h$ with a constant $c_{stab} > 0$. Then the stabilized saddle-point problem has a unique solution $(u_h, p_h) \in V_h \times Q_h$ for every $\ell \in V_h'$.

Proof We show that $\widehat{\Gamma}_h$ satisfies the conditions of the generalized Lax–Milgram lemma. Boundedness is an immediate consequence of the boundedness of a, b, and

c_h. Since $\widehat{\Gamma}_h$ is symmetric, its nondegeneracy follows from the inf-sup condition. To prove the inf-sup condition, let $(u_h, p_h) \in V_h \times Q_h$. We construct a pair $(\hat{v}_h, \hat{q}_h)$ such that

$$\frac{\widehat{\Gamma}_h((u_h, p_h), (\hat{v}_h, \hat{q}_h))}{\|\nabla \hat{v}_h\|_{L^2(\Omega)} + \|\hat{q}_h\|_{L^2(\Omega)}} \geq \gamma \left(\|\nabla u_h\|_{L^2(\Omega)} + \|p_h\|_{L^2(\Omega)} \right).$$

According to Lemma 7.3 there exists $w_h \in V_h$ with

$$\int_\Omega p_h \operatorname{div} w_h \, dx \geq \beta' \|p_h\|_{L^2(\Omega)} \|\nabla w_h\|_{L^2(\Omega)} - c_p h \|\nabla p_h\|_{L^2(\Omega)} \|\nabla w_h\|_{L^2(\Omega)}.$$

Replacing w_h by sw_h with an appropriate number $s > 0$, we may assume that

$$\|\nabla w_h\|_{L^2(\Omega)} = \|p_h\|_{L^2(\Omega)}.$$

We then have that

$$\int_\Omega p_h \operatorname{div} w_h \, dx \geq \beta' \|p_h\|_{L^2(\Omega)}^2 - c_p h \|\nabla p_h\|_{L^2(\Omega)} \|p_h\|_{L^2(\Omega)}.$$

Noting that

$$\widehat{\Gamma}_h((u_h, p_h), (u_h, -p_h)) = \mu \|\nabla u_h\|_{L^2(\Omega)}^2 + |p_h|_{c_h}^2,$$

and, using $\|\nabla w_h\|_{L^2(\Omega)} = \|p_h\|_{L^2(\Omega)}$, we have

$$\widehat{\Gamma}_h((u_h, p_h), (-w_h, 0)) = -\mu \int_\Omega \nabla u_h \cdot \nabla w_h \, dx + \int_\Omega p_h \operatorname{div} w_h \, dx$$

$$\geq -\mu \|\nabla u_h\|_{L^2(\Omega)} \|p_h\|_{L^2(\Omega)} + (\beta' \|p_h\|_{L^2(\Omega)}^2 - c_p h \|\nabla p_h\|_{L^2(\Omega)} \|p_h\|_{L^2(\Omega)});$$

and we deduce that for $(\hat{v}_h, \hat{q}_h) = (u_h - \alpha w_h, -p_h)$ and every $\alpha \geq 0$ we have

$$\widehat{\Gamma}_h((u_h, p_h), (\hat{v}_h, \hat{q}_h)) \geq \mu \|\nabla u_h\|_{L^2(\Omega)}^2 + |p_h|_{c_h}^2 - \mu \|\nabla u_h\|_{L^2(\Omega)} \|p_h\|_{L^2(\Omega)}$$

$$+ \alpha \beta' \|p_h\|_{L^2(\Omega)}^2 - \alpha c_p h \|\nabla p_h\|_{L^2(\Omega)} \|p_h\|_{L^2(\Omega)} - \alpha \mu \|\nabla u_h\|_{L^2(\Omega)} \|p_h\|_{L^2(\Omega)}.$$

We use Young's inequality $ab \leq \varepsilon a^2 + b^2/(4\varepsilon)$ to check that

$$\mu \|\nabla u_h\|_{L^2(\Omega)} \|p_h\|_{L^2(\Omega)} \leq \frac{\mu^2}{\beta'} \|\nabla u_h\|_{L^2(\Omega)}^2 + \frac{\beta'}{4} \|p_h\|_{L^2(\Omega)}^2,$$

and

$$c_p h \|\nabla p_h\|_{L^2(\Omega)} \|p_h\|_{L^2(\Omega)} \leq \frac{c_p^2 h^2}{\beta'} \|\nabla p_h\|_{L^2(\Omega)}^2 + \frac{\beta'}{4} \|p_h\|_{L^2(\Omega)}^2.$$

Together with $|p_h|_{c_h} \geq c_{stab}^{1/2} h \|\nabla p_h\|_{L^2(\Omega)}$, this leads to

$$\widehat{\Gamma}_h\big((u_h, p_h), (\hat{v}_h, \hat{q}_h)\big) \geq \mu \left(1 - \frac{\alpha \mu}{\beta'}\right) \|\nabla u_h\|_{L^2(\Omega)}^2$$

$$+ \frac{\alpha \beta'}{2} \|p_h\|_{L^2(\Omega)}^2 + \left(1 - \frac{\alpha c_p^2}{\beta' c_{stab}}\right) c_{stab} h^2 \|\nabla p_h\|_{L^2(\Omega)}^2.$$

We choose $\alpha > 0$ sufficiently small, so that the brackets in the first and third term on the right-hand side are greater than $1/2$, i.e., we have

$$\widehat{\Gamma}_h\big((u_h, p_h), (\hat{v}_h, \hat{q}_h)\big) \geq \frac{\mu}{2} \|\nabla u_h\|_{L^2(\Omega)}^2 + \frac{\alpha \beta'}{2} \|p_h\|_{L^2(\Omega)}^2.$$

With the triangle inequality and $\|\nabla w_h\|_{L^2(\Omega)} = \|p_h\|_{L^2(\Omega)}$, we verify that

$$\|\nabla \hat{v}_h\|_{L^2(\Omega)} + \|\hat{q}_h\|_{L^2(\Omega)} = \|\nabla(u_h - \alpha w_h)\|_{L^2(\Omega)} + \|p_h\|_{L^2(\Omega)}$$

$$\leq \|\nabla u_h\|_{L^2(\Omega)} + (1 + \alpha) \|p_h\|_{L^2(\Omega)}.$$

The combination of the last two estimates proves the inf-sup condition and implies that $\widehat{\Gamma}_h$ satisfies the conditions of the generalized Lax–Milgram lemma.   □

The Stokes system is related to the bilinear form

$$\Gamma\big((u, p), (v, q)\big) = a(u, v) + b(v, p) + b(u, q).$$

Since the discrete problem uses a different bilinear form, we cannot apply the generalized Céa lemma directly to deduce a best-approximation result. A modification of the proof leads to the following result.

Theorem 7.4 (Error Estimate) *Assume that the conditions of Proposition 7.5 are satisfied. For the solution* $(u, p) \in H_0^1(\Omega; \mathbb{R}^d) \times L_0^2(\Omega)$ *of the weak formulation of the Stokes system, and the unique solution* $(u_h, p_h) \in V_h \times Q_h$ *of the stabilized discrete Stokes system, we have*

$$\|\nabla(u - u_h)\|_{L^2(\Omega)} + \|p - p_h\|_{L^2(\Omega)}$$

$$\leq c_\Gamma \inf_{(v_h, q_h) \in V_h \times Q_h} \big(\|\nabla(u - v_h)\|_{L^2(\Omega)} + \|p - q_h\|_{L^2(\Omega)} + c' |q_h|_{c_h}\big).$$

Proof Exercise.   □

Remarks 7.10

(i) The choice

$$c_h(p_h, q_h) = h^2 \int_\Omega \nabla p_h \cdot \nabla q_h \, dx$$

satisfies the requirements of the theorem and leads to an error contribution $\mathcal{O}(h)$, which is of the same order as the other terms in case of a regular solution $(u, p) \in H^2(\Omega; \mathbb{R}^d) \times H^1(\Omega)$.

(ii) The stabilized saddle-point formulation is similar to a perturbed saddle-point problem. For this we do not require an inf-sup condition for b.

(iii) The method discussed is based on an inconsistent stabilization of the discrete system. By making appropriate use of residuals, consistent stabilization methods can be constructed.

7.2.5 Stable Nonconforming Stokes Element

The stable conforming Stokes elements require higher-order finite elements. Allowing for nonconformity of the numerical method, i.e., that we do not have $V_h \subset V$, enables us to work with elementwise affine vector fields and elementwise constant pressure approximations.

Definition 7.8 The *broken Sobolev space* associated with triangulation $\mathcal{T}_h$ of $\Omega \subset \mathbb{R}^d$ is defined by

$$H^1(\mathcal{T}_h) = \{v \in L^2(\Omega) : v|_T \in H^1(T) \text{ for all } T \in \mathcal{T}_h\}.$$

For $v \in H^1(\mathcal{T}_h)$ we define $\nabla_{\mathcal{T}} v \in L^2(\Omega; \mathbb{R}^d)$ for every $T \in \mathcal{T}_h$ by

$$(\nabla_{\mathcal{T}} v)|_T = \nabla(v|_T).$$

The space has the norm $\|v\|_{H^1(\mathcal{T}_h)} = \|v\|_{L^2(\Omega)} + \|\nabla_{\mathcal{T}} v\|_{L^2(\Omega)}$.

Note that $H^1(\Omega) \subseteq H^1(\mathcal{T}_h)$ but $H^1(\mathcal{T}_h) \not\subseteq H^1(\Omega)$ unless $\mathcal{T}_h$ consists of only one element. With each side $S \in \mathcal{S}_h$, we associate its midpoint

$$x_S = |S|^{-1} \int_S s \, ds = \frac{1}{d} \sum_{z \in \mathcal{N}_h \cap S} z.$$

The nonconforming Crouzeix–Raviart finite element imposes continuity only at the midpoints of the sides of elements.

Fig. 7.11 Basis function
associated with a side $S \in \mathscr{S}_h$
(*left*); schematical description
of the Crouzeix–Raviart finite
element (*right*)

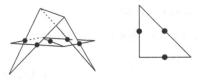

Definition 7.9 The *Crouzeix–Raviart finite element space* subordinated to the
triangulation $\mathscr{T}_h$ is defined by

$$\mathscr{S}^{1,cr}(\mathscr{T}_h) = \big\{v_h \in L^\infty(\Omega) : v_h|_T \in \mathscr{P}_1(T) \text{ for all } T \in \mathscr{T}_h,$$

$$v_h \text{ continuous at } x_S \text{ for all } S \in \mathscr{S}_h\big\}.$$

The subspace $\mathscr{S}_D^{1,cr}(\mathscr{T}_h)$ consists of all $v_h \in \mathscr{S}^{1,cr}(\mathscr{T}_h)$ with $v_h(z_S) = 0$ for all
$S \in \mathscr{S}_h \cap \Gamma_D$. If $\Gamma_D = \partial\Omega$, we also write $\mathscr{S}_0^{1,cr}(\mathscr{T}_h)$ instead of $\mathscr{S}_D^{1,cr}(\mathscr{T}_h)$.

Proposition 7.6 (Definiteness) *The seminorm* $v \mapsto \|\nabla_{\mathscr{T}} v\|_{L^2(\Omega)}$ *defines a norm on*
$\mathscr{S}_D^{1,cr}(\mathscr{T}_h)$.

Proof If $\|\nabla_{\mathscr{T}} v_h\|_{L^2(\Omega)} = 0$, then v_h is elementwise constant, and the continuity at
the midpoints of sides, together with the boundary condition imply $v_h = 0$. □

We construct a basis for $\mathscr{S}^{1,cr}(\mathscr{T}_h)$ consisting of functions that are associated
with sides of elements, cf. Fig. 7.11.

Lemma 7.4 (Crouzeix–Raviart Basis) *For* $S \in \mathscr{S}_h$, *let* $\varphi_S \in \mathscr{S}^{1,cr}(\mathscr{T}_h)$ *be the
uniquely defined function in* $\mathscr{S}^{1,cr}(\mathscr{T}_h)$ *with*

$$\varphi_S(x_R) = \delta_{SR}$$

for all $R \in \mathscr{S}_h$. *Then* $(\varphi_S : S \in \mathscr{S}_h)$ *is a basis for* $\mathscr{S}^{1,cr}(\mathscr{T}_h)$.

Proof The condition $\varphi_S(x_R) = \delta_{SR}$ defines the elementwise affine function uniquely
on every element $T \in \mathscr{T}_h$. Due to continuity at x_R for every $R \in \mathscr{S}_h$, we have
$\varphi_S \in \mathscr{S}^{1,cr}(\mathscr{T}_h)$. If $u_h \in \mathscr{S}^{1,cr}(\mathscr{T}_h)$, let $\alpha_S = u_h(x_S)$ for every $S \in \mathscr{S}_h$. Then the
function

$$\tilde{u}_h = \sum_{S \in \mathscr{S}_h} \alpha_S \varphi_S$$

coincides with u_h. The property $\varphi_S(x_R) = \delta_{SR}$ implies that $\alpha_S = 0$ for all $S \in \mathscr{S}_h$ if
$\tilde{u}_h = 0$. □

It is useful for implementing the Crouzeix–Raviart method to have a local
representation of the basis functions in terms of the standard nodal basis.

Remark 7.11 On every element $T \in \mathcal{T}_h$, the basis function $\varphi_S \in \mathcal{S}^{1,cr}(\mathcal{T}_h)$ associated with $S \subset \partial T$ satisfies

$$\varphi_S = \varphi_{z_1} + \cdots + \varphi_{z_d} - (d-1)\varphi_{z_{d+1}} = 1 - d\varphi_{z_{d+1}},$$

if $S = \mathrm{conv}\{z_1, z_2, \ldots, z_d\}$ and z_{d+1} is the vertex of T opposite S.

For a vector field $v_h \in \mathcal{S}^{1,cr}(\mathcal{T}_h)^d$, we define

$$\mathrm{div}_{\mathcal{T}} v_h = \mathrm{tr}\left[\nabla_{\mathcal{T}} v_h\right],$$

which coincides with the elementwise application of the divergence operator. We assume that $u_D = 0$ in what follows.

Definition 7.10 The *Crouzeix–Raviart method* for the Stokes system consists in finding $(u_h, p_h) \in \mathcal{S}_D^{1,cr}(\mathcal{T}_h)^d \times \mathcal{L}^0(\mathcal{T}_h)$ such that

$$\mu \int_\Omega \nabla_{\mathcal{T}} u_h : \nabla_{\mathcal{T}} v_h \, dx - \int_\Omega p_h \, \mathrm{div}_{\mathcal{T}} v_h \, dx = \int_\Omega f \cdot v_h \, dx + \int_{\Gamma_N} g \cdot v_h \, ds,$$

$$-\int_\Omega q_h \, \mathrm{div}_{\mathcal{T}} u_h \, dx = 0,$$

for all $(v_h, q_h) \in \mathcal{S}_D^{1,cr}(\mathcal{T}_h)^d \times \mathcal{L}^0(\mathcal{T}_h)$.

The nonconforming element is schematically described in Fig. 7.12.

To establish the existence of a unique solution, we construct a Fortin interpolant.

Proposition 7.7 (Fortin Interpolant) *There exists a bounded linear operator* $I_F : H_D^1(\Omega; \mathbb{R}^d) \to \mathcal{S}_D^{1,cr}(\mathcal{T}_h)^d$ *such that*

$$\|\nabla_{\mathcal{T}} I_F v\|_{L^2(\Omega)} \leq c_F \|\nabla v\|_{L^2(\Omega)}$$

and

$$b(v - I_F v, q_h) = -\int_\Omega q_h \, \mathrm{div}_{\mathcal{T}}(v - I_F v) \, dx = 0$$

for all $v \in H_D^1(\Omega; \mathbb{R}^d)$ *and* $q_h \in \mathcal{L}^0(\mathcal{T}_h)$.

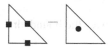

Fig. 7.12 Degrees of freedom for the nonconforming Stokes element

Proof

(i) For $v \in H_D^1(\Omega; \mathbb{R}^d)$, let $v_S \in \mathbb{R}^d$ for every $S \in \mathscr{S}_h$ be defined by

$$v_S = |S|^{-1} \int_S v \, ds.$$

Note that $v_S = 0$ if $S \subset \Gamma_D$, and define $I_F v \in \mathscr{S}_D^{1,cr}(\mathscr{T}_h)^d$ by

$$I_F v = \sum_{S \in \mathscr{S}_h} v_S \varphi_S.$$

On an element $T \in \mathscr{T}_h$ we have, using $\|\nabla \varphi_S\|_{L^\infty(T)} \le c h_T^{-1}$, that

$$\|\nabla I_F v\|_{L^2(T)} \le \sum_{S \in \mathscr{S}_h, S \subset \partial T} |v_S| \|\nabla \varphi_S\|_{L^2(T)} \le c \sum_{S \in \mathscr{S}_h, S \subset \partial T} |v_S| h_T^{-1} |T|^{1/2}.$$

The trace inequality

$$\|v\|_{L^2(S)} \le c\big(h_T^{-1/2} \|v\|_{L^2(T)} + h_T^{1/2} \|\nabla v\|_{L^2(T)}\big),$$

Hölder's inequality, and $|S| \sim h_T^{d-1}$ imply that

$$|v_S| \le |S|^{-1/2} \|v\|_{L^2(S)} \le c h_T^{-(d-1)/2} \big(h_T^{-1/2} \|v\|_{L^2(T)} + h_T^{1/2} \|\nabla v\|_{L^2(T)}\big)$$

$$\le c h_T^{-d/2} \big(\|v\|_{L^2(T)} + h_T \|\nabla v\|_{L^2(T)}\big).$$

Combining the last estimates and noting $|T| \sim h_T^d$ yield that

$$\|\nabla I_F v\|_{L^2(T)} \le c h_T^{-1} \big(\|v\|_{L^2(T)} + h_T \|\nabla v\|_{L^2(T)}\big).$$

We note that we may replace v by $v - \overline{v}_T$, where $\overline{v}_T = |T|^{-1} \int_T v \, dx$. With the local Poincaré inequality

$$\|v - \overline{v}_T\|_{L^2(T)} \le c h_T \|\nabla v\|_{L^2(T)},$$

we thus deduce that

$$\|\nabla I_F v\|_{L^2(T)} \le c \|\nabla v\|_{L^2(T)}.$$

Summation over $T \in \mathscr{T}_h$ proves the boundedness of the operator I_F.

(ii) It remains to show that

$$\int_\Omega q_h \operatorname{div}_{\mathscr{T}}(v - I_F v)\, dx = 0$$

for all $q_h \in \mathscr{L}^0(\mathscr{T}_h)$. An elementwise integration-by-parts proves that

$$\int_\Omega q_h \operatorname{div}_{\mathscr{T}}(v - I_F v)\, dx = \sum_{T \in \mathscr{T}_h} q_h|_T \int_T \operatorname{div}(v - I_F v)\, dx$$

$$= \sum_{T \in \mathscr{T}_h} q_h|_T \int_{\partial T} (v - I_F v) \cdot n\, ds.$$

The midpoint rule is exact for affine functions, and this shows that for $R, S \in \mathscr{S}_h$, we have

$$\int_S \varphi_R|_T\, ds = |S|\delta_{RS},$$

where the trace of φ_R is taken from the element T. Therefore by definition of the coefficients $(v_S)_{S \in \mathscr{S}_h}$ we have that

$$\int_S (v - I_F v)|_T\, ds = \int_S v\, ds - \sum_{R \in \mathscr{S}_h} v_R \int_S \varphi_R|_T\, ds = 0.$$

This implies that

$$\int_\Omega q_h \operatorname{div}_{\mathscr{T}}(v - I_F v)\, dx = 0$$

which finishes the proof. □

Remark 7.12 Due to the nonconformity of the method, the Fortin criterion and the generalized Céa lemma are not directly applicable. They can be modified and lead to the well-posedness of the discrete problem and the error estimate

$$\|\nabla_{\mathscr{T}}(u - u_h)\|_{L^2(\Omega)} + \|p - p_h\|_{L^2(\Omega)} = \mathcal{O}(h),$$

provided that we have the regularity property $u \in H^2(\Omega; \mathbb{R}^d)$ and $p \in H^1(\Omega)$.

7.2.6 Implementation

To describe a MATLAB implementation of the stable nonconforming finite element method, we first discuss realization of the Crouzeix–Raviart finite element approximation of the Poisson problem, i.e., the implementation of the problem of determining $u_h \in \mathscr{S}_D^{1,cr}(\mathscr{T}_h)$ such that

$$\int_\Omega \nabla_\mathscr{T} u_h \cdot \nabla_\mathscr{T} v_h \, \mathrm{d}x = \int_\Omega f v_h \, \mathrm{d}x + \int_{\Gamma_N} g v_h \, \mathrm{d}s - \int_\Omega \nabla_\mathscr{T} u_{D,h} \cdot \nabla_\mathscr{T} v_h \, \mathrm{d}x$$

for all $v_h \in \mathscr{S}_D^{1,cr}(\mathscr{T}_h)$. In this formulation, Dirichlet boundary conditions have been included on the right-hand side. The array n4s defines the sides of elements, with a local ordering such that the j-th side S_j of T is opposite the j-th vertex z_j of T for $j = 0, 1, \ldots, d$. This allows us to make use of the formula

$$\nabla_\mathscr{T} \varphi_{S_j} = -d \nabla \varphi_{z_j},$$

which is valid on T. The remaining parts of the code displayed in Fig. 7.15 are similar to the implementation of the P_1-finite element method, where the role of the vertices of elements is taken in this case by the sides of the elements. A visualization of the solution is provided by the routine show_cr.m shown in Fig. 7.17. Examples of numerical solutions of the Poisson problem are plotted in Fig. 7.13.

Figure 7.14 shows a numerical solution for the Stokes problem computed with the nonconforming method. The underlying implementation requires defining a basis for Crouzeix–Raviart vector fields. With the canonical basis vectors

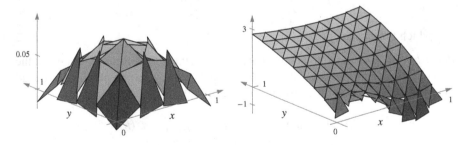

Fig. 7.13 Crouzeix–Raviart finite element approximations of two-dimensional Poisson problems

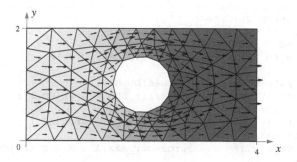

Fig. 7.14 Nonconforming finite element approximation of a Stokes problem

$e_1, e_2, \ldots, e_d \in \mathbb{R}^d$, we use that

$$\mathcal{S}^{1,cr}(\mathcal{T}_h)^d = \operatorname{span} \{\varphi_S e_j : S \in \mathcal{S}_h, j = 1, 2, \ldots, d\}.$$

For the scalar pressure variable, we use that

$$\mathcal{L}^0(\mathcal{T}_h) = \operatorname{span} \{\chi_T : T \in \mathcal{T}_h\}.$$

The matrix of the resulting discrete saddle-point problem has the structure

$$\begin{bmatrix} A & B^\mathsf{T} \\ B & 0 \end{bmatrix}.$$

The matrix A is a vectorial version of the stiffness matrix for approximating the Poisson problem with the Crouzeix–Raviart element. Its entries are

$$A_{(S,p),(R,q)} = a(\varphi_S e_p, \varphi_R e_q) = \mu \, e_p \cdot e_q \int_\Omega \nabla_{\mathcal{T}} \varphi_S \cdot \nabla_{\mathcal{T}} \varphi_R \, \mathrm{d}x.$$

for sides $R, S \in \mathcal{S}_h$ and $p, q = 1, 2, \ldots, d$, cf. Figs. 7.15 and 7.16. Note that we only obtain a contribution for $p = q$. The matrix B has the entries

$$B_{(S,p),T} = b(\varphi_S e_p, \chi_T) = -\int_\Omega \chi_T \operatorname{div}_{\mathcal{T}} (\varphi_S e_p) \, \mathrm{d}x = -|T| e_p \cdot \nabla_{\mathcal{T}} \varphi_S|_T.$$

Computing the right-hand side is realized with appropriate midpoint rules. We restricted parts of the implementation shown in Fig. 7.16 to the case $d = 2$, and we assume that $\Gamma_N \neq \emptyset$. Routines for visualizing numerical solutions are shown in Fig. 7.17.

```
function poisson_cr(d,red)
[c4n,n4e,Db,Nb] = triang_cube(d);
for j = 1:red
    [c4n,n4e,Db,Nb] = red_refine(c4n,n4e,Db,Nb);
end
[s4e,~,n4s,s4Db,s4Nb] = sides(n4e,Db,Nb);
nS = size(n4s,1); nE = size(n4e,1); nNb = size(Nb,1);
fSides = setdiff(1:nS,s4Db);
u = zeros(nS,1); tu_D = zeros(nS,1); b = zeros(nS,1);
ctr = 0; ctr_max = (d+1)^2*nE;
I = zeros(ctr_max,1); J = zeros(ctr_max,1); X = zeros(ctr_max,1);
for j = 1:nE
    X_T = [ones(1,d+1);c4n(n4e(j,:),:)'];
    grads_T = X_T\[zeros(1,d);eye(d)];
    vol_T = det(X_T)/factorial(d);
    mp_T = sum(c4n(n4e(j,:),:),1)/(d+1);
    for m = 1:d+1
        b(s4e(j,m)) = b(s4e(j,m))+(1/(d+1))*vol_T*f(mp_T);
        for n = 1:d+1
            ctr = ctr+1; I(ctr) = s4e(j,m); J(ctr) = s4e(j,n);
            X(ctr) = d^2*vol_T*grads_T(m,:)*grads_T(n,:)';
        end
    end
end
s = sparse(I,J,X,nS,nS);
for j = 1:nNb
    if d == 2
        vol_S = norm(c4n(Nb(j,1),:)-c4n(Nb(j,2),:));
    elseif d == 3
        vol_S = norm(cross(c4n(Nb(j,3),:)-c4n(Nb(j,1),:),...
            c4n(Nb(j,2),:)-c4n(Nb(j,1),:)),2)/2;
    end
    mp_S = sum(c4n(Nb(j,:),:),1)/d;
    for k = 1:d
        b(s4Nb(j)) = b(s4Nb(j))+(1/d)*vol_S*g(mp_S);
    end
end
for j = 1:nS
    mp_S = sum(c4n(n4s(j,:),:),1)/d;
    tu_D(j) = u_D(mp_S);
end
b = b-s*tu_D; u(fSides) = s(fSides,fSides)\b(fSides); u = u+tu_D;
show_cr(c4n,n4e,s4e,u)

function val = f(x); val = 1;
function val = g(x); val = 1;
function val = u_D(x); val = sin(2*pi*x(:,1));
```

Fig. 7.15 MATLAB implementation of the Crouzeix–Raviart finite element method for approximating the Poisson problem

```
function stokes_cr
d = 2; mu = 1;
load triang_cyl_w_hole_2d;
idx = (min([c4n(Db(:,1),1),c4n(Db(:,2),1)],[],2)>=2.0);
Nb = Db(idx,:); Db(idx,:) = [];
[s4e,~,n4s,s4Db,s4Nb] = sides(n4e,Db,Nb);
nS = size(n4s,1); nE = size(n4e,1); nNb = size(Nb,1);
fNodes = setdiff(1:2*nS+nE, [d*(s4Db-1)+1;d*(s4Db-1)+2]);
u = zeros(d*nS,1); tu_D = u; mp_T = zeros(nE,d);
x = zeros(d*nS+nE,1); tx = x; b = x;
ctr_A = 0; ctr_A_max = (d+1)^2*nE;
X = zeros(ctr_A_max,1); I = X; J = X;
ctr_B = 0; ctr_B_max = (d+1)*nE;
Y = zeros(ctr_B_max,1); K = Y; L = Y;
for j = 1:nE
    X_T = [ones(1,d+1);c4n(n4e(j,:),:)'];
    grads_T = X_T\[zeros(1,d);eye(d)];
    vol_T = det(X_T)/factorial(d);
    mp_T(j,:) = sum(c4n(n4e(j,:),:),1)/(d+1);
    for m = 1:d+1
      b(d*(s4e(j,m)-1)+(1:d)) = b(d*s4e(j,m)-1)...
          +(1/(d+1))*vol_T*f(mp_T(j,:));
      for n = 1:d+1
        ctr_A = ctr_A+1; I(ctr_A) = s4e(j,m); J(ctr_A) = s4e(j,n);
        X(ctr_A) = mu*d^2*vol_T*grads_T(m,:)*grads_T(n,:)';
      end
      K(ctr_B+(1:d)) = j; L(ctr_B+(1:d)) = d*(s4e(j,m)-1)+(1:d);
      Y(ctr_B+(1:d)) = d*vol_T*grads_T(m,:);
      ctr_B = ctr_B+d;
    end
end
A = sparse([2*I-1;2*I],[2*J-1;2*J],[X;X],d*nS,d*nS);      % d=2
B = sparse(K,L,Y,nE,d*nS);
for j = 1:nNb
    vol_S = norm(c4n(Nb(j,1),:)-c4n(Nb(j,2),:));            % d=2
    mp_S = sum(c4n(n4s(j,:),:),1)/d;
    b(d*(s4Nb(j)-1)+(1:d)) = b(d*(s4Nb(j)-1)+(1:d))+vol_S*g(mp_S);
end
for j = 1:nS
    mp_S = sum(c4n(n4s(j,:),:),1)/d;
    tu_D(d*(j-1)+(1:d)) = u_D(mp_S);
end
tx(1:d*nS) = tu_D;
G = [A,B';B,sparse(nE,nE)]; b = b-G*tx;
x(fNodes) = G(fNodes,fNodes)\b(fNodes);
u = x(1:d*nS); p = x(d*nS+(1:nE));
show_stokes_cr(c4n,n4e,s4e,u,p,mp_T)

function val = f(x); d = size(x,2); val = zeros(d,1);
function val = g(x); d = size(x,2); val = zeros(d,1);
function val = u_D(x); d = size(x,2); val = zeros(d,1);
if x(1)<=-2; val(1) = 1; end
```

Fig. 7.16 MATLAB implementation of the nonconforming finite element method for the Stokes system

```
function show_cr(c4n,n4e,s4e,u)
nE = size(n4e,1); d = size(c4n,2);
Signum = ones(d+1)-d*eye(d+1);
if d == 2
    E = reshape(1:3*nE,3,nE)'; n4e_t = n4e';
    X = c4n(n4e_t(:),1); Y = c4n(n4e_t(:),2); Z = Signum*u(s4e)';
    trisurf(E,X,Y,Z(:));
else
    tetramesh(n4e,c4n,sum(u(s4e),2)/(d+1));
end
```

```
function show_stokes_cr(c4n,n4e,s4e,u,p,mp_T)
[nC,d] = size(c4n);
u1_T = sum(u(d*(s4e-1)+1),2)/(d+1);
u2_T = sum(u(d*(s4e-1)+2),2)/(d+1);
if d == 2
    trisurf(n4e,c4n(:,1),c4n(:,2),zeros(nC,1)',p);
    hold on; view(0,90);
    quiver(mp_T(:,1),mp_T(:,2),u1_T,u2_T,'k');
    shading flat; hold off;
else
    tetramesh(n4e,c4n,p); hold on;
    u3_T = sum(u(d*s4e-1)+3,2)/(d+1);
    quiver3(mp_T(:,1),mp_T(:,2),mp_T(:,3),u1_T,u2_T,u3_T,'k');
    shading flat; hold off;
end
colorbar;
```

Fig. 7.17 Visualization of a Crouzeix–Raviart finite element function (*top*); routine to display the nonconforming finite element solution of a Stokes problem (*bottom*)

7.3   Convection-Dominated Problems

7.3.1   Boundary Layers and Subcharacteristics

The convection of a substance by the flow of another substance leads to first-order terms in partial differential equations. To understand their effect, we consider the model boundary value problem that determines $u : \Omega \to \mathbb{R}$ as the solution of

$$-\varepsilon \Delta u + b \cdot \nabla u = f \text{ in } \Omega, \quad u|_{\Gamma_\mathrm{D}} = u_\mathrm{D}, \quad \partial_n u|_{\Gamma_\mathrm{N}} = g$$

for a given vector field $b : \Omega \to \mathbb{R}^d$, which may be the solution of the Stokes problem. We follow [12] and consider the case that $0 < \varepsilon \ll 1$.

Fig. 7.18 Solutions of a
convection-diffusion equation
with $\varepsilon = 1, 1/10, 1/100$
(*from left to right*)

Example 7.6 If $\Omega = (0, 1)$, $\Gamma_D = \{0, 1\}$, and $b = 1$, then the unique solution $u \in C^2([0, 1])$ of

$$-\varepsilon u'' + u' = 0 \text{ in } \Omega, \quad u(0) = 0, u(1) = 1,$$

is for every $\varepsilon > 0$ given by

$$u(x) = \frac{1 - e^{x/\varepsilon}}{1 - e^{1/\varepsilon}}.$$

The solution is sketched for $\varepsilon = 1, 1/10, 1/100$ in Fig. 7.18.

As $\varepsilon \to 0$, the solutions do not converge uniformly. Instead, they develop a *boundary layer*, i.e., large gradients occur in a small neighborhood of the boundary $x = 1$. This behavior can be explained by considering the formal limit problem for $\varepsilon = 0$.

Example 7.7 The functions $u(x) = 0$ and $u(x) = 1$ for $x \in [0, 1]$ satisfy $u' = 0$ but only one of the boundary conditions $u(0) = 0$ and $u(1) = 1$.

The formal limit problem of the multidimensional problem with $f = 0$ leads to the equation

$$b \cdot \nabla u = b_1 \partial_{x_1} u + \cdots + b_d \partial_{x_d} u = 0 \text{ in } \Omega,$$

which has similarities with the transport equation $\partial_t u + a \partial_x u = 0$. Their solutions are determined via characteristics by the boundary conditions that can only be imposed at $t = 0$ and at $x = 0$ if $a > 0$ or at $x = 1$ if $a < 0$.

Definition 7.11

(i) We say that the convection-diffusion equation

$$-\varepsilon \Delta u + b \cdot \nabla u = f$$

is *convection-dominated* if the *Péclet number*

$$Pe = \varepsilon^{-1} \|b\|_{L^\infty(\Omega)} \operatorname{diam}(\Omega)$$

is significantly larger than 1, i.e., if $Pe \gg 1$.

(ii) A *subcharacteristic* of a convection-dominated equation is a curve $\phi : (\alpha, \beta) \to \Omega$ with

$$\phi'(s) = b(\phi(s))$$

for all $s \in (\alpha, \beta)$.

In the limit $\varepsilon \to 0$, information is transported along subcharacteristics, which are the integral curves of b.

Remarks 7.13

(i) If $u : \Omega \to \mathbb{R}$ solves the formal limit problem $b \cdot \nabla u = 0$ and $\phi : (\alpha, \beta) \to \mathbb{R}$ is a subcharacteristic, then

$$\frac{d}{ds} u \circ \phi(s) = \nabla u(\phi(s)) \cdot \phi'(s) = b(\phi(s)) \cdot \nabla u(\phi(s)) = 0,$$

i.e., u is constant along subcharacteristics.
(ii) The convection-diffusion equation is well-posed but the conditioning of the problem depends critically on $\varepsilon > 0$. The problem is therefore called a *singularly perturbed* problem.

With the subcharacteristics we can construct a solution of the formal limit problem. To obtain a meaningful solution, we partition the boundary of Ω, cf. Fig. 7.19.

Definition 7.12 Given $b : \Omega \to \mathbb{R}^d$, the *inflow*, *outflow*, and *parabolic boundary* of Ω is defined by

$$\Gamma_{\text{in}} = \{x \in \partial\Omega : b \cdot n < 0\},$$

$$\Gamma_{\text{out}} = \{x \in \partial\Omega : b \cdot n > 0\},$$

$$\Gamma_{\text{par}} = \{x \in \partial\Omega : b \cdot n = 0\},$$

respectively, where n denotes the outer unit normal on $\partial\Omega$.

Fig. 7.19 Inflow, outflow, and parabolic boundaries of Ω; subcharacteristics are integral curves of the vector field b

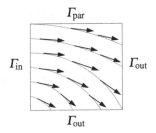

Along the subcharacteristics that intersect Γ_{in}, a solution u of $b \cdot \nabla u = 0$ is determined by the boundary data on Γ_{in}. In particular, discontinuities are propagated along subcharacteristics. The inflow boundary also appears when establishing the existence of solutions for the convection dominated problem with $\varepsilon > 0$ within the Lax–Milgram framework. For simplicity we restrict to homogeneous Dirichlet boundary conditions.

Proposition 7.8 (Existence and Uniqueness) *If $b \in H(\mathrm{div}; \Omega) \cap L^\infty(\Omega; \mathbb{R}^d)$ satisfies*

$$-\frac{1}{2} \mathrm{div}\, b \geq 0 \text{ in } \Omega, \quad b \cdot n \geq 0 \text{ on } \Gamma_{N},$$

i.e., if $\Gamma_N \subset \Gamma_{out} \cup \Gamma_{par}$, then the bilinear form

$$a_\varepsilon(u, v) = \varepsilon \int_\Omega \nabla u \cdot \nabla v \, dx + \int_\Omega b \cdot (\nabla u) v \, dx$$

is bounded and coercive on $H_D^1(\Omega)$. In particular, for every $\ell \in H_D^1(\Omega)'$, there exists a unique solution $u \in H_D^1(\Omega)$ with

$$a_\varepsilon(u, v) = \ell(v)$$

for all $v \in H_D^1(\Omega)$.

Proof Exercise.   □

We can approximate the solution by a Galerkin method and control the error with a refinement of Céa's lemma.

Proposition 7.9 (Galerkin Approximation) *Assume that the conditions of Proposition 7.8 are satisfied, and let $u_h \in \mathscr{S}_D^1(\mathscr{T}_h)$ be the Galerkin approximation of the convection-dominated problem. If $u \in H^2(\Omega) \cap H_D^1(\Omega)$, we have*

$$\|\nabla(u - u_h)\|_{L^2(\Omega)} \leq \big(1 + \varepsilon^{-1} h \|b\|_{L^\infty(\Omega)}\big) h \|D^2 u\|_{L^2(\Omega)}.$$

Proof By coercivity of a with constant ε and Galerkin orthogonality, we have

$$\varepsilon \|\nabla(u - u_h)\|_{L^2(\Omega)}^2 \leq a_\varepsilon(u - u_h, u - v_h)$$

$$= \varepsilon \int_\Omega \nabla(u - u_h) \cdot \nabla(u - v_h) \, dx + \int_\Omega b \cdot \nabla(u - u_h)(u - v_h) \, dx$$

$$\leq \varepsilon \|\nabla(u - u_h)\|_{L^2(\Omega)} \big(\|\nabla(u - v_h)\|_{L^2(\Omega)} + \varepsilon^{-1} \|b\|_{L^\infty(\Omega)} \|u - v_h\|_{L^2(\Omega)}\big).$$

Choosing $v_h = \mathscr{I}_h u$ as the nodal interpolant of u implies the estimate.   □

Remark 7.14 Note that $\|D^2u\|_{L^2(\Omega)}$ also depends on ε, i.e., we typically have that $\|D^2u\|_{L^2(\Omega)} \le c\varepsilon^{-2}$.

The upper bound of the error estimate is optimal in the sense of the following example.

Example 7.8 A finite difference discretization of the one-dimensional convection-diffusion equation

$$-\varepsilon u'' + u' = 0 \text{ in } (0, 1), \quad u(0) = 0, u(1) = 1,$$

leads with grid points $x_i = i/M$, $i = 0, 1, \dots, M$, to the scheme

$$-\varepsilon \frac{U_{i-1} - 2U_i + U_{i+1}}{h^2} + \frac{U_{i+1} - U_{i-1}}{2h} = 0, \quad U_0 = 0, \quad U_M = 1,$$

for $i = 1, 2, \dots, M - 1$ with $h = 1/M$. Interpreting the scheme as a three-term recursion shows that

$$U_i = \left(1 - \lambda^i\right)/\left(1 - \lambda^M\right).$$

for $i = 0, 1, \dots, M$, and with $\lambda = (2\varepsilon + h)/(2\varepsilon - h)$. For $h > 2\varepsilon$, the numerical solution oscillates rapidly. If $h = \varepsilon$, then at $x_{M-1} = (M - 1)h$ we have

$$u(x_{M-1}) = \frac{1 - e^{(M-1)h/\varepsilon}}{1 - e^{1/\varepsilon}} \to \frac{1}{e}, \quad U_{M-1} = \frac{1 - \lambda^{M-1}}{1 - \lambda^M} \to \frac{1}{\lambda} = \frac{1}{3},$$

as $M \to \infty$.

The example does not contradict convergence of the numerical method as $h \to 0$ when ε is fixed. In the calculation, we set $\varepsilon = h$ to obtain an understanding of the behavior when $h \sim \varepsilon$ and $0 < \varepsilon \ll 1$.

Remark 7.15 The CFL condition motivates discretizing the convection term by an upwinding method, i.e.,

$$b(x_i)u'(x_i) \approx \begin{cases} b(x_i)(U_i - U_{i-1})/h & \text{if } b(x_i) \ge 0, \\ b(x_i)(U_{i+1} - U_i)/h & \text{if } b(x_i) < 0. \end{cases}$$

In the above example this implies a discrete maximum principle. The generalization to multidimensional settings is difficult.

7.3.2 Streamline Diffusion Method

We assume that $b \in H_N(\mathrm{div}; \Omega) \cap L^\infty(\Omega; \mathbb{R}^d)$ with $\mathrm{div}\, b = 0$, and consider the boundary value problem

$$-\varepsilon \Delta u + b \cdot \nabla u = f \text{ in } \Omega, \quad u|_{\Gamma_D} = 0, \quad \partial_n u|_{\Gamma_N} = g.$$

To improve the suboptimal performance of standard Galerkin methods, the *streamline-diffusion method* adds the discrete element residual

$$r|_T = \left(-\varepsilon \Delta u_h + b \cdot \nabla u_h - f \right)|_T$$

weighted by a parameter δ_T, and tested with $b \cdot \nabla v_h$ on the weak formulation. We thus consider the bilinear form

$$a_{\mathrm{sd}}(u_h, v_h) = a_\varepsilon(u_h, v_h) + \sum_{T \in \mathcal{T}_h} \delta_T \int_T (-\varepsilon \Delta u_h + b \cdot \nabla u_h)(b \cdot \nabla v_h)\, dx$$

and the right-hand side functional

$$\ell_{\mathrm{sd}}(v_h) = \int_\Omega f v_h\, dx + \varepsilon \int_{\Gamma_N} g v_h\, ds + \sum_{T \in \mathcal{T}_h} \delta_T \int_T f(b \cdot \nabla v_h)\, dx.$$

Noting $\Delta u_h|_T = 0$ for $u_h \in \mathscr{S}_D^1(\mathcal{T}_h)$ and $T \in \mathcal{T}_h$, the bilinear form a_{sd} may be regarded as a discretization of the differential operator

$$-\mathrm{div}\left([\varepsilon I + \delta b^\top b] \nabla u \right) + b \cdot \nabla u,$$

i.e., diffusion in the direction of b is added. The formulation is *consistent* in the sense that the exact solution $u \in H_D^1(\Omega)$ satisfies

$$a_{\mathrm{sd}}(u, v_h) = \ell_{\mathrm{sd}}(v_h)$$

for all $v_h \in \mathscr{S}_D^1(\mathcal{T}_h)$. This implies a Galerkin orthogonality.

Lemma 7.5 (Streamline Diffusion Approximation) *There exists a unique function $u_h \in \mathscr{S}_D^1(\mathcal{T}_h)$ satisfying*

$$a_{\mathrm{sd}}(u_h, v_h) = \ell_{\mathrm{sd}}(v_h)$$

for all $v_h \in \mathscr{S}_D^1(\mathscr{T}_h)$. With the exact weak solution $u \in H_D^1(\Omega)$, we have

$$a_{\mathrm{sd}}(u - u_h, v_h) = 0$$

for all $v_h \in \mathscr{S}_D^1(\mathscr{T}_h)$.

Proof The result is an immediate consequence of the Lax–Milgram lemma and the Galerkin orthogonality.

We assume for simplicity that $\delta_T = \delta$ for all $T \in \mathscr{T}_h$, and introduce the streamline diffusion norm

$$\|v\|_{\mathrm{sd}}^2 = \varepsilon \|\nabla v\|_{L^2(\Omega)}^2 + \delta \|b \cdot \nabla v\|_{L^2(\Omega)}^2.$$

For all $v_h \in \mathscr{S}_D^1(\mathscr{T}_h)$, we have that

$$a_{\mathrm{sd}}(v_h, v_h) \geq \|v_h\|_{\mathrm{sd}}^2,$$

where due to the assumptions $\operatorname{div} b = 0$ and $\Gamma_N = \emptyset$ in fact equality holds. We follow [12] and consider discretizations with $h \geq \varepsilon$.

Theorem 7.5 (Streamline Diffusion Error) *Assume $\Gamma_D = \partial\Omega$ and let $u_h \in \mathscr{S}_0^1(\mathscr{T}_h)$ satisfy*

$$a_{\mathrm{sd}}(u_h, v_h) = \ell_{\mathrm{sd}}(v_h)$$

for all $v_h \in \mathscr{S}_0^1(\mathscr{T}_h)$. If $u \in H^2(\Omega) \cap H_0^1(\Omega)$, and if $h \geq \varepsilon$ and $\delta = h$, then we have

$$\|u - u_h\|_{\mathrm{sd}} \leq c_{\mathrm{sd}} h^{3/2} \|D^2 u\|_{L^2(\Omega)}.$$

Proof With the nodal interpolant $\mathscr{I}_h u$ of u, we decompose the approximation error by setting

$$u - u_h = (u - \mathscr{I}_h u) + (\mathscr{I}_h u - u_h) = e_I + d_h,$$

where e_I is the interpolation error and $d_h \in \mathscr{S}_0^1(\mathscr{T}_h)$. By Galerkin orthogonality we have

$$\|d_h\|_{\mathrm{sd}}^2 \leq a_{\mathrm{sd}}(\mathscr{I}_h u - u_h, d_h) = a_{\mathrm{sd}}(\mathscr{I}_h u - u, d_h) = a_{\mathrm{sd}}(e_I, d_h).$$

Using $\Delta u_h|_T = 0$, it follows that

$$\|d_h\|_{sd}^2 \leq \varepsilon \int_\Omega \nabla e_I \cdot \nabla d_h \, dx + \int_\Omega (b \cdot \nabla e_I) d_h \, dx + \delta \int_\Omega (b \cdot \nabla e_I)(b \cdot \nabla d_h) \, dx$$

$$+ \sum_{T \in \mathcal{T}_h} \delta \int_T (-\varepsilon \Delta u)(b \cdot \nabla d_h) \, dx$$

$$= I + II + III + IV.$$

For the first term on right-hand side, we use Hölder's inequality to verify that

$$I \leq \varepsilon \|\nabla e_I\|_{L^2(\Omega)} \|\nabla d_h\|_{L^2(\Omega)} \leq \varepsilon^{1/2} \|\nabla e_I\|_{L^2(\Omega)} \|d_h\|_{sd}.$$

Integration-by-parts with $\text{div } b = 0$ and $e_I = 0$ on $\partial\Omega$ proves that

$$II = -\int_\Omega e_I b \cdot \nabla d_h \, dx \leq \|e_I\|_{L^2(\Omega)} \|b \cdot \nabla d_h\|_{L^2(\Omega)} \leq \delta^{-1/2} \|e_I\|_{L^2(\Omega)} \|d_h\|_{sd}.$$

With Hölder's inequality we find that

$$III \leq \delta \|b \cdot \nabla e_i\|_{L^2(\Omega)} \|b \cdot \nabla d_h\|_{L^2(\Omega)} \leq \|e_I\|_{sd} \|d_h\|_{sd}.$$

We use Hölder inequalities and the Cauchy–Schwarz inequality to estimate

$$IV \leq \delta\varepsilon \left(\sum_{T \in \mathcal{T}_h} \|\Delta u\|_{L^2(T)}^2 \right)^{1/2} \left(\sum_{T \in \mathcal{T}_h} \|b \cdot \nabla d_h\|_{L^2(T)}^2 \right)^{1/2}$$

$$\leq \delta^{1/2}\varepsilon \|\Delta u\|_{L^2(\Omega)} \|d_h\|_{sd}.$$

Upon combining the previous estimates, we find that

$$\|d_h\|_{sd} \leq \varepsilon^{1/2} \|\nabla e_I\|_{L^2(\Omega)} + \delta^{-1/2} \|e_I\|_{L^2(\Omega)} + \|e_I\|_{sd} + \delta^{1/2}\varepsilon \|\Delta u\|_{L^2(\Omega)}.$$

With the triangle inequality we deduce that

$$\|u - u_h\|_{sd} \leq \|e_I\|_{sd} + \|d_h\|_{sd}$$

$$\leq 2\|e_I\|_{sd} + \varepsilon^{1/2} \|\nabla e_I\|_{L^2(\Omega)} + \delta^{-1/2} \|e_I\|_{L^2(\Omega)} + \delta^{1/2}\varepsilon \|\Delta u\|_{L^2(\Omega)}.$$

By nodal interpolation estimates, we have

$$h^{-1}\|e_I\|_{L^2(\Omega)} + \|\nabla e_I\|_{L^2(\Omega)} \le c_{\mathscr{I}} h \|D^2 u\|_{L^2(\Omega)}.$$

This implies that

$$\|e_I\|_{sd} \le c_{\mathscr{I}} h \big(\varepsilon + \delta\|b\|_{L^\infty(\Omega)}^2\big)^{1/2} \|D^2 u\|_{L^2(\Omega)}.$$

Using $\varepsilon \le h$, we thus have that

$$\|u - u_h\|_{sd} \le c\big(\varepsilon^{1/2} + \delta^{1/2} c_b + \varepsilon^{1/2} + \delta^{-1/2} h + \delta^{1/2}\big) h \|D^2 u\|_{L^2(\Omega)}.$$

Incorporating $\varepsilon \le h$ and $\delta = h$ proves the estimate.   □

Remarks 7.16

(i) The approximation error of the streamline diffusion method in H^1 is of the order $\mathcal{O}(\varepsilon^{-1/2} h^{3/2})$, whereas for the standard Galerkin method we obtain $\mathcal{O}(\varepsilon^{-1} h^2)$, i.e., we obtain a factor $(h/\varepsilon)^{1/2}$ instead of h/ε, which is an improvement in the regime $h \ge \varepsilon$.

(ii) To allow for locally refined triangulations, an elementwise definition of δ_T is used, e.g., for $T \in \mathscr{T}_h$, one sets

$$\delta_T = \max\{0, h_T(1 - Pe_T^{-1})\},$$

where $Pe_T = \|b\|_{L^\infty(T)} h_T / \varepsilon$ is the *element Péclet number*. Hence no stabilization is used if $Pe_T < 1$ and a stabilization with $\delta_T \sim h_T$ occurs otherwise.

Figure 7.20 displays a MATLAB realization of the streamline diffusion method. The stabilization parameter is chosen elementwise as $\delta_T = h_T = |T|^{1/d}$. Figure 7.21 compares numerical solutions obtained with and without the stabilizing term.

7.3.3 Finite Volume Method

If $\operatorname{div} b = 0$, we may rewrite the convection-diffusion equation as

$$\operatorname{div} q(u) = f \text{ in } \Omega, \quad q(u) = -\varepsilon\nabla u + bu,$$

```
function streamline_diffusion(red)
d = 2; eps = 1e-5;
[c4n,n4e,Nb,Db] = triang_cube(d);
for j = 1:red
    [c4n,n4e,Db,Nb] = red_refine(c4n,n4e,Db,Nb);
end
[nC,d] = size(c4n); nE = size(n4e,1);
dNodes = unique(Db); fNodes = setdiff(1:nC,dNodes);
u = zeros(nC,1); tu_D = zeros(nC,1);
ctr = 0; ctr_max = (d+1)^2*nE;
I = zeros(ctr_max,1); J = zeros(ctr_max,1); X = zeros(ctr_max,1);
for j = 1:nE
    X_T = [ones(1,d+1);c4n(n4e(j,:),:)'];
    grads_T = X_T\[zeros(1,d);eye(d)];
    vol_T = det(X_T)/factorial(d);
    mp_T = sum(c4n(n4e(j,:),:),1)/(d+1);
    b_T = b_field(mp_T);
    delta_T = vol_T^(1/d);
    for m = 1:d+1
        for n = 1:d+1
            ctr = ctr+1; I(ctr) = n4e(j,m); J(ctr) = n4e(j,n);
            X(ctr) = vol_T*(eps*grads_T(m,:)*grads_T(n,:)'...
                +delta_T*(b_T*grads_T(m,:)')*(b_T*grads_T(n,:)')...
                +b_T*grads_T(n,:)'/(d+1));
        end
    end
end
s = sparse(I,J,X,nC,nC);
for j = 1:nC
    tu_D(j) = u_D(c4n(j,:));
end
b = -s*tu_D; u(fNodes) = s(fNodes,fNodes)\b(fNodes); u = u+tu_D;
show_p1(c4n,n4e,Db,Nb,u);

function val = b_field(x)
[phi,~] = cart2pol(x(1),x(2));
val = [sin(phi),-cos(phi)];

function val = u_D(x)
val = 0;
if (x(1)==0 && x(2)<=1/2);
    val = 1;
end
```

Fig. 7.20 MATLAB implementation of the streamline diffusion method with elementwise defined stabilization parameter $\delta_T = h_T = |T|^{1/d}$

where $q(u)$ is called a *flux*. We follow [12], restrict to the case $d = 2$, and let $(\Omega_j)_{j=1,\dots,J}$ be a nonoverlapping partition of Ω into polyhedral sets Ω_j called *control volumes*, cf. Fig. 7.22. We then require the flux equation to be satisfied in average

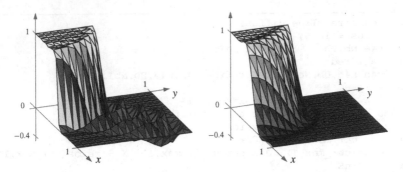

Fig. 7.21 Finite element approximations of a convection dominated boundary value problem without (*left*) and with (*right*) a stabilizing streamline diffusion term

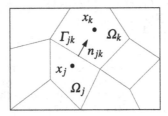

Fig. 7.22 Control volumes $\Omega_j, j = 1, 2, \ldots, J$, and interfaces $\Gamma_{jk}, j, k = 1, 2, \ldots, J$

on every set Ω_j, i.e., the *local conservation property*

$$\int_{\Omega_j} \operatorname{div} q \, dx = \int_{\Omega_j} f \, dx$$

for $j = 1, 2, \ldots, J$. Applying Gauss's theorem yields that

$$\int_{\Omega_j} f \, dx = \int_{\partial \Omega_j} q(u) \cdot n \, ds = \sum_{k=1}^{J} \int_{\Gamma_{jk}} n_{jk} \cdot q(u) \, ds,$$

where the interfaces $\Gamma_{jk} = \partial \Omega_j \cap \partial \Omega_k$, are assumed to be flat components of $\partial \Omega_j$ with outer unit normal n_{jk}. We set $\Gamma_{jj} = \emptyset$ for $j = 1, 2, \ldots, J$.

The term $\nabla u \cdot n_{jk}$ is the derivative of u in the direction of n_{jk} which can be approximated by an appropriate difference quotient. Assuming that we are given points $x_j \in \Omega_j$ for $j = 1, 2, \ldots, J$, such that their difference vectors are normal to the nontrivial interfaces Γ_{jk}, i.e.,

$$x_j - x_k \parallel n_{jk},$$

cf. Fig. 7.22, a first-order approximation is given by

$$\nabla u \cdot n_{jk}|_{\Gamma_{jk}} \approx \frac{u(x_k) - u(x_j)}{|x_k - x_j|}.$$

An essential question is whether it is possible to find appropriate points $(x_j)_{j=1,\dots,J}$. One strategy is to first fix nodes $(x_j)_{j=1,\dots,J} \subset \overline{\Omega}$, and then construct appropriate control volumes $\Omega_j, j = 1, 2, \dots, J$.

Definition 7.13 The *Voronoi polygons* $\Omega_j, j = 1, 2, \dots, J$, associated with disjoint points $(x_j)_{j=1,2,\dots,J} \in \overline{\Omega}$, are defined by

$$\Omega_j = \{x \in \Omega : |x - x_j| < |x - x_k| \text{ for all } k = 1, 2, \dots, J, \ k \neq j\}.$$

The *Voronoi diagram* $(\Omega_j)_{j=1,\dots,J}$ is called *regular* if at most three Voronoi polygons intersect at one point.

The Voronoi polygons are separated by hyperplanes Γ_{jk} with normals n_{jk} that are parallel to the vectors $x_j - x_k$.

Remark 7.17 It can be shown that at the corner of a Voronoi polygon at least three polygons intersect. If four or more polygons intersect at a corner, then all the corresponding points lie on a circle.

With the construction of the sets Ω_j, the *finite volume method* for the Poisson problem consists in solving the equations

$$-\sum_{k=1}^{J} |\Gamma_{jk}| \frac{U_k - U_j}{|x_k - x_j|} = |\Omega_j| f(x_j)$$

for $j = 1, 2, \dots, J$, subject to appropriate boundary conditions. Unique solvability can be shown if the Voronoi diagram is regular. An equivalent P_1-finite element method can be constructed.

Definition 7.14 The *dual* or *Delaunay triangulation* of a Voronoi diagram is obtained by connecting those points in $(x_j)_{j=1,\dots,J}$, whose Voronoi polygons are neighbors.

An example of a Delaunay triangulation is shown in Fig. 7.23.

Fig. 7.23 Voronoi polygons defined by points $(x_j)_{j=1,\dots,J}$ (*left*); Delaunay triangulation associated with the Voronoi diagram (*right*)

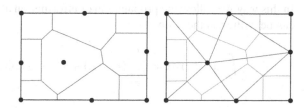

Remarks 7.18

(i) The Delaunay triangulation obtained from a regular Voronoi diagram has the property that the sum of angles opposite an inner edge is always bounded by π, i.e., the triangulation is *weakly acute*.

(ii) If $\mathscr{T}_h$ is a triangulation of Ω such that all inner angles of triangles are bounded by $\pi/2$, then a corresponding Voronoi diagram can be reconstructed by considering perpendicular edge bisectors.

The equivalence of the finite volume method on a Voronoi diagram for the Poisson problem to the finite element method on the corresponding Delaunay triangulation is a consequence of the following lemma.

Lemma 7.6 (Cotangens Formula) *Assume that the nodes z, y are the endpoints of an inner edge $S = T_1 \cap T_2$ in a Delaunay triangulation. For the associated nodal basis functions $\varphi_z, \varphi_y \in \mathscr{S}^1(\mathscr{T}_h)$, we have*

$$\int_\Omega \nabla\varphi_z \cdot \nabla\varphi_y \, dx = -\frac{|m_{T_1} - m_{T_2}|}{|z - y|} = -\frac{1}{2}(\cot\alpha_1 + \cot\alpha_2),$$

where m_{T_1}, m_{T_2} are the circumcenters of T_1, T_2, and α_1, α_2 are the inner angles of T_1, T_2 that are opposite S.

Proof Exercise.   □

Remark 7.19 The formula implies that the finite element method for the Poisson problem on a Delaunay triangulation satisfies a discrete maximum principle.

For a finite volume discretization of a convection term, we note that by Gauss's theorem we have

$$\int_{\Omega_j} \mathrm{div}(bu) \, dx = \int_{\partial\Omega_j} b \cdot nu \, ds \approx \sum_{k=1}^{J} |\Gamma_{jk}| b(x_{jk}) \cdot n_{jk}\hat{u}_{jk},$$

where x_{jk} is a point on Γ_{jk} and $\hat{u}_{jk}$ is an appropriately chosen approximation of $u(x_{jk})$. To achieve numerical stability, an upwinding is realized by

$$\hat{u}_{jk} = \begin{cases} u(x_j) & \text{if } b(x_{jk}) \cdot n_{jk} \geq 0, \\ u(x_k) & \text{if } b(x_{jk}) \cdot n_{jk} < 0. \end{cases}$$

In this way we follow the propagation of information along subcharacteristics. This can be written in the form

$$\int_{\Omega_j} \mathrm{div}(bu) \, dx = \sum_{k=1}^{J} \left(r_{jk}u(x_j) + (1 - r_{jk})u(x_k)\right)b(x_{jk}) \cdot n_{jk}|\Gamma_{jk}|,$$

where

$$r_{jk} = \frac{1}{2}\,\mathrm{sign}\,(Pe_{jk}) + \frac{1}{2}, \quad Pe_{jk} = \varepsilon^{-1}|\Gamma_{jk}|b(x_{jk}) \cdot n_{jk},$$

with the local discrete Péclet number Pe_{jk}. Note that for the treatment of the convective term, we do not assume that the vector $x_j - x_k$ is perpendicular to Γ_{jk}. Hence, more general partitions are possible.

Definition 7.15 Given a triangulation $\mathscr{T}_h$ of Ω into triangles, the *Donald diagram* associates with each node $z \in \mathscr{N}_h$ the set

$$\Omega_z = \{x \in \Omega : \varphi_z(x) > \varphi_y(x) \text{ for all } y \in \mathscr{N}_h \setminus \{z\}\}.$$

If $z \in T$ for $T \in \mathscr{T}_h$, then we have that $\Omega_z \cap T$ is the convex hull of the midpoint x_T of T, the midpoints of the edges of T that contain z, and the point z, cf. Fig. 7.24. Note that the boundary $\Gamma_{zy} = \Omega_z \cap \Omega_y$ is in general not flat, and hence the vector $z - y$ may not be normal to Γ_{zy}. We choose an enumeration $(z_1, z_2, \dots, z_J)$ of the nodes in $\mathscr{N}_h$, and let x_{jk} and n_{jk} be a point on Γ_{jk} and an approximate unit normal to Γ_{jk} at x_{jk}, respectively.

Definition 7.16 The *upwinding finite element method* for a convection-diffusion equation consists in finding $u_h \in \mathscr{S}_D^1(\mathscr{T}_h)$ such that

$$\varepsilon \int_{\Omega} \nabla u_h \cdot \nabla v_h \,dx + \sum_{j,k=1}^{J} |\Gamma_{jk}| v_h(z_k) \big[r_{jk} u_h(z_j) + (1 - r_{jk}) u_h(z_k)\big] b(x_{jk}) \cdot n_{jk}$$

$$= \int_{\Omega} f v_h \,dx + \int_{\Gamma_N} g v_h \,ds$$

for all $v_h \in \mathscr{S}_D^1(\mathscr{T}_h)$.

Remarks 7.20

(i) Advantages of the finite volume method are its flexibility with respect to geometry, its simple implementation, and dimension independence. Drawbacks are difficulties in its mathematical analysis and the development of higher-order versions.

Fig. 7.24 Two Donald cells in a triangulation

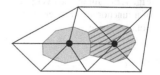

(ii) The finite volume method is of particular importance for hyperbolic conservation laws that do not involve a diffusion term, i.e., equations of the form $\partial_t u - \operatorname{div} f(u) = 0$ with a function $f : \mathbb{R} \to \mathbb{R}^n$.

7.4　Discontinuous Galerkin Methods

7.4.1　Jumps and Averages

For a triangulation $\mathcal{T}_h$ of a bounded Lipschitz domain $\Omega \subset \mathbb{R}^d$, $d = 2, 3$, we let $\mathcal{S}_h$ denote the set of sides of elements.

Definition 7.17 For $s \geq 0$ the *broken Sobolev space* $H^s(\mathcal{T}_h)$ subordinated to $\mathcal{T}_h$ is defined by

$$H^s(\mathcal{T}_h) = \{v \in L^2(\Omega) : v|_T \in H^s(T) \text{ for all } T \in \mathcal{T}_h\}.$$

The *elementwise gradient*, defined by the application of the weak gradient on every element, defines the vector field $\nabla_{\mathcal{T}} v \in L^2(\Omega; \mathbb{R}^d)$ via

$$\left(\nabla_{\mathcal{T}} v\right)|_T = \nabla\left(v|_T\right)$$

for all $T \in \mathcal{T}_h$.

Sobolev functions with noninteger order of differentiability can be defined via interpolation or Fourier transform. For the following results, the integer cases $s = 1$ and $s = 2$ are sufficient. Note that functions in $H^s(\mathcal{T}_h)$ may jump across the sides of elements, and that the inclusion $H^1(\Omega) \subset H^1(\mathcal{T}_h)$ is strict unless $\mathcal{T}_h$ consists of only one element.

To measure the discontinuity of functions in the broken Sobolev space, we define jumps and averages across sides, cf. Fig. 7.25.

Definition 7.18

(i) Let $s > 1/2$ and $v \in H^s(\mathcal{T}_h)$, and let $S \in \mathcal{S}_h$ be an interior side with $S = T_+ \cap T_-$ for $T_-, T_+ \in \mathcal{T}_h$, and unit normal n_S pointing from T_- into T_+. The

Fig. 7.25 Jump and average of a function in a broken Sobolev space

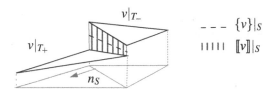

jump of v across S is for almost every $x \in S$ defined by

$$\llbracket v \rrbracket|_S(x) = v|_{T_+}(x) - v|_{T_-}(x).$$

The *average of v across S* is for almost every $x \in S$ defined by

$$\{v\}|_S(x) = \left(v|_{T_+}(x) + v|_{T_-}(x)\right)/2.$$

(ii) For $s > 1/2$, $v \in H^s(\mathcal{T}_h)$, and a boundary side $S \in \mathcal{S}_h \cap \partial\Omega$, we let $n_S = n|_S$, and define

$$\llbracket v \rrbracket|_S = -v|_S, \quad \{v\}|_S = v|_S.$$

Remarks 7.21

(i) The condition $s > 1/2$ guarantees that the trace operator is well defined as a mapping $\mathrm{Tr} : H^s(T) \to L^2(\partial T)$; in particular, we have that $\llbracket v \rrbracket, \{v\} \in L^2(\cup\mathcal{S}_h)$.
(ii) If $v \in H^s(\mathcal{T}_h)$ such that $v|_T \in C(T)$ for all $T \in \mathcal{T}_h$, then we have for every interior side $S \in \mathcal{S}_h$ and $x \in S$ that

$$\llbracket v \rrbracket|_S(x) = \lim_{\varepsilon \to 0} \left(v(x + \varepsilon n_S) - v(x - \varepsilon n_S)\right),$$
$$\{v\}|_S(x) = \lim_{\varepsilon \to 0} \left(v(x + \varepsilon n_S) + v(x - \varepsilon n_S)\right)/2.$$

For functions that are globally sufficiently regular, the jumps of a function and its gradient vanish.

Lemma 7.7

(i) If $v \in H^s(\Omega)$, $s > 1/2$, then we have for every interior side $S \in \mathcal{S}_h$ that

$$\llbracket v \rrbracket|_S = 0, \quad \{v\}|_S = v|_S.$$

(ii) If $v \in H^s(\mathcal{T}_h) \cap H^1(\Omega)$ with $s > 3/2$, and $\nabla v \in H(\mathrm{div}; \Omega)$, then for every interior side $S \in \mathcal{S}_h$ we have

$$\llbracket \nabla v \cdot n_S \rrbracket|_S = 0,$$

where n_S is extended constantly to a neighborhood of S.

Proof Exercise. □

7.4.2 DG Methods

Similar to the derivation of the finite volume method, we test a partial differential equation on subsets of its domain. In the case of the Poisson problem with homogeneous Dirichlet conditions, i.e.,

$$-\Delta u = f \text{ in } \Omega, \quad u = 0 \text{ on } \partial\Omega,$$

we multiply the equation by $v \in H^1(\mathcal{T}_h)$, integrate over $T \in \mathcal{T}_h$, and integrate-by-parts, i.e., we have

$$\int_T \nabla u \cdot \nabla v \, dx - \int_{\partial T} (\nabla u \cdot n_T) v \, ds = -\int_T (\Delta u) v \, dx = \int_T f v \, dx,$$

where n_T is the outer unit normal to ∂T. Summing over all elements leads to

$$\int_\Omega \nabla_{\mathcal{T}} u \cdot \nabla_{\mathcal{T}} v \, dx - \sum_{T \in \mathcal{T}_h} \int_{\partial T} (\nabla u \cdot n_T) v \, ds = \int_\Omega f v \, dx.$$

Since every interior side S occurs twice with opposite normals, we have, assuming that $u \in H^{3/2+\varepsilon}(\Omega)$, so that traces of ∇u are well defined,

$$\sum_{T \in \mathcal{T}_h} \int_{\partial T} (\nabla u \cdot n_T) v \, ds = \sum_{S \in \mathcal{S}_h} \int_S \left(\nabla u|_{T_-} n_S v|_{T_-} - \nabla u|_{T_+} n_S v|_{T_+} \right) ds$$

$$= -\sum_{S \in \mathcal{S}_h} \int_S [\![(\nabla u \cdot n_S) v]\!] \, ds,$$

where $T_+ = \emptyset$ if $S \subset \partial\Omega$. We use the identity

$$ab - cd = \frac{1}{2}(a+c)(b-d) + \frac{1}{2}(a-c)(b+d)$$

to deduce that

$$[\![(\nabla u \cdot n_S) v]\!]|_S = \{\nabla u \cdot n_S\}|_S [\![v]\!]|_S + [\![\nabla u \cdot n_S]\!]|_S \{v\}|_S.$$

Under the assumption that $u \in H^{3/2+\varepsilon}(\Omega)$, we have $[\![\nabla u \cdot n_S]\!]|_S = 0$ for all interior sides S, and hence

$$\int_\Omega \nabla_{\mathcal{T}} u \cdot \nabla_{\mathcal{T}} v \, dx + \sum_{S \in \mathcal{S}_h} \int_S \{\nabla u \cdot n_S\} [\![v]\!] \, ds = \int_\Omega f v \, dx$$

for all $v \in H^1(\mathcal{T}_h)$. To obtain either a symmetric- or the sum of a symmetric- and an antisymmetric bilinear form on the left-hand side, we note that $[\![u]\!]|_S = 0$ and for $\sigma \in \{-1, 0, 1\}$ add the vanishing integrals

$$\sigma \int_S \{\nabla v \cdot n_S\} [\![u]\!] \, ds$$

to the left-hand side. Coercivity on $H^s(\mathcal{T}_h)$ cannot be expected for the resulting bilinear form for any $s > 3/2$, and therefore we also add the bilinear form

$$J_{\beta,\gamma} : H^1(\mathcal{T}_h) \times H^1(\mathcal{T}_h) \to \mathbb{R}, \quad J_{\beta,\gamma}(v, w) = \sum_{S \in \mathscr{S}_h} \frac{\beta_S}{h_S^\gamma} \int_S [\![v]\!] [\![w]\!] \, ds,$$

where $h_S = \operatorname{diam}(S)$, $\beta_S > 0$, and $\gamma > 0$. Note that we have $J_{\beta,\gamma}(v, w) = 0$ if $w \in H^1(\Omega)$, and that $J_{\beta,\gamma}(v, v) = 0$ implies $v \in H^1(\Omega)$. We thus consider the bilinear form

$$a_{\mathrm{dG}}(u, v) = \int_\Omega \nabla_{\mathcal{T}} u \cdot \nabla_{\mathcal{T}} v \, dx + \sum_{S \in \mathscr{S}_h} \int_S \{\nabla u \cdot n_S\} [\![v]\!] \, ds$$

$$+ \sigma \sum_{S \in \mathscr{S}_h} \int_S \{\nabla v \cdot n_S\} [\![u]\!] \, ds + J_{\beta,\gamma}(u, v).$$

The construction of a_{dG} shows that if the solution $u \in H_0^1(\Omega)$ of the Poisson problem satisfies $u \in H^{3/2+\varepsilon}(\Omega)$, then we have

$$a_{\mathrm{dG}}(u, v) = \ell(v) = \int_\Omega f v \, dx$$

for all $v \in H^s(\mathcal{T}_h)$ with $s > 3/2$. Since $H^r(\Omega) \subset H^s(\mathcal{T}_h)$ for every $r \geq s$, we also have the converse implication.

Proposition 7.10 (Consistency) *The function $u \in H_0^1(\Omega) \cap H^s(\mathcal{T}_h)$ with $s > 3/2$ is a weak solution of the Poisson problem, if and only if*

$$a_{\mathrm{dG}}(u, v) = \ell(v)$$

for all $v \in H^s(\mathcal{T}_h)$ and $\nabla u \in H(\operatorname{div}; \Omega)$.

Proof Exercise. □

The discontinuous Galerkin method consists in discretizing the variational formulation with discontinuous, piecewise polynomial functions.

Definition 7.19 Let $k \in \mathbb{N}_0$ and set

$$\mathscr{S}^{k,\mathrm{dG}}(\mathscr{T}_h) = \{v_h \in H^1(\mathscr{T}_h) : v_h|_T \in \mathscr{P}_k(T) \text{ for all } T \in \mathscr{T}_h\}.$$

For parameters σ, β, γ, the *interior penalty discontinuous Galerkin method (IPdG)* seeks $u_h \in \mathscr{S}^{k,\mathrm{dG}}(\mathscr{T}_h)$ such that

$$a_{\mathrm{dG}}(u_h, v_h) = \ell(v_h)$$

for all $v_h \in \mathscr{S}^{k,\mathrm{dG}}(\mathscr{T}_h)$. For $\sigma = 1$, $\sigma = -1$, or $\sigma = 0$, the method is called *symmetric (SIPdG)*, *nonsymmetric (NIPdG)*, or *incomplete (IIPdG)*, respectively.

The terminology *interior penalty* corresponds to the occurrence of the bilinear form $J = J_{\beta,\gamma}$, i.e., it is always assumed that $\beta_S > 0$ for all $S \in \mathscr{S}_h$.

Remarks 7.22

(i) For the SIPdG-method with $\sigma = 1$, we have that every solution $u_h \in \mathscr{S}^{k,\mathrm{dG}}(\mathscr{T}_h)$ minimizes the functional

$$u_h \mapsto \frac{1}{2} a_{\mathrm{dG}}(u_h, u_h) - \ell(u_h).$$

In this case, the quadratic term $J_{\beta,\gamma}(u_h, u_h)$ penalizes jumps of u_h.

(ii) The bilinear form a_{dG} for the NIPdG-method with $\sigma = -1$ is unconditionally coercive, i.e., for all $u_h \in \mathscr{S}^{k,\mathrm{dG}}(\mathscr{T}_h)$, we have

$$a_{\mathrm{dG}}(u_h, u_h) = \int_\Omega |\nabla_{\mathscr{T}} u_h|^2 \, dx + \sum_{S \in \mathscr{S}_h} \frac{\beta_S}{h_S^\gamma} \int_S |\llbracket u_h \rrbracket|^2 \, ds.$$

In particular, the NIPdG-method admits a unique solution if $\beta_S > 0$ for all $S \in \mathscr{S}_h$.

(iii) Note that Dirichlet conditions on u_h are only imposed implicitly.

7.4.3 Well-Posedness

The previous estimates motivate defining a *dG-norm* by

$$\|v\|_{\mathrm{dG}}^2 = \|\nabla_{\mathscr{T}} v\|_{L^2(\Omega)}^2 + \sum_{S \in \mathscr{S}_h} \frac{\beta_S}{h_S^\gamma} \int_S |\llbracket v \rrbracket|^2 \, ds,$$

which is a norm provided that $\beta_S > 0$ for all $S \in \mathscr{S}_h$. To identify sufficient conditions for coercivity for a_{dG}, we note that for $\sigma \in \{1, 0\}$, we have

$$a_{dG}(v, v) = \int_\Omega |\nabla_{\mathscr{T}} v|^2 \, dx + (\sigma + 1) \sum_{S \in \mathscr{S}_h} \int_S \{\nabla v \cdot n_S\} [\![v]\!] \, ds + \sum_{S \in \mathscr{S}_h} \frac{\beta_S}{h_S^\gamma} \int_S [\![v]\!]^2 \, ds.$$

We thus have to bound the middle term on the right-hand side from below by the other two terms.

Lemma 7.8 (Jump-Average Bound) *For every $v_h \in \mathscr{S}^{1,dG}(\mathscr{T}_h)$ and $S \in \mathscr{S}_h$ with $S = T_+ \cap T_-$ for $T_+, T_- \in \mathscr{T}_h$, or $S \subset T_- \cap \partial\Omega$ and $T_+ = \emptyset$, we have*

$$\int_S \{\nabla v_h \cdot n_S\} [\![v_h]\!] \, ds \leq c_1^2 \frac{h_S^{\gamma-1}}{\beta_S} \left(\|\nabla v_h\|_{L^2(T_+)} + \|\nabla v_h\|_{L^2(T_-)} \right)^2 + \frac{\beta_S}{4h_S^\gamma} \|[\![v_h]\!]\|_{L^2(S)}^2.$$

Proof Let $v_h \in \mathscr{S}^{1,dG}(\mathscr{T}_h)$ and $S = T_+ \cap T_-$. With Hölder's inequality and the estimate $|S|^{1/2} \leq c_{usr} h_S^{-1/2} |T|^{1/2}$, we find that

$$\int_S \{\nabla v_h \cdot n_S\} [\![v_h]\!] \, ds \leq \|\{\nabla v_h \cdot n_S\}\|_{L^2(S)} \|[\![v_h]\!]\|_{L^2(S)}$$

$$\leq \left(\|\nabla v_h|_{T_+}\|_{L^2(S)} + \|\nabla v_h|_{T_-}\|_{L^2(S)} \right) \|[\![v_h]\!]\|_{L^2(S)}$$

$$\leq c_{usr} h_S^{-1/2} \left(\|\nabla v_h\|_{L^2(T_+)} + \|\nabla v_h\|_{L^2(T_-)} \right) \|[\![v_h]\!]\|_{L^2(S)}.$$

With Young's inequality $ab \leq a^2 + b^2/4$, we deduce the estimate. $\qquad \square$

By summation of the estimate over all sides $S \in \mathscr{S}_h$, we obtain the coercivity of a_{dG} under appropriate assumptions.

Proposition 7.11 (Existence and Uniqueness) *The discontinuous Galerkin method, defined by determining $u_h \in \mathscr{S}^{1,dG}(\mathscr{T}_h)$ such that*

$$a_{dG}(u_h, v_h) = \ell(v_h)$$

for all $v_h \in \mathscr{S}^{1,dG}(\mathscr{T}_h)$, has a unique solution if

- $\sigma = -1$ *and $\beta_S > 0$ for all $S \in \mathscr{S}_h$, or*
- $\sigma \in \{1, 0\}$ *and $\beta_S \geq c_3$ with $c_3 > 0$ sufficiently large, and $\gamma \geq 1$.*

In particular, under these conditions, the bilinear form a_{dG} is bounded and coercive on $\mathscr{S}^{1,dG}(\mathscr{T}_h)$.

Proof The bilinear form a_{dG} is coercive if $\sigma = -1$, and we show that this is also true for $\sigma \in \{0, 1\}$ under the conditions stated. A summation of the jump-average bound of Lemma 7.8 shows that, using that every element has $d + 1$ sides,

$$(\sigma + 1) \sum_{S \in \mathscr{S}_h} \int_S \{\nabla v_h \cdot n_S\} [\![v_h]\!] \, ds$$

$$\geq -2(d+1)c_1^2 \big(\max_{S \in \mathscr{S}_h} h_S^{\gamma-1} \beta_S^{-1} \big) \|\nabla_{\mathscr{T}} v_h\|_{L^2(\Omega)}^2 - \frac{1}{2} \sum_{S \in \mathscr{S}_h} \frac{\beta_S}{h_S^\gamma} \int_S |[\![v_h]\!]|^2 \, ds.$$

If $\gamma \geq 1$, then $h_S^{\gamma-1} \leq c_2$ and assuming that $\beta_S \geq c_3$ for all $S \in \mathscr{S}_h$, then

$$(\sigma + 1) \sum_{S \in \mathscr{S}_h} \int_S \{\nabla v_h \cdot n_S\} [\![v_h]\!] \, ds$$

$$\geq -\frac{2(d+1)c_1^2 c_2}{c_3} \|\nabla_{\mathscr{T}} v_h\|_{L^2(\Omega)}^2 - \frac{1}{2} \sum_{S \in \mathscr{S}_h} \frac{\beta_S}{h_S^\gamma} \int_S |[\![v_h]\!]|^2 \, ds.$$

If c_3 is sufficiently large so that $2(d+1)c_1^2 c_2/c_3 \leq 1/2$, then we may deduce that

$$a_{\mathrm{dG}}(v_h, v_h) \geq \frac{1}{2} \|\nabla_{\mathscr{T}} v_h\|_{L^2(\Omega)}^2 + \frac{1}{2} \sum_{S \in \mathscr{S}_h} \frac{\beta_S}{h_S^\gamma} \int_S |[\![v_h]\!]|^2 \, ds = \frac{1}{2} \|v_h\|_{\mathrm{dG}}^2,$$

i.e., that a_{dG} is coercive on $\mathscr{S}^{1,\mathrm{dG}}(\mathscr{T}_h)$. An exercise shows that a_{dG} is also continuous, i.e.,

$$a_{\mathrm{dG}}(v_h, w_h) \leq k_a \|v_h\|_{\mathrm{dG}} \|w_h\|_{\mathrm{dG}}$$

for all $w_h \in \mathscr{S}^{1,\mathrm{dG}}(\mathscr{T}_h)$. The Lax–Milgram-lemma implies the statement. □

Remark 7.23 The bilinear form a_{dG} is in general neither bounded nor coercive on $H^s(\mathscr{T}_h)$ with respect to the norm $\|\cdot\|_{\mathrm{dG}}$.

7.4.4 Error Analysis

To derive an error estimate, we have to control the error on sides. For this, certain trace inequalities are needed.

Lemma 7.9 (Trace Inequalities) *For $v \in H^1(\mathscr{T}_h)$ and a side $S \subset \partial T$, we have*

$$\|v\|_{L^2(S)} \leq c_{\mathrm{Tr}} \big(h_S^{1/2} \|\nabla v\|_{L^2(T)} + h_S^{-1/2} \|v\|_{L^2(T)} \big).$$

For $v_h \in \mathscr{S}^{k,\mathrm{dG}}(\mathscr{T}_h)$, we have

$$\|v_h\|_{L^2(S)} \le c_{\mathrm{Tr},k} h_S^{-1/2} \|v_h\|_{L^2(T)},$$

where $c_{\mathrm{Tr},k} \ge 0$ depends on the polynomial degree $k \ge 0$.

Proof Exercise. $\qquad\qquad\qquad\qquad\qquad\qquad\qquad\qquad\qquad\qquad\qquad\qquad$ $\square$

We prove an optimal error estimate for the dG-method in the dG-norm, under the assumption of a regular solution.

Theorem 7.6 (Error Estimate) *Assume that the exact solution of the Poisson problem satisfies $u \in H^2(\Omega) \cap H_0^1(\Omega)$, and that the conditions of Proposition 7.11 are satisfied. We then have*

$$\|u - u_h\|_{\mathrm{dG}} \le c_{\mathrm{dG}} h \|D^2 u\|_{L^2(\Omega)}.$$

Proof With the nodal interpolant $\mathscr{I}_h u \in \mathscr{S}_0^1(\mathscr{T}_h) \subset \mathscr{S}^{1,\mathrm{dG}}(\mathscr{T}_h)$, we note that

$$\|u - u_h\|_{\mathrm{dG}} \le \|u - \mathscr{I}_h u\|_{\mathrm{dG}} + \|\mathscr{I}_h u - u_h\|_{\mathrm{dG}} = \|e_I\|_{\mathrm{dG}} + \|d_h\|_{\mathrm{dG}},$$

where $e_I = u - \mathscr{I}_h u \in H_0^1(\Omega)$ and $d_h = u_h - \mathscr{I}_h u$. Since $[\![e_I]\!]|_S = 0$ for every $S \in \mathscr{S}_h$, we have by nodal interpolation estimates that

$$\|e_I\|_{\mathrm{dG}} = \|\nabla e_I\|_{L^2(\Omega)} \le ch \|D^2 u\|_{L^2(\Omega)}.$$

The consistency of the method implies the Galerkin orthogonality

$$a_{\mathrm{dG}}(u_h - u, v_h) = 0$$

for all $v_h \in \mathscr{S}^{1,\mathrm{dG}}(\mathscr{T}_h)$. Therefore,

$$a_{\mathrm{dG}}(d_h, v_h) = a_{\mathrm{dG}}(u_h - \mathscr{I}_h u, v_h) = a_{\mathrm{dG}}(u - \mathscr{I}_h u, v_h) = a_{\mathrm{dG}}(e_I, v_h).$$

For $v_h = d_h$, we deduce with the coercivity of a_{dG} and the fact that $[\![e_I]\!] = 0$ that

$$\alpha \|d_h\|_{\mathrm{dG}}^2 \le a_{\mathrm{dG}}(d_h, d_h) = a_{\mathrm{dG}}(e_I, d_h)$$

$$= \sum_{T \in \mathscr{T}_h} \int_T \nabla e_I \cdot \nabla d_h \, dx + \sum_{S \in \mathscr{S}_h} \int_S \{\nabla e_I \cdot n_S\} [\![d_h]\!] \, ds$$

$$\le \sum_{T \in \mathscr{T}_h} \|\nabla e_I\|_{L^2(T)} \|\nabla d_h\|_{L^2(T)} + \sum_{S \in \mathscr{S}_h} \|\{\nabla e_I \cdot n_S\}\|_{L^2(S)} \|[\![d_h]\!]\|_{L^2(S)}.$$

The trace inequality, the interpolation estimate $\|\nabla e_I\|_{L^2(T)} \leq ch\|D^2u\|_{L^2(T)}$, and $D^2\mathscr{I}_h u|_T = 0$ for every $T \in \mathscr{T}_h$, yield for every $S \in \mathscr{S}_h$ that

$$\|\{\nabla e_I \cdot n_S\}\|_{L^2(S)} \leq c \sum_{T=T_+,T_-} \left(h_S^{1/2}\|D^2 e_I\|_{L^2(T)} + h_S^{-1/2}\|\nabla e_I\|_{L^2(T)}\right)$$

$$\leq 2ch_S^{1/2}\|D^2u\|_{L^2(T_S)},$$

where $T_S \in \{T_+, T_-\}$ is the element with the larger contribution to the sum. With this and the Cauchy–Schwarz inequality we verify that

$$\alpha\|d_h\|_{dG}^2 \leq \sum_{T\in\mathscr{T}_h} \|\nabla e_I\|_{L^2(T)}\|\nabla d_h\|_{L^2(T)} + c\sum_{S\in\mathscr{S}_h} h_S^{1/2}\|D^2u\|_{L^2(T_S)}\|[\![d_h]\!]\|_{L^2(S)}$$

$$\leq \frac{1}{2\alpha}\sum_{T\in\mathscr{T}_h}\|\nabla e_I\|_{L^2(T)}^2 + \frac{\alpha}{2}\sum_{T\in\mathscr{T}_h}\|\nabla d_h\|_{L^2(T)}^2$$

$$+ \frac{c^2}{2\alpha}\sum_{S\in\mathscr{S}_h} h_S\frac{h_S^\gamma}{\beta_S}\|D^2u\|_{L^2(T_S)}^2 + \frac{\alpha}{2}\sum_{S\in\mathscr{S}_h}\frac{\beta_S}{h_S^\gamma}\|[\![d_h]\!]\|_{L^2(S)}^2.$$

The sum of the second and fourth term on the right-hand side coincides with $(\alpha/2)\|d_h\|_{dG}^2$. Using again the interpolation estimate $\|\nabla e_I\|_{L^2(T)} \leq ch\|D^2u\|_{L^2(T)}$, we thus find that

$$\frac{\alpha}{2}\|d_h\|_{dG}^2 \leq \frac{c^2}{2\alpha}\sum_{T\in\mathscr{T}_h}h^2\|D^2u\|_{L^2(T)}^2 + \frac{c^2}{2\alpha}\sum_{S\in\mathscr{S}_h}h_S\frac{h_S^\gamma}{\beta_S}\|D^2u\|_{L^2(T_S)}^2$$

$$\leq ch^2\|D^2u\|_{L^2(\Omega)}^2,$$

which implies the error estimate.   □

Remarks 7.24

(i) Note that we may choose $\gamma \geq 1$ arbitrarily for the coercivity and error estimate. It is preferable to choose $\gamma = 1$ since this leads to the smallest condition number.

(ii) Since $[\![u]\!]|_S = 0$ for all interior sides $S \in \mathscr{S}_h$, the error estimate implies that

$$\sum_{S\in\mathscr{S}_h} h_S^{-\gamma}\beta_S \int_S |[\![u_h]\!]|^2 \, ds \leq ch^2\|D^2u\|_{L^2(\Omega)}^2,$$

i.e., the error estimate controls the failure of continuity, and as $h \to 0$, the discontinuous Galerkin approximation becomes continuous. The parameters γ and $(\beta_S)_{S\in\mathscr{S}_h}$ define for a fixed triangulation the amount of discontinuity of the discrete solution.

(iii) The dG-method is flexible with respect to the used partitions, e.g., hanging nodes or more general partitions $\mathcal{T}_h$ can be considered.

7.4.5 Convection Terms

The importance of dG-methods is related to their flexibility, in particular when dealing with convection-dominated equations. To illustrate the main ideas of how to discretize a convection term, we consider a vector field $b \in C^1(\overline{\Omega}; \mathbb{R}^d)$ with $\operatorname{div} b = 0$ in Ω. For $u \in H^1(\Omega)$ and $v \in H^1(\mathcal{T}_h)$, we have

$$
\int_\Omega b \cdot \nabla u \, v \, dx = -\sum_{T \in \mathcal{T}_h} \int_T bu \cdot \nabla v \, dx + \sum_{T \in \mathcal{T}_h} \int_{\partial T} b \cdot n_T uv \, ds
$$

$$
= -\int_\Omega bu \cdot \nabla_{\mathcal{T}} v \, dx - \sum_{S \in \mathcal{S}_h} \int_S b \cdot n_S \{u\} [\![v]\!] \, ds.
$$

To improve numerical stability, we add the term

$$
K_\varrho(u, v) = \sum_{S \in \mathcal{S}_h} \varrho_S \int_S [\![u]\!] [\![v]\!] \, ds.
$$

This motivates defining the bilinear form to be

$$
c_{\mathrm{dG}}(u, v) = -\int_\Omega bu \cdot \nabla_{\mathcal{T}} v \, dx - \sum_{S \in \mathcal{S}_h} b_S \int_S \{u\} [\![v]\!] \, ds + \sum_{S \in \mathcal{S}_h} \varrho_S \int_S [\![u]\!] [\![v]\!] \, ds,
$$

where we set $b_S = b(x_S) \cdot n_S$ with the midpoint x_S of S. The choice of the coefficients ϱ_S is intended to capture the underlying physical behavior. Setting

$$
\varrho_S = \frac{1}{2} |b_S|,
$$

we have, for every interior side $S \in \mathcal{S}_h$ with adjacent elements T_- and T_+ so that n_S points from T_- into T_+, that

$$
b_S \{u\}|_S - \varrho_S [\![u]\!]|_S = \frac{1}{2} b_S \big(u|_{T_-} + u|_{T_+}\big) - \frac{1}{2} |b_S| \big(u|_{T_+} - u|_{T_-}\big)
$$

$$
= \begin{cases} b_S u|_{T_-} & \text{if } b_S \geq 0, \\ b_S u|_{T_+} & \text{if } b_S \leq 0. \end{cases}
$$

Therefore the choice of the coefficients $\varrho_S = |b(x_S) \cdot n_S|/2$ realizes an upwinding scheme.

Remark 7.25 When we add the diffusion bilinear form a_{dG} with $\gamma = 1$, weighted by a small factor ε, i.e., discretizing

$$-\varepsilon \Delta u + b \cdot \nabla u$$

with the discontinuous Galerkin method, and the bilinear form c_{dG}, then the penalty and the stabilization terms $J_{\beta,\gamma}$ and K_ϱ sum up to

$$\sum_{S \in \mathscr{S}_h} \left(\varepsilon \frac{\beta_S}{h_S} + \frac{1}{2} |b \cdot n_S| \right) \int_S [\![u]\!] [\![v]\!] \, ds.$$

We again observe the relevance of a local mesh Péclet number defined by $Pe_S = h_S |b \cdot n_S| / (2\varepsilon \beta_S)$.

7.4.6   Implementation

To obtain system matrices with few entries, an appropriate basis of the space $\mathscr{S}^{1,dG}(\mathscr{T}_h)$ has to be chosen. We associate with every element $T \in \mathscr{T}_h$ with vertices $z_1^T, z_2^T, \ldots, z_{d+1}^T$, the functions $\varphi_{T,j} \in \mathscr{S}^{1,dG}(\mathscr{T}_h)$ defined by

$$\varphi_{T,j}(x) = \begin{cases} 1 - d\varphi_{z_j^T}(x), & x \in T, \\ 0, & x \notin T. \end{cases}$$

A typical basis function is shown in Fig. 7.26.

Fig. 7.26   Basis function for the space $\mathscr{S}^{1,dG}(\mathscr{T}_h)$

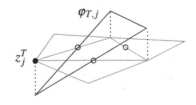

The basis functions are thus restrictions of Crouzeix–Raviart basis functions to particular elements. With the midpoints of sides $(x_S : S \in \mathscr{S}_h)$, we have the property that

$$\varphi_{T,j}|_T(x_{S_k}) = \delta_{jk},$$

for the sides $S_1, \ldots, S_{d+1} \subset \partial T$ that are opposite the vertices $z_1^T, \ldots, z_{d+1}^T$. Since the basis functions are elementwise affine, this implies that the integral of the basis functions over sides of elements either vanish or coincide with the surface area of the side. We also note that

$$[\![\varphi_{T_\pm,j}]\!]|_S = \pm\left(1 - d\varphi_{z_j^T}\right)\big|_S,$$

$$\{\varphi_{T_\pm,j}\}|_S = \frac{1}{2}\left(1 - d\varphi_{z_j^T}\right)\big|_S,$$

$$\{\nabla\varphi_{T_\pm,j} \cdot n_S\}|_S = -\frac{d}{2}\nabla\varphi_{z_j^T}|_T \cdot n_S.$$

Here the sign of $T_\pm$ is chosen according to the direction of the normal n_S. In our implementation, we use that the chosen normal points from the element with the higher element number into the one with lower element number. Hence, the adjacent element with the higher number is the negative one with respect to the side, written $\sigma_T^S = -1$. With these conventions, for a side S and adjacent elements T_k, T_ℓ, which do not have to be disjoint, we have that

$$\int_S \{\nabla\varphi_{T_k,m} \cdot n_S\}[\![\varphi_{T_k,n}]\!]\, ds = \begin{cases} (-d|S|/2)\nabla\varphi_{z_m}^{T_k} \cdot n_S\sigma_{T_\ell}^S & \text{if } z_n^{T_\ell} \notin S, \\ 0 & \text{otherwise.} \end{cases}$$

The discretization of the penalty bilinear form leads to contributions given by

$$\int_S [\![\varphi_{T_k,m}]\!][\![\varphi_{T_\ell,n}]\!]\, ds = (-1)^{\delta_{T_k,T_\ell}} \begin{cases} |S| & \text{if } z_m^{T_k}, z_n^{T_\ell} \notin S, \\ d|S|(\delta_{mn} - 1)/(d+1) & \text{if } z_m^{T_k}, z_n^{T_\ell} \in S. \end{cases}$$

The main program that realizes low-order discontinuous Galerkin methods for the Poisson problem is shown in Fig. 7.27. The assembly of the matrices representing the bilinear form associated with the sides of the triangulation is realized in the

```
function poisson_dg(d,red)
global beta gamma;
sigma = 1; beta = 10; gamma = 1;
[c4n,n4e,Db,Nb] = triang_cube(d); Db = [Db;Nb]; Nb = [];
for j = 1:red
    [c4n,n4e,Db,Nb] = red_refine(c4n,n4e,Db,Nb);
end
[s4e,~,n4s,~,~,e4s] = sides(n4e,Db,Nb);
nS = size(n4s,1); nE = size(n4e,1);
b = zeros((d+1)*nE,1);
ctr = 0; ctr_max = (d+1)^2*nE;
grads = zeros((d+1)*nE,d);
normals_S = zeros(nS,d); vol_S = zeros(nS,1);
I = zeros(ctr_max,1); J = zeros(ctr_max,1); X = zeros(ctr_max,1);
for j = 1:nE
    X_T = [ones(1,d+1);c4n(n4e(j,:),:)'];
    grads_T = X_T\[zeros(1,d);eye(d)];
    vol_T = det(X_T)/factorial(d);
    mp_T = sum(c4n(n4e(j,:),:),1)/(d+1);
    for m = 1:d+1
        b((j-1)*(d+1)+m) = (1/(d+1))*vol_T*f(mp_T);
        for n = 1:d+1
            ctr = ctr+1;
            I(ctr) = (j-1)*(d+1)+m; J(ctr) = (j-1)*(d+1)+n;
            X(ctr) = d^2*vol_T*grads_T(m,:)*grads_T(n,:)';
        end
    end
    grads((j-1)*(d+1)+(1:d+1),:) = grads_T;
    heights = 1./sqrt(sum(grads_T.^2,2));
    vol_S(s4e(j,:)) = factorial(d)*vol_T./heights;
    normals_S(s4e(j,:),:) = -grads_T.*(heights*ones(1,d));
end
s_elements = sparse(I,J,X,(d+1)*nE,(d+1)*nE);
[s_sides,s_penal] = ...
    dg_side_matrices(n4e,e4s,n4s,vol_S,grads,normals_S);
s = s_elements+s_sides+sigma*s_sides'+s_penal;
u = s\b;
show_dg(c4n,n4e,u)

function val = f(x); val = 1;
```

Fig. 7.27 MATLAB implementation of low-order discontinuous Galerkin methods for the Poisson problem

MATLAB program dg_side_matrices.m shown in Fig. 7.28. Discontinuous Galerkin approximations of a two-dimensional Poisson problem are shown in Fig. 7.29.

```
function [s_sides,s_penal] = ...
    dg_side_matrices(n4e,e4s,n4s,vol_S,grads,normals_S)
global beta gamma;
nE = size(n4e,1); [nS,d] = size(n4s);
ctr1 = 0; ctr2 = 0; ctr_max = 4*d^2*nS;
I = zeros(ctr_max,1); J = zeros(ctr_max,1); X = zeros(ctr_max,1);
K = zeros(ctr_max,1); L = zeros(ctr_max,1); Y = zeros(ctr_max,1);
shift = [2,1];
for j = 1:nS
    h_S = vol_S(j)^(1/(d-1));
    for k = 1:2
        for ell = 1:2
            sigma_T_ell = 2*(e4s(j,ell)<e4s(j,shift(ell)))-1;
            if (e4s(j,k)&&e4s(j,ell))
                [~,n] = setdiff(n4e(e4s(j,ell),:),n4s(j,:));
                for m = 1:d+1
                    ctr1 = ctr1+1;
                    I(ctr1) = (d+1)*(e4s(j,k)-1)+m;
                    J(ctr1) = (d+1)*(e4s(j,ell)-1)+n;
                    X(ctr1) = (-d*vol_S(j)/2) ...
                        *grads((d+1)*(e4s(j,k)-1)+m,:) ...
                        *normals_S(j,:)'*sigma_T_ell;
                end
                [~,m] = setdiff(n4e(e4s(j,k),:),n4s(j,:));
                [~,n] = setdiff(n4e(e4s(j,ell),:),n4s(j,:));
                ctr2 = ctr2+1;
                K(ctr2) = (d+1)*(e4s(j,k)-1)+m;
                L(ctr2) = (d+1)*(e4s(j,ell)-1)+n;
                Y(ctr2) = (-1)^(k-ell)*beta*h_S^(-gamma)*vol_S(j);
                [~,ind1] = intersect(n4e(e4s(j,k),:),n4s(j,:));
                [~,ind2] = intersect(n4e(e4s(j,ell),:),n4s(j,:));
                for m = reshape(ind1,1,d)
                    for n = reshape(ind2,1,d)
                        ctr2 = ctr2+1;
                        delta = (n4e(e4s(j,k),m)==n4e(e4s(j,ell),n));
                        K(ctr2) = (d+1)*(e4s(j,k)-1)+m;
                        L(ctr2) = (d+1)*(e4s(j,ell)-1)+n;
                        Y(ctr2) = (-1)^(k-ell)*beta*h_S^(-gamma) ...
                            *vol_S(j)*(d*delta-1)/(d+1);
                    end
                end
            end
        end
    end
end
s_sides = sparse(I(1:ctr1),J(1:ctr1),X(1:ctr1),(d+1)*nE,(d+1)*nE);
s_penal = sparse(K(1:ctr2),L(1:ctr2),Y(1:ctr2),(d+1)*nE,(d+1)*nE);
```

Fig. 7.28 Assembly of the matrices representing bilinear forms defined on sides of elements in the discontinuous Galerkin method

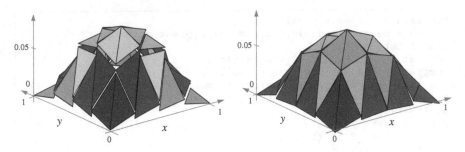

Fig. 7.29 Discontinuous Galerkin finite element approximations of a two-dimensional Poisson problem obtained with the nonsymmetric (*left*) and the symmetric (*right*) interior penalty discontinuous Galerkin method

References

Original articles containing the construction and analysis of the methods discussed in this chapter are the references [3, 7, 8, 15, 16]. Textbooks on mixed and discontinuous finite element methods are the references [5, 6, 9, 10, 17]. Singularly perturbed equations are analyzed in [12, 18]. Numerical methods for conservation laws are the subject of [13, 14]. Further aspects of mixed and nonstandard methods are contained in [1, 2, 4]. A proof of the mapping properties of the divergence operator is contained in [11].

1. Arnold, D.N., Brezzi, F., Cockburn, B., Marini, L.D.: Unified analysis of discontinuous Galerkin methods for elliptic problems. SIAM J. Numer. Anal. **39**(5), 1749–1779 (2001/02). URL http://dx.doi.org/10.1137/S0036142901384162
2. Bahriawati, C., Carstensen, C.: Three MATLAB implementations of the lowest-order Raviart-Thomas MFEM with a posteriori error control. Comput. Methods Appl. Math. **5**(4), 333–361 (electronic) (2005). URL http://dx.doi.org/10.2478/cmam-2005-0016
3. Baker, G.A.: Finite element methods for elliptic equations using nonconforming elements. Math. Comp. **31**(137), 45–59 (1977)
4. Bochev, P.B., Dohrmann, C.R., Gunzburger, M.D.: Stabilization of low-order mixed finite elements for the Stokes equations. SIAM J. Numer. Anal. **44**(1), 82–101 (electronic) (2006). URL http://dx.doi.org/10.1137/S0036142905444482
5. Boffi, D., Brezzi, F., Fortin, M.: Mixed finite element methods and applications. Springer Series in Computational Mathematics, vol. 44. Springer, Heidelberg (2013). URL http://dx.doi.org/10.1007/978-3-642-36519-5
6. Braess, D.: Finite Elements, 3rd edn. Cambridge University Press, Cambridge (2007). URL http://dx.doi.org/10.1017/CBO9780511618635
7. Brooks, A.N., Hughes, T.J.R.: Streamline upwind/Petrov-Galerkin formulations for convection dominated flows with particular emphasis on the incompressible Navier-Stokes equations. Comput. Methods Appl. Mech. Eng. **32**(1–3), 199–259 (1982). URL http://dx.doi.org/10.1016/0045-7825(82)90071-8
8. Crouzeix, M., Raviart, P.A.: Conforming and nonconforming finite element methods for solving the stationary Stokes equations. I. Rev. Française Automat. Informat. Recherche Opérationnelle Sér. Rouge **7**(R-3), 33–75 (1973)
9. Di Pietro, D.A., Ern, A.: Mathematical aspects of discontinuous Galerkin methods. Mathématiques & Applications (Berlin) [Mathematics & Applications], vol. 69. Springer, Heidelberg (2012). URL http://dx.doi.org/10.1007/978-3-642-22980-0

10. Ern, A., Guermond, J.L.: Theory and practice of finite elements. Applied Mathematical Sciences, vol. 159. Springer, New York (2004). URL http://dx.doi.org/10.1007/978-1-4757-4355-5

11. Girault, V., Raviart, P.A.: Finite element methods for Navier-Stokes equations. Springer Series in Computational Mathematics, vol. 5. Springer, Berlin (1986). URL http://dx.doi.org/10.1007/978-3-642-61623-5

12. Knabner, P., Angermann, L.: Numerical methods for elliptic and parabolic partial differential equations. Texts in Applied Mathematics, vol. 44. Springer, New York (2003)

13. Kröner, D.: Numerical schemes for conservation laws. Wiley-Teubner Series. Advances in Numerical Mathematics. Wiley, Chichester; B. G. Teubner, Stuttgart (1997)

14. LeVeque, R.J.: Numerical methods for conservation laws. Lectures in Mathematics ETH Zürich, 2nd edn. Birkhäuser Verlag, Basel (1992). URL http://dx.doi.org/10.1007/978-3-0348-8629-1

15. Raviart, P.A., Thomas, J.M.: A mixed finite element method for 2nd order elliptic problems. Lecture Notes in Math., Vol. 606, pp. 292–315. Springer, Berlin (1977)

16. Reed, W., Hill, T.: Triangular mesh methods for the neutron transport equation (1973). URL http://www.osti.gov/scitech/servlets/purl/4491151

17. Rivière, B.: Discontinuous Galerkin methods for solving elliptic and parabolic equations. Frontiers in Applied Mathematics, vol. 35. Society for Industrial and Applied Mathematics (SIAM), Philadelphia, PA (2008). URL http://dx.doi.org/10.1137/1.9780898717440

18. Roos, H.G., Stynes, M., Tobiska, L.: Robust numerical methods for singularly perturbed differential equations. Springer Series in Computational Mathematics, vol. 24, 2nd edn. Springer, Berlin (2008)

Chapter 8
Applications

8.1 Linear Elasticity

8.1.1 Navier–Lamé Equations

Let $\Omega \subset \mathbb{R}^d$, $d = 2, 3$, be the domain occupied by a solid body. We assume that the body behaves *elastically*, i.e., the body returns to its reference configuration when sufficiently small forces stop acting, e.g., as in the case of a sponge or a network of springs. We want to specify a mapping $\phi : \Omega \to \mathbb{R}^d$ called a *deformation* that describes the elastic deformation of the body subject to such forces, cf. Fig. 8.1.

Small changes in the distance of material points due to the deformation ϕ are measured by

$$\|\phi(x + z) - \phi(x)\|^2 \approx z^\mathsf{T} \nabla\phi(x)^\mathsf{T} \nabla\phi(x)z.$$

The matrix $C = \nabla\phi(x)^\mathsf{T} \nabla\phi(x)$ is called the *right Cauchy–Green strain tensor* and $E = (C - I)/2$ is the corresponding *strain*. For small *displacements* $u = \phi - \mathrm{id}$, i.e., $|\nabla u| \ll 1$, we have

$$E = \frac{1}{2}\big((\nabla u)^\mathsf{T} + \nabla u\big) + \frac{1}{2}(\nabla u)^\mathsf{T}\nabla u \approx \frac{1}{2}\big((\nabla u)^\mathsf{T} + \nabla u\big) = \varepsilon(u)$$

with the *linearized strain* or *symmetric gradient* $\varepsilon(u) = (\nabla u^\mathsf{T} + \nabla u)/2$. A *rigid body motion* is a deformation ϕ with $E = 0$. A theorem due to Liouville states that the only rigid body motions are rotations and translations, i.e., if $E = 0$, then we have $\phi(x) = Qx + b$ for all $x \in \Omega$ with $Q \in SO(d)$ and $b \in \mathbb{R}^d$. *Linearized rigid body motions* are characterized by $\varepsilon(u) = 0$ and given by $u(x) = Ax + b$ with $A \in so(d)$ and $b \in \mathbb{R}^d$, where

$$so(d) = \{X \in \mathbb{R}^{d\times d} : X^\mathsf{T} = -X\},$$

© Springer International Publishing Switzerland 2016
S. Bartels, *Numerical Approximation of Partial Differential Equations*,
Texts in Applied Mathematics 64, DOI 10.1007/978-3-319-32354-1_8

Fig. 8.1 Deformation of an
elastic body

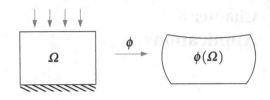

which is the tangent space of $SO(d)$ at the identity matrix. The restoring elastic forces generated by a displacement are described by the *Cauchy stress tensor* σ : $\Omega \to \mathbb{R}^{d\times d}_{\text{sym}}$ with

$$\mathbb{R}^{d\times d}_{\text{sym}} = \{X \in \mathbb{R}^{d\times d} : X^{\top} = X\}.$$

Mathematically, *linearly elastic* materials are defined by an invertible linear relation between stresses and linearized strains, i.e., the *Hookean law*

$$\sigma = \mathbb{C}\varepsilon,$$

with a linear, positive definite, and symmetric fourth order tensor $\mathbb{C} : \mathbb{R}^{d\times d}_{\text{sym}} \to \mathbb{R}^{d\times d}_{\text{sym}}$, i.e., for all $A, B \in \mathbb{R}^{d\times d}_{\text{sym}}$, we have

$$\mathbb{C}A : B = A : \mathbb{C}B, \quad \mathbb{C}A : A \geq c_{\mathbb{C}}|A|^2,$$

where $A : B = \sum_{i,j=1}^{d} A_{ij}B_{ij}$ denotes the canonical inner product on $\mathbb{R}^{d\times d}$. The concepts of *frame indifference* and *material isotropy* imply that $\mathbb{C}$ is described by two parameters $\lambda, \mu > 0$ called the *Lamé constants*, i.e., for all $A \in \mathbb{R}^{d\times d}_{\text{sym}}$, we have

$$\mathbb{C}A = \lambda \operatorname{tr}(A)I + 2\mu A.$$

The *bulk modulus* λ defines the resistance of the solid to compression, and the *shear modulus* μ the resistance to shear forces. Note that we have $I : A = \operatorname{tr}(A)$ and hence $c_{\mathbb{C}} = 2\mu$. We postulate that the actual displacement $u : \Omega \to \mathbb{R}^d$ of the body subject to the *volume* and *surface forces*

$$f : \Omega \to \mathbb{R}^d, \quad g : \Gamma_{\text{N}} \to \mathbb{R}^d$$

minimizes the elastic energy

$$I(u) = \frac{1}{2}\int_{\Omega} \mathbb{C}\varepsilon(u) : \varepsilon(u)\,\mathrm{d}x - \int_{\Omega} f \cdot u\,\mathrm{d}x - \int_{\Gamma_{\text{N}}} g \cdot u\,\mathrm{d}s$$

in the set of all possible displacements $u \in H_D^1(\Omega; \mathbb{R}^d)$; displacements thus tend to align with the forces. A minimizer satisfies the optimality conditions

$$\int_\Omega \mathbb{C}\varepsilon(u) : \varepsilon(v)\, dx = \int_\Omega f \cdot v\, dx + \int_{\Gamma_N} g \cdot v\, ds$$

for all $v \in H_D^1(\Omega; \mathbb{R}^d)$. Noting that $\mathbb{C}\varepsilon(u) : \varepsilon(v) = \mathbb{C}\varepsilon(u) : \nabla v$, we obtain a strong form of the equations.

Definition 8.1 The *Navier–Lamé equations* define a displacement $u \in H_D^1(\Omega; \mathbb{R}^d)$ as a weak solution of the boundary value problem

$$-\operatorname{div} \mathbb{C}\varepsilon(u) = f \text{ in } \Omega, \quad u|_{\Gamma_D} = 0, \quad (\mathbb{C}\varepsilon(u))n|_{\Gamma_N} = g.$$

The optimality conditions for the minimum-energy principle thus coincide with the weak formulation of the Navier–Lamé equations.

Remark 8.1 The case $d = 2$ arises as a simplification of three-dimensional situations, i.e., *plane strain* and *plane stress* approximations that describe certain deformations of long cylinders and thin plates, respectively.

8.1.2 Well-Posedness and Approximation

To establish the existence of solutions with the Lax–Milgram lemma, we have to show that the bilinear form

$$a(u, v) = \int_\Omega \mathbb{C}\varepsilon(u) : \varepsilon(v)\, dx$$

is coercive on $H_D^1(\Omega; \mathbb{R}^d)$. Its definiteness is a consequence of the definiteness of $\mathbb{C}$, and the fact that linearized rigid body motions are trivial due to the boundary conditions on Γ_D; the induced seminorm is denoted

$$\|\mathbb{C}^{1/2}\varepsilon(u)\|_{L^2(\Omega)} = a(u, u)^{1/2}$$

for $u \in H_D^1(\Omega; \mathbb{R}^d)$. The following lemma shows that the symmetric part of the gradient controls the full gradient for vector fields in $H_D^1(\Omega; \mathbb{R}^d)$.

Lemma 8.1 (Korn Inequality, See [5, Chap. 6]) *For all $v \in H_D^1(\Omega; \mathbb{R}^d)$, we have*

$$\|\varepsilon(v)\| \geq c_K \|\nabla v\|.$$

Proof (Sketched) We prove the estimate for the special case when $\Gamma_D = \partial\Omega$. For every $v \in C^2(\overline{\Omega}; \mathbb{R}^d)$, we have

$$2 \operatorname{div} \varepsilon(v) = \Delta v + \nabla \operatorname{div} v,$$

where $\operatorname{div} \varepsilon(v)$ is understood to be row- and Δv component-wise. Therefore, if $w \in H_0^1(\Omega; \mathbb{R}^d)$, we have that

$$\int_\Omega \varepsilon(v) : \varepsilon(w) \, dx = - \int_\Omega \operatorname{div} \varepsilon(v) \cdot w \, dx$$

$$= \frac{1}{2} \int_\Omega \nabla v : \nabla w \, dx + \frac{1}{2} \int_\Omega \operatorname{div}(v) \operatorname{div}(w) \, dx.$$

For $v = w$, we conclude that

$$\|\varepsilon(v)\|_{L^2(\Omega)}^2 \geq \frac{1}{2} \|\nabla v\|_{L^2(\Omega)}^2.$$

The result for $v \in H_0^1(\Omega; \mathbb{R}^d)$ follows from a density argument.   □

Remark 8.2 The full proof of the lemma uses mapping properties of the divergence operator, in particular an inf-sup condition.

Corollary 8.1 (Existence and Uniqueness) *For $f \in L^2(\Omega; \mathbb{R}^d)$ and $g \in L^2(\Gamma_N; \mathbb{R}^d)$ there exists a unique weak solution $u \in H_D^1(\Omega; \mathbb{R}^d)$ of the Navier–Lamé equations.*

Proof The result follows with the Korn inequality, i.e., the fact that

$$a(u, u) = \|\mathbb{C}^{1/2} \varepsilon(u)\|_{L^2(\Omega)}^2 \geq (2\mu) c_K^2 \|\nabla u\|_{L^2(\Omega)}^2$$

for all $u \in H_D^1(\Omega; \mathbb{R}^d)$, from the Lax–Milgram lemma.   □

To obtain convergence rates for numerical approximation schemes for the Navier–Lamé equations, the regularity of solutions is required.

Theorem 8.1 (H^2-Regularity, See [4, Chap. 11]) *If $\Omega \subset \mathbb{R}^d$ is convex and $\Gamma_D = \partial\Omega$, then we have $u \in H^2(\Omega; \mathbb{R}^d)$ and*

$$\|u\|_{H^2(\Omega)} + \lambda \|\nabla \operatorname{div} u\|_{L^2(\Omega)} \leq c_{NL} \|f\|_{L^2(\Omega)},$$

with a constant $c_{NL} > 0$ that is independent of λ.

Remark 8.3 For $\lambda \gg 1$ it follows that $\|\nabla \operatorname{div} u\|_{L^2(\Omega)}$ is small.

Via Céa's lemma we obtain error estimates for conforming finite element approximations.

Proposition 8.1 (Galerkin Approximation) *Let* $u \in H^1_D(\Omega; \mathbb{R}^d)$ *be the weak solution of the Navier–Lamé equations for a right-hand side functional* $\ell \in H^1_D(\Omega; \mathbb{R}^d)'$, *and let* $u_h \in \mathscr{S}^1_D(\mathscr{T}_h)^d$ *satisfy*

$$a(u_h, v_h) = \ell(v_h)$$

for all $v_h \in \mathscr{S}^1_D(\mathscr{T}_h)$. *If* $u \in H^2(\Omega; \mathbb{R}^d)$, *we have*

$$\|\mathbb{C}^{1/2}\varepsilon(u - u_h)\|_{L^2(\Omega)} \leq c_{\mathscr{I}}(\lambda d + 2\mu)^{1/2}h\|D^2 u\|_{L^2(\Omega)}.$$

Proof By Galerkin orthogonality and the Cauchy–Schwarz inequality, we have

$$\|\mathbb{C}^{1/2}\varepsilon(u - u_h)\|^2_{L^2(\Omega)} = a(u - u_h, u - v_h)$$

$$\leq \|\mathbb{C}^{1/2}\varepsilon(u - u_h)\|_{L^2(\Omega)}\|\mathbb{C}^{1/2}\varepsilon(u - v_h)\|_{L^2(\Omega)}.$$

With $\operatorname{tr}\varepsilon(w) = I : \varepsilon(w) = \operatorname{div}(w)$ for every $w \in H^1(\Omega; \mathbb{R}^d)$, we note that

$$\|\mathbb{C}^{1/2}\varepsilon(w)\|^2_{L^2(\Omega)} = \int_\Omega \mathbb{C}\varepsilon(w) : \varepsilon(w)\,dx$$

$$= \lambda \int_\Omega \operatorname{tr}\varepsilon(w)I : \varepsilon(w)\,dx + 2\mu \int_\Omega \varepsilon(w) : \varepsilon(w)\,dx$$

$$= \lambda \int_\Omega (\operatorname{div} w)^2\,dx + 2\mu \int_\Omega |\varepsilon(w)|^2\,dx$$

$$\leq (\lambda d + 2\mu) \int_\Omega |\nabla w|^2\,dx.$$

The choice $v_h = \mathscr{I}_h u$ implies the estimate.   □

8.1.3 Incompressibility Locking

Nearly incompressible materials like rubber are characterized by a large Lamé constant $\lambda \gg 1$. In this case the error estimate for the conforming Galerkin approximation may be of limited use; numerical experiments reveal that they provide sub-optimal approximations for moderate mesh-sizes. This is due to the estimate

$$\lambda^{1/2}\|\operatorname{div}(u - \mathscr{I}_h u)\|_{L^2(\Omega)} \leq (\lambda d)^{1/2}\|\nabla(u - \mathscr{I}_h u)\|_{L^2(\Omega)}$$

$$\leq c_{\mathscr{I}}(\lambda d)^{1/2}h\|D^2 u\|_{L^2(\Omega)}.$$

This estimate cannot be improved as the example

$$u(x, y) = [\sin(x)\sin(y), \cos(x)\cos(y)]^{\top}$$

shows, for which we have $\operatorname{div} u = 0$ and $\|\operatorname{div} \mathscr{I}_h u\|_{L^2(T_h)} \geq ch\|D^2 u\|_{L^2(T_h)}$ with a constant $c > 0$ and $T_h = \operatorname{conv}\{(0,0),(h,0),(0,h)\}$. If we could construct an interpolant $I_F u$ that allows us to make use of the regularity result, i.e., such that

$$\lambda^{1/2}\|\operatorname{div}(u - I_F u)\|_{L^2(\Omega)} \leq c_F h \lambda^{1/2}\|\nabla \operatorname{div} u\|_{L^2(\Omega)}$$
$$\leq c_{\mathrm{NL}} c_F \lambda^{-1/2} h \|f\|_{L^2(\Omega)},$$

then we could obtain error estimates that are robust in the limit $\lambda \to \infty$. A key to such estimates are reformulations of the weak form of the Navier–Lamé equations as perturbed saddle-point problems.

Example 8.1 Introducing $p = \lambda \operatorname{div} u$ and using

$$\int_{\Omega} \mathbb{C}\varepsilon(u) : \varepsilon(v)\, dx = 2\mu \int_{\Omega} \varepsilon(u) : \varepsilon(v)\, dx + \lambda \int_{\Omega} \operatorname{div} u \operatorname{div} v\, dx,$$

the variational formulation is equivalent to determining $(u, p) \in H_{\mathrm{D}}^1(\Omega; \mathbb{R}^d) \times L^2(\Omega)$ such that

$$2\mu \int_{\Omega} \varepsilon(u) : \varepsilon(v)\, dx + \int_{\Omega} p \operatorname{div} v\, dx = \ell(v),$$

$$\int_{\Omega} q \operatorname{div} u\, dx \quad - \lambda^{-1} \int_{\Omega} pq\, dx = 0,$$

for all $(v, q) \in H_{\mathrm{D}}^1(\Omega; \mathbb{R}^d) \times L^2(\Omega)$. The formulation is qualitatively similar to a Stokes system with penalty term.

An alternative is the *Hellinger–Reissner* principle.

Example 8.2 Introducing $\sigma = \mathbb{C}\varepsilon(u)$, we obtain the equivalent system

$$-\operatorname{div} \sigma = f, \quad \varepsilon(u) = \mathbb{C}^{-1}\sigma, \quad u|_{\Gamma_{\mathrm{D}}} = 0, \quad \sigma n|_{\Gamma_{\mathrm{N}}} = g.$$

A corresponding weak formulation seeks $(\sigma, u) \in L^2(\Omega; \mathbb{R}_{\mathrm{sym}}^{d \times d}) \times H_{\mathrm{D}}^1(\Omega; \mathbb{R}^d)$ such that

$$\int_{\Omega} \mathbb{C}^{-1}\sigma : \tau\, dx - \int_{\Omega} \tau : \varepsilon(u)\, dx = 0,$$

$$\int_{\Omega} \sigma : \varepsilon(v)\, dx \qquad\qquad = \ell(v),$$

for all $(\tau, v) \in L^2(\Omega; \mathbb{R}^{d \times d}_{sym}) \times H^1_D(\Omega; \mathbb{R}^d)$. This formulation is similar to the primal mixed formulation of the Poisson problem. A dual mixed version is obtained via an integration-by-parts. Robustness in $\lambda \to \infty$ follows from the fact that coercivity is only required on the kernel of

$$b(\tau, v) = -\int_\Omega \tau : \varepsilon(v) \, dx = \int_\Omega \operatorname{div} \tau \cdot v \, dx,$$

which consists of all $\tau \in L^2(\Omega; \mathbb{R}^{d \times d}_{sym})$ with $\operatorname{div} \tau = 0$ and $\tau n = 0$ on Γ_N.

The development of stable numerical methods is difficult because of the possible failure of a discrete Korn inequality and the proof of an inf-sup condition for symmetric matrix fields.

8.1.4 Crouzeix–Raviart Approximation

A discretization of the saddle-point formulation of Example 8.1 with a Crouzeix–Raviart method leads to the problem of determining $(u_h, p_h) \in \mathscr{S}^{1,cr}_D(\mathscr{T}_h)^d \times \mathscr{L}^0(\mathscr{T}_h)$ such that

$$2\mu \int_\Omega \varepsilon_\mathscr{T}(u_h) : \varepsilon_\mathscr{T}(v_h) \, dx + \int_\Omega p_h \operatorname{div}_\mathscr{T} v_h \, dx = \ell(v_h),$$

$$\int_\Omega q_h \operatorname{div} u_h \, dx \qquad - \lambda^{-1} \int_\Omega p_h q_h \, dx \ = 0,$$

for all $(v_h, q_h) \in \mathscr{S}^{1,cr}_D(\mathscr{T}_h)^d \times \mathscr{L}^0(\mathscr{T}_h)$. We recall that

$$\mathscr{L}^0(\mathscr{T}_h) = \{q_h \in L^1(\Omega) : q_h|_T \in \mathscr{P}_0(T) \text{ for all } T \in \mathscr{T}_h\},$$

$$\mathscr{S}^{1,cr}_D(\mathscr{T}_h) = \{v_h \in L^1(\Omega) : v_h|_T \in \mathscr{P}_1(T) \text{ for all } T \in \mathscr{T}_h,$$

$$v_h \text{ continuous at } x_S \text{ for all } S \in \mathscr{S}_h,$$

$$v_h(x_S) = 0 \text{ for all } S \in \mathscr{S}_h \cap \Gamma_D\},$$

where x_S is the midpoint of a side $S \in \mathscr{S}_h$. Moreover, the differential operators $\nabla_\mathscr{T}$, $\varepsilon_\mathscr{T}$, and $\operatorname{div}_\mathscr{T}$ denote the elementwise application of ∇, ε, and div, respectively. We let $(\varphi_S : S \in \mathscr{S}_h)$ denote the standard basis of $\mathscr{S}^{1,cr}(\mathscr{T}_h)$ and recall the properties of the Fortin interpolant.

Lemma 8.2 (Fortin Interpolant) *The operator* $I_F : H^1_D(\Omega; \mathbb{R}^d) \to \mathscr{S}^{1,cr}_D(\mathscr{T}_h)^d$,

$$I_F v = \sum_{S \in \mathscr{S}_h} v_S \varphi_S, \quad v_S = |S|^{-1} \int_S v \, ds,$$

satisfies $\operatorname{div} I_F v = |T|^{-1} \int_T \operatorname{div} v \, dx$ *and*

$$\|v - I_F v\|_{L^2(T)} + h_T \|\nabla_{\mathscr{T}}(v - I_F v)\|_{L^2(T)} \le c_F h_T^2 \|D^2 v\|_{L^2(T)}$$

and

$$\| \operatorname{div}_{\mathscr{T}}(v - I_F v)\|_{L^2(T)} \le c_F' h_T \|\nabla \operatorname{div} v\|_{L^2(T)}$$

for all $v \in H_D^1(\Omega; \mathbb{R}^d) \cap H^2(\Omega; \mathbb{R}^d)$ *and every* $T \in \mathscr{T}_h$.

Proof Exercise.   □

Remark 8.4 Note that $p_h \in \mathscr{L}^0(\mathscr{T}_h)$ can be eliminated from the discrete saddle-point problem, since we have $p_h = \lambda \operatorname{div}_{\mathscr{T}} u_h$. It follows that $u_h \in \mathscr{S}_D^{1,cr}(\mathscr{T}_h)^d$ satisfies

$$2\mu \int_\Omega \varepsilon_{\mathscr{T}}(u_h) : \varepsilon_{\mathscr{T}}(v_h) \, dx + \lambda \int_\Omega \operatorname{div}_{\mathscr{T}} u_h \operatorname{div}_{\mathscr{T}} v_h \, dx = \ell(v_h)$$

for all $v_h \in \mathscr{S}_D^{1,cr}(\mathscr{T}_h)^d$.

The Fortin interpolant leads to a robust error estimate, provided that a discrete Korn inequality holds.

Proposition 8.2 (Error Estimate) *Assume that a discrete Korn inequality is satisfied, i.e., for all* $v_h \in \mathscr{S}_D^{1,cr}(\mathscr{T}_h)^d$, *we have*

$$\|\nabla_{\mathscr{T}} v_h\|_{L^2(\Omega)} \le c_{dK} \|\varepsilon_{\mathscr{T}}(v_h)\|_{L^2(\Omega)},$$

and assume that $\|u\|_{H^2(\Omega)} + \lambda \|\nabla \operatorname{div} u\|_{L^2(\Omega)} \le c_{NL} \|f\|_{L^2(\Omega)}$. *Then we have*

$$\|\mathbb{C}^{1/2} \varepsilon_{\mathscr{T}}(u - u_h)\|_{L^2(\Omega)} \le c_{cr} \big((2\mu)^{1/2} + \lambda^{-1/2}\big) h \|f\|_{L^2(\Omega)}.$$

Proof Exercise.   □

If $\Gamma_D = \partial\Omega$, then a discrete Korn inequality holds, but in general this is not the case.

Example 8.3 Let $\Omega = (-1, 1)^2$ with $\Gamma_D = \{-1\} \times [-1, 1] \cup [-1, 1] \times \{-1\}$. For the triangulation $\mathscr{T}_h = \{T_1, T_2\}$ indicated in Fig. 8.2, let $v_h \in \mathscr{S}_D^{1,cr}(\mathscr{T}_h)^2$ for an arbitrary matrix $A \in so(2)$ be defined by

$$v_h(x) = \begin{cases} 0 & \text{for } x \in T_1, \\ Ax & \text{for } x \in T_2. \end{cases}$$

We then have $\varepsilon_{\mathscr{T}}(v_h) = 0$ in Ω, but $\nabla v_h|_{T_2} = A$.

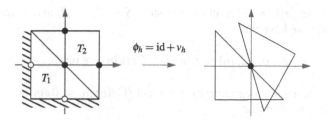

Fig. 8.2 Failure of a discrete Korn inequality for Crouzeix–Raviart elements

Remark 8.5 Higher order Crouzeix–Raviart elements have more than one degree of freedom on every side and satisfy a discrete Korn inequality.

8.1.5 Stabilized Crouzeix–Raviart Approximation

To circumvent the problem of the failure of a discrete Korn inequality for the low order Crouzeix–Raviart method, we modify the variational formulation following [13]. For this, we consider the bilinear form

$$a_h(u_h, v_h) = \int_\Omega \mathbb{C}\varepsilon_{\mathscr{T}}(u_h) : \varepsilon_{\mathscr{T}}(v_h)\,dx + \sum_{S \in \mathscr{S}_h \setminus \partial\Omega} \frac{\beta_S}{h_S} \int_S \llbracket u_h \rrbracket \llbracket v_h \rrbracket\,ds,$$

where $\llbracket w_h \rrbracket|_S$ denotes the jump of w_h across the side S. The bilinear form is well defined on $H^1(\mathscr{T}_h; \mathbb{R}^d)$, and defines a norm $\|\!|\cdot|\!\|_h$ on $\mathscr{S}_D^{1,cr}(\mathscr{T}_h)^d$. Hence, there exists a unique solution $u_h \in \mathscr{S}_D^{1,cr}(\mathscr{T}_h)^d$ such that

$$a_h(u_h, v_h) = \ell(v_h)$$

for all $v_h \in \mathscr{S}_D^{1,cr}(\mathscr{T}_h)^d$. In general, we do not have that $a_h(u, v_h) = \ell(v_h)$ for all $v_h \in \mathscr{S}_D^{1,cr}(\mathscr{T}_h)^d$. In general a Galerkin orthogonality is false. Instead, the following consistency estimate holds.

Lemma 8.3 (Consistency) *Assume that $u \in H^2(\Omega; \mathbb{R}^d) \cap H_D^1(\Omega; \mathbb{R}^d)$ solves the Navier–Lamé problem. For the consistency error $\mathscr{C}_h(u; v_h) = a_h(u, v_h) - \ell(v_h)$, we have that*

$$\left|\mathscr{C}_h(u; v_h)\right| \le ch(2\mu)^{-1/2}\left(2\mu\|D^2 u\|_{L^2(\Omega)} + \lambda\|\nabla \operatorname{div} u\|_{L^2(\Omega)}\right)\|\!|v_h|\!\|_h$$

for all $v_h \in \mathscr{S}_D^{1,cr}(\mathscr{T}_h)^d$.

Proof We have $[\![u]\!]|_S = 0$ for all interior sides $S \in \mathscr{S}_h \setminus \partial\Omega$, and $\varepsilon_{\mathscr{T}}(u) = \varepsilon(u)$ in Ω. Moreover, we have that

$$-\operatorname{div}\mathbb{C}\varepsilon(u) = f \text{ in } \Omega, \quad \mathbb{C}\varepsilon(u)n = g \text{ on } \Gamma_{\mathrm{N}}.$$

With an elementwise integration-by-parts and $[\![\mathbb{C}\varepsilon(u)]\!]|_S = 0$ for all interior sides $S \in \mathscr{S}_h \setminus \partial\Omega$, we find that

$$
\begin{aligned}
a_h(u, v_h) &= \sum_{T \in \mathscr{T}_h} \int_\Omega \mathbb{C}\varepsilon(u) : \nabla_{\mathscr{T}} v_h \, dx \\
&= -\sum_{T \in \mathscr{T}_h} \int_T \operatorname{div}\mathbb{C}\varepsilon(u) \cdot v_h \, dx + \sum_{T \in \mathscr{T}_h} \int_{\partial T} (\mathbb{C}\varepsilon(u)n) \cdot v_h \, ds \\
&= -\int_\Omega f \cdot v_h \, dx - \int_{\Gamma_{\mathrm{N}}} g \cdot v_h \, ds - \sum_{S \in \mathscr{S}_h \setminus \overline{\Gamma}_{\mathrm{N}}} \int_S (\mathbb{C}\varepsilon(u)n_S) \cdot [\![v_h]\!] \, ds.
\end{aligned}
$$

The first two terms on the right-hand side coincide with $\ell(v_h)$. Since $[\![v_h]\!]|_S$ is affine and vanishes at the midpoint x_S for every $S \in \mathscr{S}_h \setminus \overline{\Gamma}_{\mathrm{N}}$, we have that

$$\int_S [\![v_h]\!] \, ds = 0.$$

For every $S \in \mathscr{S}_h \setminus \overline{\Gamma}_{\mathrm{N}}$ with adjacent element $T_S \in \mathscr{T}_h$, we use that $\nabla I_F u|_{T_S}$ is constant and hence also $\mathbb{C}\varepsilon(I_F u)|_{T_S}$. This implies that

$$
\begin{aligned}
\int_S \mathbb{C}\varepsilon(u)n_S \cdot [\![v_h]\!] \, ds &= \int_S (\mathbb{C}\varepsilon(u - I_F u)n_S) \cdot [\![v_h]\!] \, ds \\
&= 2\mu \int_S (\varepsilon(u - I_F u)n_S) \cdot [\![v_h]\!] \, ds + \lambda \int_S (\operatorname{div}(u - I_F u)n_S) \cdot [\![v_h]\!] \, ds.
\end{aligned}
$$

The trace inequality yields that

$$
\begin{aligned}
\int_S &\mathbb{C}\varepsilon(u)n_S \cdot [\![v_h]\!] \, ds \\
&\leq 2\mu c \big(h_S^{-1/2}\|\varepsilon(u - I_F u)\|_{L^2(T_S)} + h_S^{1/2}\|D^2 u\|_{L^2(T_S)}\big)\|[\![v_h]\!]\|_{L^2(S)} \\
&\quad + \lambda c \big(h_S^{-1/2}\|\operatorname{div}(u - I_F u)\|_{L^2(T_S)} + h_S^{1/2}\|\nabla \operatorname{div} u\|_{L^2(T_S)}\big)\|[\![v_h]\!]\|_{L^2(S)}.
\end{aligned}
$$

With the approximation properties of I_F, we deduce that

$$\int_S (\mathbb{C}\varepsilon(u)n_S) \cdot [\![v_h]\!] \, ds \leq ch_S^{1/2}\big(2\mu\|D^2 u\|_{L^2(T_S)} + \lambda\|\nabla \operatorname{div} u\|_{L^2(T_S)}\big)\|[\![v_h]\!]\|_{L^2(S)}.$$

We sum the estimate over all sides $S \in \mathscr{S}_h \setminus \overline{\Gamma}_N$ and apply the Cauchy–Schwarz inequality to deduce that

$$\sum_{S \in \mathscr{S}_h \setminus \overline{\Gamma}_N} \int_S (\mathbb{C}\varepsilon(u)n_S) \cdot [\![v_h]\!] \, ds$$

$$\leq \Big(\sum_{S \in \mathscr{S}_h} \frac{ch_S^2}{2\mu} \big(2\mu \|D^2 u\|_{L^2(T_S)} + \lambda \|\nabla \operatorname{div} u\|_{L^2(T_S)}\big)^2 \Big)^{1/2} \Big(\sum_{S \in \mathscr{S}_h} \frac{2\mu}{h_S} \|[\![v_h]\!]\|_{L^2(S)}^2 \Big)^{1/2}.$$

Noting that every $T \in \mathscr{T}_h$ occurs at most $d+1$ times in the first factor on the right-hand side, and that the second factor is bounded by $\| v_h \|_h$, we deduce the estimate. □

To derive an error estimate, we have to control the interpolation error in the norm induced by a_h.

Lemma 8.4 (Stabilization Error) *If $u \in H^2(\Omega, \mathbb{R}^d)$ and $\beta_S = 2\mu$ for all $S \in \mathscr{S}_h$, then we have*

$$\sum_{S \in \mathscr{S}_h \setminus \partial\Omega} \frac{\beta_S}{h_S} \|[\![u - I_F u]\!]\|_{L^2(S)}^2 \leq c_{\text{stab}}(2\mu)h^2 \|D^2 u\|_{L^2(\Omega)}^2.$$

Proof The trace inequality and the interpolation estimates of Lemma 8.2 imply that for every $S \in \mathscr{S}_h \setminus \partial\Omega$ with adjacent elements $T_+, T_- \in \mathscr{T}_h$ with $e_F = u - I_F u$, we have

$$\frac{\beta_S}{h_S} \|[\![e_F]\!]\|_{L^2(S)}^2 \leq c\beta_S \sum_{T = T_+, T_-} \big(h_S^{-2} \|e_F\|_{L^2(T)}^2 + \|\nabla e_F\|_{L^2(T)}^2\big)$$

$$\leq c\beta_S \sum_{T = T_+, T_-} \big(h_S^{-2}h_T^4 + \beta_S h_T^2\big)\|D^2 u\|_{L^2(T)}^2.$$

The choice of β_S implies the estimate. □

This interpolation result implies an error estimate in the norm $\| \cdot \|_h^2 = a_h(\cdot, \cdot)$. Remarkably, the stabilization does not reduce the order of convergence and is robust for $\lambda \to \infty$.

Theorem 8.2 (Optimal Convergence) *Assume that $\beta_S = 2\mu$ for all $S \in \mathscr{S}_h$. For the solution $u_h \in \mathscr{S}_D^{1,cr}(\mathscr{T}_h)^d$ of the stabilized Crouzeix–Raviart discretization of the Navier–Lamé equations, we have that*

$$\| u - u_h \|_h \leq c_{\text{crs}}h \big((2\mu)^{1/2} \|D^2 u\|_{L^2(\Omega)} + (\lambda^{1/2} + (2\mu)^{-1/2}\lambda)\|\nabla \operatorname{div} u\|_{L^2(\Omega)}\big).$$

In particular, if $\|u\|_{H^2(\Omega)} + \lambda\|\nabla\operatorname{div}u\|_{L^2(\Omega)} \le c_{NL}\|f\|_{L^2(\Omega)}$, *then*

$$\|\mathbb{C}^{1/2}\varepsilon_{\mathscr{T}}(u - u_h)\|_{L^2(\Omega)} \le c_{crs}h\left((2\mu)^{1/2} + \lambda^{-1/2} + (2\mu)^{-1/2}\right)c_{NL}\|f\|_{L^2(\Omega)}.$$

Proof The triangle inequality for the norm $\|\cdot\|_h$ shows that

$$\|u - u_h\|_h \le \|u_h - I_F u\|_h + \|u - I_F u\|_h.$$

The consistency estimate and the Cauchy–Schwarz inequality imply that

$$\begin{aligned}\|u_h - I_F u\|_h^2 &= a_h(u - I_F u, u_h - I_F u) + a_h(u_h - u, u_h - I_F u)\\ &\le \|u - I_F u\|_h\|u_h - I_F u\|_h + |\mathscr{C}_h(u; u_h - I_F u)|.\end{aligned}$$

Noting that

$$\begin{aligned}\|u - I_F u\|_h^2 &= \lambda\|\operatorname{div}_{\mathscr{T}}(u - I_F u)\|_{L^2(\Omega)}^2 + 2\mu\|\varepsilon_{\mathscr{T}}(u - I_F u)\|_{L^2(\Omega)}^2\\ &\quad + \sum_{S\in\mathscr{S}_h\setminus\partial\Omega}\frac{2\mu}{h_S}\|[\![u - I_F u]\!]\|_{L^2(S)}^2,\end{aligned}$$

the estimate follows from Lemmas 8.2, 8.3, and 8.4.   □

8.1.6 Implementation

The implementation of the Crouzeix–Raviart approximation of the Navier–Lamé equations extends the MATLAB program for the nonconforming approximation of the Poisson problem. With the standard basis $(\varphi_S : S \in \mathscr{S}_h)$ of $\mathscr{S}^{1,cr}(\mathscr{T}_h)$, we have for $S, R, R' \in \mathscr{S}_h$ and $T_+, T_- \in \mathscr{T}_h$, such that $S = T_+ \cap T_-$ and $R, R' \subset T_+ \cup T_-$, that

$$\int_S [\![\varphi_R]\!][\![\varphi_{R'}]\!]\,ds = \begin{cases}0, & R = S \text{ or } R' = S,\\ |S|\dfrac{d\delta_{RR'} - 1}{d + 1}, & R, R' \subset T_+ \text{ or } R, R' \subset T_-,\\ -|S|\dfrac{d\delta_{RR'} - 1}{d + 1}, & R \subset T_+, R' \subset T_- \text{ or } R \subset T_-, R' \subset T_+.\end{cases}$$

Here, we used $[\![\varphi_S]\!]|_S = 0$, $\int_S \varphi_z\varphi_y\,ds = |S|(1 + \delta_{zy})/(d(d + 1))$ for $z, y \in S \cap \mathscr{N}_h$, and $\varphi_R = 1 - d\varphi_z$ for some $z \in S \cap \mathscr{N}_h$ if $R \ne S$. A basis for the Crouzeix–Raviart vector fields $\mathscr{S}^{1,cr}(\mathscr{T}_h)^d$ is defined by

$$(e_j\varphi_S : j = 1, 2, \ldots, d, S \in \mathscr{S}_h)$$

```
function cr_elast_stabilized(d,red)
global lambda mu;
lambda = 1000; mu = 1; beta = 2*mu;
[c4n,n4e,Db,Nb] = triang_cube(d);
for j = 1:red
    [c4n,n4e,Db,Nb] = red_refine(c4n,n4e,Db,Nb);
end
[s4e,~,n4s,s4Db,~,e4s] = sides(n4e,Db,Nb);
nS = size(n4s,1);
fSides = setdiff(1:nS,s4Db);
FSides = d*(repmat(fSides,d,1)-1)+(1:d)'*ones(1,size(fSides,2));
FSides = FSides(:);
u = zeros(d*nS,1);
[S_cr,b,vol_S] = cr_stiffness_elast(c4n,n4e,s4e,nS,d);
ctr = 0; ctr_max = 4*d^3*nS;
I = zeros(ctr_max,1); J = zeros(ctr_max,1); X = zeros(ctr_max,1);
for j = 1:nS
    h_S = vol_S(j)^(1/(d-1));
    if (e4s(j,1) && e4s(j,2))
        for k = 1:2
            for ell = 1:2
                [~,ind1] = intersect(n4e(e4s(j,k),:),n4s(j,:));
                [~,ind2] = intersect(n4e(e4s(j,ell),:),n4s(j,:));
                for m = reshape(ind1,1,d)
                    for n = reshape(ind2,1,d)
                        for p = 1:d
                            ctr = ctr+1;
                            delta = (n4e(e4s(j,k),m) ...
                                ==n4e(e4s(j,ell),n));
                            I(ctr) = d*(s4e(e4s(j,k),m)-1)+p;
                            J(ctr) = d*(s4e(e4s(j,ell),n)-1)+p;
                            X(ctr) = (-1)^(k-ell)*beta*h_S^(-1) ...
                                *vol_S(j)*(d*delta-1)/(d+1);
                        end
                    end
                end
            end
        end
    end
end
Z_stab = sparse(I(1:ctr),J(1:ctr),X(1:ctr),d*nS,d*nS);
S_full = S_cr+Z_stab;
u(FSides) = S_full(FSides,FSides)\b(FSides);
show_cr_def(c4n,n4e,s4e,u)
```

Fig. 8.3 Stabilized Crouzeix–Raviart finite element approximation of the Navier–Lamé equations

with the canonical basis vector $e_1, e_2, \ldots, e_d \in \mathbb{R}^d$. The matrix defined by the stabilizing bilinear form is assembled in a loop over all sides. For this, appropriate data structures, provided by the routine sides.m, are used. For simplicity, the code shown in Fig. 8.3 only realizes homogeneous Dirichlet and Neumann boundary conditions. The routines cr_stiffness_elast.m and show_cr_def.m shown in Fig. 8.4 provide the vectorial Crouzeix–Raviart stiffness matrix, the

```
function [S_cr,b,vol_S] = cr_stiffness_elast(c4n,n4e,s4e,nS,d)
global lambda mu;
nE = size(n4e,1); ctr = 0; ctr_max = d^2*(d+1)^2*nE;
b = zeros(d*nS,1); vol_S = zeros(nS,1); zz = zeros(d,d);
I = zeros(ctr_max,1); J = zeros(ctr_max,1); X = zeros(ctr_max,1);
for j = 1:nE
    X_T = [ones(1,d+1);c4n(n4e(j,:),:)'];
    grads_T = X_T\[zeros(1,d);eye(d)];
    vol_T = det(X_T)/factorial(d);
    f_mp_T = f(sum(c4n(n4e(j,:),:),1)/(d+1));
    heights = 1./sqrt(sum(grads_T.^2,2));
    vol_S(s4e(j,:)) = factorial(d)*vol_T./heights;
    for m = 1:d+1
        for p = 1:d
            d_phi_mp = zz; d_phi_mp(p,:) = -d*grads_T(m,:);
            Ceps_phi_mp = lambda*sum(diag(d_phi_mp))*eye(d)...
                +2*mu*(d_phi_mp'+d_phi_mp)/2;
            b(d*(s4e(j,m)-1)+p) = b(d*(s4e(j,m)-1)+p)...
                +(1/(d+1))*vol_T*f_mp_T(p);
            for n = 1:d+1
                for q = 1:d
                    ctr = ctr+1;
                    d_phi_nq = zz; d_phi_nq(q,:) = -d*grads_T(n,:);
                    I(ctr) = d*(s4e(j,m)-1)+p;
                    J(ctr) = d*(s4e(j,n)-1)+q;
                    X(ctr) = vol_T*sum(sum(Ceps_phi_mp.*d_phi_nq));
                end
            end
        end
    end
end
S_cr = sparse(I,J,X,d*nS,d*nS);

function val = f(x); d = size(x,2); val = zeros(d,1); val(d) = -1;
```

```
function show_cr_def(c4n,n4e,s4e,u)
nE = size(n4e,1); d = size(c4n,2);
Signum = ones(d+1)-d*eye(d+1); X = zeros((d+1)*nE,d);
E = reshape(1:3*nE,3,nE)'; n4e_t = n4e';
for k = 1:d
    u_p1_disc_k = Signum*u(d*(s4e-1)+k)';
    X(:,k) = c4n(n4e_t(:),k)+u_p1_disc_k(:);
end
if d == 2
    trisurf(E,X(:,1),X(:,2),zeros((d+1)*nE,1)); view(0,90);
else
    for j = 1:nE
        tetramesh(1:d+1,X((d+1)*(j-1)+(1:d+1),:),0); hold on;
    end
    hold off; s = .1; axis([-s 1+s -s 1+s -s 1+s]); view(44,14)
end
```

Fig. 8.4 Assembly of a vectorial Crouzeix–Riravient stiffness matrix (*top*) and visualization (*bottom*)

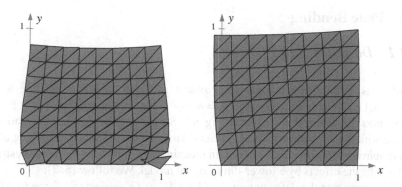

Fig. 8.5 Accurate and inaccurate approximation of a nearly incompressible deformation obtained with the stabilized Crouzeix–Raviart method for a moderate penalization (*left*) and an overpenalization (*right*) of jumps, respectively

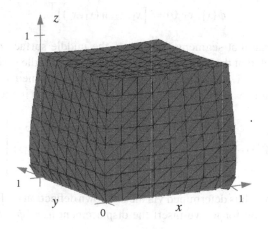

Fig. 8.6 Stabilized Crouzeix–Raviart finite element approximation of a nearly incompressible three-dimensional elastic deformation

right-hand side vector, and the surface areas of the sides, and we can visualize a Crouzeix–Raviart deformation, respectively. Results of two- and three-dimensional experiments are shown in Figs. 8.5 and 8.6. The results show that for an over-penalization of the jumps, which corresponds to a $P1$-finite element method, the approximations do not reflect the correct physical behavior.

8.2 Plate Bending

8.2.1 Description of Plates

A *plate* is a three-dimensional solid object of small thickness relative to the
diameter, i.e., it occupies a domain $\Omega_t = \omega \times (-t/2, t/2)$ with a domain $\omega \subset \mathbb{R}^2$ and
a thickness parameter $0 < t \ll 1$, cf. Fig. 8.7. Simulating the elastic deformation of
a plate with the three-dimensional Navier–Lamé equations leads to a locking effect
and requires mesh-sizes $h \ll t$ to obtain meaningful results. It is therefore desirable
to describe the effects by a lower-dimensional model. We follow [6, 7].

We assume that the deformation $\phi : \Omega_t \to \mathbb{R}^3$ of Ω_t subject to a force $f : \Omega_t \to$
$\mathbb{R}^3$ is entirely described by the *deflection* $w : \omega \to \mathbb{R}$ of the middle surface ω, i.e.,
for $(x_1, x_2) \in \omega$, we have

$$\phi(x_1, x_2, 0) = \left[x_1, x_2, w(x_1, x_2)\right]^\top.$$

We further assume that segments, normal to the middle surface ω, are deformed
isometrically such that they are normal to the deformed middle surface, cf. Fig. 8.7.
Hence if $|\nabla w| \ll 1$, then the unit normal to the deformed middle surface is
approximated via $\nu \approx \partial_1 \phi \times \partial_2 \phi = [-\nabla w, 1]^\top$. This results in the representation

$$\phi(x_1, x_2, x_3) = \begin{bmatrix} x_1 \\ x_2 \\ w(x_1, x_2) \end{bmatrix} + x_3 \begin{bmatrix} -\partial_1 w(x_1, x_2) \\ -\partial_2 w(x_1, x_2) \\ 1 \end{bmatrix}.$$

The deformation is thus determined via the unknown deflection w. To derive a partial
differential equation for w, we insert the displacement $u = \phi - \mathrm{id}$ in the energy
functional

$$I(u) = \frac{1}{2} \int_{\Omega_t} \mathbb{C}\varepsilon(u) : \varepsilon(u) \, dx - \int_{\Omega_t} f \cdot u \, dx.$$

Fig. 8.7 A plate $\Omega_t = \omega \times (-t/2, t/2)$ (*left*); illustration of the Kirchhoff assumptions (*right*)

Abbreviating $x' = (x_1, x_2)$, we have $u(x', x_3) = [-x_3 \nabla w(x'), w(x')]^\top$ and hence

$$\varepsilon(u) = \frac{1}{2}(\nabla u + \nabla u^\top) = -x_3 \begin{bmatrix} D^2 w(x') & 0 \\ 0 & 0 \end{bmatrix} = -x_3 [D^2 w(x')]_{3\times3},$$

where $[B]_{3\times3}$ denotes the canonical embedding of a matrix $B \in \mathbb{R}^{2\times2}$ into $\mathbb{R}^{3\times3}$. This implies that

$$\mathbb{C}\varepsilon(u) = -\lambda x_3 \Delta w I - 2\mu x_3 [D^2 w]_{3\times3}$$

and

$$\mathbb{C}\varepsilon(u) : \varepsilon(u) = \lambda x_3^2 |\Delta w|^2 + 2\mu x_3^2 |D^2 w|^2.$$

We assume that f only depends on x' and deduce that

$$f \cdot u = -x_3 f_{1,2}(x') \cdot \nabla w(x') + f_3(x') w(x').$$

Using these expressions in the functional I leads to

$$
\begin{aligned}
I(u) &= \frac{\lambda}{2} \int_\omega \int_{-t/2}^{t/2} x_3^2 |\Delta w(x')|^2 \, dx_3 \, dx' + \mu \int_\omega \int_{-t/2}^{t/2} x_3^2 |D^2 w(x')|^2 \, dx_3 \, dx' \\
&\quad + \int_\omega \int_{-t/2}^{t/2} x_3 f_{1,2}(x') \cdot \nabla w(x') \, dx_3 \, dx' - \int_\omega \int_{-t/2}^{t/2} f_3(x') w(x') \, dx_3 \, dx' \\
&= \frac{\lambda t^3}{24} \int_\omega |\Delta w|^2 \, dx' + \frac{2\mu t^3}{24} \int_\omega |D^2 w|^2 \, dx' - t \int_\omega f_3 w \, dx'.
\end{aligned}
$$

Defining $f = 12(\lambda + 2\mu)^{-1} f_3 / t^2$, we thus find that w minimizes the *Kirchhoff bending energy*

$$I_{Ki}(w) = \frac{\theta}{2} \int_\omega |\Delta w|^2 \, dx' + \frac{1 - \theta}{2} \int_\omega |D^2 w|^2 \, dx' - \int_\omega f w \, dx',$$

where we denote $\theta = \lambda / (\lambda + 2\mu)$. If the deflection w vanishes on $\partial \omega$, but the boundary segments $\{x\} \times (-t/2, t/2)$ for $x \in \partial \omega$ are free to rotate, then we impose the boundary condition

$$w|_{\partial\omega} = 0,$$

Fig. 8.8 A simply supported plate (*left*), and a clamped plate (*right*), leading to different boundary conditions

which is called a *simple support*. If the lateral boundary $\partial\omega \times (-t/2, t/2) \subset \partial\Omega_t$ is fixed, then we obtain the boundary conditions

$$w|_{\partial\omega} = 0, \quad \nabla w|_{\partial\omega} = 0,$$

which are referred to as *clamped plate* boundary conditions, cf. Fig. 8.8. The condition $w|_{\partial\omega} = 0$ implies that the tangential derivative of w along $\partial\omega$ vanishes and hence the condition $\nabla w|_{\partial\omega} = 0$ can be replaced by $\partial_n w|_{\partial\omega} = 0$.

Remark 8.6 Our assumptions about the actual deformation lead to a model that describes bending phenomena. So-called *membrane phenomena*, such as shearing or stretching effects, are not covered by the Kirchhoff model.

8.2.2   Euler–Lagrange Equations

The Euler–Lagrange equations for the Kirchhoff bending energy take a simple form due to certain relations for squares of second order derivatives. We write x instead of x' for the variable on the middle surface ω, and u for its deflection in what follows.

Lemma 8.5 (Hessian Versus Laplacian) *For every $u \in H^2(\omega)$, we have*

$$\int_\omega |D^2 u|^2 - |\Delta u|^2 \, dx = 2 \int_\omega (\partial_1 \partial_2 u)^2 - \partial_1^2 u \partial_2^2 u \, dx.$$

In particular, we have

$$I_{Ki}(u) = \frac{1}{2} \int_\omega |\Delta u|^2 \, dx + (1 - \theta) \int_\omega (\partial_1 \partial_2 u)^2 - \partial_1^2 u \partial_2^2 u \, dx - \int_\omega fu \, dx.$$

Proof Exercise. $\square$

With the lemma we obtain a simple form of the optimality conditions.

Proposition 8.3 (Bilinear Form) *Let $V \subset H^2(\omega)$ be a linear subspace and assume that $u \in V$ is minimal for I_{Ki} in V. With the bilinear form $a : H^2(\omega) \times H^2(\omega) \to \mathbb{R}$,*

$$a(u, v) = \int_\omega \Delta u \, \Delta v \, dx + (1 - \theta) \int_\omega 2\partial_1 \partial_2 u \, \partial_1 \partial_2 v - \partial_1^2 u \, \partial_2^2 v - \partial_2^2 u \, \partial_1^2 v \, dx,$$

and with the linear functional $\ell : H^2(\omega) \to \mathbb{R}$,

$$\ell(v) = \int_\omega fv \, dx,$$

we have $a(u, v) = \ell(v)$ for all $v \in V$.

Proof Exercise. □

The second integral in the definition of the bilinear form a can be expressed as a boundary integral.

Lemma 8.6 (Boundary Terms) *For* $u, v \in H^3(\omega)$, *we have*

$$\int_\omega 2\partial_1\partial_2 u \, \partial_1\partial_2 v - \partial_1^2 u \, \partial_2^2 v - \partial_2^2 u \, \partial_1^2 v \, dx = \int_{\partial\omega} \partial_n\partial_\tau u \, \partial_\tau v - \partial_\tau^2 u \, \partial_n v \, ds,$$

where $\tau = [-n_2, n_1]^\mathsf{T}$ *is a unit tangent vector on* $\partial\omega$, *and*

$$\partial_\tau^2 u = \tau^\mathsf{T} D^2 u \, \tau, \qquad \partial_n\partial_\tau u = n^\mathsf{T} D^2 u \, \tau.$$

Proof Assume that $u, v \in C^\infty(\overline{\omega})$. Repeated integration-by-parts shows that

$$2\int_\omega \partial_1\partial_2 u \partial_1\partial_2 v \, dx = -\int_\omega \partial_1^2\partial_2 u \partial_2 v \, dx + \int_{\partial\omega} \partial_1\partial_2 u \partial_2 v n_1 \, ds$$

$$-\int_\omega \partial_1\partial_2^2 u \partial_1 v \, dx + \int_{\partial\omega} \partial_1\partial_2 u \partial_1 v n_2 \, ds$$

$$= \int_\omega \partial_1^2 u \partial_2^2 v \, dx - \int_{\partial\omega} \partial_1^2 u \partial_2 v n_2 \, ds + \int_{\partial\omega} \partial_1\partial_2 u \partial_2 v n_1 \, ds$$

$$+ \int_\omega \partial_2^2 u \partial_1^2 v \, dx - \int_{\partial\omega} \partial_2^2 u \partial_1 v n_1 \, ds + \int_{\partial\omega} \partial_1\partial_2 u \partial_1 v n_2 \, ds.$$

For the boundary terms $I_{\partial\omega}$ on the right-hand side we have

$$I_{\partial\omega} = \int_{\partial\omega} \tau_1 \partial_1^2 u \partial_2 v + \tau_2 \partial_1\partial_2 u \partial_2 v - \tau_2 \partial_2^2 u \partial_1 v - \tau_1 \partial_1\partial_2 u \partial_1 v \, ds$$

$$= \int_{\partial\omega} \begin{bmatrix} -\tau_2 \partial_2^2 u - \tau_1 \partial_1\partial_2 u \\ \tau_1 \partial_1^2 u + \tau_2 \partial_1\partial_2 u \end{bmatrix} \cdot \nabla v \, ds$$

$$= \int_{\partial\omega} \begin{bmatrix} -\nabla\partial_2 u \cdot \tau \\ \nabla\partial_1 u \cdot \tau \end{bmatrix} \cdot \left[(n \cdot \nabla v)n + (\tau \cdot \nabla v)\tau\right] ds$$

$$= \int_{\partial\omega} \left(-n_1 \nabla \partial_2 u + n_2 \nabla \partial_1 u \right) \cdot \tau \partial_n v + \left(-\tau_1 \nabla \partial_2 u + \tau_2 \nabla \partial_1 u \right) \cdot \tau \partial_\tau v \, ds$$

$$= \int_{\partial\omega} \left(-\tau^\top D^2 u \tau \right) \partial_n v + (n^\top D^2 u \tau) \partial_\tau v \, ds.$$

This proves the lemma for smooth functions; a density argument implies the general result.   □

The second tangential derivative $\partial_\tau^2 u$ can be expressed in terms of the curvature of $\partial\omega$ and the normal derivative $\partial_n u$.

Remark 8.7 If $\phi : (-\delta, \delta) \to \mathbb{R}^2$ is a local C^2-parametrization of $\partial\omega$ with $\phi'(s) = \tau(s)$ for $s \in (-\delta, \delta)$, and if $u \in H^2(\omega)$ with $u|_{\partial\omega} = 0$, then we have $u \circ \phi(s) = 0$ for $s \in (-\delta, \delta)$, and hence

$$\partial_\tau u(\phi(s)) = \nabla u(\phi(s)) \cdot \phi'(s) = 0.$$

Differentiating once again with respect to s leads to

$$\phi'(s)^\top D^2 u(\phi(s)) \phi'(s) + \nabla u(\phi(s)) \cdot \phi''(s) = 0.$$

Since $|\phi'(s)|^2 = 1$, we have $\phi'' \cdot \tau = 0$ and hence $\phi'' = \kappa n$ with the curvature $\kappa : \partial\omega \to \mathbb{R}$. This implies that

$$\partial_\tau^2 u = -\kappa \partial_n u.$$

Note that for $\omega = B_r(0)$ and $\phi(s) = r[\cos(s), \sin(s)]$ we have $\kappa = -1/r$.

To derive the Euler–Lagrange equations for the Kirchhoff bending energy, we use the following integration-by-parts formula.

Lemma 8.7 (Bilaplacian)   *For $u \in H^4(\omega)$ and $v \in H^2(\omega)$ we have*

$$\int_\omega \Delta u \, \Delta v \, dx = \int_\omega \Delta^2 u \, v \, dx - \int_{\partial\omega} (\partial_n \Delta u) \, v \, ds + \int_{\partial\omega} \Delta u \, \partial_n v \, ds.$$

Proof Exercise.   □

The Euler–Lagrange equations define a fourth order partial differential equation with two boundary conditions. Note that for clamped boundary conditions we have $\partial_\tau u = 0$ and $\partial_n u = \nabla u \cdot n = 0$ on $\partial\omega$, so that all boundary integrals identified above disappear.

Proposition 8.4 (Euler–Lagrange Equations) *Assume that $u \in H^4(\omega)$ is a minimizer for the Kirchhoff energy I_{Ki} in the set of functions in $H^2(\omega)$ satisfying simple support or clamped plate boundary conditions.*

(i) *In the case of clamped plate conditions we have*

$$\Delta^2 u = f \text{ in } \omega, \quad u|_{\partial\omega} = 0, \quad \nabla u|_{\partial\omega} = 0.$$

(ii) *In the case of simple support conditions we have*

$$\Delta^2 u = f \text{ in } \omega, \quad u|_{\partial\omega} = 0, \quad \Delta u + (1-\theta)\kappa\partial_n u = 0 \text{ on } \partial\omega.$$

Proof Let $u \in H^4(\omega)$ be a minimizer of the Kirchhoff energy subject to clamped or simple support boundary conditions, so that

$$a(u, v) = \ell(v)$$

for all $v \in H^2(\omega) \cap H_0^1(\omega)$ satisfying also $\nabla v|_{\partial\omega} = 0$ in the case of clamped plate conditions. For every compactly supported function $v \in C_0^\infty(\omega)$, we thus have by Proposition 8.3 and Lemma 8.6 that

$$\int_\omega \Delta u \, \Delta v \, dx = \int_\omega f v \, dx.$$

Lemma 8.7 then shows that

$$\int_\omega \Delta^2 u \, v \, dx = \int_\omega f v \, dx,$$

which implies the partial differential equation via the fundamental lemma of the calculus of variations. The boundary conditions for the clamped plate coincide with the imposed essential boundary conditions. In the case of a simply supported plate we have the essential boundary condition $u|_{\partial\omega} = 0$, and it remains to derive the natural boundary condition $\Delta u + (1-\theta)\kappa\partial_n u = 0$ on $\partial\omega$. We let $v \in C^\infty(\overline{\omega})$ with $v|_{\partial\omega} = 0$ and note that $\partial_\tau u = \partial_\tau v = 0$ on $\partial\omega$. Hence, using that $\Delta^2 u = f$, the relation $a(u, v) = \ell(v)$ and Lemmas 8.6 and 8.7 imply that

$$\int_{\partial\omega} \Delta u \, \partial_n v \, ds - (1-\theta) \int_{\partial\omega} \partial_\tau^2 u \, \partial_n v \, ds = 0.$$

Letting $x_0 \in \partial\omega$ and $\varepsilon > 0$, and choosing v such that $\partial_n v(x_0) = 1$ and $\text{supp } v \subset B_\varepsilon(x_0) \cap \omega$, we find that $\Delta u - (1-\theta)\partial_\tau^2 u = 0$. Incorporating the relation $\partial_\tau^2 u = -\kappa\partial_n u$ implies the asserted condition. $\qquad\square$

Remarks 8.8

(i) The fourth order problem is called *plate bending* or *biharmonic problem*.
(ii) If the boundary of ω is piecewise flat, i.e., if $\kappa = 0$ almost everywhere, or if clamped boundary conditions are imposed, then the material parameter θ does

not enter the Euler–Lagrange equations, i.e., all clamped plates deform in the
same way.

(iii) The appearance of the curvature of the boundary implies that the bending
problem on a domain with curved boundary cannot be approximated by a
sequence of problems on domains with polyhedral boundaries, known as the
Babuška paradox for simply supported plates.

(iv) The bilinear form defined by the bilaplacian, i.e., the bilinear form a with $\theta = 1$, is coercive on $H^2(\omega) \cap H_0^1(\omega)$ if and only if the Laplace operator on ω is H^2-regular.

8.2.3 Well-Posedness

To establish the existence of minimizers for Kirchhoff bending energy, we show that
the bilinear form a is coercive.

Proposition 8.5 (Poincaré Inequality) *There exists $c_P > 0$ such that for all $u \in H^2(\omega) \cap H_0^1(\omega)$, we have that*

$$\|u\|_{H^2(\omega)} \le c_P \|D^2 u\|_{L^2(\omega)}.$$

Proof Assume that the statement is false, so that there exists a sequence $(u_j)_{j \in \mathbb{N}} \subset H^2(\omega)$ with $u_j|_{\partial \omega} = 0$ for all $j \in \mathbb{N}$ and

$$\|u_j\|_{H^2(\omega)} > j \|D^2 u_j\|_{L^2(\omega)}.$$

By a rescaling we may assume that $\|u_j\|_{H^2(\omega)} = 1$ for all $j \in \mathbb{N}$. We thus have
that $D^2 u_j \to 0$ in $L^2(\omega)$. The Eberlein–Smuljan theorem implies that there exists a
subsequence $(u_{j_k})_{k \in \mathbb{N}}$ that converges weakly to some $u \in H^2(\omega)$. The compactness
of the embedding id : $H^2(\omega) \to H^1(\omega)$ proves that we have $u_{j_k} \to u$ in $H^1(\omega)$
as $k \to \infty$. But then $u_{j_k} \to u$ converges strongly in $H^2(\omega)$ with $D^2 u = 0$ and
$\|u\|_{H^2(\omega)} = 1$. Since $D^2 u = 0$ we have that u is affine. By continuity of the trace
operator tr : $H^2(\omega) \to L^2(\partial \omega)$, we also have that $u|_{\partial \omega} = 0$, which yields that $u = 0$
and contradicts $\|u\|_{H^2(\omega)} = 1$.   □

The coercivity guaranteed by the Poincaré inequality implies the existence and
uniqueness of solutions for the plate bending problem.

Theorem 8.3 (Well-Posedness) *If $\mu > 0$ or if the Poisson problem in ω is H^2-regular, then there exists a unique solution $u \in H^2(\omega)$ for the plate bending problem
with clamped plate or simple support boundary conditions.*

Proof The result follows from the Lax–Milgram lemma.   □

A best approximation result is another immediate observation.

Proposition 8.6 (Approximation) *Assume that $V_h \subset H^2(\omega)$ such that functions in V_h satisfy clamped plate or simple support boundary conditions. For the corresponding solution of the bending problem, we have*

$$\|u - u_h\|_{H^2(\omega)} \leq c_{\text{bend}} \inf_{v_h \in V_h} \|u - v_h\|_{H^2(\omega)}.$$

Proof Exercise. □

8.2.4 Plate Elements

The construction of finite element spaces that are contained in $H^2(\omega)$ is difficult, since continuity of the derivatives across interelement edges has to be guaranteed.

Example 8.4 The *Argyris element* is the finite element $(T, \mathscr{P}_5(T), \mathscr{K}_T)$ for a triangle $T = \text{conv}\{z_0, z_1, z_2\}$, the space of fifth order polynomials $\mathscr{P}_5(T)$ on T, and the set of node functionals in $\mathscr{K}_T$ which are for $v \in H^4(T)$ given by

$$\chi_{i,\alpha}(v) = D^\alpha v(z_i)$$

for $i = 0, 1, 2$, and $\alpha \in \mathbb{N}_0^2$ with $|\alpha| \leq 2$, and

$$\chi_{i,n}(v) = \nabla v(x_{S_i}) \cdot n_{S_i}$$

for the sides $S_i \subset \partial T$ with unit normals n_{S_i} and midpoints x_{S_i}, $i = 0, 1, 2$, cf. Fig. 8.9.

The Argyris element is a C^1-conforming finite element with 21 degrees of freedom.

Proposition 8.7 (Argyris Element)

(i) *The Argyris element is a finite element, i.e., for every $v \in H^4(T)$ there exists a uniquely defined $q_T \in \mathscr{P}_5(T)$ such that $\chi_{i,\alpha}(v - q_T) = 0$ and $\chi_{i,n}(v - q_T) = 0$ for $i = 0, 1, 2$ and $|\alpha| \leq 2$.*

(ii) *The Argyris element is a C^1 element, i.e., if $v \in H^4(\omega)$, then the elementwise defined interpolant $\mathscr{I}_{\text{Arg}} v|_T = q_T$ for all $T \in \mathscr{T}_h$ satisfies $\mathscr{I}_{\text{Arg}} v \in C^1(\overline{\omega})$.*

Fig. 8.9 Schematical description of the Argyris element (*left*); assumed situation in the proof of Proposition 8.7 (*right*)

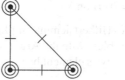

Proof

(i) To prove that the Argyris element is a finite element, we have to show that if
$p \in \mathscr{P}_5(T)$ with $\chi_{i,\alpha}(v) = 0$ and $\chi_{i,n}(v) = 0$ for $i = 0, 1, 2$ and $|\alpha| \le 2$, it
follows that $p = 0$. For this, we note the conditions imply $p|_{S_i} = 0$ and $\nabla p|_{S_i} =$
0 for $i = 0, 1, 2$. Indeed, assuming without loss of generality that $S_i = [0, 1]$,
we have that the fifth-order polynomial $p_i = p|_{S_i}$, i.e., $p_i(t) = p(t, 0)$, satisfies
$p_i(t) = p_i'(t) = p_i''(t) = 0$ for $t \in \{0, 1\}$, and hence $p_i(t) = p_i'(t) = 0$ for
all $t \in S_i$. The function $q_i = \nabla p|_{S_i} \cdot n_{S_i}$ is a polynomial of degree four. Since
$q_i(t) = 0$ for $t \in \{0, 1\}$, $q_i(x_{S_i}) = 0$ with $x_{S_i} = (1/2, 0)$, and $q_i'(t) = 0$ for
$t \in \{0, 1\}$, we also have that $q_i = 0$. Hence, $\nabla p|_{S_i} = 0$. To prove that $p = 0$, we
show that we have

$$p = \varphi_{z_0}^2 \varphi_{z_1}^2 \varphi_{z_2}^2 q$$

with the nodal basis functions $\varphi_{z_i} \in \mathscr{P}_1(T)$, and a polynomial q. Since $p \in$
$\mathscr{P}_5(T)$ this implies that $q = 0$ and hence $p = 0$. Without loss of generality,
we assume that $S_2 \subset \{0\} \times \mathbb{R}$ and $z_2 = (1, 0)$, cf. Fig. 8.9, so that $\varphi_{z_2}(x) = x_1$.
Noting that we may write

$$p(x_1, x_2) = \sum_{i=0}^{5} x_1^i r_i(x_2)$$

with polynomials $r_i \in \mathscr{P}_{5-i}(\mathbb{R})$, and that $p(0, x_2) = 0$ and $\partial_1 p(0, x_2) = 0$, we
find that $r_0 = r_1 = 0$, and hence

$$p(x_1, x_2) = x_1^2 \sum_{i=2}^{5} x_1^{i-2} r_i(x_2) = \varphi_{z_2}^2(x_1, x_2) r(x_1, x_2).$$

Repeating the argument with the sides S_0 and S_1 proves the statement.

(ii) Assume that $T_1, T_2 \in \mathscr{T}_h$ are neighboring triangles with common side $S =$
$[0, 1] \times \{0\}$, and that for $v_h \in L^1(\omega)$, we have $v_h|_{T_\ell} \in \mathscr{P}_5(T_\ell)$, $\ell = 1, 2$,
$D^\alpha v_h(x)$ is continuous at $x = (0, 0)$ and $x = (1, 0)$, and $\nabla v_h(x_S) \cdot n_S = \partial_2 v_h(x_S)$
is continuous. We have to show that v_h and ∇v_h are continuous on S. We set
$v_\ell = v_h|_{T_\ell}$ for $\ell = 1, 2$. Then $v_\ell|_S$ is for $\ell = 1, 2$, a polynomial of degree five.
Since $\partial_1^r v_1(x) = \partial_1^r v_2(x)$ for $r = 0, 1, 2$ and $x \in \{(0, 0), (1, 0)\}$, we have that
the polynomials coincide on S. It remains to show that $\partial_2 v_1 = \partial_2 v_2$ on S. We
have that $\partial_2 v_\ell|_S$ is for $\ell = 1, 2$ a polynomial of degree four. Since $\partial_1^r \partial_2 v_1(x) =$
$\partial_1^r \partial_2 v_2(x)$ for $r = 0, 1$ and $x \in \{(0, 0), (1, 0)\}$ and $\partial_2 v_1(x_S) = \partial_2 v_2(x_S)$, we find
that $\partial_2 v_1 = \partial_2 v_2$ on S. $\square$

The Bramble–Hilbert lemma implies an error estimate for approximating the
plate bending problem with the Argyris element. We denote the finite element space
defined by the Argyris element by

$$\mathscr{S}_0^{5,1}(\mathscr{T}_h) = \{v_h \in C^1(\overline{\omega}) : v_h|_{\partial\omega} = 0, \ v_h|_T \in \mathscr{P}_5(T) \text{ for all } T \in \mathscr{T}_h\}.$$

We state an error estimate for boundary conditions of simple support. The case of clamped boundary conditions is analogous.

Proposition 8.8 (Error Estimate) *There exists a unique function $u_h \in \mathscr{S}_0^{5,1}(\mathscr{T}_h)$ such that*

$$a(u_h, v_h) = \ell(v_h)$$

for all $v_h \in \mathscr{S}_0^{5,1}(\mathscr{T}_h)$. If the exact solution of the plate bending problem satisfies $u \in H^{s+1}(\omega)$ for $s \in \{3, 4, 5\}$, then we have

$$\|u - u_h\|_{H^2(\omega)} \leq c_{\mathrm{Arg}} h^{s-1} \|u\|_{H^{s+1}(\omega)}.$$

Proof Exercise. $\qquad\qquad\qquad\qquad\qquad\qquad\qquad\qquad\qquad\qquad\qquad\qquad\qquad\qquad\qquad$ □

The use of higher-order polynomials can be avoided by using nonconforming or nonpolynomial elements.

Example 8.5 For a triangle $T = \mathrm{conv}\{z_0, z_1, z_2\}$ with midpoint x_T and sides S_0, S_1, S_2, let K_0, K_1, K_2 be the subtriangles with vertex x_T and sides S_0, S_1, S_2, respectively. Define

$$\widetilde{P}_T = \{v \in C^1(T) : v|_{K_i} \in \mathscr{P}_3(K_i), \ i = 0, 1, 2\},$$

and let $\mathscr{K}_T$ be the set of node functionals given for $v \in C^1(T)$ by

$$\chi_{i,\alpha}(v) = \partial^\alpha v(z_i), \quad \chi_{i,n}(v) = \nabla v(x_{S_i}) \cdot n_{S_i},$$

for $i = 0, 1, 2$ and $\alpha \in \mathbb{N}_0^2$ with $|\alpha| \leq 1$, with the midpoints x_{S_i} and normals n_{S_i} for the sides S_i, $i = 0, 1, 2$. Then $(T, \widetilde{P}_T, \mathscr{K}_T)$ defines a (generalized) finite element called the *Hsieh–Clough–Tocher (HCT-) element* with 12 degrees of freedom.

The HCT-element is depicted in Fig. 8.10. It is easier to implement but leads to lower convergence rates.

Fig. 8.10 Schematical description of the nonpolynomial HCT-element

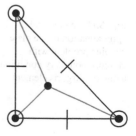

8.2.5 *Implementation*

Figure 8.11 shows the output of a numerical experiment for a plate bending problem
with clamped boundary conditions. The underlying simple MATLAB implementation of the Argyris element shown in Fig. 8.12 uses on every triangle representations
of the nodal basis $(N_m^T : m = 1, 2, \ldots, 21)$ defined by the condition $\chi_j^T(N_m^T) = \delta_{jm}$
for $j, m = 1, 2, \ldots, 21$, in terms of the monomial basis $(p_1, p_2, \ldots, p_{21})$ of $\mathscr{P}_5(\mathbb{R}^2)$,
given by

$$1,\ x,\ y,\ x^2,\ xy,\ y^2, \ldots, x^5,\ x^4y,\ x^3y^2,\ xy^4,\ y^5.$$

An appropriate sign has to be chosen for the nodal basis functions corresponding to
the edge midpoints in order to guarantee continuity. For computing the corresponding coefficients, the node functionals χ_j^T have to be evaluated at the monomials,
which is realized with the first routine shown in Fig. 8.13. This defines a coefficient
matrix $C^{(T)} \in \mathbb{R}^{21 \times 21}$ such that

$$N_m^T = \sum_{j=1}^{21} C_{jm}^{(T)} p_j$$

and allows for an assembly of the elementwise system matrix via

$$a_T(N_m, N_n) = \theta \int_T \Delta N_m \Delta N_n \, dx + (1 - \theta) \int_T D^2 N_m : D^2 N_n \, dx$$

$$= \sum_{j,k=1}^{21} C_{jm}^{(T)} C_{kn}^{(T)} a_T(p_j, p_k).$$

The integrals $a_T(p_j, p_k)$ are evaluated with a 12-point Gaussian quadrature rule; the
evaluation of the monomials and their derivatives at the quadrature points is done
in the second routine displayed in Fig. 8.13. The quadrature points on the element
T are obtained via a transformation from the reference element. This is realized

Fig. 8.11 Numerical
approximation of the plate
bending problem with
clamped boundary conditions
using the Argyris element

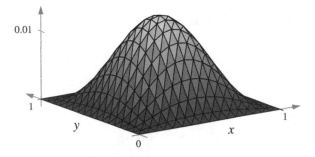

```
function argyris(red)
theta = 1/2;
[c4n,n4e,Db,Nb] = triang_cube(2); Db = [Db;Nb];
for j = 1:red
    [c4n,n4e,Db,Nb] = red_refine(c4n,n4e,Db,Nb);
end
[s4e,sign_s4e,n4s,s4Db] = sides(n4e,Db,Nb);
nE = size(n4e,1); nS = size(n4s,1); nC = size(c4n,1);
dNodes_tmp = 6*(reshape(unique(Db),length(unique(Db)),1)-1);
dNodes = repmat(dNodes_tmp,1,3)+ones(size(dNodes_tmp,1),1)*(1:3);
fNodes = setdiff(1:6*nC+nS,[dNodes(:);6*nC+s4Db]);
A = sparse(6*nC+nS,6*nC+nS);
u = zeros(6*nC+nS,1); b = zeros(6*nC+nS,1);
for j = 1:nE
    loc_c4n = c4n(n4e(j,:),:);
    N = argyris_monomials_at_dofs(loc_c4n);       % conditioning ok?
    E = eye(21); E(19:21,19:21) = diag(sign_s4e(j,:));
    C = N\E;
    [xi,kappa] = argyris_gauss_quad_ref_deg_6(loc_c4n);
    X_T = [ones(1,3);loc_c4n'];
    [p,p_xx,p_yy,p_xy] = argyris_monomials_at_qps(xi);
    f_qp = f(xi);
    A_loc = zeros(21,21);
    b_loc = zeros(21,1);
    for i = 1:size(kappa,1);
        A_loc = A_loc+kappa(i)*det(X_T)*...
            ((p_xx(i,:)'*p_xx(i,:)+p_yy(i,:)'*p_yy(i,:)...
            +p_xx(i,:)'*p_yy(i,:)+p_yy(i,:)'*p_xx(i,:))...
            +(1-theta)*(2*p_xy(i,:)'*p_xy(i,:)...
            -p_xx(i,:)'*p_yy(i,:)-p_xx(i,:)'*p_yy(i,:)));
        b_loc = b_loc+kappa(i)*det(X_T)*f_qp(i)*p(i,:)';
    end
    I_tmp = 6*(n4e(j,:)-1)';
    I = repmat(I_tmp,1,6)+ones(3,1)*(1:6);
    I = [I(:);6*nC+s4e(j,:)'];
    A(I,I) = A(I,I)+C'*A_loc*C;
    b(I) = b(I)+C'*b_loc;
end
u(fNodes) = A(fNodes,fNodes)\b(fNodes);
u_p1 = u(1:6:6*nC);
trisurf(n4e,c4n(:,1),c4n(:,2),u_p1);

function val = f(x)
val = 10*ones(size(x,1),1);
```

Fig. 8.12 Simple MATLAB implementation of the Argyris finite element for the plate bending problem

in the routine shown in Fig. 8.14. The implementation only allows for clamped boundary conditions and is only useful for triangulations of moderate mesh-sizes. For finer triangulations, the linear systems of equations that define the coefficients

```
function N = argyris_monomials_at_dofs(loc_c4n)
J = [0,-1;1,0];
x = loc_c4n(:,1); y = loc_c4n(:,2);
e = ones(3,1); o = zeros(3,1);
p = [e,x,y,x.^2,x.*y,y.^2,x.^3,x.^2.*y,x.*y.^2,...
     y.^3,x.^4,x.^3.*y,x.^2.*y.^2,x.*y.^3,y.^4,...
     x.^5,x.^4.*y,x.^3.*y.^2,x.^2.*y.^3,x.*y.^4,y.^5];
p_x = [o,e,o,2*x,y,o,3*x.^2,2*x.*y,y.^2,o,...
       4*x.^3,3*x.^2.*y,2*x.*y.^2,y.^3,o,...
       5*x.^4,4*x.^3.*y,3*x.^2.*y.^2,2*x.*y.^3,y.^4,o];
p_y = [o,o,e,o,x,2*y,o,x.^2,2*x.*y,3*y.^2,...
       o,x.^3,2*x.^2.*y,3*x.*y.^2,4*y.^3,...
       o,x.^4,2*x.^3.*y,3*x.^2.*y.^2,4*x.*y.^3,5*y.^4];
p_xx = [o,o,o,2*e,o,o,6*x,2*y,o,o,...
        12*x.^2,6*x.*y,2*y.^2,o,o,...
        20*x.^3,12*x.^2.*y,6*x.*y.^2,2*y.^3,o,o];
p_yy = [o,o,o,o,o,2*e,o,o,2*x,6*y,...
        o,o,2*x.^2,6*x.*y,12*y.^2,...
        o,o,2*x.^3,6*x.^2.*y,12*x.*y.^2,20*y.^3];
p_xy = [o,o,o,o,e,o,o,2*x,2*y,o,o,3*x.^2,4*x.*y,3*y.^2,o,...
        o,4*x.^3,6*x.^2.*y,6*x.*y.^2,4*y.^3,o];
N = [p;p_x;p_y;p_xx;p_yy;p_xy];
shift1 = [2,3,1]; shift2 = [3,1,2];
for k = 1:3
    z_a = loc_c4n(shift2(k),:); z_b = loc_c4n(shift1(k),:);
    m_S = (z_a+z_b)/2; n_S = J*(z_b-z_a)'/norm(z_b-z_a);
    x = m_S(1); y = m_S(2); e = 1; o = 0;
    p_x = [o,e,o,2*x,y,o,3*x.^2,2*x.*y,y.^2,o,...
           4*x.^3,3*x.^2.*y,2*x.*y.^2,y.^3,o,...
           5*x.^4,4*x.^3.*y,3*x.^2.*y.^2,2*x.*y.^3,y.^4,o];
    p_y = [o,o,e,o,x,2*y,o,x.^2,2*x.*y,3*y.^2,...
           o,x.^3,2*x.^2.*y,3*x.*y.^2,4*y.^3,...
           o,x.^4,2*x.^3.*y,3*x.^2.*y.^2,4*x.*y.^3,5*y.^4];
    N(18+k,:) = n_S(1)*p_x+n_S(2)*p_y;
end
```

```
function [p,p_xx,p_yy,p_xy] = argyris_monomials_at_qps(qp)
x = qp(:,1); y = qp(:,2);
e = ones(size(qp,1),1); o = zeros(size(qp,1),1);
p = [e,x,y,x.^2,x.*y,y.^2,x.^3,x.^2.*y,x.*y.^2,...
     y.^3,x.^4,x.^3.*y,x.^2.*y.^2,x.*y.^3,y.^4,...
     x.^5,x.^4.*y,x.^3.*y.^2,x.^2.*y.^3,x.*y.^4,y.^5];
p_xx = [o,o,o,2*e,o,o,6*x,2*y,o,o,...
        12*x.^2,6*x.*y,2*y.^2,o,o,...
        20*x.^3,12*x.^2.*y,6*x.*y.^2,2*y.^3,o,o];
p_yy = [o,o,o,o,o,2*e,o,o,2*x,6*y,...
        o,o,2*x.^2,6*x.*y,12*y.^2,...
        o,o,2*x.^3,6*x.^2.*y,12*x.*y.^2,20*y.^3];
p_xy = [o,o,o,o,e,o,o,2*x,2*y,o,o,3*x.^2,4*x.*y,3*y.^2,o,...
        o,4*x.^3,6*x.^2.*y,6*x.*y.^2,4*y.^3,o];
```

Fig. 8.13 Auxiliary routines for the implementation of the Argyris element

```
function [xi,kappa] = argyris_gauss_quad_ref_deg_6(loc_c4n)
G = [0.24928674517091    0.24928674517091    0.11678627572638;...
     0.24928674517091    0.50142650965818    0.11678627572638;...
     0.50142650965818    0.24928674517091    0.11678627572638;...
     0.06308901449150    0.06308901449150    0.05084490637021;...
     0.06308901449150    0.87382197101700    0.05084490637021;...
     0.87382197101700    0.06308901449150    0.05084490637021;...
     0.31035245103378    0.63650249912140    0.08285107561837;...
     0.63650249912140    0.05314504984482    0.08285107561837;...
     0.05314504984482    0.31035245103378    0.08285107561837;...
     0.63650249912140    0.31035245103378    0.08285107561837;...
     0.31035245103378    0.05314504984482    0.08285107561837;...
     0.05314504984482    0.63650249912140    0.08285107561837];
Phi = [1-G(:,1)-G(:,2),G(:,1),G(:,2)];
xi = Phi*loc_c4n;
kappa = G(:,3)/2;
```

Fig. 8.14 Affine transformation of a Gaussian quadrature rule with 12 quadrature points which is exact for polynomials of degree 6. The first two columns in the matrix G contain the coordinates on the reference element and the third column the corresponding normalized weights

of the nodal basis becomes ill-conditioned. An appropriate elementwise scaling of the monomials may improve this deficiency.

8.3 Electromagnetism

8.3.1 Maxwell Equations

When electric and magnetic fields undergo temporal changes, they influence each other, reflecting the physical laws of electromagnetic induction. The *Maxwell system* comprises four partial differential equations that describe the interaction of electric and magnetic fields E and B, and current and charge densities j and ϱ, via

$$\operatorname{curl} B = \mu j + \mu \varepsilon \partial_t E,$$

$$\operatorname{curl} E = -\partial_t B,$$

$$\operatorname{div} E = \varrho/\varepsilon,$$

$$\operatorname{div} B = 0.$$

Fig. 8.15 Illustration of
Faraday's law of electric
induction

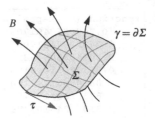

The constants $\varepsilon, \mu > 0$ are the permittivity and permeability of the considered
material, respectively. For a vector field $F : \Omega \to \mathbb{R}^3$, with $\Omega \subset \mathbb{R}^3$, we have

$$\operatorname{curl} F = \nabla \times F = \begin{bmatrix} \partial_2 F_3 - \partial_3 F_2 \\ \partial_3 F_1 - \partial_1 F_3 \\ \partial_1 F_2 - \partial_2 F_1 \end{bmatrix}.$$

The curl of a vector field is proportional to the angular velocity of a body transported
by the vector field and aligned with the axis about which the body rotates. The first
equation in the Maxwell system is *Ampère's law*, which states that electric currents
and temporally varying electric fields are surrounded by magnetic fields. The second
equation is *Faraday's law* of electric induction; its integral form states that for every
surface Σ and every closed loop γ such that $\gamma = \partial \Sigma$, we have, using *Stokes's
theorem*,

$$\int_{\partial \Sigma} E \cdot \tau \, dr = \int_{\Sigma} \operatorname{curl} E \cdot n \, ds = -\frac{d}{dt} \int_{\Sigma} B \cdot n \, ds,$$

i.e., a changing magnetic flux through the surface Σ generates a current in the closed
wire γ, cf. Fig. 8.15. The third and fourth equations are *Gauss's laws*, modeling
that electric charges are sources of electric fields, whereas magnetic field lines are
closed.

Remark 8.9 The Maxwell system predicts the existence of electromagnetic waves
with wave speed $c = (\varepsilon \mu)^{-1/2}$, which coincides with the speed of light in the case
of free space.

We consider a situation without electric charges, i.e., $\varrho = 0$, and eliminate the
magnetic field from the system by taking the curl of the second equation, i.e., we
have

$$\operatorname{curl} \operatorname{curl} E = -\partial_t \operatorname{curl} B = -\mu \partial_t j - \mu \varepsilon \partial_t^2 E.$$

We have thus obtained a system of equations for E, given by

$$\mu\varepsilon\partial_t^2 E + \operatorname{curl}\operatorname{curl} E = -\mu\partial_t j,$$

$$\operatorname{div} E = 0,$$

which is supplemented by appropriate boundary conditions, e.g., that the boundary of the domain under consideration is *perfectly conducting*, so that according to *Ohm's law* the electric field E has no tangential component, i.e., on the boundary $\partial\Omega$ we have

$$n \times E = 0.$$

In applications it is often sufficient to consider a time-harmonic situation, i.e., the field E is given by

$$E(t, x) = \operatorname{Re}\left(e^{-i\omega t} u(x)\right),$$

with a fixed frequency $\omega \in \mathbb{R}_{\geq 0}$, and a time-independent vector field $u : \Omega \to \mathbb{R}^3$. Mathematically, this results from a Fourier transformation in time. The unknown variable u then solves the stationary boundary value problem

$$\operatorname{curl}\operatorname{curl} u - \omega^2 u = f \quad \text{in } \Omega,$$

$$\operatorname{div} u = 0 \quad \text{in } \Omega,$$

$$u \times n = 0 \quad \text{on } \partial\Omega.$$

The right-hand side function f results from a Fourier transformation of $\partial_t j$, and is assumed to be divergence-free; we set for simplicity $\mu\varepsilon = 1$.

Remarks 8.10

(i) If $\operatorname{div} f = 0$ and $\omega \neq 0$, then $\operatorname{div} u = 0$ and the second equation is redundant.
(ii) The system is ill-posed if ω^2 is an eigenvalue of the operator $A = \operatorname{curl}\operatorname{curl}$ on the space of functions satisfying the boundary condition $v \times n = 0$.

8.3.2 The Space $H(\operatorname{curl}; \Omega)$

To derive a weak formulation for the time-harmonic Maxwell system, we have to identify an appropriate function space related to the curl operator. For this we note that the integration-by-parts formula implies that for $\phi, w \in C^1(\overline{\Omega}; \mathbb{R}^3)$, we have

$$\int_\Omega w \cdot \operatorname{Curl}\phi \, dx = \int_\Omega \operatorname{curl} w \cdot \phi \, dx - \int_{\partial\Omega} \phi \cdot (w \wedge n) \, ds,$$

where Curl $=$ curl and $\phi \wedge n = \phi \times n$. By defining

$$J = \begin{bmatrix} 0 & 1 \\ -1 & 0 \end{bmatrix},$$

which realizes a clockwise rotation by $\pi/2$, noting that $J^\top = -J$, and setting

$$\operatorname{curl} w = \operatorname{div} Jw = \partial_1 w_2 - \partial_2 w_1,$$

$$\operatorname{Curl} \phi = J \nabla \phi = \begin{bmatrix} \partial_2 \phi \\ -\partial_1 \phi \end{bmatrix},$$

$$w \wedge n = w \cdot Jn = -w \cdot \tau,$$

where $\tau = -Jn$ is a tangent vector on $\partial \Omega$, the integral formula also holds for $d = 2$, $w \in C^1(\overline{\Omega}; \mathbb{R}^2)$, and $\phi \in C^1(\overline{\Omega})$. We let $\ell_d = 3$ for $d = 3$ and $\ell_d = 1$ for $d = 2$ and let $\Omega \subset \mathbb{R}^d$ be a bounded Lipschitz domain in what follows.

Definition 8.2 The space $H(\operatorname{curl}; \Omega)$ consists of all vector fields $v \in L^2(\Omega; \mathbb{R}^d)$ that have a *weak curl* in $L^2(\Omega; \mathbb{R}^{\ell_d})$, i.e., there exists $g \in L^2(\Omega; \mathbb{R}^{\ell_d})$ denoted $g = \operatorname{curl} v$ such that

$$\int_\Omega v \cdot \operatorname{Curl} \phi \, dx = \int_\Omega g \cdot \phi \, dx$$

for all $\phi \in C_0^\infty(\overline{\Omega}; \mathbb{R}^{\ell_d})$. It is equipped with the norm

$$\|v\|_{H(\operatorname{curl};\Omega)} = \left(\|v\|_{L^2(\Omega)}^2 + \|\operatorname{curl} v\|_{L^2(\Omega)}^2 \right)^{1/2}.$$

The subspace $H_0(\operatorname{curl}; \Omega)$ consists of all $v \in H(\operatorname{curl}; \Omega)$ such that $v \wedge n = 0$ on $\partial \Omega$ in the sense that

$$\langle v \wedge n, \phi \rangle_{\partial \Omega} = \int_\Omega v \cdot \operatorname{Curl} \phi \, dx - \int_\Omega \operatorname{curl} v \cdot \phi \, dx = 0$$

for all $\phi \in C^\infty(\overline{\Omega}; \mathbb{R}^{\ell_d})$.

The space $H_0(\operatorname{curl}; \Omega)$ consists of all vector fields in $L^2(\Omega; \mathbb{R}^d)$ whose distributional curl belongs to $L^2(\Omega; \mathbb{R}^{\ell_d})$, and whose tangential components vanish on $\partial \Omega$ in a distributional sense. It coincides with the closure of the set $C_0^\infty(\Omega; \mathbb{R}^d)$ in the space $H(\operatorname{curl}; \Omega)$.

Remarks 8.11

(i) The space $H_0(\operatorname{curl}; \Omega)$ is a Banach space, and we have $\nabla H_0^1(\Omega) \subset H_0(\operatorname{curl}; \Omega)$, in particular, $\operatorname{curl} \nabla \phi = 0$ for all $\phi \in H_0^1(\Omega)$.

(ii) The three-dimensional *de Rham complex* states that we have the mapping properties

$$\mathbb{R} \xrightarrow{\text{id}} H^1(\Omega) \xrightarrow{\nabla} H(\text{curl}; \Omega) \xrightarrow{\text{curl}} H(\text{div}; \Omega) \xrightarrow{\text{div}} L^2(\Omega) \xrightarrow{0} 0.$$

If Ω is simply connected and $\partial\Omega$ is connected, then the sequence is exact, i.e., the range of every indicated operator is the kernel of the subsequent operator. In this case there exist for every $v \in L^2(\Omega; \mathbb{R}^3)$ uniquely defined functions $\phi \in H^1(\Omega)/\mathbb{R}$ and $\psi \in H_0(\text{curl}; \Omega) \cap H(\text{div}; \Omega)$ or alternatively $\phi \in H_0^1(\Omega)$ and $\psi \in H(\text{curl}; \Omega) \cap H_0(\text{div}; \Omega)$ with $\text{div}\,\psi = 0$ such that we have the *Helmholtz decomposition*

$$v = \nabla\phi + \text{curl}\,\psi.$$

If Ω is not simply connected, e.g., a torus, or $\partial\Omega$ is not connected, e.g., $\Omega = B_1(0) \setminus \overline{B_{1/2}(0)}$, then *harmonic fields* have to be incorporated into the decomposition.

(iii) Assume that $\partial\Omega$ is connected and let $K = H_0(\text{curl}; \Omega) \cap H(\text{div}^0; \Omega)$, where div^0 stands for divergence-free vector fields. Then the embedding $K \to L^2(\Omega; \mathbb{R}^d)$ is compact. As a consequence, there exists a constant $c_M > 0$ such that the *Maxwell inequality*

$$\|v\|_{L^2(\Omega)} \le c_M \|\text{curl}\,v\|_{L^2(\Omega)}$$

holds for all $v \in K$.

8.3.3 Weak Formulation

With the space $H_0(\text{curl}; \Omega)$ we consider the following weak formulation of the time-harmonic Maxwell system. We assume that $\partial\Omega$ is connected, so that the Maxwell inequality is available on the set K. We follow [3] to establish the well-posedness of a weak formulation for the time-harmonic Maxwell system.

Definition 8.3 The *weak formulation of the time-harmonic Maxwell system* with $\omega \ne 0$ consists in determining for $f \in H(\text{div}; \Omega)$ satisfying $\text{div}f = 0$ a vector field $u \in H_0(\text{curl}; \Omega)$ such that

$$\int_\Omega \text{curl}\,u \cdot \text{curl}\,v \, dx - \omega^2 \int_\Omega u \cdot v \, dx = \int_\Omega f \cdot v \, dx$$

for all $v \in H_0(\text{curl}; \Omega)$.

By choosing $v = \nabla\phi$ for arbitrary $\phi \in H_0^1(\Omega)$, we see that the formulation implies div $u = 0$. We may thus consider an equivalent saddle-point problem, which also includes the case $\omega = 0$.

Definition 8.4 The *saddle-point formulation of the time-harmonic Maxwell system* consists in determining $(u, p) \in H_0(\text{curl}; \Omega) \times H_0^1(\Omega)$ such that

$$\int_\Omega \text{curl}\, u \cdot \text{curl}\, v \, dx - \omega^2 \int_\Omega uv \, dx + \int_\Omega \nabla p \cdot v \, dx = \int_\Omega f \cdot v \, dx,$$

$$\int_\Omega \nabla q \cdot u \, dx \qquad\qquad\qquad\qquad = 0$$

for all $(v, q) \in H_0(\text{curl}; \Omega) \times H_0^1(\Omega)$.

To establish the existence of solutions, it is preferable to consider the saddle-point formulation.

Remark 8.12 The bilinear form $b(v, q)$ that realizes the divergence constraint satisfies an inf-sup condition, i.e., the choice $v = \nabla q$ leads to

$$\sup_{v \in H_0(\text{curl};\Omega)} \frac{b(v, q)}{\|v\|_{H(\text{curl};\Omega)}} = \sup_{v \in H_0(\text{curl};\Omega)} \frac{\int_\Omega \nabla q \cdot v \, dx}{\|v\|_{H(\text{curl};\Omega)}} \geq \|\nabla q\|_{L^2(\Omega)}$$

for all $q \in H_0^1(\Omega)$, where we used that $\|\nabla q\|_{H(\text{curl};\Omega)} = \|\nabla q\|_{L^2(\Omega)}$.

The divergence-free vector fields in $H_0(\text{curl}; \Omega)$ define the kernel of the bilinear form b, and we have to show that the continuous, symmetric bilinear form

$$a(u, v) = \int_\Omega \text{curl}\, u \cdot \text{curl}\, v \, dx - \omega^2 \int_\Omega u \cdot v \, dx$$

satisfies an inf-sup condition on this kernel, i.e., on

$$K = H_0(\text{curl}; \Omega) \cap H(\text{div}^0; \Omega).$$

If $\omega^2 = 0$, then the Maxwell inequality $\|v\|_{L^2(\Omega)} \leq c_M \|\text{curl}\, v\|_{L^2(\Omega)}$ for $v \in K$ implies that a is coercive on K and the saddle-point formulation admits a unique solution $u_f \in K$ for every $f \in H(\text{div}^0; \Omega)$. Since K is compactly embedded into $L^2(\Omega; \mathbb{R}^d)$, and since the composition of a compact and a bounded linear operator is compact, we have that the solution operator $L_0 : H(\text{div}^0; \Omega) \to H(\text{div}^0; \Omega), f \mapsto u_f$, is compact. Since L_0 is also bounded, self-adjoint, and positive definite, there exists a countable orthonormal family of eigenfunctions in $H(\text{div}^0; \Omega)$ associated with positive eigenvalues. We let $(\lambda_j, u_j)_{j \in \mathbb{N}}$ be such an eigenbasis, i.e., we have $u_j \in K$ and

$$\int_\Omega \operatorname{curl} u_j \cdot \operatorname{curl} v \, dx = \lambda_j \int_\Omega u_j \cdot v \, dx$$

for all $v \in K$ and $j \in \mathbb{N}$. Noting $\|u_j\|_{L^2(\Omega)} = 1$ and choosing $v = u_j$, we find that

$$\|\operatorname{curl} u_j\|^2_{L^2(\Omega)} = \lambda_j, \quad \|u_j\|^2_{H(\operatorname{curl};\Omega)} = 1 + \lambda_j$$

for all $j \in \mathbb{N}$. For all $u, v \in K$ there exist $(a_j)_{j \in \mathbb{N}}, (b_j)_{j \in \mathbb{N}} \in \ell^2(\mathbb{N})$ with

$$u = \sum_{j \in \mathbb{N}} a_j u_j, \quad v = \sum_{j \in \mathbb{N}} b_j u_j.$$

To verify an inf-sup condition for $\omega^2 \neq 0$, we let $u \in K$ with coefficients $(a_j)_{j \in \mathbb{N}}$, and define $v \in K$ by setting $b_j = \sigma_j a_j$, with $\sigma_j = \operatorname{sign}(\lambda_j - \omega^2)$. Then $\|v\|_{H(\operatorname{curl};\Omega)} = \|u\|_{H(\operatorname{curl};\Omega)}$ and due to the L^2-orthonormality of the eigenfunctions, we have

$$a(u, v) = \sum_{j \in \mathbb{N}} a_j b_j \left(\|\operatorname{curl} u_j\|^2_{L^2(\Omega)} - \omega^2 \|u_j\|^2_{L^2(\Omega)} \right)$$

$$= \sum_{j \in \mathbb{N}} a_j^2 |\lambda_j - \omega^2|$$

$$\geq \min_{j \in \mathbb{N}} \frac{|\lambda_j - \omega^2|}{1 + \lambda_j} \|u\|^2_{H(\operatorname{curl};\Omega)}.$$

Thus for all $u \in K$ we have

$$\sup_{v \in K} \frac{a(u, v)}{\|v\|_{H(\operatorname{curl};\Omega)}} \geq c_\omega \|u\|_{H(\operatorname{curl};\Omega)}$$

with $c_\omega = \min_{j \in \mathbb{N}} |\lambda_j - \omega^2|/(1 + \lambda_j)$. This implies the following result.

Proposition 8.9 (Well-Posedness) *If $\partial\Omega$ is connected, ω^2 is not an eigenvalue of the operator* Curl curl *in the weak sense, and $f \in H(\operatorname{div}^0; \Omega)$, then there exists a unique solution $(u, p) \in H_0(\operatorname{curl}; \Omega) \times H_0^1(\Omega)$ of the saddle-point formulation of the time-harmonic Maxwell system.*

Proof The result follows from the abstract saddle-point theory since b satisfies an inf-sup condition, and a defines an invertible operator on the kernel K of b. □

8.3.4 Finite Element Approximation

A finite element approximation of the time-harmonic Maxwell problem seeks for a given finite element space $V_h \subset H_0(\mathrm{curl}; \Omega)$ a vector field $u_h \in V_h$ such that

$$a(u_h, v_h) = \ell(v_h),$$

for all $v_h \in V_h$. A discrete saddle-point formulation that incorporates a discrete version of the constraint $\mathrm{div}\, u = 0$ determines a pair $(u_h, p_h) \in V_h \times Q_h$ such that

$$a(u_h, v_h) + b(v_h, p_h) = \ell(v_h),$$
$$b(u_h, q_h) \qquad\qquad = 0,$$

for all $(v_h, q_h) \in V_h \times Q_h$ with a space $Q_h \subset H_0^1(\Omega)$.

Remark 8.13 If $\nabla Q_h \subset V_h$, then the two formulations are equivalent with $p_h = 0$. The mixed formulation involves more degrees of freedom but has better stability properties, e.g., if $|\omega| \ll 1$.

   The arguments that showed the well-posedness of the continuous problem can be carried over to the discrete problem, provided that the discrete eigenvalues $\lambda_{h,j}$, $j = 1, 2, \ldots, N = \dim V_h$, for the discrete eigenvalue problem

$$\int_\Omega \mathrm{curl}\, u_{h,j} \cdot \mathrm{curl}\, v_h \, \mathrm{d}x = \lambda_{h,j} \int_\Omega u_{h,j} \cdot v_h \, \mathrm{d}x$$

are different from ω^2. In this case, and under the assumption $\nabla Q_h \subset V_h$, the abstract saddle-point theory yields that

$$\|u - u_h\|_{H(\mathrm{curl};\Omega)} \leq c_{\omega,h}^{-1} \inf_{v_h \in V_h} \|u - v_h\|_{H(\mathrm{curl};\Omega)},$$

with $c_{\omega,h} = c \min_{j=1,\ldots,N} |\lambda_{h,j} - \omega^2| / (1 + \lambda_{h,j})$. To obtain useful approximations, the continuous spectrum thus has to be accurately approximated by the discrete one related to the space V_h.

Remark 8.14 Depending on Ω, the space $H^1(\Omega; \mathbb{R}^d) \cap H_0(\mathrm{curl}; \Omega)$ may be a closed and proper subspace of $H_0(\mathrm{curl}; \Omega) \cap H(\mathrm{div}; \Omega)$, e.g., if Ω is a nonconvex polyhedral domain. Since every piecewise polynomial vector field in $H_0(\mathrm{curl}; \Omega) \cap H(\mathrm{div}; \Omega)$ belongs to $H^1(\Omega; \mathbb{R}^d)$, we deduce that conforming polynomial finite element methods with $V_h \subset H_0(\mathrm{curl}; \Omega) \cap H(\mathrm{div}; \Omega)$ cannot be convergent in general.

   Finite element spaces in $H(\mathrm{curl}; \Omega)$ are closely related to finite element spaces in $H(\mathrm{div}; \Omega)$. For $d = 2$ they can be defined as rotated versions of Raviart–Thomas finite element spaces.

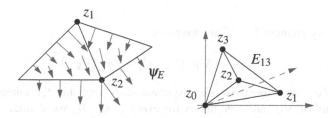

Fig. 8.16 Basis function in the Nédélec space $\mathcal{N}ed^0(\mathcal{T}_h)$ associated with the edge $E = \text{conv}\{z_1, z_2\}$ (*left*). Tetrahedron with edges $E_{ij} = \text{conv}\{z_i, z_j\}$, $i, j = 0, 1, \ldots, d$, and $z_0 = 0$ (*right*)

Definition 8.5 The *Nédélec finite element space* $\mathcal{N}ed^0(\mathcal{T}_h)$ subordinated to a triangulation $\mathcal{T}_h$ of Ω consists of all $u_h \in H(\text{curl}; \Omega)$, such that for every $T \in \mathcal{T}_h$ there exist $a_T \in \mathbb{R}^d$ and $b_T \in \mathbb{R}^{\ell_d}$ with

$$u_h(x) = a_T + b_T \wedge x = a_T + \begin{cases} b_T J x & \text{if } d = 2, \\ b_T \times x & \text{if } d = 3, \end{cases}$$

for all $x \in T$. The subspace with vanishing tangential components on $\partial\Omega$ is given by

$$\mathcal{N}ed^0_{\mathrm{T}}(\mathcal{T}_h) = \mathcal{N}ed^0(\mathcal{T}_h) \cap H_0(\text{curl}; \Omega).$$

Functions in $\mathcal{N}ed^0(\mathcal{T}_h)$ are exactly those piecewise affine vector fields whose tangential component is continuous across element sides and whose elementwise gradient is skew-symmetric. As a consequence, the tangential component $v_h|_E \cdot \tau_E$ is constant on every edge $E \in \mathcal{E}_h$, since for $x \in E$ we have $x = z_i + r\tau_E$ and $(A\tau_E) \cdot \tau_E = 0$ for every skew-symmetric matrix A. We construct basis functions for $\mathcal{N}ed^0(\mathcal{T}_h)$ which are associated with one-dimensional edges of triangles or tetrahedra, i.e., line segments connecting two vertices, cf. Fig. 8.16.

Lemma 8.8 (Edge Basis) *For every edge* $E \in \mathcal{E}_h$ *with* $E = \text{conv}\{z_1, z_2\}$ *for* $z_1, z_2 \in \mathcal{N}_h$, *let*

$$\psi_E = \varphi_{z_1} \nabla \varphi_{z_2} - \varphi_{z_2} \nabla \varphi_{z_1}.$$

We then have that $\psi_E \in \mathcal{N}ed^0(\mathcal{T}_h)$ *and*

$$L_{E'}(\psi_E) = \int_{E'} \psi_E \cdot \tau_{E'} \, \mathrm{d}r = \delta_{E',E}$$

for all $E, E' \in \mathcal{E}_h$, *provided that* τ_E *points from* z_2 *to* z_1. *In particular, we have that* $(\psi_E : E \in \mathcal{E}_h)$ *is a basis for* $\mathcal{N}ed^0(\mathcal{T}_h)$.

Proof

(i) On every element $T \in \mathcal{T}_h$ we have that

$$\nabla \psi_E = \nabla \varphi_{z_1} (\nabla \varphi_{z_2})^{\mathsf{T}} - \nabla \varphi_{z_2} (\nabla \varphi_{z_1})^{\mathsf{T}},$$

   i.e., $\nabla \psi_E$ is elementwise skew-symmetric. Noting that the tangential components of $\nabla \varphi_z$ are continuous for every $z \in \mathcal{N}_h$, we deduce that $\psi_E \in \mathcal{N}ed^0(\mathcal{T}_h)$.

(ii) If $z \in \mathcal{N}_h$ and $E' \in \mathcal{E}_h$ such that $z \notin E'$, then we have that $\nabla \varphi_z|_{E'}$ is normal to E' or vanishes. If $z \in E$, then $\nabla \varphi_z|_E \cdot \tau_E$ is the tangential derivative of φ_z along E and equals $\pm 1/|E|$. Since $\varphi_{z_1} + \varphi_{z_2} = 1$ on E, this implies that we have

$$L_{E'}(\psi_E) = \int_{E'} \psi_E \cdot \tau_{E'} \, dr = \delta_{E',E}$$

   for all $E, E' \in \mathcal{E}_h$.

(iii) To prove that the family $(\psi_E : E \in \mathcal{E}_h)$ defines a basis for $\mathcal{N}ed^0(\mathcal{T}_h)$, it suffices to show that if $v_h \in \mathcal{N}ed^0(\mathcal{T}_h)$ is such that $L_E(v_h) = 0$ for all $E \in \mathcal{E}_h$, then we have that $v_h = 0$. We consider the case $d = 3$, an element $T \in \mathcal{T}_h$, and for $a_T, b_T \in \mathbb{R}^3$ the function

$$v_T(x) = v_h|_T(x) = a_T + b_T \times x.$$

   We have to show that if $L_E(v_T) = 0$ for all $E \in \mathcal{E}_h \cap T$, then we have $v_T = 0$. The condition is equivalent to $v_T|_E \cdot \tau_E = 0$ for all $E \in \mathcal{E}_h \cap T$. Without loss of generality we may assume that $T = \mathrm{conv}\{z_0, z_1, z_2, z_3\}$ with $z_0 = 0$, cf. Fig. 8.16. Then the vectors $\tau_i = z_i/|z_i|$, $i = 1, 2, 3$, are linearly independent and the condition $v_T(0) \cdot \tau_i = 0$ implies that $a_T = 0$. For $i = 1, 2, 3$ we have $(b_T \times z_i) \cdot (z_k - z_i) = 0$ for $k = 0, 1, 2, 3$ with $k \neq i$. Since the vectors $z_k - z_i$ are linearly independent, it follows that $b_T \times z_i = 0$ for $i = 1, 2, 3$. The vectors z_i, $i = 1, 2, 3$, are linearly independent, which implies that $b_T = 0$. Hence the functionals $L_{E'}$, $E' \in \mathcal{E}_h$, span the dual space of $\mathcal{N}ed^0(\mathcal{T}_h)$. The orthogonality $L_{E'}(\psi_E) = \delta_{EE'}$ implies that $(\psi_E : E \in \mathcal{E}_h)$ is a basis for $\mathcal{N}ed^0(\mathcal{T}_h)$.   □

The nodal basis allows us to define an interpolation operator.

Proposition 8.10 (Nédélec Interpolant) *For* $v \in C(\overline{\Omega}; \mathbb{R}^d)$ *define*

$$\mathscr{I}_{\mathcal{N}ed} v = \sum_{E \in \mathcal{E}_h} L_E(v) \psi_E.$$

If $u \in H^2(\Omega; \mathbb{R}^d)$ *we have*

$$\|u - \mathscr{I}_{\mathcal{N}ed} u\|_{L^2(\Omega)} \leq c_{\mathcal{N}ed} h \big(\|\nabla u\|_{L^2(\Omega)} + \|D^2 u\|_{L^2(\Omega)} \big).$$

If $u \in H(\mathrm{curl}; \Omega)$ *such that* $\mathrm{curl}\, u \in H^1(\Omega; \mathbb{R}^{\ell_d})$ *we have*

$$\| \mathrm{curl}(u - \mathscr{I}_{\mathcal{N}ed} u)\|_{L^2(\Omega)} \leq c'_{\mathcal{N}ed} h \| \mathrm{curl}\, u\|_{H^1(\Omega)}.$$

Proof

(i) The first estimate is a consequence of the fact that $\mathscr{I}_{\mathcal{N}ed}$ is exact for constant functions and the Bramble–Hilbert lemma.

(ii) To prove the second estimate we first note that $\operatorname{curl}\mathscr{I}_{\mathcal{N}ed}u \in \mathscr{R}T^0(\mathscr{T}_h)$. This follows from the identity

$$\operatorname{curl}\psi_E = 2\nabla\varphi_{z_1} \wedge \nabla\varphi_{z_2}$$

for $E = \operatorname{conv}\{z_1, z_2\}$, which implies that the normal component of every function $\operatorname{curl} v_h$ with $v_h \in \mathcal{N}ed^0(\mathscr{T}_h)$ is continuous across element sides. By definition of $\mathscr{I}_{\mathcal{N}ed}(u)$, we have for every side $S \in \mathscr{S}_h$ with edges $E \in \mathscr{E}_h \cap S$ whose union is ∂S by Stokes's theorem that

$$\int_S \operatorname{curl} u \cdot n_S \, ds = \int_{\partial S} u \cdot \tau \, dr = \sum_{E \in \mathscr{E}_h \cap S} \sigma_{S,E} L_E(u)$$

$$= \sum_{E \in \mathscr{E}_h \cap S} \sigma_{S,E} L_E(\mathscr{I}_{\mathcal{N}ed}u) = \int_{\partial S} \mathscr{I}_{\mathcal{N}ed}u \cdot \tau \, dr = \int_S \operatorname{curl}\mathscr{I}_{\mathcal{N}ed}u \cdot n_S \, ds,$$

where $\sigma_{S,E} = \pm 1$ are appropriate signs. This shows that $\operatorname{curl}\mathscr{I}_{\mathcal{N}ed}u = \mathscr{I}_{\mathscr{R}T}\operatorname{curl} u$ with the Raviart–Thomas interpolation operator $\mathscr{I}_{\mathscr{R}T}$, for which we have

$$\|w - \mathscr{I}_{\mathscr{R}T}w\|_{L^2(\Omega)} \leq ch\|\nabla w\|_{L^2(\Omega)}.$$

With $w = \operatorname{curl} u$ we obtain the second estimate. □

Remark 8.15 The proof of the proposition reveals a commutativity property which applies in other situations as well. The diagram shown in Fig. 8.17 relates the finite element spaces $\mathscr{S}^1(\mathscr{T}_h)$, $\mathscr{R}T^0(\mathscr{T}_h)$, $\mathcal{N}ed^0(\mathscr{T}_h)$, and $\mathscr{L}^0(\mathscr{T}_h)$, and the corresponding interpolation and differential operators. In the case of elementwise constant functions, the interpolation operator is given by

$$\pi_h v|_T = |T|^{-1}\int_T v \, dx$$

for all $T \in \mathscr{T}_h$.

$$
\begin{array}{ccccccc}
H^1(\Omega) \cap C^\infty & \xrightarrow{\nabla} & H(\operatorname{curl};\Omega) \cap C^\infty & \xrightarrow{\operatorname{curl}} & H(\operatorname{div};\Omega) \cap C^\infty & \xrightarrow{\operatorname{div}} & L^2(\Omega) \cap C^\infty \\
\mathscr{I}_h \downarrow & & \mathscr{I}_{\mathcal{N}ed} \downarrow & & \mathscr{I}_{\mathscr{R}T} \downarrow & & \pi_h \downarrow \\
\mathscr{S}^1(\mathscr{T}_h) & \xrightarrow{\nabla} & \mathcal{N}ed^0(\mathscr{T}_h) & \xrightarrow{\operatorname{curl}} & \mathscr{R}T^0(\mathscr{T}_h) & \xrightarrow{\operatorname{div}} & \mathscr{L}^0(\mathscr{T}_h)
\end{array}
$$

Fig. 8.17 The interpolation operators indicated in the diagram commute with the application of the indicated differential operators, e.g., $\nabla\mathscr{I}_h\phi = \mathscr{I}_{\mathcal{N}ed}\nabla\phi$ for $\phi \in H^1(\Omega) \cap C^\infty(\overline{\Omega})$. The finite element spaces are related to vertices, edges, sides, and elements from left to right

The interpolation properties and quasi-best approximation results lead to an error estimate for the approximation of the time-harmonic Maxwell problem.

Theorem 8.4 (Error Estimate) *Assume that* $\lambda_j^h \neq \omega^2$ *for* $j = 1, 2, \ldots, N$, *and* $\lambda_j \neq \omega^2$ *for* $j \in \mathbb{N}$. *If* $u \in H^2(\Omega)$, *we have*

$$\|u - u_h\|_{H(\mathrm{curl};\Omega)} \leq c_{\omega,h}^{-1} c' h \|D^2 u\|_{L^2(\Omega)}.$$

Proof The result follows from the abstract saddle-point theory, i.e., the generalized Céa Lemma, and the interpolation properties of the Nédélec interpolation operator.

$\square$

8.3.5 Implementation

The edge basis functions for $\mathcal{N}ed^0(\mathcal{T}_h)$ are only defined up to signs. To work with compatible, elementwise defined functions, we define a sign σ_E for an edge $E \in \mathcal{E}_h$ with $E = \mathrm{conv}\{z_1, z_2\}$ as positive if the global node number of z_1 is larger than the global node number of z_2, and as negative otherwise. We thus use the basis $(\psi_E : E \in \mathcal{E}_h)$ of the Nédélec finite element space given by the functions

$$\psi_E = \sigma_E\big(\varphi_{z_1}\nabla\varphi_{z_2} - \varphi_{z_2}\nabla\varphi_{z_1}\big).$$

We have $\psi_E \in H(\mathrm{curl}; \Omega)$ with

$$\mathrm{curl}\,\psi_E = 2\sigma_E\nabla\varphi_{z_1} \wedge \nabla\varphi_{z_2},$$

where $a \wedge b = \det[a, b]$ if $d = 2$ and $a \wedge b = a \times b$ if $d = 3$. The realization of the method also requires a representation of the L^2-inner product of the edge basis functions. For this, for sides $E_{k\ell} = \mathrm{conv}\{z_k, z_\ell\}$ and $E_{mn} = \mathrm{conv}\{z_m, z_n\}$ that belong to the boundary of the same element T, we use the identity

$$\int_T \psi_{E_{k\ell}} \cdot \psi_{E_{mn}}\, dx$$

$$= \sigma_{E_{k\ell}}\sigma_{E_{mn}} \int_T (\varphi_k\nabla\varphi_\ell - \varphi_\ell\nabla\varphi_k) \cdot (\varphi_m\nabla\varphi_n - \varphi_n\nabla\varphi_m)\, dx$$

$$= \sigma_{E_{k\ell}}\sigma_{E_{mn}}\big(m_{km}s_{\ell n} - m_{\ell m}s_{kn} - m_{kn}s_{\ell m} + m_{\ell n}s_{km}\big),$$

where $m_{km} = \int_T \varphi_k\varphi_m\, dx$ and $s_{\ell n} = \int_T \nabla\varphi_\ell \cdot \nabla\varphi_n\, dx$. Figures 8.18 and 8.19 show the components of a MATLAB realization. The output of a numerical experiment is shown in Fig. 8.20.

```
function maxwell(d,red)
[c4n,n4e,Db,Nb] = triang_cube(d); o_sq = 1;
for j = 1:red
    [c4n,n4e,Db,Nb] = red_refine(c4n,n4e,Db,Nb);
end
nE = size(n4e,1); d = size(c4n,2);
switch d
    case 2; edgeInd = [1 2,1 3,2 3];
    case 3; edgeInd = [1 2,1 3,1 4,2 3,2 4,3 4];
end
[edges,el2edges,Db2edges] = edge_data(n4e,Db,Nb);
nEdges = size(edges,1); fEdges = setdiff(1:nEdges,Db2edges);
u = zeros(nEdges,1); b = zeros(nEdges,1);
idx = [1,3,6]; edgeInd = reshape(edgeInd,2,idx(d))';
m_loc = (eye(d+1)+ones(d+1,d+1))/((d+2)*(d+1));
max_ctr = idx(d)*nE; ctr = 0;
I = zeros(max_ctr,1); J = zeros(max_ctr,1);
X_C = zeros(max_ctr,1); X_M = zeros(max_ctr,1);
for j = 1:nE
    X_T = [ones(1,d+1);c4n(n4e(j,:),:)'];
    grads_T = X_T\[zeros(1,d);eye(d)];
    vol_T = det(X_T)/factorial(d);
    mp_T = sum(c4n(n4e(j,:),:),1)/(d+1);
    for m = 1:idx(d)
        m1 = edgeInd(m,1); m2 = edgeInd(m,2);
        s_m = 2*(n4e(j,m1)<n4e(j,m2))-1;
        curl_psi_m = 2*s_m*wedge(grads_T(m1,:),grads_T(m2,:),d);
        for n = 1:idx(d)
            ctr = ctr+1;
            I(ctr) = el2edges(j,m); J(ctr) = el2edges(j,n);
            n1 = edgeInd(n,1); n2 = edgeInd(n,2);
            s_n = 2*(n4e(j,n1)<n4e(j,n2))-1;
            curl_psi_n = 2*s_n*wedge(grads_T(n1,:),grads_T(n2,:),d);
            X_C(ctr) = vol_T*dot(curl_psi_m,curl_psi_n);
            X_M(ctr) = vol_T*s_m*s_n*...
                (m_loc(m1,n1)*grads_T(m2,:)*grads_T(n2,:)'...
                -m_loc(m2,n1)*grads_T(m1,:)*grads_T(n2,:)'...
                -m_loc(m1,n2)*grads_T(m2,:)*grads_T(n1,:)'...
                +m_loc(m2,n2)*grads_T(m1,:)*grads_T(n1,:)');
        end
        b(el2edges(j,m))= b(el2edges(j,m))+vol_T*f(mp_T,o_sq,d)'...
            *s_m*(grads_T(m2,:)-grads_T(m1,:))'/(d+1);
    end
end
C = sparse(I,J,X_C); M = sparse(I,J,X_M); X = C-o_sq*M;
u(fEdges) = X(fEdges,fEdges)\b(fEdges);
show_nedelec(c4n,n4e,el2edges,edgeInd,u);

function val = f(x,o_sq,d)
val = (pi^2-o_sq)*[sin(pi*x(2));sin(pi*x(1));zeros(d-2,1)];

function val = wedge(v,w,d)
if d == 2; val = det([v;w]); else val = cross(v,w); end
```

Fig. 8.18 MATLAB implementation of the Nédélec finite element method for the time-harmonic Maxwell system

```
function show_nedelec(c4n,n4e,element2edges,edgeInd,u)
d = size(c4n,2); nE = size(n4e,1); idx = [1,3,6];
u_T = zeros(nE,d); mp_T = zeros(nE,d);
for j = 1:nE
    X_T = [ones(1,d+1);c4n(n4e(j,:),:)'];
    grads_T = X_T\[zeros(1,d);eye(d)];
    mp_T(j,:) = sum(c4n(n4e(j,:),:),1)/(d+1);
    for m = 1:idx(d)
        m1 = edgeInd(m,1); m2 = edgeInd(m,2);
        s_m = 2*(n4e(j,m1)<n4e(j,m2))-1;
        vec = s_m*(grads_T(m2,:)-grads_T(m1,:))/(d+1);
        u_T(j,:) = u_T(j,:)+u(element2edges(j,m))*vec;
    end
end
if d == 2
    triplot(n4e,c4n(:,1),c4n(:,2),'k'); hold on;
    quiver(mp_T(:,1),mp_T(:,2),u_T(:,1),u_T(:,2)); hold off;
else
    quiver3(mp_T(:,1),mp_T(:,2),mp_T(:,3),...
        u_T(:,1),u_T(:,2),u_T(:,3));
end
```

```
function [edges,el2edges,Db2edges,Nb2edges] = edge_data(n4e,Db,Nb)
[nE,nV] = size(n4e); d = nV-1;
idx = [0,1,3,6]; Bdy = [Db;Nb]; nEdges = idx(d+1)*nE;
nDb = idx(d)*size(Db,1); nNb = idx(d)*size(Nb,1);
switch d
 case 1; edges = n4e;
 case 2; edges = [reshape(n4e(:,[1 2,1 3,2 3])',2,[])';Bdy];
 case 3; edges = reshape(n4e(:,[1 2,1 3,1 4,2 3,2 4,3 4])',2,[])';
     edges = [edges;reshape(Bdy(:,[1 2,1 3,2 3])',2,[])'];
end
[edges,~,edgeNumbers] = unique(sort(edges,2),'rows','first');
el2edges = reshape(edgeNumbers(1:nEdges),idx(d+1),[])';
Db2edges = reshape(edgeNumbers(nEdges+(1:nDb))',idx(d),[])';
Nb2edges = reshape(edgeNumbers(nEdges+nDb+(1:nNb))',idx(d),[])';
```

Fig. 8.19 Visualization of a Nédélec finite element vector field (*top*); auxiliary data structures related to edges in a triangulation (*bottom*)

Fig. 8.20 Numerical solution of a Maxwell problem computed with the Nédélec finite element method

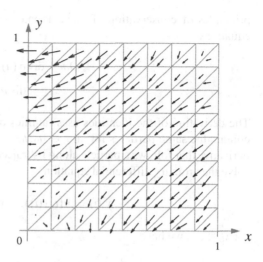

8.4 Viscous, Incompressible Fluids

8.4.1 Navier–Stokes Equations

The mathematical description of incompressible viscous fluids such as water, air, or certain oils, has similarities to the modeling of elastic solids. Viscosity here refers to the property that friction between particles matters, and these may accelerate or decelerate each other. This is specified by an appropriate relation between stresses and strains. For isotropic, homogeneous viscous fluids, this relation introduces the pressure p and is given by

$$\sigma(u) = 2\mu\varepsilon(u) + \left(\lambda \operatorname{tr}\varepsilon(u) - p\right)I,$$

for a velocity field u and its symmetric gradient $\varepsilon(u) = (\nabla u + \nabla u^\top)/2$, and the Lamé constants $\lambda, \mu > 0$. Fluids that allow for a linear stress-strain relation are called *Newtonian fluids*, and the particular expression follows from the conservation of angular momentum. In contrast to solids where a fixed number of particles is considered, processes of interest in fluid mechanics typically involve inflow and outflow effects and therefore an unbounded number of particles. Instead of describing where a particle is mapped, one considers a position x in the domain of interest and lets $u(t, x)$ be the velocity of the particle that occupies this position at time t. This is called a *Eulerian description*, whereas the one used for solids is called a *Lagrangian description*. Assuming a constant mass density $\varrho > 0$, the

principles of conservation of linear momentum and mass lead to the system of
equations

$$\varrho D_t u - \operatorname{div} \sigma(u) = f,$$
$$\operatorname{div} u = 0.$$

The derivation of the identities uses families of control volumes $(\omega_t)_{t \in [t_0, t_0 + \delta]}$ that
contain the same particles for all $t \in [t_0, t_0 + \delta]$. Accordingly, $D_t u$ is the material
derivative that follows a particle along its trajectory, i.e., letting $y : [t_0, t_0 + \delta] \to \mathbb{R}^d$
solve the ordinary differential equation

$$y'(t) = u(t, y(t)), \quad y(t_0) = x_0,$$

for $x_0 \in \omega_0$, we have

$$D_t u(t_0, x_0) = \frac{d}{dt} u(t, y(t))\big|_{t=t_0} = \partial_t u(t_0, x_0) + u(t_0, x_0) \cdot \nabla u(t_0, x_0).$$

The combination of the identities leads to the *(incompressible) Navier–Stokes
equations*

$$\varrho \partial_t u + \varrho u \cdot \nabla u - \mu \Delta u + \nabla p = f,$$
$$\operatorname{div} u = 0.$$

The equations are supplemented by the initial condition $u(0) = u_0$, and appropriate
boundary conditions, e.g., the *no-slip condition*

$$u(t, x) = 0 \quad \text{for } x \in \Gamma_{ns},$$

which models that adhesive forces dominate cohesive ones at a fluid-solid interface,
so that the velocity vanishes on the boundary Γ_{ns}; the natural or *do-nothing*
condition,

$$\big(\mu \nabla u(t, x) - p(t, x)I\big)n(x) = 0 \quad \text{for } x \in \Gamma_N,$$

which can be used to model a transparent outflow boundary, or an *inflow* boundary
condition

$$u(t, x) = u_D(t, x) \quad \text{for } x \in \Gamma_{in}$$

which prescribes an inflow on a part of the boundary.

8.4.2 Weak Formulation

Dividing by the constant density ϱ, and redefining p and f, we obtain the equations

$$\partial_t u - \nu \Delta u + u \cdot \nabla u + \nabla p = f \quad \text{in } (0, T) \times \Omega,$$

$$\text{div } u = 0 \quad \text{in } (0, T) \times \Omega,$$

$$u = u_{\mathrm{D}} \quad \text{on } (0, T) \times \Gamma_{\mathrm{D}},$$

$$(\nu \nabla u - p I) n = 0 \quad \text{on } (0, T) \times \Gamma_{\mathrm{N}},$$

$$u(0) = u_0 \quad \text{in } \Omega,$$

with *kinematic viscosity* $\nu = \mu/\varrho > 0$. The convective term, given by

$$u \cdot \nabla u = (u \cdot \nabla) u = u_1 \partial_1 \begin{bmatrix} u_1 \\ \vdots \\ u_d \end{bmatrix} + u_2 \partial_2 \begin{bmatrix} u_1 \\ \vdots \\ u_d \end{bmatrix} + \cdots + u_d \partial_d \begin{bmatrix} u_1 \\ \vdots \\ u_d \end{bmatrix},$$

is nonlinear and makes the analysis of the Navier–Stokes equations complicated. It is nevertheless possible to define a weak formulation by multiplying the equations with appropriate functions and integrating over Ω. This leads to

$$\int_\Omega \left[\partial_t u \cdot v + \nu \nabla u : \nabla v + (u \cdot \nabla u) \cdot v \right] dx - \int_\Omega p \, \text{div } v \, dx = \int_\Omega f \cdot v \, dx,$$

$$\int_\Omega q \, \text{div } u \, dx = 0,$$

for $v \in H_{\mathrm{D}}^1(\Omega; \mathbb{R}^d)$ and $q \in L^2(\Omega)$. We are thus motivated to define the bilinear forms

$$(w, v) = \int_\Omega w \cdot v \, dx,$$

$$a(u, v) = \nu \int_\Omega \nabla u : \nabla v \, dx,$$

$$b(v, q) = -\int_\Omega q \, \text{div } v \, dx,$$

and the linear functional

$$\ell(v) = \int_\Omega f \cdot v \, dx.$$

Moreover, we define the trilinear form

$$n(z; u, v) = \int_\Omega (z \cdot \nabla u) \cdot v \, dx.$$

While it is obvious that a and b are bounded bilinear forms, and ℓ is a bounded linear functional, showing that

$$n : H_D^1(\Omega; \mathbb{R}^d) \times H_D^1(\Omega; \mathbb{R}^d) \times H_D^1(\Omega; \mathbb{R}^d) \to \mathbb{R}$$

is bounded requires using Sobolev embedding results. With Hölder inequalities, we find that

$$\begin{aligned}
|n(z; u, v)| &\leq \big\| |z| |v| \big\|_{L^2(\Omega)} \| \nabla u \|_{L^2(\Omega)} \\
&\leq \| z \|_{L^4(\Omega)} \| v \|_{L^4(\Omega)} \| \nabla u \|_{L^2(\Omega)} \\
&\leq c_S^2 \| \nabla z \|_{L^2(\Omega)} \| \nabla v \|_{L^2(\Omega)} \| \nabla u \|_{L^2(\Omega)},
\end{aligned}$$

where we used that $\| w \|_{L^4(\Omega)} \leq c_S \| \nabla w \|_{L^2(\Omega)}$ for all $w \in H_D^1(\Omega; \mathbb{R}^d)$. An important property of the trilinear form n is that it is skew-symmetric in the second and third argument, provided that $\operatorname{div} z = 0$ and $z \cdot n = 0$ on $\partial\Omega$, i.e.,

$$n(z; u, v) = -n(z; v, u)$$

for all $u, v \in H_0^1(\Omega; \mathbb{R}^d)$. This implies that for $\Gamma_D = \partial\Omega$ and fixed $z \in H_0^1(\Omega; \mathbb{R}^d)$ with $\operatorname{div} z = 0$, the bilinear form

$$a_z(u, v) = a(u, v) + n(z; u, v)$$

is coercive. We note that the pressure p is only defined up to a constant if $\Gamma_D = \partial\Omega$ and in this case $L^2(\Omega)$ is replaced by $L_0^2(\Omega)$. The inner product in $L^2(\Omega)$ is denoted by $(\cdot, \cdot)$.

Definition 8.6 The *weak formulation of the Navier–Stokes equations* determines $u : [0, T] \to H_D^1(\Omega; \mathbb{R}^d)$ and $p : [0, T] \to L^2(\Omega)$, such that $u(0) = u_0$, $u(t) \in H^1(\Omega; \mathbb{R}^d)$, and $u(t, \cdot)|_{\Gamma_D} = u_D(t)$ for almost every $t \in (0, T]$, and

$$\begin{aligned}
(\partial_t u, v) + a(u, v) + n(u; u, v) + b(v, p) &= \ell(v), \\
b(u, q) &= 0,
\end{aligned}$$

for almost every $t \in (0, T]$ and all $v \in H_D^1(\Omega; \mathbb{R}^d)$ and $q \in L^2(\Omega)$.

Note that we omitted time-dependence in the variational formulation, i.e., we write $a(u, v)$ for $a(u(t), v)$.

Remarks 8.16

(i) The stationary Navier–Stokes equations consider situations in which $\partial_t u = 0$. In this case, and if $\Gamma_D = \partial\Omega$, the existence of solutions can be proved using Brouwer's fixed-point theorem. Uniqueness holds if $u_D = 0$, and $\|f\|_{L^2(\Omega)} \leq \nu^2/(c_P c_S^2)$.

(ii) Existence results for time-dependent Navier–Stokes equations require an appropriate definition of a weak time derivative, and are only known for $d = 2$.

(iii) The limit $\nu \to 0$ leads to the incompressible Euler equations, which describe certain frictionless gases. With a different scaling of the pressure, a time-dependent Stokes system can be motivated for large viscosities. Characteristic of the qualitative behavior of solutions is the *Reynold's number* $R = UL/\nu$, with a characteristic length and velocity L and U.

8.4.3 Approximation Via Stokes Problems

If a good initial guess of a solution is available, then the stationary Navier–Stokes equations can be solved with a Newton iteration. This is in general not the case, and fixed-point iterations are more likely to converge. A semi-explicit treatment of the convective term leads to the Picard iteration for the stationary system.

Algorithm 8.1 (Picard Iteration, Stationary Case) *Let $u^0 \in H^1(\Omega; \mathbb{R}^d)$ with $u^0|_{\Gamma_D} = u_D$ and div $u^0 = 0$. Let $\varepsilon_{\text{stop}} > 0$ and set $k = 0$.*

(1) Compute $u^{k+1} \in H^1(\Omega; \mathbb{R}^d)$ with $u^{k+1}|_{\Gamma_D} = u_D$ and $p^{k+1} \in L^2(\Omega)$, such that

$$a(u^{k+1}, v) + n(u^k; u^{k+1}, v) + b(v, p^{k+1}) = \ell(v),$$
$$b(u^{k+1}, q) = 0,$$

for all $v \in H_D^1(\Omega; \mathbb{R}^d)$ and $q \in L^2(\Omega)$.

(2) Stop if $\|\nabla(u^{k+1} - u^k)\|_{L^2(\Omega)} \leq \varepsilon_{\text{stop}}$; continue with (1) otherwise.

The linear system of equations in Step (1) of the Picard iteration is called an *Oseen system*.

Remarks 8.17

(i) Due to the skew-symmetry of the trilinear form n, the Oseen systems can be analyzed and approximated analogous to the Stokes system if $\Gamma_D = \partial\Omega$.

(ii) The radius of the ball of convergence and the rate of convergence depend critically on ν. If $\Gamma_D = \partial\Omega$ and $u_D = 0$, the Picard iteration is globally convergent.

The treatment of time-dependent Navier–Stokes equations can be realized in a similar way, replacing the time derivative by a backward difference quotient.

Algorithm 8.2 (Semi-Implicit Time-Stepping) *Set $u^0 = u_0$, $k = 0$, and let $\tau > 0$.*

(1) Compute $u^{k+1} \in H^1(\Omega; \mathbb{R}^d)$ with $u^{k+1}|_{\Gamma_D} = u_D$ and $p^{k+1} \in L^2(\Omega)$, such that

$$\tau^{-1}(u^{k+1} - u^k, v) + a(u^{k+1}, v) + n(u^k; u^{k+1}, v) + b(v, p^{k+1}) = \ell(v),$$

$$b(u^{k+1}, q) = 0,$$

for all $v \in H_D^1(\Omega; \mathbb{R}^d)$ and $q \in L^2(\Omega)$.
(2) Stop if $(k + 1)\tau \geq T$; set $k \to k + 1$ and continue with (1) otherwise.

Note that we have to impose the *compatibility condition* $\operatorname{div} u_0 = 0$ in order to establish the existence of a solution in the first time step.

8.4.4 Projection Methods

For deriving a simple numerical scheme we assume that the convective term can be neglected in the time-dependent Navier–Stokes equations, i.e., we consider the Stokes flow

$$\partial_t u - \nu \Delta u + \nabla p = f, \quad \operatorname{div} u = 0, \quad u|_{\partial\Omega} = u_D,$$

subject to an initial condition $u(0) = u_0$. Ignoring the incompressibility constraint for a moment, and using a backward difference quotient for discretizing the time derivative lead to the problem of determining the solution $\tilde{u}^{k+1}$ of the boundary value problem

$$\tau^{-1}(\tilde{u}^{k+1} - u^k) - \nu \Delta \tilde{u}^{k+1} = f \text{ in } \Omega, \quad \tilde{u}^{k+1}|_{\partial\Omega} = u_D(t_{k+1}).$$

This is a simple elliptic boundary value problem that can be approximated with standard finite element methods. To incorporate the incompressibility constraint, we replace $\tilde{u}^{k+1}$ in the first term by $u^{k+1} + \tau \nabla p^{k+1}$, with u^{k+1} and p^{k+1} obtained as the solution of the system,

$$u^{k+1} + \tau \nabla p^{k+1} = \tilde{u}^{k+1},$$

$$\operatorname{div} u^{k+1} = 0,$$

subject to the boundary condition $u^{k+1} \cdot n = u_D(t_{k+1}) \cdot n$ on $\partial\Omega$. Note that here we only have $u^{k+1} \in H(\operatorname{div}; \Omega)$, so that we can only impose a boundary condition on the normal component of u^{k+1}. The correction step is similar to a mixed formulation of the Poisson problem. It is called a *projection step*, since u^{k+1} is the L^2-projection of $\tilde{u}^{k+1}$ onto the set $\{v \in H(\operatorname{div}; \Omega) : \operatorname{div} v = 0\}$ subject to the boundary condition.

The correction step can be simplified by taking the divergence and the normal component of the first equation, i.e., noting that

$$-\tau \Delta p^{k+1} = -\operatorname{div} \tilde{u}^{k+1} \text{ in } \Omega, \quad \nabla p^{k+1} \cdot n = 0 \text{ on } \partial\Omega.$$

Hence p^{k+1} is the solution of a Poisson problem with Neumann boundary condition, which can be approximated using standard finite element methods. Having determined p^{k+1}, we obtain the divergence-free vector field $u^{k+1} \in H(\operatorname{div}; \Omega)$ via

$$u^{k+1} = \tilde{u}^{k+1} - \tau \nabla p^{k+1}.$$

The vector fields u^{k+1}, $\tilde{u}^{k+1}$, and the function p^{k+1} then satisfy

$$\tau^{-1}\left(u^{k+1} - u^k\right) - \nu \Delta \tilde{u}^{k+1} + \nabla p^{k+1} = f,$$
$$\operatorname{div} u^{k+1} = 0,$$

and the boundary conditions

$$\tilde{u}^{k+1} = 0, \quad u^{k+1} \cdot n = 0, \quad \partial_n p^{k+1} = 0$$

on $\partial\Omega$. Incorporating the convective term leads to the following time-stepping method for the Navier–Stokes equations.

Algorithm 8.3 (Chorin Projection Scheme) *Set $u^0 = u_0$ and $k = 0$. Let $\tau > 0$.*

(1) Compute $\tilde{u}^{k+1} \in H^1(\Omega; \mathbb{R}^d)$ such that $\tilde{u}^{k+1}|_{\partial\Omega} = u_D(t_{k+1})$ and

$$\tau^{-1}\left(\tilde{u}^{k+1} - u^k, v\right) + a\left(\tilde{u}^{k+1}, v\right) + n\left(u^k; \tilde{u}^{k+1}, v\right) = \ell(v)$$

for all $v \in H_0^1(\Omega; \mathbb{R}^d)$.
(2) Compute $p^{k+1} \in H^1(\Omega) \cap L_0^2(\Omega)$ such that and

$$\tau\left(\nabla p^{k+1}, \nabla q\right) = \left(\tilde{u}^{k+1}, \nabla q\right) - \int_{\partial\Omega} u_D(t_{k+1}) \cdot nq \, ds$$

for all $q \in H^1(\Omega) \cap L_0^2(\Omega)$.
(3) Set

$$u^{k+1} = \tilde{u}^{k+1} - \tau \nabla p^{k+1}.$$

(4) Stop if $(k+1)\tau \geq T$; set $k \to k+1$ and continue with (1) otherwise.

The algorithm can be discretized with standard finite element methods and every step is well defined.

Remarks 8.18

(i) In general, only the velocity field $\tilde{u}^{k+1}$ satisfies the correct boundary conditions, and only u^{k+1} is divergence-free.

(ii) The pressure variable satisfies the artificial boundary condition $\nabla p^{k+1} \cdot n = 0$, which may lead to the occurrence of nonphysical boundary layers.

An error analysis of the projection scheme can be based on the observations that the difference between u^{k+1} and $\tilde{u}^{k+1}$, which satisfies

$$\tilde{u}^{k+1} - u^{k+1} = \tau \nabla p^{k+1},$$

is small, provided that τ is small and ∇p^{k+1} remains bounded as $\tau \to 0$. Then, by replacing $u^k = \tilde{u}^k - \tau \nabla p^k$ in the equation that determines $\tilde{u}^{k+1}$, we find that in a strong form without a convection term we have

$$\tau^{-1}\left(\tilde{u}^{k+1} - \tilde{u}^k\right) - \nu \Delta \tilde{u}^{k+1} + \nabla p^k = f,$$
$$\operatorname{div} \tilde{u}^{k+1} - \tau \Delta p^{k+1} = 0.$$

This is a pressure-stabilized, decoupled discretization of the Navier–Stokes equations. In the case of a smooth solution, one can derive the suboptimal error estimate

$$\max_{k=0,\dots,K} \|u(t_k) - u^k\|_{H^1(\Omega)} + \|p(t_k) - p^k\|_{L^2(\Omega)} = \mathcal{O}(\tau^{1/2}).$$

Various modifications are available to improve error estimates for projection schemes.

8.4.5 Implementation

A snapshot of a two-dimensional simulation carried out with an implementation in MATLAB is shown in Fig. 8.21. Figure 8.22 shows the underlying MATLAB code

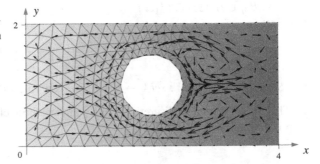

Fig. 8.21 Snapshot of finite element pressure and velocity approximations obtained with the Chorin projection scheme in a two-dimensional Navier–Stokes problem with vortices

```
function chorin_projection(d_tmp,red)
global d; d = d_tmp;
tau = 2^(-red)/10; nu = .01; T = 10; K = ceil(T/tau);
str = strcat('load triang_cyl_w_hole_ ',num2str(d),'d');
eval(str);
for j = 1:red
    [c4n,n4e,Db,Nb] = red_refine(c4n,n4e,Db,Nb);
end
nC = size(c4n,1);
dNodes = unique(Db); fNodes = setdiff(1:nC,dNodes);
FNodes = repmat(d*(fNodes-1),d,1)+(1:d)'*ones(1,size(fNodes,2),1);
FNodes = FNodes(:);
[s,m] = fe_matrices(c4n,n4e);
S = sparse(d*nC,d*nC); M = sparse(d*nC,d*nC);
for j = 1:d
    S(j:d:d*nC,j:d:d*nC) = s; M(j:d:d*nC,j:d:d*nC) = m;
end
[D,Grads_T,Vol_T] = chorin_div_matrix(c4n,n4e);
u_old = u_D(0,c4n);
for k = 1:K
    t = k*tau
    tu_new = u_D(t,c4n); tu_new(FNodes) = 0;
    % W = chorin_conv_matrix(c4n,n4e,Grads_T,Vol_T,u_old);
    W = chorin_conv_matrix_vec(c4n,n4e,Grads_T,Vol_T,u_old);
    A = M+tau*nu*S+tau*W;
    b = tau*M*f(t,c4n)+M*u_old-A*tu_new;
    tu_new(FNodes) = A(FNodes,FNodes)\b(FNodes);
    c = (1/tau)*D*(tu_new-u_D(t,c4n))-(1/tau)*m*div_u_D(t,c4n);
    p = zeros(nC,1);
    p(2:nC) = s(2:nC,2:nC)\c(2:nC);
    Pi_nabla_p = M\(D'*p);
    u_new = tu_new-tau*Pi_nabla_p;
    show_chorin(c4n,n4e,u_new,p);
    u_old = u_new;
end

function val = u_D(t,x)
global d; val = zeros(d,size(x,1)); idx = find(abs(x(:,1))>1);
val(1,idx) = sin(t)*(abs(x(idx,1))-1).*(sum(x(idx,2:d).^2,2)-1);
val = val(:);

function val = div_u_D(t,x)
global d; val = zeros(size(x,1),1); idx = find(abs(x(:,1))>1);
val(idx) = sin(t)*sign(x(idx,1)).*(sum(x(idx,2:d).^2,2)-1);

function val = f(t,x)
global d; val = zeros(d*size(x,1),1);
```

Fig. 8.22 MATLAB implementation of the Chorin projection scheme

of the projection method with P_1 discretizations for velocity and pressure. The routine `chorin_div_matrix.m` shown in Fig. 8.23 provides the matrix D that represents the bilinear form

$$(q_h, v_h) \rightarrow \int_\Omega \nabla q_h \cdot v_h \, dx$$

in the standard nodal basis. Its transpose allows us to compute the projection $\Pi[\nabla p_h]$ of the gradient of a pressure field $p_h \in \mathscr{S}^1(\mathscr{T}_h)$ onto the $\mathscr{S}^1(\mathscr{T}_h)^d$, i.e., to compute $\Pi[\nabla p_h] \in \mathscr{S}^1(\mathscr{T}_h)^d$ with

$$\int_\Omega \Pi[\nabla p_h] \cdot v_h \, dx = \int_\Omega \nabla p_h \cdot v_h \, dx$$

for all $v_h \in \mathscr{S}^1(\mathscr{T}_h)^d$. With this projection we compute the update

$$u_h^{k+1} = \tilde{u}_h^{k+1} - \tau \Pi[\nabla p_h^k].$$

```
function [D,Grads_T,Vol_T] = chorin_div_matrix(c4n,n4e)
[nC,d] = size(c4n); nE = size(n4e,1);
ctr = 0; ctr_max = d*(d+1)^2*nE;
I = zeros(ctr_max,1); J = zeros(ctr_max,1);
X_D = zeros(ctr_max,1);
Vol_T =zeros(nE,1);
Grads_T = zeros((d+1)*nE,d);
for j = 1:nE
    X_T = [ones(1,d+1);c4n(n4e(j,:),:)'];
    grads_T = X_T\[zeros(1,d);eye(d)];
    vol_T = det(X_T)/factorial(d);
    for m = 1:d+1
        for n = 1:d+1
            for p = 1:d
                ctr = ctr+1;
                I(ctr) = n4e(j,m); J(ctr) = d*(n4e(j,n)-1)+p;
                X_D(ctr) = vol_T*grads_T(m,p)'/(d+1);
            end
        end
    end
    Vol_T(j) = vol_T;
    Grads_T((d+1)*(j-1)+(1:d+1),:) = grads_T;
end
D = sparse(I,J,X_D,nC,d*nC);
```

Fig. 8.23 Assembly of the bilinear form $(q_h, v_h) \mapsto (\nabla q_h, v_h)$

```
function show_chorin(c4n,n4e,u,p)
d = size(c4n,2);
if d == 2
    trisurf(n4e,c4n(:,1),c4n(:,2),p-max(p)-1); hold on;
    quiver(c4n(:,1),c4n(:,2),u(1:2:end),u(2:2:end),'k.');
    hold off; view(0,90); shading flat; drawnow;
elseif d == 3
    p_T = sum(p(n4e),2)/(d+1);
    idx = find(c4n(n4e(:,1),2)>0);
    tetramesh(n4e(idx,:),c4n,p_T(idx)); hold on;
    quiver3(c4n(:,1),c4n(:,2),c4n(:,3),...
        u(1:3:end),u(2:3:end),u(3:3:end),'k');
    view(-26,14); hold off; drawnow;
end
```

Fig. 8.24 Visualization of a velocity and a pressure field

We only consider the case $\Gamma_D = \partial\Omega$, and for a simple implementation we use the normalization $p_h(z_1) = 0$ instead of the equivalent normalization

$$\int_\Omega p_h \, dx = 0.$$

The visualization of the solution in every time step is done by the routine displayed in Fig. 8.24. While the above matrix D, as well as the matrices representing the inner products in $L^2(\Omega)$ and $H_0^1(\Omega)$, only have to be computed once, the matrix related to the convective term, i.e., the bilinear form

$$(u_h, v_h) \mapsto n(u_h^k; u_h, v_h) = \int_\Omega (u_h^k \cdot \nabla u_h) \cdot v_h \, dx$$

has to be assembled in every time step. Figure 8.25 shows two different MATLAB realizations. The first one uses a standard loop over all elements in the triangulation. The second one is based on vector arithmetic and avoids the global loop. Since vector operations make efficient use of multi-core processors, the second implementation is significantly faster. This is important since in general many time steps have to be considered.

```
function W = chorin_conv_matrix(c4n,n4e,Grads_T,Vol_T,u)
[nC,d] = size(c4n); nE = size(n4e,1);
ctr = 0; ctr_max = (d+1)^2*nE;
I = zeros(ctr_max,1); J = zeros(ctr_max,1);
m_loc = (ones(d+1,d+1)+eye(d+1))/((d+1)*(d+2));
X_W = zeros(ctr_max,1);
for j = 1:nE
    for m = 1:d+1
        for n = 1:d+1
            val = 0;
            for i = 1:d
                val = val+Vol_T(j)*Grads_T((d+1)*(j-1)+m,i)...
                    *u(d*(n4e(j,:)-1)+i)'*m_loc(:,n);
            end
            ctr = ctr+1;
            I(ctr) = n4e(j,m); J(ctr) = n4e(j,n); X_W(ctr) = val;
        end
    end
end
w = sparse(I,J,X_W,nC,nC);
W = sparse(d*nC,d*nC);
for p = 1:d
    W(p:d:end,p:d:end) = w;
end
```

```
function W = chorin_conv_matrix_vec(c4n,n4e,Grads_T,Vol_T,u)
[nC,d] = size(c4n); nE = size(n4e,1); n4e_t = n4e';
m_loc = (ones(d+1,d+1)+eye(d+1))/((d+1)*(d+2));
X_w = zeros((d+1)^2,nE);
for i = 1:d
    X1 = repmat(Vol_T,1,(d+1)^2);
    X2 = reshape(repmat(Grads_T(:,i),1,d+1)',(d+1)^2,nE)';
    X3 = repmat(u(d*(n4e-1)+i)*m_loc,1,d+1);
    X_w = X_w+(X1.*X2.*X3)';
end
I = repmat(reshape(n4e_t,1,(d+1)*nE),d+1,1);
J = repmat(n4e_t,d+1,1);
if d == 2
    W = sparse([2*I-1,2*I],[2*J-1,2*J],[X_w,X_w],d*nC,d*nC);
elseif d == 3
    W = sparse([3*I-2,3*I-1,3*I],[3*J-2,3*J-1,3*J],...
        [X_w,X_w,X_w],d*nC,d*nC);
end
```

Fig. 8.25 Two implementations of the matrix representing the bilinear form related to the convective term

References

The use of discontinuous Galerkin and Crouzeix–Raviart finite element methods for nearly incompressible elastic materials is due to [13]. Various aspects of finite element methods for fourth order problems can be found in [2, 4, 6, 7, 10]. Finite element methods for Maxwell equations are the subject of [1, 3, 14, 16, 17, 21]. Numerical methods for Navier–Stokes equations are discussed in [9, 11, 12, 18–20, 22]. Mathematical models in elasticity, electromagnetism, and fluid mechanics are derived in [5, 8, 23]. Analytical properties of various initial boundary value problems are the subject of [15].

1. Arnold, D.N., Falk, R.S., Winther, R.: Finite element exterior calculus: from Hodge theory to numerical stability. Bull. Am. Math. Soc. (N.S.) **47**(2), 281–354 (2010). URL http://dx.doi.org/10.1090/S0273-0979-10-01278-4
2. Babuška, I., Pitkäranta, J.: The plate paradox for hard and soft simple support. SIAM J. Math. Anal. **21**(3), 551–576 (1990). URL http://dx.doi.org/10.1137/0521030
3. Boffi, D., Brezzi, F., Fortin, M.: Mixed finite element methods and applications. Springer Series in Computational Mathematics, vol. 44. Springer, Heidelberg (2013). URL http://dx.doi.org/10.1007/978-3-642-36519-5
4. Brenner, S.C., Scott, L.R.: The mathematical theory of finite element methods. Texts in Applied Mathematics, vol. 15, 3rd edn. Springer, New York (2008). URL http://dx.doi.org/10.1007/978-0-387-75934-0
5. Ciarlet, P.G.: Mathematical elasticity, Vol. I. Studies in Mathematics and Its Applications, vol. 20. North-Holland Publishing, Amsterdam (1988)
6. Ciarlet, P.G.: The finite element method for elliptic problems. Classics in Applied Mathematics, vol. 40. Society for Industrial and Applied Mathematics (SIAM), Philadelphia, PA (2002). URL http://dx.doi.org/10.1137/1.9780898719208
7. Dziuk, G.: Theorie und Numerik partieller Differentialgleichungen. Walter de Gruyter GmbH & Co. KG, Berlin (2010). URL http://dx.doi.org/10.1515/9783110214819
8. Eck, C., Garcke, H., Knabner, P.: Mathematische Modellierung. Springer-Lehrbuch. Springer, Berlin-Heidelberg-New York (2011)
9. Elman, H.C., Silvester, D.J., Wathen, A.J.: Finite elements and fast iterative solvers: with applications in incompressible fluid dynamics. Numerical Mathematics and Scientific Computation, 2nd edn. Oxford University Press, Oxford (2014). URL http://dx.doi.org/10.1093/acprof:oso/9780199678792.001.0001
10. Falk, R.S.: Approximation of the biharmonic equation by a mixed finite element method. SIAM J. Numer. Anal. **15**(3), 556–567 (1978)
11. Girault, V., Raviart, P.A.: Finite element methods for Navier-Stokes equations. Springer Series in Computational Mathematics, vol. 5. Springer, Berlin (1986). URL http://dx.doi.org/10.1007/978-3-642-61623-5
12. Guermond, J.L., Minev, P., Shen, J.: An overview of projection methods for incompressible flows. Comput. Methods Appl. Mech. Engrg. **195**(44–47), 6011–6045 (2006). URL http://dx.doi.org/10.1016/j.cma.2005.10.010
13. Hansbo, P., Larson, M.G.: Discontinuous Galerkin and the Crouzeix-Raviart element: application to elasticity. M2AN Math. Model. Numer. Anal. **37**(1), 63–72 (2003). URL http://dx.doi.org/10.1051/m2an:2003020
14. Hiptmair, R.: Finite elements in computational electromagnetism. Acta Numer. **11**, 237–339 (2002). URL http://dx.doi.org/10.1017/S0962492902000041
15. Leis, R.: Initial-boundary value problems in mathematical physics. B. G. Teubner, Stuttgart; Wiley, Chichester (1986). URL http://dx.doi.org/10.1007/978-3-663-10649-4

16. Monk, P.: Finite element methods for Maxwell's equations. Numerical Mathematics and Scientific Computation. Oxford University Press, New York (2003). URL http://dx.doi.org/10.1093/acprof:oso/9780198508885.001.0001

17. Nédélec, J.C.: Mixed finite elements in $\mathbf{R}^3$. Numer. Math. **35**(3), 315–341 (1980). URL http://dx.doi.org/10.1007/BF01396415

18. Prohl, A.: Projection and quasi-compressibility methods for solving the incompressible Navier-Stokes equations. Advances in Numerical Mathematics. B. G. Teubner, Stuttgart (1997). URL http://dx.doi.org/10.1007/978-3-663-11171-9

19. Quarteroni, A., Valli, A.: Numerical approximation of partial differential equations. Springer Series in Computational Mathematics, vol. 23. Springer, Berlin (1994)

20. Rannacher, R.: Finite element methods for the incompressible Navier-Stokes equations. In: Fundamental Directions in Mathematical Fluid Mechanics, Adv. Math. Fluid Mech., pp. 191–293. Birkhäuser, Basel (2000)

21. Schöberl, J.: Numerical methods for Maxwell equations (2009). Lecture Notes, Vienna University of Technology, Austria

22. Temam, R.: Navier-Stokes equations. Studies in Mathematics and Its Applications, vol. 2, 3rd edn. North-Holland Publishing, Amsterdam (1984)

23. Temam, R., Miranville, A.: Mathematical Modeling in Continuum Mechanics, 2nd edn. Cambridge University Press, Cambridge (2005). URL http://dx.doi.org/10.1017/CBO9780511755422

Appendix A
Problems and Projects

A.1 Finite Difference Method

A.1.1 Transport Equation

Exercise A.1.1 Derive a partial differential equation that describes the transport of a substance through a long, thin tube that allows for the injection of a substance at any time $t \in [0, T]$ and any position $x \in \mathbb{R}$ described through a function $f(t, x)$ that specifies the number of injected particles per unit volume.

Exercise A.1.2

(i) Let $f \in C(\mathbb{R})$ and such that

$$\int_{x_1}^{x_2} f(x)\, dx = 0$$

for all $x_1 \leq x_2$. Show that $f = 0$.

(ii) Show that in the derivation of the transport equation, we have that

$$\int_{x_1}^{x_2} u(t, x)\, dx = \int_{x_1}^{x_2} u(t + \tau, x + a\tau)\, dx,$$

and conclude that $u(t, x) = u(t + \tau, x + a\tau)$ for all $t \in [0, T]$, $x \in \mathbb{R}$, and $\tau > 0$.

© Springer International Publishing Switzerland 2016

S. Bartels, *Numerical Approximation of Partial Differential Equations*,
Texts in Applied Mathematics 64, DOI 10.1007/978-3-319-32354-1

Exercise A.1.3

(i) Prove the following estimates for difference quotients:

$$|\partial^{\pm} u(x_j) - u'(x_j)| \leq \frac{\Delta x}{2} \|u''\|_{C([0,1])},$$

$$|\widehat{\partial} u(x_j) - u'(x_j)| \leq \frac{\Delta x^2}{6} \|u'''\|_{C([0,1])},$$

$$|\partial^+ \partial^- u(x_j) - u''(x_j)| \leq \frac{\Delta x^2}{12} \|u^{(4)}\|_{C([0,1])}.$$

Show that these estimates do not hold if u does not satisfy the required differentiability properties.

(ii) Show that $\partial^+ \partial^- = \partial^- \partial^+$.

(iii) Prove an error estimate for the difference $\partial^+ \partial^+ u(x_j) - u''(x_j)$.

Exercise A.1.4 Let $a < 0$ and consider the numerical scheme $\partial_t^+ U_j^k + a \partial_x^+ U_j^k = 0$. Show that the scheme is stable under appropriate conditions on Δt and Δx and prove an error estimate.

Exercise A.1.5 Let $u_0 \in C^2([0,1])$ and let $\tilde{u}_0$ denote its trivial extension by zero to $\mathbb{R}$.

(i) Find conditions on u_0 that guarantee $\tilde{u}_0 \in C^2(\mathbb{R})$.

(ii) Show that the solution of the transport problem $\partial_t u + a \partial_x u = 0$ with $u(t, 0) = 0$ and $u(0, x) = u_0(x)$ satisfies $u \in C^2([0, T] \times [0, 1])$, and that

$$\|\partial_x^2 u(t, \cdot)\|_{C([0,1])} = a^{-2} \|\partial_t^2 u(t, \cdot)\|_{C([0,1])} = \|u_0''\|_{C([0,1])}$$

for all $t \in [0, T]$.

Exercise A.1.6 Show that the upwinding scheme for the transport equation is equivalent to the scheme

$$\partial_t^+ U_j^k + a_j^k \widehat{\partial}_x U_j^k = |a_j^k| \Delta x \partial_x^+ \partial_x^- U_j^k,$$

and discuss the incorporation of boundary conditions.

Exercise A.1.7

(i) Show by constructing appropriate initial data that the difference scheme $U_j^{k+1} = U_j^k + \mu(U_j^k - U_{j-1}^k)$ with $\mu = a\Delta t / \Delta x$ is unstable if $\mu > 1$.

(ii) Check the CFL condition and the estimate $\sup_{j=0,\dots,J} |U_j^{k+1}| \leq \sup_{j=0,\dots,J} |U_j^k|$ of the following difference schemes for the transport equation:

$$\partial_t^+ U_j^k - \partial_x^- U_j^k = 0, \quad \partial_t^+ U_j^k + \partial_x^+ U_j^k = 0, \quad \partial_t^+ U_j^k + \widehat{\partial}_x U_j^k = 0.$$

Exercise A.1.8

(i) Show that the functions $\phi_\ell(x) = e^{ikx}$, $x \in [-\pi, \pi]$, $\ell \in \mathbb{Z}$, define an orthonormal system in $L^2(-\pi, \pi)$, i.e., for all $\ell, m \in \mathbb{Z}$, we have

$$\frac{1}{2\pi} \int_{-\pi}^{\pi} \phi_\ell(x)\overline{\phi_m(x)} \, dx = \delta_{\ell m}.$$

(ii) For $f \in L^2(-\pi, \pi)$ and $\ell \in \mathbb{Z}$ set

$$f_\ell = \frac{1}{2\pi} \int_{-\pi}^{\pi} f(x)\overline{\phi_\ell(x)} \, dx.$$

Prove that

$$\frac{1}{2\pi} \int_{-\pi}^{\pi} |f|^2 \, dx = \sum_{\ell \in \mathbb{Z}} |f_\ell|^2.$$

Exercise A.1.9 Let $g \in C([-\pi, \pi])$ be such that

$$\int_{-\pi}^{\pi} |fg| \, dx \leq \int_{-\pi}^{\pi} |f| \, dx$$

for all $f \in L^1(-\pi, \pi)$. Show that $|g(x)| \leq 1$ for all $x \in [-\pi, \pi]$. Is it sufficient to assume that $g \in L^2(-\pi, \pi)$?

Exercise A.1.10 Let u solve the partial differential equation $\partial_t u + a(t, x)\partial_x u = 0$.

(i) Show that u is constant along curves $(t, y(t))$ for solutions of the initial boundary value problems $y'(t) = a(t, y(t))$, $y(0) = x_0$, called characteristics.
(ii) Determine the characteristics for the equation $\partial_t u + tx\partial_x u = 0$, i.e., for $a(t, x) = tx$, sketch them, and determine the solution for the initial condition $u_0(x) = \sin(x)$.

Quiz A.1.1 Decide for each of the following statements whether it is true or false. You should be able to justify your decision.

The transport equation describes the motion of a substance in a motionless fluid	
The total amount of substance in the transport problem is conserved	
For a C^2 function u, the central difference quotient $\widehat{\partial}$ provides a more accurate approximation of the derivative than the one-sided difference quotients $\partial^{\pm}$	
The implementation of the difference scheme $\partial_t^+ U_j^k + a\partial_x^- U_j^k = 0$ requires the solution of linear systems of equations in every time step	
The CFL condition is a necessary and sufficient condition for stability of a finite difference scheme	

A.1.2   Heat Equation

Exercise A.1.11 Let $u \in C^2([0, T] \times [\alpha, \beta])$ solve the heat equation $\partial_t u - \kappa \partial_x^2 u = 0$. Show that for appropriate $\tau, L, x_0 > 0$, the function $\tilde{u}(s, y) = u(\tau s, Ly + x_0)$ solves $\partial_s \tilde{u} - \partial_y^2 \tilde{u} = 0$ in $(0, T') \times (0, 1)$.

Exercise A.1.12 Derive a mathematical model for a diffusion process that includes sinks and sources of the diffusing substance, described by a function $f \in C([0, T] \times [0, 1])$.

Exercise A.1.13 Let $u \in C^2([0, T] \times [0, 1])$ solve the heat equation $\partial_t u - \partial_x^2 u = 0$ with homogeneous Dirichlet boundary conditions. Prove that

$$\frac{d}{dt} \frac{1}{2} \int_0^1 \left(\partial_x u(t, x) \right)^2 dx \leq 0$$

and deduce the uniqueness of solutions for the heat equation with general Dirichlet boundary conditions.

Exercise A.1.14 The construction of a solution via a *separation of variables* consists in finding functions $u_n(t, x) = v_n(t) w_n(x)$ that solve the heat equation and the prescribed boundary conditions. A solution of the initial boundary value problem is then obtained by determining coefficients $(\alpha_n)_{n \in \mathbb{N}}$ such that

$$u(t, x) = \sum_{n=1}^{\infty} \alpha_n v_n(t) w_n(x)$$

converges in an appropriate sense and satisfies $u(0, x) = u_0(x)$.

(i) Construct pairs (v_n, w_n) such that $u_n(t, x) = v_n(t) w_n(x)$ satisfies $\partial_t u_n - \partial_x^2 u_n = 0$ in $(0, T) \times (0, 1)$ and $u_n(t, 0) = u_n(t, 1) = 0$ for all $t \in (0, T)$.
(ii) Assume that the function $u_0 \in C([0, 1])$ is given as

$$u_0(x) = \sum_{n=1}^{\infty} \gamma_n \sin(n \pi x).$$

Construct the solution of the corresponding initial boundary value problem for the heat equation.

Remark It can be shown that every function $u_0 \in C([0, 1])$ can be represented in the specified form.

Exercise A.1.15

(i) Show that the explicit Euler scheme is unstable if $\Delta t > \Delta x^2 / 2$ by constructing appropriate initial data.

(ii) Show that numerical solutions obtained with the Crank–Nicolson scheme do in general not satisfy a discrete maximum principle.

Exercise A.1.16 Show that the discretization in the space of the heat equation leads to a stiff initial value problem $\partial_t U + AU = 0$, $U(0) = U_0$, which admits a unique solution on every time interval $[0, T]$.

Exercise A.1.17 For $a, b \in \mathbb{R}$ and $n \in \mathbb{N}$, let $A \in \mathbb{R}^{n \times n}$ be the bandmatrix

$$
A = \begin{bmatrix}
a & b & & & \\
b & \ddots & \ddots & & \\
& \ddots & \ddots & b \\
& & b & a
\end{bmatrix}.
$$

Show that A has the eigenvalues $\lambda_p = a + 2b \cos(p\pi/(n+1))$, $p = 1, 2, \ldots, n$.
Hint: Show that for $a = 0$, corresponding eigenvectors $v_p \in \mathbb{R}^n$ are given by $v_{p,j} = \sin(pj\pi/(n+1))$, $j = 1, 2, \ldots, n$.

Exercise A.1.18 Let $J \in \mathbb{N}$ and set $\Delta x = 1/J$.

(i) Prove that the vectors $\varphi_p \in \mathbb{R}^{J+1}$, $p = 1, \ldots, J - 1$, given by $\varphi_{p,j} = \sqrt{2} \sin(pj\Delta x)$, $j = 0, 1, \ldots, J$, define an orthonormal basis for $\ell_{0,\Delta x}^2 = \{V \in \mathbb{R}^{J+1} : V_0 = V_{J+1} = 0\}$ with respect to the inner product

$$
(V, W)_{\Delta x} = \Delta x \sum_{j=0}^{J} V_j W_j.
$$

(ii) Show that the vectors φ_p are eigenvectors of the operator $-\partial_x^+ \partial_x^- : \mathbb{R}^{J+1} \to \mathbb{R}^{J+1}$, defined by

$$
\left(-\partial_x^+ \partial_x^- V \right)_j = \begin{cases} 0 & \text{for } j = 0, J, \\ -\partial_x^+ \partial_x^- V_j & \text{for } j = 1, 2, \ldots, J - 1. \end{cases}
$$

Hint: Use that $\varphi_{p,j} = \sqrt{2} \,\mathrm{Im}(\omega^{pj})$ with $\omega = e^{i\pi \Delta x}$.

Exercise A.1.19

(i) Show formally that the function

$$
u(t, x) = \frac{1}{(4\pi t)^{1/2}} \int_{\mathbb{R}} e^{-|x-y|^2/(4t)} u_0(y) \, dy
$$

solves the heat equation $\partial_t u - \partial_x^2 u = 0$ in $(0, T) \times \mathbb{R}$ for every $T > 0$.

(ii) Explain why we can expect that $u(t, x) \to u_0(x)$ as $t \to 0$, e.g., for piecewise constant initial data u_0 and $x = 0$.

(iii) Let $u_0(x) = 1$ for $x \geq 0$ and $u_0(x) = 0$ for $x < 0$. Show that $u(t, x)$ is positive for all $t \in (0, T)$ and $x \in \mathbb{R}$, and conclude that information is propagated with infinite speed.

(iv) How does the formula in (i) have to be modified to provide a solution of the heat equation $\partial_t u - \kappa \partial_x^2 u = 0$?

Exercise A.1.20

(i) Show that the θ-method is well defined for every choice of θ and every choice of $\Delta t, \Delta x > 0$.

(ii) Show that the θ-method is unstable if $\theta < 1/2$ and $\lambda = \Delta t / \Delta x^2 > 1/2$.

Quiz A.1.2 Decide for each of the following statements whether it is true or false. You should be able to justify your decision.

The larger the constant $\kappa > 0$ in the heat equation $\partial_t u - \kappa \partial_x^2 u = 0$, the faster is the diffusion process	
The explicit Euler scheme is stable if $\Delta x / \Delta t^2 \leq 1/2$	
The θ-method is explicit for $\theta < 1/2$ and implicit for $\theta \geq 1/2$	
The implicit Euler scheme requires the solution of a linear system of equations in every time step, whose system matrix is diagonally dominant and irreducible	
The Crank–Nicolson scheme approximates the exact solution of the heat equation with an error of order $\mathcal{O}(\Delta t^2 + \Delta x^2)$ if $u \in C^3([0, T] \times [0, 1])$	

A.1.3   Wave Equation

Exercise A.1.21

(i) Determine functions $u_n(t, x) = v_n(t) w_n(x)$, $n \in \mathbb{N}$, that satisfy the wave equation in $(0, T) \times (0, 1)$ subject to homogeneous Dirichlet boundary conditions.

(ii) Assume that $u_0, v_0 \in C([0, 1])$ satisfy

$$u_0(x) = \sum_{n \in \mathbb{N}} \alpha_n \sin(n\pi x), \quad v_0(x) = \sum_{n \in \mathbb{N}} \beta_n \sin(n\pi x)$$

with given sequences $(a_n)_{n \in \mathbb{N}}, (b_n)_{n \in \mathbb{N}}$. Derive a representation formula for the solution of the wave equation $\partial_t^2 u - c^2 \partial_x^2 u = 0$ in $(0, T) \times (0, 1)$ with homogeneous Dirichlet boundary conditions and initial conditions $u(0, x) = u_0(x)$ and $\partial_t u(0, x) = v_0(x)$ for all $x \in [0, 1]$.

Exercise A.1.22 Let $u \in C^2([0, T] \times \mathbb{R})$ solve the wave equation $\partial_t^2 u - c^2 \partial_x^2 u = 0$ with initial conditions $u(0, x) = u_0(x)$ and $\partial_t u(0, x) = v_0(x)$ for all $x \in \mathbb{R}$.

(i) By introducing the variables $\xi = x + ct$ and $\eta = x - ct$, show that the function $\tilde{u}(\xi, \eta) = u(t, x)$ satisfies $\partial_\xi \partial_\eta \tilde{u} = 0$ and deduce that $\tilde{u}(\xi, \eta) = f(\xi) + g(\eta)$.

(ii) Conclude that there exist functions $f, g \in C(\mathbb{R})$ such that $u(t, x) = f(x + ct) + g(x - ct)$.

(iii) Determine f and g in terms of u_0 and v_0.

Exercise A.1.23

(i) We consider the wave equation $\partial_t^2 u - c^2 \partial_x^2 u = 0$ in $(0, T) \times \mathbb{R}_{>0}$ subject to the boundary condition $u(t, 0) = 0$ for all $t \in (0, T)$. Use d'Alembert's formula to represent the solution for the initial conditions $u(0, x) = u_0(x)$ and $\partial_t u(0, x) = 0$ for $x \in \mathbb{R}_{\geq 0}$, where $u_0 \in C(\mathbb{R}_{\geq 0})$ satisfies $u_0(0) = 0$.

(ii) Specify the solution for $u_0(x) = \max\{0, 1 - |x - 2|\}$ and sketch it for $t = 0, 5/6, 2, 7/3, 15/6$.

Exercise A.1.24

(i) Prove the energy conservation principle for the wave equation with homogeneous Neumann boundary conditions.

(ii) Deduce uniqueness of solutions for solutions of the wave equation with homogeneous Dirichlet or homogeneous Neumann boundary conditions.

Exercise A.1.25 Show that the explicit finite difference scheme for the wave equation is unstable if $\mu = c\Delta t / \Delta x > 1$.

Exercise A.1.26 Let $(\xi_k)_{k \in \mathbb{N}_0}$ be a sequence of real numbers that for $\alpha, \beta \in \mathbb{R}$ and all $k \in \mathbb{N}$ satisfies the recursion

$$\begin{bmatrix} \xi_k \\ \xi_{k+1} \end{bmatrix} = A \begin{bmatrix} \xi_{k-1} \\ \xi_k \end{bmatrix}, \quad A = \begin{bmatrix} 0 & 1 \\ \alpha & \beta \end{bmatrix}.$$

(i) Show that if the eigenvalues $\lambda_1, \lambda_2 \in \mathbb{C}$ of A satisfy $|\lambda_i| < 1$, $i = 1, 2$, then there exists $c > 0$ such that $|\xi_k| \leq c$ for all $k \in \mathbb{N}_0$.

(ii) Show that if the eigenvalues $\lambda_1, \lambda_2 \in \mathbb{C}$ of A coincide and satisfy $|\lambda_i| = 1$, $i = 1, 2$, or if $\max_{i=1,2} |\lambda_i| > 1$, then there exist unbounded sequences $(\xi_k)_{k \in \mathbb{N}_0}$ that satisfy the recursion.

Exercise A.1.27 For $J \in \mathbb{N}$, let $\Delta x = 1/J$ and let $V, W \in \mathbb{R}^{J+1}$.

(i) Prove the discrete product rule

$$\partial_x^-(W_j V_j) = W_j(\partial_x^- V_j) + (\partial_x^+ W_{j-1})V_{j-1}.$$

(ii) Deduce the summation-by-parts formula

$$\Delta x \sum_{j=0}^{J-1} (\partial_x^+ W_j) V_j = -\Delta x \sum_{j=1}^{J} W_j (\partial_x^- V_j) + W_J V_J - W_0 V_0,$$

and explain its relation to the integration-by-parts formula.

Exercise A.1.28 Let $J \in \mathbb{N}$, $\Delta x = 1/J$, and $(U_j^k) \in \mathbb{R}^{(K+1) \times (J+1)}$. Prove the identities

$$(\partial_t^+ \partial_t^- U_j^k)(\widehat{\partial}_t U_j^k) = \frac{1}{2} \partial_t^- (U_j^k)^2,$$

and

$$\frac{c^2}{4} \partial_x^+ \partial_x^- (U_j^{k+1} + 2U_j^k + U_j^{k-1})(\widehat{\partial}_t U_j^k)$$

$$= \frac{c^2}{2\Delta t} (\partial_x^+ \partial_x^- (U_j^{k+1/2} + U_j^{k-1/2}))(U_j^{k+1/2} - U_j^{k-1/2}),$$

for $1 \le j \le J - 1$ and $1 \le k \le K - 1$, where $U_j^{k \pm 1/2} = (U_j^k + U_j^{k \pm 1})/2$.

Exercise A.1.29

(i) Prove that the implicit difference scheme for the wave equation is well defined, i.e., leads to regular linear systems of equations in all time steps.

(ii) Show that the implicit difference scheme for the wave equation has a consistency error $\mathcal{O}(\Delta t^2 + \Delta x^2)$.

Exercise A.1.30 The wave equation $\partial_t^2 u = \partial_x^2 u$ can be written as the system $\partial_t u = v$, $\partial_t v = \partial_x^2 u$. Discretize this system with backward difference quotients in time and a central difference quotient in space and analyze the stability of the resulting scheme.

Quiz A.1.3 Decide for each of the following statements whether it is true or false. You should be able to justify your decision.

The total kinetic energy of a solution for the wave equation is constant	
The explicit difference scheme for the wave equation is stable if $c\Delta t \le \Delta x$	
The implicit scheme for the wave equation unconditionally satisfies a discrete maximum principle	
The larger the constant c in the wave equation $\partial_t^2 - c^2 \partial_x^2 u = 0$, the smaller is the wave speed	
The discretization of the initial condition $\partial_t u(0, x) = v_0(x)$ with a central difference quotient leads to a consistency error $\mathcal{O}(\Delta t^2)$	

A.1.4 Poisson Equation

Exercise A.1.31 Let $\Omega = (0,1)^2$ and let $f \in C(\overline{\Omega})$ be given by

$$f(x_1, x_2) = \sum_{m,n \in \mathbb{N}} \alpha_{m,n} \sin(m\pi x_1) \sin(n\pi x_2).$$

Compute $-\Delta u_{m,n}$ for $u_{m,n}(x_1, x_2) = \sin(\pi m x_1) \sin(\pi n x_2)$ and construct the solution of the Poisson problem $-\Delta u = f$ in Ω and $u = 0$ on $\partial\Omega$.

Exercise A.1.32

(i) Show that it is sufficient to assume that $-\Delta u \leq 0$ to prove the maximum principle $\max_{x \in \overline{\Omega}} u(x) \leq \max_{x \in \partial\Omega} u(x)$.
(ii) Let $u \in C^2(\overline{\Omega})$ solve $-\Delta u = f$ in Ω and $u = 0$ on $\partial\Omega$. Apply the maximum principle to an appropriately defined function $v = u + \|f\|_{C(\overline{\Omega})} w$ to prove that

$$\|u\|_{C(\overline{\Omega})} \leq \max_{x \in \partial\Omega} \frac{|x|^2}{2d} \|f\|_{C(\overline{\Omega})}.$$

Is it possible to improve this estimate?

Exercise A.1.33

(i) Use Gauss's theorem to show that for $u, v \in C^2(\overline{\Omega})$, we have

$$\int_{\partial\Omega} v \nabla u \cdot n \, ds = \int_{\Omega} (\nabla u \cdot \nabla v \, dx + v \Delta u) \, dx,$$

$$\int_{\Omega} (u \Delta v - v \Delta u) \, dx = \int_{\partial\Omega} (u \nabla v \cdot n - v \nabla u \cdot n) \, ds.$$

(ii) Let $u_1, u_2 \in C^2(\overline{\Omega})$ be solutions of the boundary value problem $-\Delta u = f$ in Ω and $u = 0$ on $\partial\Omega$. Show that

$$\int_{\Omega} |\nabla(u_1 - u_2)|^2 \, dx = 0$$

and conclude that $u_1 = u_2$.

Exercise A.1.34 Let $x_0 \in \mathbb{R}^d$ for $d \in \{2, 3\}$, $a > 0$, and $u \in C^1(\overline{B_a(x_0)})$.

(i) Show that in polar coordinates with respect to x_0, we have

$$\nabla u \cdot n = \partial_r u$$

on $\partial B_{a'}(x_0)$ for every $0 < a' \leq a$.

(ii) Show that

$$\lim_{r \to 0} \frac{1}{|\partial B_r(x_0)|} \int_{\partial B_r(x_0)} u(s)\, ds = u(x_0),$$

where $|\partial B_r(x_0)|$ denotes the surface measure of $\partial B_r(x_0)$.

Exercise A.1.35 Let $w : \mathbb{R}^2 \to \mathbb{R}$ be a quadratic polynomial and $\Delta x = 1/J$ for some $J \in \mathbb{N}$. For $j, m \in \mathbb{Z}^2$, let $x_{j,m} = (j,m)\Delta x$ and $W_{j,m} = w(x_{j,m})$. Show that

$$\Delta_h W_{j,m} = \partial_{x_1}^+ \partial_{x_1}^- W_{j,m} + \partial_{x_2}^+ \partial_{x_2}^- W_{j,m} = \Delta w(x_{j,m})$$

for all $j, m \in \mathbb{Z}^2$.

Exercise A.1.36 Let $J \in \mathbb{N}$, set $L = (J-1)^2$, and let $X \in \mathbb{R}^{(J-1) \times (J-1)}$ and $A \in \mathbb{R}^{L \times L}$ be defined by

$$X = \begin{bmatrix} 2 & -1 & & \\ -1 & \ddots & \ddots & \\ & \ddots & \ddots & -1 \\ & & -1 & 2 \end{bmatrix}, \quad A = \begin{bmatrix} X & -I & & \\ -I & \ddots & \ddots & \\ & \ddots & \ddots & -I \\ & & -I & X \end{bmatrix},$$

where $I \in \mathbb{R}^{(J-1) \times (J-1)}$ denotes the identity matrix. Show that A is diagonally dominant and irreducible.

Exercise A.1.37 Let $AU = F$ be the linear system of equations corresponding to the discretized Poisson problem $-\Delta u = f$ in $\Omega = (0, 1)^2$ with homogeneous Dirichlet boundary conditions. Show that the Richardson scheme for the iterative solution of the linear system can be identified with an explicit discretization of the heat equation.

Exercise A.1.38 Let $A \in \mathbb{R}^{n \times n}$ be the system matrix corresponding to the discretization of the Poisson problem. Use the discrete maximum principle to show that for the matrix $B = A^{-1}$, we have $b_{ij} \geq 0$, $i, j = 1, 2, \ldots, n$.

Exercise A.1.39 We consider a swimming pool that has the horizontal shape of an annulus and assume that the stationary temperature distribution is independent of the vertical direction. Moreover, we assume that the temperature is prescribed at the boundary. We thus consider the two-dimensional boundary value problem

$$-\Delta u = 0 \text{ in } \Omega = B_{r_2}(0) \setminus \overline{B_{r_1}(0)}, \quad u = u_1 \text{ on } \partial B_{r_1}(0), \quad u = u_2 \text{ on } \partial B_{r_2}(0)$$

for given real numbers $0 < r_1 < r_2$ and $u_1, u_2 \in \mathbb{R}$.

(i) Show that for $g \in C^2(\mathbb{R}_{\geq 0})$ and $r(x_1, x_2) = (x_1^2 + x_2^2)^{1/2}$, we have

$$\Delta(g \circ r) = g''(r) + r^{-1}g'(r) = r^{-1}\big(rg'(r)\big)'.$$

(ii) Justify the assumption $u = \tilde{u} \circ r$ and solve the Poisson problem for the swimming pool with $r_1 = 10$, $r_2 = 20$, and $u_1 = 20$, $u_2 = 40$.

(iii) On which radius do you have to swim to be surrounded by water of $30°C$?

Exercise A.1.40

(i) Let $L > 1$ and for $\alpha_\ell, p_\ell \in \mathbb{R}$, $0 \leq \ell \leq L$, assume that $\alpha_\ell < 0$ for $\ell = 1, 2, \ldots, L$, and

$$\sum_{\ell=0}^{L} \alpha_\ell \geq 0, \quad \sum_{\ell=0}^{L} \alpha_\ell p_\ell \leq 0.$$

Suppose further that $p_0 \geq 0$ or $\sum_{\ell=0}^{L} \alpha_\ell = 0$. Show that $p_0 \geq \max_{1 \leq \ell \leq L} p_\ell$ implies $p_0 = p_1 = \cdots = p_L$.

(ii) Let $(U_{j,m})_{0 \leq j, m \leq J}$ be the finite difference approximation of the Poisson problem $-\Delta u = f$ in $\Omega = (0,1)^2$ and $u = u_D$ on $\partial \Omega$. Assume that $f \leq 0$ and show that

$$\max_{1 \leq j, m \leq J-1} U_{j,m} \leq \max_{x_{j,m} \in \partial \Omega} u_D(x_{j,m}).$$

Quiz A.1.4 Decide for each of the following statements whether it is true or false. You should be able to justify your decision.

If f is constant, then the solution of the Poisson problem $-\Delta u = f$ in Ω, $u\|_{\partial \Omega} = 0$, is constant
If u_1 and u_2 are harmonic functions, then $u_1 - u_2$ is also a harmonic function
The finite difference discretization of the Poisson problem has a consistency error of order $\mathcal{O}(\Delta x^2)$
If $-\Delta u \geq 0$, then $\max_{x \in \overline{\Omega}} u(x) \geq \max_{x \in \partial \Omega} u(x)$
If U_i, $i = 1, 2$, are the coefficient vectors of finite difference solutions of Poisson problems $-\Delta_h U_i = F_i$ with homogeneous Dirichlet boundary conditions, then we have $\|U_1 - U_2\|_\infty \leq c\|F_1 - F_2\|_\infty$ with a constant $c > 0$

A.1.5   General Concepts

Exercise A.1.41 Write the initial boundary value problem for the wave equation as an abstract boundary value problem $F(u) = 0$ in U and $G(u) = 0$ on ∂U by defining appropriate mappings F and G.

Exercise A.1.42 Discuss the well-posedness of the heat equation with homogeneous boundary conditions. In particular, discuss the effect of perturbations of initial data.

Exercise A.1.43

(i) Let $A \in \mathbb{R}^{n \times n}$, $b \in \mathbb{R}^n$, $c \in \mathbb{R}$, and $f \in C(\overline{U})$. Assume that $u \in C^2(\overline{U})$ satisfies

$$\sum_{i,j=1}^{n} a_{ij} \partial_{z_i} \partial_{z_j} u(z) + \sum_{j=1}^{n} b_j \partial_{z_j} u(z) + c\, u(z) = f(z)$$

for all $z \in U$. Suppose that $A = Q^{\mathsf{T}} \Lambda Q$ is diagonalizable and define $\tilde{u}(\xi) = u(Q\xi)$. Determine the partial differential equation satisfied by $\tilde{u}$.

(ii) Determine the type of the following partial differential equations:

$$\partial_t u + \Delta u = f \quad \text{in } (0, T) \times \Omega \subset \mathbb{R}_{\geq 0} \times \mathbb{R}^d,$$

$$\partial_{x_1}^2 u - 3 \partial_{x_1} \partial_{x_2} u + \partial_{x_2}^2 u = 0 \quad \text{in } \Omega \subset \mathbb{R}^2,$$

$$\partial_t u - \partial_{x_1}^2 u + \partial_{x_2} u = f \quad \text{in } (0, T) \times \Omega \subset \mathbb{R}_{\geq 0} \times \mathbb{R}^2.$$

Exercise A.1.44 Formulate the stability and consistency of the difference scheme for the transport equation in an abstract framework and apply the Lax–Richtmyer theorem to derive an error estimate.

Exercise A.1.45 Formulate the stability and consistency of the implicit difference scheme for the wave equation in an abstract framework and apply the Lax–Richtmyer theorem to derive an error estimate.

Exercise A.1.46 Let $L_h(U_h) = \ell_h$ be the linear system of equations resulting from the discretization of a boundary value problem. Here, boundary nodes have been eliminated appropriately. Suppose that the discretization is stable in the sense that $\|U_h\|_{\ell,N_h} \leq c \|\ell_h\|_{r,N_h}$ for all $\ell_h \in \mathbb{R}^{N_h}$ and $U_h \in \mathbb{R}^{N_h}$ with $L_h(U_h) = \ell_h$ and with norms $\| \cdot \|_{\ell,N_h}$ and $\| \cdot \|_{r,N_h}$ on $\mathbb{R}^{N_h}$. Show that $L_h : \mathbb{R}^{N_h} \to \mathbb{R}^{N_h}$ is an isomorphism and conclude that the discrete problem admits a unique solution.

Exercise A.1.47 For $u \in C^2([0, 1]^2)$ and grid points $x_{j,m} = (j, m) \Delta x$, $0 \leq j, m \leq J$, with $\Delta x = 1/J$, define the interpolant of u by

$$\mathscr{I}_h u = \left(u(x_{j,m}) \right)_{0 \leq j,m \leq J} \in \mathbb{R}^{(J+1)^2}.$$

Show that with the norm $\|V\|_\infty = \max_{0 \leq j,m \leq J} |V_{j,m}|$ on $\mathbb{R}^{(J+1)^2}$, we have for $\Delta x \to 0$ that

$$\|\mathscr{I}_h u\|_\infty \to \|u\|_{C([0,1]^2)}.$$

Exercise A.1.48 Discuss the discretization and numerical solution of the three-dimensional Poisson problem $-\Delta u = f$ in $\Omega = (0, 1)^3$ and $u = 0$ on $\partial\Omega$. Provide a stability estimate, determine the consistency error, and specify the resulting linear system of equations.

Exercise A.1.49

(i) Let $J \in \mathbb{N}$ and $\Delta x = 1/J$. Let $(\varphi_p : p = 1, \ldots, J - 1)$ be the eigenvectors of $-\partial_x^+ \partial_x^-$ given by $\varphi_{p,j} = \sqrt{2} \sin(pj\pi\Delta x)$, $0 \le j \le J$. Show that the vectors $\psi_{(p,q)} \in \mathbb{R}^{(J+1)^2}$, defined by

$$\psi_{(p,q),(j,m)} = \varphi_{p,j}\varphi_{q,m} = 2\sin(pj\pi\Delta x)\sin(qm\pi\Delta x)$$

are eigenvectors of the operator $-\Delta_h = -\partial_{x_1}^+\partial_{x_1}^- - \partial_{x_2}^+\partial_{x_2}^-$ and that they define an orthonormal basis of the space of grid functions with vanishing boundary conditions with respect to the inner product

$$(V, W)_{\Delta x} = \Delta x^2 \sum_{j,m=0}^{J} V_{j,m}W_{j,m}.$$

(ii) Carry out a stability analysis of the θ-method for approximating the two-dimensional heat equation.

Exercise A.1.50 Assume that $F_h(U_h) = L_h U_h - \ell_h$ is a discretization of a linear boundary value problem which is bounded and convergent in the sense that we have the estimates

$$\|LV_h\|_{r,N_h} \le c_1 \|V_h\|_{\ell,N_h},$$

$$\|\mathscr{I}_h u - U_h\|_{\ell,N_h} \le c_2 h^\alpha \|u\|_{C^{k+s}(\overline{U})},$$

for every $V_h \in \mathbb{R}^{N_h}$ and continuous and discrete solutions $u \in C^{k+s}(\overline{U})$ and $U_h \in \mathbb{R}^{N_h}$. Show that the numerical scheme is consistent of order α.

Quiz A.1.5 Decide for each of the following statements whether it is true or false. You should be able to justify your decision.

Every initial value problem $y'(t) = f(t, y(t))$ for $t \in (0, T]$, $y(0) = y_0$, with a continuous function f defines a well-posed boundary value problem	
The implicit Euler scheme for the two-dimensional heat equation is unconditionally stable	
Every partial differential equation admits solutions	
The implicit scheme for the two-dimensional wave equation is stable if $\Delta t \le c\Delta x$	
The equation $\partial_x^2 u - 4\partial_y u + \partial_z^2 u = 0$ is elliptic	

A.1.6   Projects

Project A.1.1

(i) Numerically solve the transport equation $\partial_t u + \partial_x u = 0$ in $(0, T) \times (0, 1)$ for $T = 1$ with boundary condition $u(t, 0) = 0$ and initial condition defined by $u_0(x) = 1$ for $0.4 \le x \le 0.6$ and $u_0(x) = 0$ otherwise, using a forward difference quotient in time and a backward difference quotient in space. Try the pairs of discretization parameters

$$(\Delta t, \Delta x) = \frac{1}{80}(2, 2), \quad (\Delta t, \Delta x) = \frac{1}{80}(2, 1), \quad (\Delta t, \Delta x) = \frac{1}{80}(1, 2).$$

Check for which of the pairs the CFL condition is satisfied, and compare the numerical solution with the exact solution of the transport equation.

(ii) Modify your code to obtain an approximation scheme for the equation

$$\partial_t u + a(x)\partial_x u = 0,$$

where $a(x) > 0$ is a given function. How should the CFL condition be formulated for nonconstant functions a? Test your code with $a(x) = (1 + 4x^2)^{1/2}$ and initial conditions $u_0(x) = 1$ if $0.05 \le x \le 0.25$ and $u_0(x) = 0$ otherwise. Compare the numerical solutions for various discretization parameters.

(iii) Run your program with $a(x) = -1$ and initial condition $u_0(x) = 1$ for $0.4 \le x \le 0.6$ and $u_0(x) = 0$ otherwise. Are there pairs of discretization parameters so that the CFL condition is satisfied?

(iv) Modify your code so that a forward difference is realized. Here the boundary condition $u(1, t) = 0$ for $t \in [0, T]$ is given. Derive the CFL condition for this approximation scheme and try different choices for Δt and Δx.

Project A.1.2   An *upwinding scheme* for the transport equation is defined by

$$U_j^{k+1} = \begin{cases} (1 - \mu_j^k)U_j^k + \mu_j^k U_{j-1}^k, & \mu_j^k \ge 0, \\ (1 + \mu_j^k)U_j^k - \mu_j^k U_{j+1}^k, & \mu_j^k < 0, \end{cases}$$

where $\mu_j^k = a(t_k, x_j)\Delta t / \Delta x$. Implement the scheme and test it with different initial conditions, discretization parameters, the function $a(x) = \sin(x)$, and boundary conditions defined by $u(0, t) = u(1, t) = 0$. Discuss your results and the validity of a CFL condition.

Project A.1.3

(i) Implement a θ-midpoint scheme to approximately solve the initial boundary value problem $\partial_t u = \kappa \partial_x^2 u$ in $(0, T) \times (0, 1)$ for $T = 1$ and $\kappa = 1/100$, $u(0, x) = \sin \pi x$ for $x \in (0, 1)$, and $u(t, 0) = u(t, 1) = 0$ for $t \in [0, T]$. Set

$\Delta x = 0.05$ and experimentally determine Δt so that the scheme is stable for $\theta = 0$.

(ii) Verify that the exact solution of the problem is given by

$$u(t, x) = \sin(\pi x) \exp\left(-\kappa \pi^2 t\right).$$

For $\theta = 1/2$, $\theta = 3/4$, and $\theta = 1$, determine the approximation error using the displaying format `long` at the point $(t, x) = (1, 0.5)$ for $\Delta x = \Delta t = 2^{-j}/10$, for $j = 2, 3, \ldots, 5$. Plot the errors in one figure using the commands `semilogy` and `hold on/off`. What is your conclusion?

(iii) Modify your code to allow for a right-hand side f, i.e., the partial differential equation $\partial_t u - \kappa \partial_x^2 u = f$, and solve the initial boundary value problem in $(0, T) \times (0, 1)$ with $T = 2$, $f(x) = (x-1/2)^2$, homogeneous Dirichlet boundary conditions at $x = 0$ and $x = 1$, and the initial condition defined by $u_0(x) = 1$ if $0.45 \le x \le 0.55$. Compare the numerical solutions for various discretization parameters and $\theta = 0, 1/2, 1$.

Project A.1.4

(i) Numerically solve the wave equation $\partial_t^2 u - \partial_x^2 u = 0$ in $(0, T) \times (0, 1)$ with homogeneous Dirichlet boundary conditions and initial conditions $v_0(x) = 0$ and $u_0(x) = \sin(\pi x)$, using an explicit difference scheme with discretization parameters

$$(\Delta t, \Delta x) = \frac{1}{40}(2, 2), \quad (\Delta t, \Delta x) = \frac{1}{40}(2, 1), \quad (\Delta t, \Delta x) - \frac{1}{40}(1, 2).$$

Compare your results and explain differences in the numerical solutions.

(ii) Change the initial conditions to

$$u_0(x) = 0, \quad v_0(x) = \begin{cases} 1 & \text{if } 0.4 \le x \le 0.6, \\ 0 & \text{otherwise,} \end{cases}$$

and run the program for different pairs of discretization parameters.

(iii) Experimentally investigate the violation of a discrete energy conservation principle by plotting the quantity

$$\Gamma^k = \frac{\Delta x}{2} \sum_{j=1}^{J-1} |\partial_t^+ U_j^k|^2 + \frac{\Delta x}{2} \sum_{j=1}^{J} |\partial_x^- U_j^k|$$

as functions of $k = 0, 1, \ldots, K - 1$.

Project A.1.5 The sound of a stringed instrument is defined by the occurrence of different overtones. To verify experimentally that the wave equation captures this effect, we consider a string of length $\ell > 0$ that is plucked at time $t = 0$ at a

Fig. A.1 Initial displacement of a string

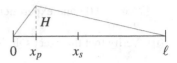

position $x_p \in (0, \ell)$ by a distance $H > 0$, so that we have

$$u_0(x) = \begin{cases} Hx/x_p & \text{for } x \leq x_p, \\ H(\ell - x)/(\ell - x_p) & \text{for } x \geq x_p, \end{cases}$$

cf. Fig. A.1. We assume that the tone is sampled at a position x_s, e.g., by a hole of an acoustic or a pickup in the case of an electric instrument. The initial velocity is assumed to vanish, and the ends of the string are fixed. Assuming for simplicity that $\ell = 1$, a separation of variables in the wave equation implies that we have

$$u(t, x) = \sum_{m=1}^{\infty} \beta_m \cos(\omega_m t) \sin(m\pi x)$$

with $\omega_m = m\pi c$ and $c = (\varrho/\sigma)^{1/2}$. Numerically solve the wave equation with $c = 2, T = 2, x_p = 1/8$, and $H = 1/100$, and use your approximations to determine coefficients $\alpha_m, m = 1, 2, \ldots, K$, such that

$$U_{j_s}^k = \sum_{m=1}^{K} \alpha_m \cos(\omega_m t_k),$$

where j_s is the index corresponding to the grid-point which equals $x_s = 1/4$. Plot the harmonics $w_m(t) = \alpha_m \cos(\omega_m t)$, $m = 1, 2, \ldots, 6$, as functions of $t \in [0, T]$, and visualize the amplitude distribution by plotting the function $m \mapsto |\alpha_m|$. Try other values for x_p and x_s and compare the results. Is it necessary to solve the wave equation in order to determine the coefficients α_m?

Project A.1.6 Implement the unconditionally stable implicit scheme

$$\partial_t^+ \partial_t^- U_j^n = \frac{1}{4} \partial_x^+ \partial_x^- (U_j^{n+1} + 2U_j^n + U_j^{n-1})$$

for approximating the wave equation $\partial_t^2 u = \partial_x^2 u$ in $(0, T) \times (0, 1)$ with homogeneous Dirichlet boundary conditions and initial conditions $u(x, 0) = u_0(x)$ and $\partial_t u(0, x) = v_0(x)$ for $x \in (0, 1)$ and given functions $u_0, v_0 \in C([0, 1])$. Use a discretization of the initial condition $\partial_t u(0, x)$ that leads to quadratic convergence. Test your program with the exact solution

$$u(t, x) = \cos(\pi t) \sin(\pi x).$$

Project A.1.7 Define functions f, g, u_D so that $u(x, y) = \sin(\pi x)\sin(\pi y)$ is the solution of the boundary value problem

$$-\Delta u = f \qquad \text{in } \Omega = (0, 1)^2,$$

$$u = u_D \qquad \text{on } \Gamma_D = [0, 1] \times \{0\},$$

$$\partial_n u = g \qquad \text{on } \Gamma_N = \partial\Omega \setminus \Gamma_D.$$

Introduce ghost points and use centered difference quotients to approximate the normal derivative on Γ_N. Experimentally verify that the scheme is quadratically convergent.

Project A.1.8 We consider an oven occupying the region $\Omega = (0, 0.4) \times (0, 0.3) \times (-\ell_z, \ell_z)$ and assume that the back of the oven is constantly heated to a temperature of $\theta = 200°C$, the front is either open or closed, and all other sides are thermally insulated, i.e., that $\partial_n\theta = 0$. When the front side is open, we assume that $\theta = 20°C$, and when it is closed we have $\partial_n\theta = 0$, cf. Fig. A.2. At time $t = 0$ we assume that the temperature inside the oven is uniformly given by $\theta = 200°C$. A mathematical model is obtained from the physical laws that heat density is proportional to temperature, i.e., $w = \varrho c_p \theta$, heat flux is proportional to the temperature gradient, i.e., $q = -\kappa\nabla\theta$, and thermal energy is conserved, i.e., $\partial_t w + \operatorname{div} q = 0$. In particular, we use the density $\varrho = 1.435\cdot10^{-3}\text{kg/m}^3$, the heat conductivity $\kappa = 0.024\,\text{W/m K}$, and the heat capacity $c_p = 1.007 \cdot 10^3\,\text{J/kg K}$. To reduce the dimension of the problem, we replace θ by its horizontal average, i.e., we consider

$$\theta'(t, x_1, x_2) = \frac{1}{2\ell_z}\int_{-\ell_z}^{\ell_z}\theta(t, x_1, x_2, x_3)\,dx_3.$$

Formulate an initial boundary value problem to describe the averaged temperature θ' in $\Omega' = (0, 0.4) \times (0, 0.3)$. Implement a Crank–Nicolson scheme and simulate the following scenarios: (i) oven open for 30s, closed for 30s, open for 30s; (ii) oven closed for 30s, open for 60s. Decide on the basis of your simulations whether it is energetically preferable to open the oven once for a long period or twice for shorter periods. Discuss limitations of the model and the numerical method.

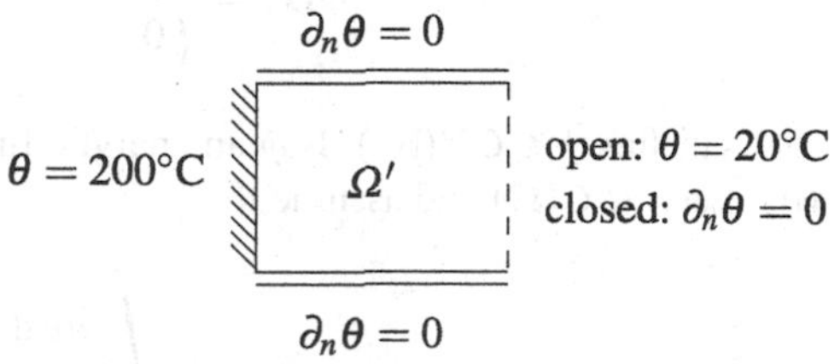

Fig. A.2 Schematical description of the cross-section of an oven

A.2 Elliptic Partial Differential Equations

A.2.1 Weak Formulation

Exercise A.2.1 For $\Omega \subset \mathbb{R}^2$ and $u \in C^2(\Omega)$, let $\tilde{u}(r, \phi) = u(r\cos\phi, r\sin\phi)$.

(i) Show that

$$\nabla u(r\cos\phi, r\sin\phi) = [\partial_r \tilde{u}(r, \phi), r^{-1}\partial_\phi \tilde{u}(r, \phi)]^\top$$

and

$$\Delta u(r\cos\phi, r\sin\phi) = \partial_r^2 \tilde{u}(r, \phi) + r^{-1}\tilde{u}(r, \phi) + r^{-2}\partial_\phi^2 \tilde{u}(r, \phi).$$

(ii) Verify that the function $\tilde{u}(r, \phi) = r^{\pi/\alpha} \sin(\phi\pi/\alpha)$ is harmonic.

Exercise A.2.2 For an open set $U \subset \mathbb{C}$, let $f : U \to \mathbb{C}$ be complex differentiable, i.e., for every $z \in U$ there exists $f'(z_0) \in \mathbb{C}$ such that

$$\lim_{h \to 0} \frac{f(z_0 + h) - f(z_0)}{h} = f'(z_0),$$

where $h \to 0$ represents an arbitrary sequence of complex numbers that converges to zero. Show that the functions $u, v : U \to \mathbb{R}$ defined by $f(x + iy) = u(x, y) + iv(x, y)$ satisfy the equations

$$\partial_x u = \partial_y v, \quad \partial_y u = -\partial_x v$$

in U and that they are harmonic, i.e., satisfy $-\Delta u = 0$ and $-\Delta v = 0$ in U.

Exercise A.2.3 Assume that Ω is connected and $\Gamma_D \neq \emptyset$. Prove that there exists at most one solution of the weak formulation of the Poisson problem.

Exercise A.2.4

(i) Show that the function

$$\phi(x) = \begin{cases} e^{-1/(1-|x|^2)} & \text{for } |x| < 1, \\ 0 & \text{for } |x| \geq 1 \end{cases}$$

satisfies $\phi \in C^\infty(\mathbb{R}^d)$. Is ϕ an analytic function?

(ii) Let $h \in C(\Omega)$ and assume that

$$\int_\Omega hv \, \mathrm{d}x = 0$$

for all $v \in C^\infty(\Omega)$ with $v = 0$ on $\partial\Omega$. Prove that $h = 0$ in Ω.

Exercise A.2.5 Sketch the nonclassical solution $u(r, \phi) = r^{\pi/\alpha} \sin(\phi\pi/\alpha)$ of the Poisson problem and its gradient for $\alpha \in \{\pi/2, \pi, 3\pi/2\}$.

Exercise A.2.6 Show that the Dirichlet energy is convex, i.e., $I((1 - t)u + tv) \leq (1 - t)I(u) + tI(v)$ for $t \in [0, 1]$.

Exercise A.2.7 Show that the functional

$$I(v) = \int_1^1 x^2 \big(u'(x)\big)^2 \, dx$$

has no minimizer in $C^1((-1, 1))$ subject to the boundary conditions $v(-1) = -1$ and $v(1) = 1$.

Exercise A.2.8 Let $\Omega \subset \mathbb{R}^d$ be open, bounded, and connected, and let $\Gamma_D \subset \partial\Omega$ be nonempty. Prove that

$$\|v\| = \left(\int_\Omega |\nabla v|^2 \, dx \right)^{1/2}$$

defines a norm on $V = \{v \in C^1(\overline{\Omega}) : v|_{\Gamma_D} = 0\}$.

Exercise A.2.9 Let $A \in \mathbb{R}^{n \times n}$ and $b \in \mathbb{R}^n$.

(i) Show that $x \in \mathbb{R}^n$ satisfies $Ax = b$ if and only if

$$(Ax) \cdot y = b \cdot y$$

for all $y \in \mathbb{R}^n$.

(ii) Assume that A is symmetric and positive definite. Show that there exists a matrix $B \in \mathbb{R}^{n \times n}$ such that the unique solution of the linear system $Ax = b$ is the unique minimizer of the mapping

$$z \mapsto \frac{1}{2}|Bz|^2 - b \cdot z.$$

Exercise A.2.10 Let $F : \mathbb{R}^n \to \mathbb{R}$ be such that there exist constants $c_1, c_2 > 0$ so that

$$F(z) \geq c_1|z| - c_2.$$

Assume that F is lower semicontinuous, i.e., whenever $z_j \to z$ as $j \to \infty$ then $F(z) \leq \liminf_{j \to \infty} F(z_j)$. Show that F has a global minimizer and provide an example of a function F that satisfies the conditions but is not continuous.

Quiz A.2.1 Decide for each of the following statements whether it is true or false. You should be able to justify your decision.

If $f \in C^1(\overline{\Omega})$, $\Gamma_{\mathrm{D}} = \partial\Omega$, and $u_{\mathrm{D}} = 0$, then the Poisson problem has a classical solution
The weak formulation of the Poisson problem specifies a function $u \in C^1(\overline{\Omega}) \cap C^2(\Omega)$
Every bilinear form $a : V \times V \to \mathbb{R}$ is symmetric
The linear system of equations $Ax = b$ can be identified with a formulation $a(x, y) = b(y)$ for all $y \in \mathbb{R}^n$
Every nonnegative functional $I : V \to \mathbb{R}$ on a Banach space V has a minimizer

A.2.2 Elementary Functional Analysis

Exercise A.2.11 Let $a : V \times V \to \mathbb{R}$ be symmetric, bilinear, and positive semidefinite. Prove that

$$a(v, w) \le \left(a(v, v)\right)^{1/2} \left(a(w, w)\right)^{1/2}.$$

Exercise A.2.12 Let $I \subset \mathbb{R}$ be a closed interval. Show that $C^1(I)$ is complete with respect to the norm

$$\|v\| = \sup_{x \in I} |v(x)| + \sup_{x \in I} |v'(x)|$$

but not with respect to the norm

$$\|v\| = \int_I |v(x)| + |v'(x)| \, \mathrm{d}x.$$

Exercise A.2.13 Let V be a Banach space and let $a : V \times V \to \mathbb{R}$ be bilinear, symmetric, and positive semidefinite. Moreover, assume that there exist $c_1, c_2 > 0$ such that

$$c_1 \|v\|_V \le \left(a(v, v)\right)^{1/2} \le c_2 \|v\|_V$$

for all $v \in V$. Show that a defines a scalar product on V and that V is a Hilbert space with this scalar product.

Exercise A.2.14 Let V, W be n- and m-dimensional linear spaces. Use the Riesz representation theorem to prove that $\mathrm{L}(V, W)$ is isomorphic to $\mathbb{R}^{n \times m}$, i.e., that linear mappings can be identified with matrices.

Exercise A.2.15 Prove that the set of square summable sequences $\ell^2(\mathbb{N}) = \{(v_j)_{j\in\mathbb{N}} : \sum_{j\in\mathbb{N}} v_j^2 < \infty\}$ is a Hilbert space.

Exercise A.2.16

(i) Show that the linear operator $A : V \to W$ is continuous if and only if it is bounded in the sense that there exists $c > 0$ such that

$$\|Av\|_W \leq c\|v\|_V$$

for all $v \in V$.

(ii) Let $A : V \to W$ be linear and bounded and let $\|A\|_{L(V,W)}$ be the infimum of all such constants $c > 0$. Show that for all $v \in V$ we have

$$\|Av\|_W \leq \|A\|_{L(V,W)}\|v\|_V.$$

(iii) Show that $A \mapsto \|A\|_{L(V,W)}$ defines a norm on the space of linear and bounded operators $L(V, W)$ such that it is a Banach space.

Exercise A.2.17 Determine all matrices $M \in \mathbb{R}^{n\times n}$ such that the bilinear mapping $a : \mathbb{R}^n \times \mathbb{R}^n \to \mathbb{R}$,

$$a(x, y) = x^\top My,$$

satisfies the conditions of the (i) Riesz representation theorem and (ii) Lax–Milgram lemma.

Exercise A.2.18 Let $(v_j)_{j\in\mathbb{N}} \subset \ell^2(\mathbb{N})$ be defined by $v_{j,n} = \delta_{j,n}$, i.e.,

$$v_j = [0,\ldots,0,1,0,\ldots].$$

Prove that the sequence converges weakly and determine the weak limit.

Exercise A.2.19 Let $(v_j)_{j\in\mathbb{N}}$ be a weakly convergent sequence in the Banach space V. Show that the sequence is bounded, that the weak limit is unique, and that the weak limit coincides with the strong limit provided it exists.

Exercise A.2.20

(i) Assume that $A : V \to W$ is a linear and compact operator between Banach spaces V and W. Show that A is bounded.
(ii) Assume that $A : V \to \mathbb{R}^n$ is linear and bounded. Show that A is compact.

Quiz A.2.2 Decide for each of the following statements whether it is true or false. You should be able to justify your decision.

Every finite-dimensional subspace of a Banach space is closed	
The scalar product of a Hilbert space defines a symmetric and positive definite bilinear form	
Every bounded sequence in a Banach space has a convergent subsequence	
The space $C([0, 1])$ equipped with the maximum norm is a Banach space	
Every linear operator $A : V \to W$ between finite-dimensional spaces is bounded	

A.2.3   Sobolev Spaces

Exercise A.2.21

(i) Prove that for $1 < p, q < \infty$ with $1/p + 1/q = 1$ and all $a, b \in \mathbb{R}_{\geq 0}$ we have

$$ab \leq \frac{1}{p}a^p + \frac{1}{q}b^q.$$

(ii) Prove Hölder's inequality

$$\int_\Omega |uv| \, dx \leq \|u\|_{L^p(\Omega)} \|v\|_{L^q(\Omega)}$$

for $u \in L^p(\Omega)$ and $v \in L^q(\Omega)$ with $1/p + 1/q = 1$.
Hint: Consider the case $\|u\|_{L^p(\Omega)} = \|v\|_{L^q(\Omega)} = 1$ first.

Exercise A.2.22 Prove that $|a + b|^p \leq |a + b|^{p-1}(|a| + |b|)$ for all $a, b \in \mathbb{R}$ and use Hölder's inequality to deduce Minkowski's inequality

$$\|u + v\|_{L^p(\Omega)} \leq \|u\|_{L^p(\Omega)} + \|v\|_{L^p(\Omega)}$$

for all $u, v \in L^p(\Omega)$.

Exercise A.2.23

(i) Let $\Omega \subset \mathbb{R}^d$ such that $\partial\Omega$ is piecewise of class C^1. Show that Gauss's theorem is equivalent to the identity

$$\int_\Omega u(\partial_i v) \, dx = -\int_\Omega (\partial_i u)v \, dx + \int_{\partial\Omega} uvn_i \, ds$$

for all $u, v \in C^1(\overline{\Omega})$ and $i = 1, 2, \ldots d$.

(ii) Let $\Omega \subset \mathbb{R}^d$ be open. Show that for all $u \in C^1(\mathbb{R}^d)$ and $\phi \in C_0^\infty(\mathbb{R}^d)$, we have

$$\int_\Omega u(\partial_i \phi)\, dx = -\int_\Omega (\partial_i u)\phi\, dx$$

for $i = 1, 2, \ldots, d$.

Exercise A.2.24 Let $d \in \mathbb{N}$, $s \in \mathbb{R}$, and $\Omega = B_1(0) \subset \mathbb{R}^d$, and define $u(x) = |x|^s$ for $x \in \Omega \setminus \{0\}$.

(i) Determine all $s \in \mathbb{R}$ so that $u \in L^p(\Omega)$.
(ii) Determine all $s \in \mathbb{R}$ so that $u \in W^{1,p}(\Omega)$.

Exercise A.2.25 Let $(\Omega_j)_{j=1,\ldots,J}$ be an open partition of Ω, i.e., the $\overline{\Omega} = \Omega_1 \cup \cdots \cup \Omega_J$, Ω_j is open for $j = 1, 2, \ldots, J$, and $\Omega_j \cap \Omega_\ell = \emptyset$ for $j \neq \ell$. Let $u \in C(\overline{\Omega})$ be such that $u|_{\Omega_j} \in C^1(\overline{\Omega}_j)$ for $j = 1, 2, \ldots, J$. Show that u is weakly differentiable and $u \in W^{1,p}(\Omega)$ for all $1 \leq p \leq \infty$.

Exercise A.2.26 Let $\Omega \subset \mathbb{R}^d$ be bounded, open, and connected, and assume that $u \in W^{1,p}(\Omega)$ satisfies $\nabla u = 0$. Use a convolution kernel and the identity $\partial_i J_\varepsilon * \phi = J_\varepsilon * \partial_i \phi$ to prove that u is constant.

Exercise A.2.27 Let $u, v \in W^{1,2}(\Omega)$. Prove that $uv \in W^{1,1}(\Omega)$ with $\nabla(uv) = u\nabla v + v\nabla u$.
Hint: Approximate u and v by smooth functions.

Exercise A.2.28 Let $\Omega = (0, 1)^2$ and $\Gamma_D = [0, 1] \times \{0\}$. Prove that for every $u \in C^\infty([0, 1]^2)$ with $u|_{\Gamma_D} = 0$, we have

$$\|u\|_{L^p(\Gamma_D)} \leq c\|\partial_2 u\|_{L^p(\Omega)}.$$

Exercise A.2.29 Let $\Omega = B_{1/2}(0) \subset \mathbb{R}^2$ and define $u(x) = \log(\log(|x|))$. Show that $u \in W^{1,2}(\Omega)$ but $u \notin L^\infty(\Omega)$ and $u \notin C(\Omega)$.
Hint: Use that for $F(r) = |\log(r)|^{-1}$, we have $F'(r) = 1/\log^2(r)$.

Exercise A.2.30

(i) Let $\mathrm{Lip}(\Omega)$ be the set of all Lipschitz continuous functions on Ω and define

$$\|u\|_{\mathrm{Lip}(\Omega)} = \|u\|_{L^\infty(\Omega)} + \sup_{x \neq y} \frac{|u(x) - u(y)|}{|x - y|}$$

for $u \in \mathrm{Lip}(\Omega)$. Show that $\mathrm{Lip}(\Omega)$ is a Banach space.
(ii) Use the Arzelà–Ascoli theorem to show that the embedding $\mathrm{Lip}(\Omega) \to C(\overline{\Omega})$ is compact.

Quiz A.2.3 Decide for each of the following statements whether it is true or false. You should be able to justify your decision.

Functions in $L^\infty(\Omega)$ can be approximated by continuous functions
Every differentiable function is weakly differentiable
The weak partial derivative ∂_i defines a bounded linear operator $\partial_i : W^{1,2}(\Omega) \to L^2(\Omega)$
If $p > q$ and Ω is bounded, then $W^{k,q}(\Omega) \subset W^{k,p}(\Omega)$
If $\Omega \subset \mathbb{R}^3$ is a bounded Lipschitz domain, then functions in $W^{1,p}(\Omega)$, $p \geq 3/2$, are continuous

A.2.4   Weak Solutions

Exercise A.2.31 For $\alpha \in (0, 2\pi)$, let

$$\Omega = \{r(\cos\phi, \sin\phi) : 0 < r < 1, 0 < \phi < \alpha\},$$

$\Gamma_D = \partial\Omega$, and $\Gamma_N = \emptyset$, and define $f = 0$ in Ω, and

$$u_D(r, \phi) = \begin{cases} 0 & \text{for } \phi \in \{0, \alpha\}, \\ \sin(\phi\pi/\alpha) & \text{for } r = 1. \end{cases}$$

Prove that

$$u(r, \phi) = r^{\pi/\alpha} \sin(\phi\pi/\alpha)$$

is a weak solution of the Poisson problem.

Exercise A.2.32

(i) Derive a weak formulation for the boundary value problem

$$\begin{cases} -\operatorname{div}(K\nabla u) + b \cdot \nabla u + cu = f & \text{in } \Omega, \\ u = u_D & \text{on } \Gamma_D, \\ (K\nabla u) \cdot n = g & \text{on } \Gamma_N. \end{cases}$$

(ii) Specify conditions on the coefficients that lead to the existence of a unique weak solution $u \in H^1(\Omega)$.

Exercise A.2.33 Let $\Omega = (0, 1)^2$ and define for $j, k = 1, 2, \ldots, N$,

$$\phi_{j,k}(x_1, x_2) = \sin(\pi x_1 j/N) \sin(\pi x_2 k/N)$$

and let V_h be the span of $(\phi_{j,k} : j, k = 1, 2, \ldots, N)$. Compute the stiffness matrix for the bilinear mapping related to the Laplace operator.

Exercise A.2.34 Prove by constructing an appropriate example that the Poisson problem is in general not H^2-regular if the domain is not convex.

Exercise A.2.35 Let $(u_j) \subset C^3(\overline{\Omega}) \cap H_0^1(\Omega)$ be such that $u_j \to u$ in $H^1(\Omega)$. Assume that for a sequence $(f_j)_{j \in \mathbb{N}} \subset L^2(\Omega)$, we have $f_j \to f$ in $L^2(\Omega)$ and

$$\int_\Omega \nabla u_j \cdot \nabla v \, dx = \int_\Omega f_j v \, dx$$

for all $j \in \mathbb{N}$ and all $v \in H_0^1(\Omega)$. Assume further that

$$\int_\Omega |D^2 u_j|^2 \, dx \le \int_\Omega |f_j|^2 \, dx$$

for all $j \in \mathbb{N}$. Show that $u \in H^2(\Omega)$ with $\|D^2 u\|_{L^2(\Omega)} \le \|f\|_{L^2(\Omega)}$.

Exercise A.2.36 Let $u \in C^3(\Omega)$. Show that we have

$$|\Delta u|^2 - |D^2 u|^2 = \text{div}\left(\nabla u \, \Delta u - \frac{1}{2} \nabla |\nabla u|^2\right).$$

Exercise A.2.37 Let $Q' \subset \mathbb{R}^{d-1}$ be open and let $h \in C^2(Q')$ be concave. Prove that for all $x' \in Q'$ we have

$$\sum_{i=1}^{d-1} \partial_i^2 h(x') \le 0.$$

Exercise A.2.38 For $f \in L^2(\Omega)$ let $u_f \in H_0^1(\Omega)$ be the unique solution of the Poisson problem with the right-hand side f, and let $L_f : L^2(\Omega) \to L^2(\Omega)$ be defined by $f \mapsto u_f$. Verify and prove whether the operator L_f is linear, bounded, injective, surjective, and compact.

Exercise A.2.39 Show that the boundary value problem

$$\Delta^2 u = f \text{ in } \Omega, \quad u = \Delta u = 0 \text{ on } \partial\Omega$$

has a unique weak solution $u \in H^2(\Omega)$.

Exercise A.2.40 Show that the Neumann problem

$$-\Delta u = f \text{ in } \Omega, \quad \partial_n u = g \text{ on } \Gamma_N = \partial\Omega$$

has a unique solution $u \in H^1(\Omega)$ satisfying $\int_\Omega u \, dx = 0$ if and only if

$$\int_\Omega f \, dx + \int_{\partial\Omega} g \, ds = 0.$$

Quiz A.2.4 Decide for each of the following statements whether it is true or false. You should be able to justify your decision.

The stiffness matrix of a Galerkin method for an elliptic boundary value problem is symmetric and positive definite	
Elliptic partial differential equations with constant coefficients define H^2-regular problems	
Every weak solution of an elliptic boundary value problem is a classical solution	
The existence and uniqueness of a weak solution for the Poisson problem is a consequence of the Riesz representation theorem	
The Galerkin approximation $u_h \in V_h$ of an elliptic boundary value problem minimizes the distance to the exact solution in the set V_h	

A.2.5   Projects

Project A.2.1 Compute approximate solutions $u_m \in \mathscr{P}_m|_{[0,1]}$ of the one-dimensional Poisson problem $-u'' = f$ in $\Omega = (0, 1)$ with boundary conditions $u(0) = u(1) = 1$, by numerically solving the system of equations

$$-u_m''(x_i) = f(x_i), \quad i = 1, 2, \ldots, m-1, \quad u_m(x_0) = u_m(x_m) = 0,$$

where $x_i = i/m$ for $i = 0, 1, \ldots, m$. Test the method for the right-hand sides $f(x) = 1$ and $f(x) = \text{sign}(x - 1/2)$. Investigate the decay of the error $\max_{i=0,\ldots,m} |u(x_i) - u_m(x_i)|$ for $m \to \infty$ and the conditioning of the linear system of equations.

Project A.2.2 For $m \in \mathbb{N}_0$ and $j = 1, 2$, consider the subspaces $V_m^{(j)} \subset H_0^1(\Omega)$ for $\Omega = (0, 1)$ defined by

$$V_m^{(1)} = \Big\{ \sum_{0 \le j+k \le m} \alpha_{j,k} x^j (1-x)^k \Big\}, \quad V_m^{(2)} = \Big\{ \sum_{0 \le j \le m} \beta_j \sin(\pi j x) \Big\}.$$

Compute the Galerkin approximations of the Poisson problem $-u'' = 1$ in $\Omega = (0, 1)$ with Dirichlet boundary conditions $u(0) = u(1) = 0$. Comparatively, investigate the convergence of the methods and the properties of the linear systems of equations.

Project A.2.3 We consider the square $Q = [0, 1]^2$ and the triangle $T = \{(x_1, x_2) \in \mathbb{R}^2 : 0 \le x_2 \le x_1 \le 1\}$, and the mapping

$$\Phi : Q \to T, \quad (\xi_1, \xi_2) \mapsto (\xi_1, \xi_1 \xi_2).$$

Explain the identity

$$\int_T f(x)\, dx = \int_0^1 \int_0^1 f(\xi_1, \xi_1 \xi_2) \xi_1\, d\xi_2\, d\xi_1$$

and use it to define a quadrature rule on T. Determine experimentally, for which $s \in \mathbb{R}$, the function $x \mapsto |x|^s$ is integrable on T.

Project A.2.4 Use MATLAB routines for the approximate solution of ordinary differential equations to numerically determine the level set $\{x \in \mathbb{R}^2 : f(x) = f(y)\}$ of a given function $f : \mathbb{R}^2 \to \mathbb{R}$ and a point $y \in \mathbb{R}^2$. Visualize the graph of the function, its gradient via arrows, and some of its level sets for the cases $f(x) = |x|^2$ and $f(x_1, x_2) = \sin(\pi x_1) \cos(\pi x_2)$.

Project A.2.5 We consider the convolution kernel $J : \mathbb{R}^2 \to \mathbb{R}$ defined by

$$J(x) = \begin{cases} c_2 e^{-1/(1-|x|^2)}, & |x| < 1, \\ 0, & |x| \ge 1. \end{cases}$$

Plot the functions $J_\varepsilon(x) = \varepsilon^{-d} J(x/\varepsilon)$ for $\varepsilon = 10^{-j}, j = 0, 1, 2$, using $c_2 = 1$. Use an iterated trapezoidal rule to determine c_2 such that $\|J\|_{L^1(\mathbb{R}^2)} = 1$. Approximate and visualize the regularizations $f_\varepsilon = J_\varepsilon * f$ for the cases $f(x_1, x_2) = |(x_1, x_2)|$ and $f(x_1, x_2) = \operatorname{sign}(x_1) \operatorname{sign}(x_2)$ with $\varepsilon = 10^{-j}, j = 0, 1, 2$.

Project A.2.6 Use an iterated trapezoidal rule to approximate the integral

$$\int_{(-1,1)^2} \frac{[x_1, x_2]^\top}{(x_1^2 + x_2^2)^{1/2}} \cdot \begin{bmatrix} (1 - x_1^2)(1 - x_2)^2 \\ \cos(\pi x_1/2) \cos(\pi x_2/2) \end{bmatrix} dx_1\, dx_2.$$

Try to improve the approximation by using integration-by-parts.

Project A.2.7 We define the two-dimensional torus $T_{R,r}$ for radii $r, R > 0$ as the image of the mapping

$$f : [0, 2\pi]^2 \to \mathbb{R}^3, \quad (\theta, \phi) \mapsto \big((R + r\cos\theta)\cos\phi, (R + r\cos\theta)\sin\phi, r\sin\theta\big).$$

Use the transformation formula and an iterated trapezoidal rule to approximate the surface integral of the function $u(x, y, z) = x^2 y^3 z^4$ on $T_{R,r}$ for $R = 1$ and $r = 1/8$. Visualize the torus and the function u by partitioning the parameter domain into triangles and using the MATLAB command `trisurf`.

Project A.2.8 Numerically compute the $W^{1,2}$ norm of the functions $u(x) = \log\log|x|$ and $u(x) = |x|^{1/2}$ in the domain $\Omega = (-1/2, 1/2)^2$. Visualize the functions and their gradients.

A.3   Finite Element Method

A.3.1   Interpolation with Finite Elements

Exercise A.3.1 Prove that the interpolant $\mathscr{I}_T v$ associated with a finite element $(T, \mathscr{P}, \mathscr{K})$ is well defined for all $v \in W^{m,p}(T)$.

Exercise A.3.2 For a triangle $T \subset \mathbb{R}^2$ with vertices $z_0, z_1, z_2 \in \mathbb{R}^2$, let $z_3, z_4, z_5 \in \mathbb{R}^2$ be the midpoints of the sides of T.

(i) Show that $(T, \mathscr{P}_2(T), \mathscr{K})$ with $\mathscr{K} = \{\chi_j : j = 0, 1, \ldots, 5\}$ for $\chi_j(\phi) = \phi(z_j)$, $j = 0, 1, \ldots, 5$, is a finite element.
(ii) Construct the dual basis for the finite element $(T, \mathscr{P}_2(T), \mathscr{K})$.

Exercise A.3.3 Let $w = (w_1, w_2, \ldots, w_d) : \mathbb{R}^d \to \mathbb{R}^d$ be a polynomial vector field of degree $m - 1$ on $\mathbb{R}^d$, and assume that $w = \nabla v$ for some function $v \in C^1(\mathbb{R}^d)$. Show that v is a polynomial of degree m.

Exercise A.3.4 Let $\omega \subset \mathbb{R}^d$ be a bounded Lipschitz domain. Provide a constructive proof for the existence of a constant $c_P > 0$, such that for all $v \in H^1(\omega)$ with

$$\int_\omega v \, dx = 0$$

we have $\|v\|_{L^2(\omega)} \le c_P \|\nabla v\|_{L^2(\omega)}$.
Hint: Use the mean-value theorem to represent $v(x)$ by an integral over ω.

Exercise A.3.5 Let $k \in \mathbb{N}$, and define $N = |\{\alpha \in \mathbb{N}_0^d : |\alpha| \le k\}|$. Show that

$$\mathscr{P}_k(T) = \Big\{ \sum_{\alpha \in \mathbb{N}_0^d, |\alpha| \le k} a_\alpha x^\alpha : a_\alpha \in \mathbb{R} \Big\},$$

and that the mapping

$$\mathscr{P}_{m-1}(T) \to \mathbb{R}^N, \quad q \mapsto \Big(\int_T \partial^\alpha q(x) \, dx \Big)_{\alpha \in \mathbb{N}_0^d, |\alpha| \le m-1}$$

is an isomorphism.

Exercise A.3.6 Let $\Phi_T : \widehat{T} \to T$ be an affine diffeomorphism. Show that

$$D\Phi_T^{-1} = \left(D\Phi_T\right)^{-1}$$

and that both matrices are independent of $x \in T$ and $\hat{x} \in \widehat{T}$.

Exercise A.3.7 Let $(\widehat{T}, \widehat{\mathscr{P}}, \widehat{\mathscr{K}})$ be a finite element, and $\Phi_T : \widehat{T} \to T$ an affine diffeomorphism.

(i) Show that the triple $(T, \mathscr{P}, \mathscr{K})$ defined by

$$T = \Phi_T(\widehat{T}), \quad \mathscr{P} = \{\hat{q} \circ \Phi_T^{-1} : \hat{q} \in \widehat{\mathscr{P}}\}, \quad \mathscr{K} = \{\widehat{\chi} \circ \Phi_T^{-1} : \widehat{\chi} \in \widehat{\mathscr{K}}\}$$

is a finite element.
(ii) Show that for the interpolants $\mathscr{I}_T$ and $\mathscr{I}_{\widehat{T}}$ of the finite elements $(\widehat{T}, \widehat{\mathscr{P}}, \widehat{\mathscr{K}})$ and $(T, \mathscr{P}, \mathscr{K})$, we have $(\mathscr{I}_T v) \circ \Phi_T = \mathscr{I}_{\widehat{T}} \hat{v}$.

Exercise A.3.8 Assume that the sequence of triangulations $(\mathscr{T}_h)_{h>0}$ of the domain $\Omega \subset \mathbb{R}^2$ satisfies a minimum angle condition, i.e., we have

$$\inf_{h>0} \min_{T \in \mathscr{T}_h} \min_{j=0,1,2} \alpha_j^T \geq c_0 > 0,$$

where for a triangle $T \subset \mathbb{R}^d$ the numbers $\alpha_0^T, \alpha_1^T, \alpha_2^T \in (0, \pi)$ are the inner angles of T. Prove that the sequence is uniformly shape regular.

Exercise A.3.9 Let $0 = x_0 < x_1 < \cdots < x_n = 1$ be a partition of $(0, 1)$. Show that the subordinated cubic spline space $\mathscr{S}^{3,2}(\mathscr{T}_h) \subset C^2([0, 1])$ is not an affine family.

Exercise A.3.10 Construct a sequence of approximating, shape regular triangulations of the ring $B_2(0) \setminus \overline{B_1(0)}$.

Quiz A.3.1 Decide for each of the following statements whether it is true or false. You should be able to justify your decision.

If $T \subset \mathbb{R}^2$ is a triangle, $x_0, x_1, x_2 \in T$ are distinct points, and $\chi_j(q) = q(x_j)$ for $j = 1, 2, 3$, then $(T, \mathscr{P}_1(T), \{\chi_0, \chi_1, \chi_2\})$ is a finite element
There exists a constant $c > 0$ such that for all piecewise polynomial functions $v \in H^1(\Omega)$, we have $\|\nabla v\|_{L^2(\Omega)} \leq c\|v\|_{L^2(\Omega)}$
If T_1 and T_2 are elements in a conforming triangulation $\mathscr{T}_h$, then we have $\mathrm{diam}(T_1) \sim \mathrm{diam}(T_2)$
For all $v \in H^3(T)$ there exists $q \in \mathscr{P}_2(T)$ such that $\|\nabla(v-q)\|_{L^2(T)} \leq ch_T^2\|D^3 v\|_{L^2(T)}$
If $\int_T v\,dx = 0$ and $\int_T Dv\,dx = 0$, then there exists a polynomial $q \in \mathscr{P}_1(T)$ with $\|v - q\|_{L^2(T)} \leq c\|D^2 v\|_{L^2(T)}$

A.3.2   P1-Approximation of the Poisson Problem

Exercise A.3.11 Let $\mathcal{T}_h$ be a triangulation of $\Omega \subset \mathbb{R}^d$ with nodes $\mathcal{N}_h$.

(i) Show that for every $z \in \mathcal{N}_h$ there exists a unique function $\varphi_z \in \mathcal{S}^1(\mathcal{T}_h)$ with
     $\varphi_z(y) = \delta_{zy}$ for all $y \in \mathcal{N}_h$.
(ii) Prove that the families $(\varphi_z : z \in \mathcal{N}_h)$ and $(\varphi_z : z \in \mathcal{N}_h \setminus \Gamma_D)$ define bases for
     the spaces $\mathcal{S}^1(\mathcal{T}_h)$ and $\mathcal{S}_D^1(\mathcal{T}_h)$.

Exercise A.3.12 Let $a : H_D^1(\Omega) \times H_D^1(\Omega) \to \mathbb{R}$ be a symmetric and coercive bilinear form, and let $\mathcal{T}_h$ be a triangulation of Ω. Let $A = (A_{zy})_{z,y \in \mathcal{N}_h \setminus \Gamma_D}$ be for $z, y \in \mathcal{N}_h \setminus \Gamma_D$ defined by

$$A_{zy} = a(\varphi_z, \varphi_y).$$

(i) Prove that A is positive definite and symmetric.
(ii) Show that for the bilinear form a induced by the Poisson problem, the resulting
     matrix A is sparse, i.e., the number of nonvanishing entries in A is proportional
     to $|\mathcal{N}_h|$.

Exercise A.3.13 Let $\Omega = (0,1)^2$, $\Gamma_D = \partial\Omega$, and $f \in C(\overline{\Omega})$. Let $\mathcal{T}_h$ be the triangulation of Ω consisting of halved squares of sidelengths $h = 1/n$, and with diagonals parallel to the vector $(1,1)$. Show that the $P1$-finite element method and the finite difference method with a five-point stencil lead to linear systems of equations with identical system matrices.

Exercise A.3.14 Let $(\mathcal{T}_h)_{h>0}$ be a family of triangulations of $\Omega \subset \mathbb{R}^d$ with maximal mesh-size $h \to 0$.

(i) Show that $\cup_{h>0}\mathcal{S}^1(\mathcal{T}_h)$ is dense in $H^1(\Omega)$.
(ii) Prove that Galerkin approximations of the Poisson problem always converge to
     the exact solution.

Exercise A.3.15 Let $S \in \mathcal{S}_h$ be an inner side in a triangulation $\mathcal{T}_h$ with endpoints $z, y \in \mathcal{N}_h$ and neighboring triangles $T_1, T_2 \in \mathcal{T}_h$. Let α_1 and α_2 be the inner angles of T_1 and T_2 opposite to S, respectively. Prove that

$$A_{zy} = \int_{T_1 \cup T_2} \nabla\varphi_z \cdot \nabla\varphi_y \, dx = -\frac{1}{2}(\cot\alpha_1 + \cot\alpha_2) = -\frac{1}{2}\frac{\sin(\alpha_1 + \alpha_2)}{\sin(\alpha_1)\sin(\alpha_2)}$$

and formulate precise conditions which imply $A_{zy} \le 0$.

Exercise A.3.16 Let $(\mathcal{T}_h)_{h>0}$ be a triangulation of $\Omega \subset \mathbb{R}^d$, and let $\Gamma_D = \partial\Omega$. Prove that if

$$A_{zy} = \int_\Omega \nabla\varphi_z \cdot \nabla\varphi_y \, dx \le 0$$

for all distinct $z, y \in \mathcal{N}_h \setminus \partial\Omega$, then the Galerkin approximation $u_h \in \mathcal{S}_0^1(\mathcal{T}_h)$ of the Poisson problem is nonnegative, whenever the right-hand side f has this property. Show that in general this is not the case.

Exercise A.3.17 Let $T \subset \mathbb{R}^d$ be a simplex with vertices $z_0, z_1, \dots, z_d \in \mathbb{R}^d$.

(i) Prove that the midpoint rule

$$Q_T(\phi) = |T|\phi(x_T), \quad x_T = (d+1)^{-1} \sum_{j=0}^d z_j$$

is an exact quadrature formula for $\phi \in \mathcal{P}_1(T)$.

(ii) Assume $d = 2$ and define $\kappa_1 = \kappa_2 = \kappa_3 = 1/3$, and

$$\xi_1 = \frac{1}{6}(4z_0 + z_1 + z_2), \quad \xi_2 = \frac{1}{6}(z_0 + 4z_1 + z_2), \quad \xi_3 = \frac{1}{6}(z_0 + z_1 + 4z_2).$$

Show that the quadrature rule $Q_T(\phi) = \sum_{j=1}^3 |T|\kappa_j\phi(\xi_j)$ is exact for polynomials of partial degree two.

Exercise A.3.18 Let $W = V + V_h$, and assume that $a : W \times W \to \mathbb{R}$ is bilinear and continuous with respect to a norm $\| \cdot \|_h$, and assume that a_h is coercive on V_h. Let $u_h \in V_h$ satisfy $a_h(u_h, v_h) = \ell_h(v_h)$ for all $v_h \in V_h$, and let $u \in V$ be such that $a(u, v) = \ell(v)$ for all $v \in V$.

(i) Show there exists $c > 0$ such that

$$c^{-1}\|u - u_h\|_h \le \inf_{v_h \in V_h} \|u - v_h\|_h + \|a_h(u, \cdot) - \ell_h\|_{V_h'}.$$

(ii) Use the estimate to control the error induced by an approximate treatment of the domain in a Poisson problem.

Exercise A.3.19 Let $(\mathcal{T}_h)_{h>0}$ be a regular family of quasiuniform triangulations of $\Omega \subset \mathbb{R}^d$.

(i) Show that there exists $c > 0$, such that for all $v_h \in \mathcal{S}^1(\mathcal{T}_h)$ we have

$$\|\nabla v_h\|_{L^2(\Omega)} \le ch^{-1}\|v_h\|_{L^2(\Omega)}.$$

(ii) Show that the quasiuniformity condition cannot be omitted in general.

(iii) Show that the estimate from (i) does not hold for functions $v \in H^1(\Omega)$.

Exercise A.3.20 Devise and analyze a $P1$-finite element method for the approximation of the boundary value problem

$$-\Delta u + c_0 u = f \text{ in } \Omega, \quad u|_{\partial\Omega} = 0.$$

Quiz A.3.2 Decide for each of the following statements whether it is true or false. You should be able to justify your decision.

We have $\mathscr{S}^1(\mathscr{T}_h) = \{v_h \in C^1(\overline{\Omega}) : v_h|_T \in \mathscr{P}_1(T) \text{ for all } T \in \mathscr{T}_h\}$
The $P1$-finite element method for the Poisson problem satisfies a discrete maximum principle
If the solution of the Poisson problem satisfies $u \in H^2(\Omega) \cap H_0^1(\Omega)$, then we have $\|u - u_h\|_{L^2(\Omega)} \leq ch^2 \|D^2 u\|_{L^2(\Omega)}$ for the Galerkin approximation $u_h \in \mathscr{S}_0^1(\mathscr{T}_h)$
For a quasiuniform triangulation $\mathscr{T}_h$ we have $\mathrm{diam}(T_1) \sim \mathrm{diam}(T_2)$ for all $T_1, T_2 \in \mathscr{T}_h$
For all $v_h \in \mathscr{S}_0^1(\mathscr{T}_h)$ we have $\|v_h\|_{L^4(\Omega)} \leq c\|\nabla v_h\|_{L^2(\Omega)}$

A.3.3   Implementation of P1- and P2-Methods

Exercise A.3.21

(i) Let $T \subset \mathbb{R}^2$ be a triangle such that two of its sides are parallel to the coordinate axes. Let δ_1, δ_2 be the lengths of these sides. Show that the components of the gradients of the nodal basis functions belong to $\{\delta_1^{-1}, \delta_2^{-1}\}$ and that

$$\sum_{z \in \mathscr{N}_h \cap T} \nabla \varphi_z = 0.$$

(ii) Let $\mathscr{T}_h$ be the triangulation of $\Omega = (0, 1)^2$ with $\Gamma_D = [0, 1] \times \{0\}$ shown in Fig. A.3. Manually compute the coefficients and right-hand side in the linear system of equations

$$Ax = b$$

that determines the nontrivial coefficients of the $P1$-Galerkin approximation of the Poisson problem $-\Delta u = 1$ in Ω, $u|_{\Gamma_D} = 0$, and $\partial_n u|_{\Gamma_N} = 2$.

Exercise A.3.22 Define arrays `edges`, `el2edges`, `Db2edges`, and `Nb2edges` that specify the edges in the triangulation, edges of elements, and edges on the Dirichlet and Neumann boundaries for the triangulation shown in Fig. A.3.

Fig. A.3 Triangulation of
$\Omega = (0, 1)^2$ with
$\Gamma_D = [0, 1] \times \{0\}$

Exercise A.3.23 Let $T \equiv (z_0, z_1, \ldots, z_d)$ be a simplex with positively oriented vertices $z_0, z_1, \ldots, z_d \in \mathbb{R}^d$ and define

$$X_T = \begin{bmatrix} 1 & 1 & \cdots & 1 \\ z_0 & z_1 & \cdots & z_d \end{bmatrix} \in \mathbb{R}^{(d+1) \times (d+1)}.$$

Prove that $\det X_T = d! \, |T|$.

Exercise A.3.24 Let $T \equiv (z_0, z_1, \ldots, z_d)$ be a simplex with positively oriented vertices $z_0, z_1, \ldots, z_d \in \mathbb{R}^d$ and define

$$X_T = \begin{bmatrix} 1 & 1 & \cdots & 1 \\ z_0 & z_1 & \cdots & z_d \end{bmatrix} \in \mathbb{R}^{(d+1) \times (d+1)}.$$

Prove that the gradients of the nodal basis functions on T satisfy

$$\left[\nabla \varphi_{z_0}|_T, \ldots, \nabla \varphi_{z_d}|_T \right]^{\mathsf{T}} = X_T^{-1} \begin{bmatrix} 0 \\ I_d \end{bmatrix}.$$

Exercise A.3.25 Let $\mathcal{T}_h$ be a triangulation.

(i) Show that for a side $S = \mathrm{conv}\{z_0, z_1, \ldots, z_{d-1}\} \in \mathcal{S}_h$, the surface area $|S|$ is given by

$$|S| = \begin{cases} 1 & \text{if } d = 1, \\ |z_1 - z_0| & \text{if } d = 2, \\ |(z_2 - z_0) \times (z_1 - z_0)|/2 & \text{if } d = 3. \end{cases}$$

(ii) Show that for $T \in \mathcal{T}_h$, $S \in \mathcal{S}_h$, and $z \in T \cap S$, we have

$$\int_T \varphi_z \, dx = \frac{|T|}{d+1}, \qquad \int_S \varphi_z \, ds = \frac{|S|}{d}.$$

Exercise A.3.26 Let $T \subset \mathbb{R}^d$ be a triangle or tetrahedron with vertices $z_0, z_1, \ldots, z_{d+1}$, and for $j = 1, 2, \ldots, d+1$, let $S_j \subset \partial T$ be the side of T opposite to the vertex z_j, with outer unit normal $n_j \in \mathbb{R}^d$. Prove that for the nodal basis function

$\varphi_j : T \to \mathbb{R}$ associated with z_j, we have

$$\nabla \varphi_j = -\varrho_j n_j,$$

where $\varrho_j > 0$ is the height of T with respect to S_j. Show that $\varrho_j = |S_j|/(d|T|)$.

Exercise A.3.27 Let $\mathcal{T}_0$ be a triangulation of $\Omega \subset \mathbb{R}^d$, and let $(\mathcal{T}_j)_{j \in \mathbb{N}}$ be the sequence of triangulations obtained from $\mathcal{T}_0$ by $j = 1, 2, \ldots$ red-refinements. Prove that the sequence $(\mathcal{T}_j)_{j \in \mathbb{N}}$ is uniformly shape regular.

Exercise A.3.28 Let $\mathcal{T}_h$ be a triangulation of the unit square $\Omega = (0,1)^2$ into halved squares. Let $A \in \mathbb{R}^{n \times n}$ be the finite element stiffness matrix corresponding to the Galerkin approximation of the Poisson problem. Determine bounds for the bandwidth of A.

Exercise A.3.29 Prove that the isoparametric P2-finite element space $\mathscr{S}^{2,iso}(\mathcal{T}_h)$ is a subspace of $C(\overline{\Omega})$.

Exercise A.3.30 Let $w \in \mathbb{R}^3$ and $t \in \mathbb{R}^5$ be defined by

$$w = \frac{1}{2400}[155 - \sqrt{15}, 155 + \sqrt{15}, 270],$$

$$t = \frac{1}{21}[6 - \sqrt{15}, 9 + 2\sqrt{15}, 6\sqrt{15}, 9 - 2\sqrt{15}, 7].$$

A quadrature rule $Q\phi = \sum_{m=1}^{M} \kappa_m \phi(\xi_m)$ on $\widehat{T} = \text{conv}\{(0,0),(1,0),(0,1)\}$ is then defined by

$$\xi_1 = [t_1, t_1], \quad \xi_2 = [t_2, t_1], \quad \xi_3 = [t_1, t_2], \quad \xi_4 = [t_3, t_4],$$
$$\xi_5 = [t_3, t_3], \quad \xi_6 = [t_4, t_3], \quad \xi_7 = [t_5, t_5],$$

and $\kappa = [w_1, w_1, w_1, w_2, w_2, w_2, w_3]$. Verify that the quadrature rule is exact of degree 5.

Quiz A.3.3 Decide for each of the following statements whether it is true or false. You should be able to justify your decision.

The P1-finite element discretization of the Poisson problem leads to a linear system of equations with symmetric and positive definite system matrix	
The P1-finite element stiffness matrix $A \in \mathbb{R}^{n \times n}$ has $\mathcal{O}(n)$ many nonvanishing entries	
The computation of the stiffness matrix $A \in \mathbb{R}^{n \times n}$ requires $\mathcal{O}(n)$ many arithmetic operations	
The linear system of equations $Ax = b$ of the P1-finite element method can be solved with $\mathcal{O}(n)$ operations	
The P1-finite element stiffness matrix is diagonally dominant	

A.3.4   P1-Approximation of Evolution Equations

Exercise A.3.31 For given $f \in C([0,T]; L^2(\Omega))$ and $u_0 \in H^1_0(\Omega)$ let $u \in C^1([0,T]; H^1_0(\Omega))$ be a weak solution of the heat equation.

(i) Show that there exists $c_P > 0$ such that

$$\sup_{t \in [0,T]} \|u(t)\|^2 + \int_0^T \|\nabla u(t)\|^2 \, dt \leq 2\|u_0\|^2 + 2c_P \int_0^T \|f(t)\|^2 \, dt.$$

Hint: Use the identity $\frac{d}{dt}\|u\|^2 = (\partial_t u, u)$.

(ii) Deduce the uniqueness of weak solutions for the heat equation.

Exercise A.3.32 Let $\mathcal{T}_h$ be a triangulation with nodes $\mathcal{N}_h$. Define the matrices M and M^h by

$$M_{zy} = (\varphi_z, \varphi_y), \quad M^h_{zy} = (\varphi_z, \varphi_y)_h$$

for $z, y \in \mathcal{N}_h$ with the nodal basis functions $\varphi_z \in \mathcal{S}^1(\mathcal{T}_h)$, $z \in \mathcal{N}_h$. Show that M^h is diagonal, has nonnegative entries, and that for all $z \in \mathcal{N}_h$ we have

$$M^h_{zz} = \sum_{y \in \mathcal{N}_h} M_{zy}.$$

Exercise A.3.33 Let $\mathcal{T}_h$ be a triangulation of $\Omega \subset \mathbb{R}^2$. Show that there exists a constant $c > 0$ such that

$$\left| (v,w) - (v,w)_h \right| \leq ch^2 \|v\|_{W^{2,2}(\Omega)} \|w\|_{W^{2,2}(\Omega)},$$

for all $v, w \in C^2(\overline{\Omega})$. Discuss weaker conditions on v and w that lead to a similar estimate.

Exercise A.3.34 Show that approximations of the heat equation obtained with the Crank–Nicolson scheme do in general not satisfy a discrete maximum principle.

Exercise A.3.35 Let the H^1-projection $Q_h : H^1_0(\Omega) \to \mathcal{S}^1_0(\mathcal{T}_h)$ be defined by

$$(\nabla Q_h v, \nabla w_h) = (\nabla v, \nabla w_h)$$

for all $w_h \in \mathcal{S}^1_0(\mathcal{T}_h)$.

(i) Show that for every $v_h \in \mathcal{S}^1_0(\mathcal{T}_h)$ we have

$$\|\nabla(v - Q_h v)\| \leq \|\nabla(v - v_h)\|.$$

(ii) Assume that the Poisson problem in Ω with $\Gamma_D = \partial\Omega$ is H^2-regular. Prove that

$$h^{-1}\|v - Q_h v\| + \|\nabla(v - Q_h v)\| \leq c_\varrho h \|D^2 v\|$$

provided that $v \in H^2(\Omega)$.

Exercise A.3.36 Let $u \in H_0^1(\Omega)$ be the weak solution of the Poisson problem $-\Delta u = f$ in Ω subject to homogeneous Dirichlet conditions on $\partial\Omega$. Let $u_h \in \mathscr{S}_0^1(\mathscr{T}_h)$ be the Galerkin approximation. Show that with the H^1-projection Q_h : $H_0^1(\Omega) \to \mathscr{S}_0^1(\mathscr{T}_h)$ we have

$$u_h = Q_h u.$$

Exercise A.3.37 Let $\mathscr{T}_h$ be a triangulation of Ω, and let $u_0 \in H^2(\Omega) \cap H_0^1(\Omega)$ and $f \in L^2(\overline{\Omega})$. A semidiscrete approximation of the heat equation seeks $u_h : [0, T] \to \mathscr{S}_0^1(\mathscr{T}_h)$ with $u_h(0) = \mathscr{I}_h u_0$ and

$$(\partial_t u_h, v_h) + (\nabla u_h, \nabla v_h) = (f, v_h)$$

for all $v_h \in \mathscr{S}_0^1(\mathscr{T}_h)$. Show that for every $T > 0$ there exists a unique solution which is bounded independently of h and T.

Exercise A.3.38 Let $\mathscr{T}_h$ be a triangulation of Ω, and let $u_0, v_0 \in H^2(\Omega) \cap H_0^1(\Omega)$, and $f \in L^2(\overline{\Omega})$. A semi-discrete approximation of the wave equation seeks u_h : $[0, T] \to \mathscr{S}_0^1(\mathscr{T}_h)$ with $u_h(0) = \mathscr{I}_h u_0$, $\partial_t u_h(0) = \mathscr{I}_h v_0$, and

$$(\partial_t^2 u_h, v_h) + (\nabla u_h, \nabla v_h) = (f, v_h)$$

for all $v_h \in \mathscr{S}_0^1(\mathscr{T}_h)$. Show that for every $T > 0$ there exists a unique solution which is bounded independently of h and T.

Exercise A.3.39 Let $(z_h^k)_{k=0,\dots,K} \subset \mathscr{S}_0^1(\mathscr{T}_h)$ and $(b_k)_{k=0,\dots,K} \subset H_0^1(\Omega)'$ satisfy

$$(d_t^2 z_h^k, v_h) + \frac{1}{4}(\nabla[z_h^k + 2z_h^{k-1} + z_h^{k-2}], \nabla v_h) = b_k(v_h)$$

for all $v_h \in cS_0^1(\mathscr{T}_h)$. Show that for $k = 1, 2, \dots, K$, we have

$$\|d_t z_h^k\|^2 + \frac{1}{2}\|\nabla z_h^{k-1/2}\|^2 \leq \|d_t z_h^1\|^2 + \|\nabla z_h^{1/2}\|^2 + \frac{\tau}{2}\sum_{k=2}^{K}\|b_k\|_{H_0^1(\Omega)'}^2.$$

Exercise A.3.40 Show that for $T \in \mathscr{T}_h$ with $T = \operatorname{conv}\{z_0, z_1, \dots, z_d\}$, we have for $0 \leq m, n \leq d$ that

$$\int_T \varphi_{z_m}\varphi_{z_n}\,dx = \frac{|T|(1 + \delta_{mn})}{(d+1)(d+2)}, \quad \int_T \mathscr{I}_h[\varphi_{z_m}\varphi_{z_n}]\,dx = \frac{|T|\delta_{mn}}{d+1}.$$

Quiz A.3.4 Decide for each of the following statements whether it is true or false. You should be able to justify your decision.

A maximum principle for the $P1$-approximation of the heat equation holds if and only if a maximum principle for the corresponding discretization of the Poisson problem is satisfied
The θ-scheme is stable if $(1/2 - \theta)\tau h_{\min}^2 \le 1$ with the minimal mesh-size $h_{\min}$
The θ-method for the approximation of the heat equation leads to a linear system of equations $AU = b$
If $\theta = 1$ and the Poisson problem is H^2-regular, then we have that $\|u(t_k) - u_h^k\| = \mathcal{O}(h^2 + \tau^2)$
The midpoint scheme for the wave equation is unconditionally stable

A.3.5 Projects

Project A.3.1

(i) Modify the MATLAB program that realizes the $P1$-finite element method for the Poisson problem to approximate the boundary value problem

$$-\operatorname{div}(K\nabla u) = f \text{ in } \Omega, \quad u = u_{\mathrm{D}} \text{ on } \Gamma_{\mathrm{D}}, \quad (K\nabla u) \cdot n = g \text{ on } \Gamma_{\mathrm{N}},$$

where $K : \Omega \to \mathbb{R}^{d \times d}$ is a given piecewise continuous mapping such that $K(x)$ is symmetric and positive definite for almost every $x \in \Omega$. Test your code with $\Omega = (0, 1) \times (0, 2)$, $\Gamma_{\mathrm{N}} = \{1\} \times (0, 2)$, $\Gamma_{\mathrm{D}} = \partial\Omega \setminus \Gamma_{\mathrm{N}}$, $u(x, y) = x^2 y$, and

$$K(x, y) = \begin{bmatrix} 2 & \sin(x) \\ \sin(x) & 2 \end{bmatrix}.$$

(ii) Modify the MATLAB program that realizes the $P1$-finite element method for the Poisson problem to approximate the boundary value problem

$$-\Delta u = f \text{ in } \Omega, \quad u + \alpha \partial_n u = g \text{ on } \partial\Omega.$$

Test your code for $\Omega = (0, 1)^2$, $\alpha = 2$, and $u(x, y) = x^2 + y^2$.

Project A.3.2 Let $\gamma \in (0, 2\pi]$ and define

$$\Omega_\gamma = (-1, 1)^2 \cap \{x = r(\cos\phi, \sin\phi) : r > 0, \ 0 < \phi < \gamma\}.$$

Let $u_{\mathrm{D}}(r, \phi) = r^{\pi/\gamma} \sin(\phi\pi/\gamma)$ for $x = r(\cos\phi, \sin\phi) \in \Gamma_{\mathrm{D}} = \partial\Omega$. The exact solution of the Poisson problem with $f = 0$ is then given by $u(r, \phi) =$

$r^{\pi/\gamma} \sin(\phi\pi/\gamma)$. Determine the experimental convergence rates on sequences of uniform triangulations of Ω_γ for $\gamma = \ell\pi/2$, $\ell = 1, 2, \ldots, 4$, for the discrete errors

$$\delta_h^{L^2} = \|u_h - \mathscr{I}_h u\|_{L^2(\Omega)}, \quad \delta_h^{L^\infty} = \|u_h - \mathscr{I}_h u\|_{L^\infty(\Omega)},$$

and for the error

$$e_h^{H^1} = \|\nabla(u_h - u)\|_{L^2(\Omega)},$$

where the latter integral can be approximated with the midpoint rule. Discuss your results.

Project A.3.3 The optimal constant $c_h > 0$ in the inverse estimate

$$\|\nabla v_h\|_{L^2(\Omega)} \le c_h h^{-1} \|v_h\|_{L^2(\Omega)}$$

for all $v_h \in \mathscr{S}^1(\mathscr{T}_h)$ can be obtained via an eigenvalue problem and the Rayleigh quotient

$$R(v_h) = \frac{\|\nabla v_h\|_{L^2(\Omega)}^2}{\|v_h\|_{L^2(\Omega)}^2}.$$

Use the power method to find approximations of c_h for different sequences of uniformly refined triangulations. Choose a function $v \in H^1(\Omega)$ and use its nodal interpolants on a sequence of uniformly refined triangulations to illustrate experimentally that an estimate of the form $\|\nabla v\|_{L^2(\Omega)} \le c\|v\|_{L^2(\Omega)}$ fails in general.

Project A.3.4 For $\varepsilon > 0$ define the triangles $T = \mathrm{conv}\{(-1,0), (1,0), (0,\varepsilon)\}$ and $T^\pm = \mathrm{conv}\{(\pm 1, 0), (0, 0), (0, \varepsilon)\}$, and the triangulations $\mathscr{T}_h^1 = \{T\}$ and $\mathscr{T}_h^2 = \{T^+, T^-\}$ of the same domain Ω_ε. Compute the interpolation errors

$$\|\nabla(u - \mathscr{I}_h u)\|_{L^2(\Omega)}$$

for the function $u(x, y) = 1 - x^2$ for $\varepsilon = 10^{-j}$, $j = 1, 2, \ldots, 5$, and comment on the relevance of the minimum angle condition.

Project A.3.5 Use a sequence of approximating triangulations of the unit disk $\Omega = B_1(0)$ to approximately solve the Poisson problem

$$-\Delta u = 1 \text{ in } \Omega, \quad u = 0 \text{ on } \Gamma_D = \partial\Omega,$$

whose exact solution is given by $u(x) = (|x|^2 - 1)/4$. and determine the experimental convergence rates in $L^2(\Omega)$ and $H^1(\Omega)$. Repeat the experiment imposing the Neumann boundary condition $\partial_n u = 1/2$ on half of the boundary.

Project A.3.6 Write a C-program `nodal_basis.c` that computes for a given triangulation of a two- or three-dimensional domain $\Omega \subset \mathbb{R}^4$ specified by arrays `c4n` and `n4e` the volumes and midpoints of elements, and the elementwise gradients of the nodal basis functions. Use the formula

$$\nabla \varphi_z|_T = -\frac{|S_z|}{d|T|} n_{S_z},$$

for an element T and a vertex $z \in T$, with opposite side $S_z \subset \partial T$, and outer unit normal n_{S_z} on S_z. Use the MATLAB interface MEX to call the routine within MATLAB, and test it for triangulations consisting of two elements.

Project A.3.7 Implement the midpoint scheme for approximating the wave equation allowing for Neumann boundary conditions. Test your implementation for meaningful initial and boundary conditions, and experimentally verify an energy conservation principle. Augment the code allowing for a damping term in the wave equation, i.e., approximating the equation $\partial_t^2 u - \alpha \Delta \partial_t u - c^2 \Delta u = f$, and test the method with an exact solution.

Project A.3.8 We consider a can in a refrigerator and want to determine the time needed to cool the liquid inside the can below a given temperature. We assume that the metal surface of the can immediately and throughout has the same temperature as its environment inside the refrigerator. To derive a model that describes the temperature changes inside the can, we use the physical laws that the heat density w is proportional to the temperature θ, i.e., $w = \varrho c_p \theta$, that the heat flux q is proportional to the temperature gradient $\nabla \theta$, i.e., $q = -\kappa \nabla \theta$, and that heat is conserved, i.e., $\partial_t w + \operatorname{div} q = 0$. Use the values

$$\varrho = 1.009 \cdot 10^{-3} \, \text{kg/m}^3, \quad \kappa = 0.597 \, \text{W/m K}, \quad c_p = 4.186 \cdot 10^3 \, \text{J/kg K},$$

assume that the can is 0.115 m high, has a diameter of 0.067 m, and stands upright in the refrigerator with an environmental temperature varying linearly from 4°C at the bottom to 5°C at the top of the can. Determine the time needed to cool the liquid from 15°C below 8°C using the Crank–Nicolson method. Discuss the reliability of your result and limitations of the mathematical model.

A.4 Adaptivity

A.4.1 Local Resolution of Corner Singularities

Exercise A.4.1 Show that the triangles in a graded grid of the reference element defined by $J \in \mathbb{N}$ and $\beta \geq 1$ satisfy a minimum angle condition which is independent of J.

Exercise A.4.2 Let $u \in C^1(\mathbb{R}^2)$ and $\tilde{u}(r, \phi) = u(r \cos \phi, r \sin \phi)$. Show that we have

$$\nabla u(x) = \begin{bmatrix} \cos \phi & \sin \phi \\ \sin \phi & -\cos \phi \end{bmatrix} \begin{bmatrix} \partial_r \tilde{u}(r, \phi) \\ r^{-1} \partial_\phi \tilde{u}(r, \phi) \end{bmatrix}$$

for $x = r(\cos \phi, \sin \phi) \in \mathbb{R}^2 \setminus \{0\}$. Conclude that $|\nabla u|^2 = (\partial_r \tilde{u})^2 + r^{-2}(\partial_\phi \tilde{u})^2$.

Exercise A.4.3

(i) For $x = (x_1, x_2) \in \mathbb{R}^2$ with $x_1, x_2 > 0$, let

$$r(x_1, x_2) = (x_1^2 + x_2^2)^{1/2}, \quad \phi(x_1, x_2) = \arctan(x_1/x_2).$$

Show that with the Frobenius norm, we have

$$|\nabla \phi| \leq \frac{1}{r}, \quad |D^2 \phi| \leq \frac{4}{r^2}, \quad |\nabla r| \leq 1, \quad |D^2 r| \leq \frac{2}{r}.$$

(ii) In polar coordinates $(r, \phi) \in \mathbb{R}_{>0} \times (0, \pi/2)$ with respect to the origin in $\mathbb{R}^2$ and for $\alpha > 0$, let $u_\alpha(r, \phi) = r^\alpha v(\alpha \phi)$ with a 2π-periodic function $v \in C^2([0, 2\pi])$. For $x = r(\cos \phi, \sin \phi)$ let $\tilde{u}_\alpha(x) = u_\alpha(r, \phi)$. Show that there exists $c_{\alpha,v} > 0$ such that $|D^2 \tilde{u}_\alpha(x)| \leq c_{\alpha,v} |x|^{\alpha-2}$.

Exercise A.4.4 For $\alpha > 0$ and polar coordinates (r, ϕ) in $\mathbb{R}^2$, let $S(x) = r^\alpha \sin(\alpha \phi)$. Show that S_α is harmonic in the slit domain $\mathbb{R}^2 \setminus \{(x_1, 0) \in \mathbb{R}^2 : x_1 \geq 0\}$.

Exercise A.4.5 For a graduation strength $\beta \geq 1$ and an integer $J \in \mathbb{N}$, construct a graded grid of the reference element $T_{\text{ref}} = \text{conv}\{(0, 0), (1, 0), (0, 1)\}$ such that every triangle T in the triangulation has a right angle, satisfies $h_T \leq c\beta/J$, and $h_T \leq c'_\beta J^{-1} |x|^{(\beta-1)/\beta}$ for all $x \in T$.

Exercise A.4.6 Let $\omega \subset \mathbb{R}^d$ be open and bounded and $f \in L^2(\omega)$. Show that for

$$\bar{f} = |\omega|^{-1} \int_\omega f \, dx$$

we have $\|f - \bar{f}\|_{L^2(\omega)}^2 = \min_{c \in \mathbb{R}} \|f - c\|_{L^2(\omega)}^2 = \|f\|_{L^2(\omega)}^2 - |\omega| |\bar{f}|^2$.

Exercise A.4.7 Let $\Omega \subset \mathbb{R}^2$ be a bounded, polygonal Lipschitz domain in $\mathbb{R}^2$ with corners $P_1, P_2, \ldots, P_L \in \mathbb{R}^2$. Devise an algorithm for the generation of a regular family of triangulations that are graded towards re-entrant corners.

Exercise A.4.8 For $s \in \mathbb{R}_{\geq 0}$ and $x \in (0, 1)$ set $f(x) = x^s$. Construct a grid $0 = x_0 < x_1 < \cdots < x_n = 1$, such that with the nodal interpolant $\mathscr{I}_n f$ of f on the grid we have $\|f - \mathscr{I}_n f\|_{L^\infty(0,1)} \leq c_s n^{-2}$.

Exercise A.4.9 Let $u \in H^1(\Omega)$ be the solution of the Poisson problem $-\Delta u = f$ in Ω with boundary condition $u|_{\partial\Omega} = u_D$ for a given function $u_D \in C(\partial\Omega)$. For a triangulation $\mathcal{T}_h$ of Ω, let $u_h \in \mathcal{S}^1(\mathcal{T}_h)$ be the Galerkin approximation with $u_h(z) = u_D(z)$ for all $z \in \mathcal{N}_h \cap \partial\Omega$. Show that

$$\|\nabla(u - u_h)\| \leq \|\nabla(u - v_h)\|$$

for every function $v_h \in \mathcal{S}^1(\mathcal{T}_h)$ satisfying $v_h(z) = u_D(z)$ for all $z \in \mathcal{N}_h \cap \partial\Omega$.

Exercise A.4.10 Derive an error estimate for approximating a singularity function $u_\alpha(r, \phi) = r^\alpha v(\alpha\phi)$ on a graded grid of the reference element with grading strength $\beta > 1/a$.

Quiz A.4.1 Decide for each of the following statements whether it is true or false. You should be able to justify your decision.

Singularity functions $S \in L^1(\Omega)$ associated with a polygonal domain $\Omega \subset \mathbb{R}^2$ satisfy $S \in H^{3/2}(\Omega)$	
The number of triangles in a graded grid with grading strength $\beta \geq 1$ and maximal mesh-size $h = 1/J$ is of order $\mathcal{O}(J^2)$	
A graded grid is a quasiuniform triangulation with a local resolution towards the origin	
The approximation error for a singularity function in H^1 is of order $\mathcal{O}(h)$	
The approximation error for a corner singularity on a graded grid is up to a constant bounded from below by $h^{\beta\alpha}(1 + |\log(h)|)$	

A.4.2 Error Control and Adaptivity

Exercise A.4.11 Let $(\mathcal{T}_h)_{h>0}$ be a sequence of uniformly shape-regular triangulations of the bounded Lipschitz domain $\Omega \subset \mathbb{R}^d$. For each node $z \in \mathcal{N}_h$, let ω_z be the node patch with diameter h_z.

(i) Prove that there exists a constant $c_{\ell oc} > 0$, such that for all $h > 0$, $z \in \mathcal{N}_h$, and $T \in \mathcal{T}_h$ such that $z \in T$, we have

$$h_T \leq h_z \leq c_{\ell oc} h_T.$$

Show that this is not true if the condition $z \in T$ is omitted.

(ii) Show that there exists a number $K \geq 0$, such that for all $h > 0$ and $z \in \mathcal{N}_h$ we have

$$|\{T \in \mathcal{T}_h : z \in T\}| \leq K.$$

Exercise A.4.12

(i) Let $Q = [0, h]^d \subset \mathbb{R}^d$ and $R = [0, h]^{d-1} \times \{0\}$. Show that there exists $c > 0$ which does not depend on $h > 0$ such that

$$\|v\|_{L^2(R)}^2 \leq c\big(h^{-1}\|v\|_{L^2(Q)}^2 + h\|\nabla v\|_{L^2(Q)}^2\big)$$

for all $v \in H^1(Q)$.
Hint: Use one-dimensional integration-by-parts.

(ii) Let $T \subset \mathbb{R}^d$ be a triangle or tetrahedron and $S \subset \partial T$ be a side of T. Show that there exists $c_{tr} > 0$ which does not depend on $h_T = \text{diam}(T)$ such that

$$\|v\|_{L^2(S)}^2 \leq c_{tr}^2\big(h_T^{-1}\|v\|_{L^2(T)}^2 + h_T\|\nabla v\|_{L^2(T)}^2\big)$$

for all $v \in H^1(T)$.

Exercise A.4.13

(i) Show that the Clément quasi-interpolant $\mathscr{J}_h : H_D^1(\Omega) \to \mathscr{S}_D^1(\mathscr{T}_h)$ is not a projection, i.e., there exists $v_h \in \mathscr{S}_D^1(\mathscr{T}_h)$ such that $v_h \neq \mathscr{J}_h v_h$.

(ii) Compare approximation results, domains of definition, and projection properties for the nodal interpolant $\mathscr{I}_h$ and the Clément quasi-interpolant $\mathscr{J}_h$ related to the finite element space $\mathscr{S}^1(\mathscr{T}_h)$.

Exercise A.4.14 Let $\mathscr{T}_h$ be the triangulation of $\Omega = (0, 1)^2$ consisting of four halved squares with diagonals parallel to the vector $(1, 1)$. Define $u(x, y) = xy$ and $u_h = \mathscr{I}_h u$. Compute the jumps $[\![\nabla u_h \cdot n_S]\!]$ for every interior side S.

Exercise A.4.15 Let $u_h \in \mathscr{S}^1(\mathscr{T}_h)$ and $S \in \mathscr{S}_h$ such that $S = T_1 \cap T_2$ for $T_1, T_2 \in \mathscr{T}_h$ and let τ_S be a tangent vector on S. Show that

$$\nabla u_h|_{T_1} \cdot \tau_S = \nabla u_h|_{T_2} \cdot \tau.$$

Exercise A.4.16 Let $v \in L^1(\Omega)$ and $\mathscr{J}_h v$ be its Clément interpolant on a regular family of triangulations $(\mathscr{T}_h)_{h>0}$. Show that $\mathscr{J}_h v \to v$ in $L^1(\Omega)$.

Exercise A.4.17 Derive an a posteriori error estimate for approximating the boundary value problem

$$-\Delta u + u = f \text{ in } \Omega, \quad \partial_n u = g \text{ on } \Gamma_N = \partial\Omega.$$

Exercise A.4.18 Show that there are constants $c_{e,1}, c_{e2} > 0$ such that for every $h > 0$ and every $T \in \mathscr{T}_h$ with $T = \text{conv}\{z_1, z_2, \ldots, z_{d+1}\}$ the function $b_T = \varphi_{z_1}\varphi_{z_2}\cdots\varphi_{z_{d+1}} \in H^1(\Omega) \cap C(\overline{\Omega})$ satisfies

$$\text{supp } b_T \subset T, \quad \int_T b_T = c_{e,1}|T|, \quad \|\nabla b_T\|_{L^2(T)} \leq c_{e,2}h_T^{d/2-1}.$$

Exercise A.4.19

(i) Let $\mathcal{T}_0$ be a triangulation of a Lipschitz domain $\Omega \subset \mathbb{R}^2$ consisting of halved squares only. Prove that every refinement obtained with the red-green-blue refinement strategy leads to a triangulation that consists of right-angled triangles.
(ii) Show that triangulations obtained from an initial triangulation $\mathcal{T}_0$ of $\Omega \subset \mathbb{R}^2$ satisfy a minimum angle condition.

Exercise A.4.20 Let $\mathcal{T}_0$ be the triangulation of $\Omega = (0, 2)^2$ consisting of four halved squares with diagonals parallel to the vector $(1, 1)$. Assume that the triangles containing the points $(2, 1)/3$ and $(5, 4)/3$ are marked for refinement. Determine the refined triangulations with the red-green-blue and the bisection strategy.

Quiz A.4.2 Decide for each of the following statements whether it is true or false. You should be able to justify your decision.

The coefficients v_z that define the Clément quasi-interpolant are obtained as the solution of a linear system of equations	
The local Poincaré inequality controls the error for approximating a function by its average	
For finite element functions the Clément interpolant coincides with the nodal interpolant	
An a posteriori error estimate bounds the approximation error by the jumps of the exact solution and computable terms	
The efficiency of an error estimator refers to a lower bound for the error up to generic constants	

A.4.3 Convergence of Adaptive Methods

Exercise A.4.21 Let $\mathcal{T}_*$ be a refinement of $\mathcal{T}_h$, i.e., every element in $\mathcal{T}_h$ is the union of elements in $\mathcal{T}_*$. Let $u_h \in \mathcal{S}_0^1(\mathcal{T}_h)$ and $u_* \in \mathcal{S}_0^1(\mathcal{T}_*)$ be the corresponding Galerkin approximations of the Poisson problem. Prove that $\mathcal{S}^1(\mathcal{T}_h) \subset \mathcal{S}^1(\mathcal{T}_*)$ and

$$\|\nabla(u - u_h)\|^2 = \|\nabla(u - u_*)\|^2 + \|\nabla(u_* - u_h)\|^2.$$

Exercise A.4.22 Let $u_h \in \mathcal{S}_0^1(\mathcal{T}_h)$ be the Galerkin approximation of the solution $u \in H_0^1(\Omega)$ of the Poisson problem with right-hand side $f \in L^2(\Omega)$. For every $v \in H_0^1(\Omega)$, define

$$\langle \mathcal{R}_{u_h}, v \rangle = \int_\Omega \nabla u_h \cdot \nabla v \, dx - \int_\Omega f v \, dx.$$

Show that for the operator norm

$$\|\mathscr{R}_{u_h}\|_* = \sup_{v \in H_0^1(\Omega)\setminus\{0\}} \frac{\langle \mathscr{R}_{u_h}, v \rangle}{\|\nabla v\|},$$

we have $\|\mathscr{R}_{u_h}\|_* = \|\nabla(u - u_h)\|$.

Exercise A.4.23 For a triangulation $\mathscr{T}_h$, let $\mathscr{T}_{h/2}$ denote the triangulation obtained from a uniform refinement of $\mathscr{T}_h$. Assume that there exists $0 < q < 1$, such that the corresponding Galerkin approximations of the Poisson problem satisfy

$$\|\nabla(u - u_{h/2})\| \le q\|\nabla(u - u_h)\|.$$

Show that the error estimator

$$\eta_{h\to h/2}(u_h) = \|\nabla(u_h - u_{h/2})\|$$

is reliable and efficient. Devise an adaptive algorithm based on this estimator.

Exercise A.4.24 Construct triangulations $\mathscr{T}_1$ and $\mathscr{T}_2$ of $\Omega = (0, 1)^2$ and a function $f \in L^2(\Omega)$, such that the Galerkin approximations u_1 and u_2 of the Poisson problem coincide but are different from the true solution. Show that the error does not decrease but the residual error estimator decays in the passage from $\mathscr{T}_1$ to $\mathscr{T}_2$.

Exercise A.4.25 Let $\varphi_z \in \mathscr{S}^1(\mathscr{T}_h)$ be the nodal basis function associated with a node $z \in \mathscr{N}_h$. Show that for every $p \in [1, \infty]$, there exists a constant $c_z > 0$ such that

$$\|\nabla^\ell \varphi_z\|_{L^p(\omega_z)} \le c_z h_z^{d/p-\ell}.$$

Exercise A.4.26 Let $z \in \mathscr{N}_h$ and $S_z \in \mathscr{S}_h$ be such that $z \in S_z$.

(i) Show that there exists $\psi_z \in \mathscr{P}_1(S_z)$, such that for all $z' \in \mathscr{N}_h$ we have

$$\int_{S_z} \psi_z \varphi_{z'} \, ds = \delta_{zz'}$$

(ii) Prove that $\|\psi_z\|_{L^\infty(S_z)} \le c h_z^{-(d-1)}$.

Exercise A.4.27 Let $\Omega \subset \mathbb{R}^d$ be a bounded Lipschitz domain, and $\mathscr{T}_h$ a triangulation of Ω. Let $f \in L^2(\Omega)$ and let $\tilde{f}$ be the elementwise constant function on $\mathscr{T}_h$ defined by the averages of f on every element $T \in \mathscr{T}_h$. Let $u, \tilde{u} \in H_0^1(\Omega)$ be the weak solutions of the Poisson problem with right-hand sides $f, \tilde{f}$, respectively. Assume that $f|_T \in H^1(T)$ for all $T \in \mathscr{T}_h$. Show that for a constant $c_P > 0$ we have

$$\|\nabla(u - \tilde{u})\| \le c_P h_{\max}^2 \|\nabla_{\mathscr{T}} f\|,$$

where $h_{max} > 0$ is the maximal mesh-size in $\mathscr{T}_h$, and $\nabla_{\mathscr{T}} f$ the elementwise gradient of f.

Exercise A.4.28 Let $\mathscr{T}_1, \mathscr{T}_2$ be triangulations of Ω, such that $\mathscr{T}_2$ is a refinement of $\mathscr{T}_1$. Assume that $\mathscr{T}_1$ and $\mathscr{T}_2$ coincide in the subdomain $\Omega' \subset \Omega$. Show that in general we have $u_1 \neq u_2$ in Ω' for related Galerkin approximations $u_j \in \mathscr{S}_0^1(\mathscr{T}_j)$, $j = 1, 2$.

Exercise A.4.29 Assume that the Poisson problem in $\Omega \subset \mathbb{R}^d$ with homogeneous Dirichlet boundary conditions is H^2-regular, i.e., $\|D^2\psi\| \leq c_2\|\Delta\psi\|$ for all $\psi \in H_0^1(\Omega) \cap H^2(\Omega)$. Prove that for the approximation error $u - u_h$ of the Poisson problem with the right-hand side $f \in L^2(\Omega)$, we have

$$\|u - u_h\| \leq c\Big(\sum_{T \in \mathscr{T}_h} \eta_{2,T}^2(u_h)\Big)^{1/2},$$

with the error indicators

$$\eta_{2,T}^2(u_h) = h_T^4\|f\|_{L^2(T)}^2 + \sum_{S \in \mathscr{S}_h, S \subset \partial T} h_T^3\|[\![\nabla u_h \cdot n_S]\!]\|_{LT^2(S)}^2.$$

Exercise A.4.30 Discuss advantages and disadvantages of finite element methods using uniform and adaptively refined triangulations.

Quiz A.4.3 Decide for each of the following statements whether it is true or false. You should be able to justify your decision.

If $\mathscr{T}_{k+1}$ is a refinement of $\mathscr{T}_k$, then we have $\|\nabla e_{k+1}\| < \|\nabla e_k\|$ for the errors $e_k = u - u_k$	
The maximal mesh-size in a sequence of adaptively refined triangulations always tends to zero	
The Scott–Zhang quasi-interpolant preserves piecewise affine boundary data	
The Scott–Zhang quasi-interpolant is well defined for functions $v \in L^\infty(\Omega)$	
For an appropriate marking strategy, the adaptive algorithm always defines a convergent sequence of approximations	

A.4.4 Adaptivity for the Heat Equation

Exercise A.4.31 Let X be a Banach space and let $f : [a, b] \to X$ be affine. Show that for all $t \in [a, b]$ we have $\|f(t) - f(a)\| \leq \|f(b) - f(a)\|$.

Exercise A.4.32

(i) Show that every weak solution of the heat equation satisfies

$$\int_0^T \|\partial_t u(t)\|^2 \, dt + \frac{1}{2}\|\nabla u(T)\|^2 = \frac{1}{2}\|\nabla u_0\|^2.$$

(ii) Prove that weak solutions for the heat equation are unique.

Exercise A.4.33 For $\tau > 0$, a fixed triangulation $\mathcal{T}_h$ of Ω, and $\theta \in [0, 1]$, the finite element version of the θ-scheme is defined by

$$\left(d_t U^j, V\right) + \left(\nabla[(1 - \theta)U^{j-1} + \theta U^j], \nabla V\right) = \left(f(t_{j-1+\theta}), V\right)$$

for $j = 1, 2, \ldots, J$, and all $V \in \mathcal{S}_0^1(\mathcal{T}_h)$, where $t_{j-1+\theta} = (j - 1 + \theta)\tau$.

(i) Show that the iterates $(U^j)_{j=0,\ldots,J}$ are well defined.
(ii) Prove that for every $\ell = 1, 2, \ldots, J$, we have

$$\frac{\tau}{2}\sum_{j=1}^{\ell}\|d_t U^j\|^2 + \frac{1}{2}\|\nabla U^\ell\|^2 + (\theta - 1/2)\tau\sum_{j=1}^{\ell}\|\nabla d_t U^j\|^2 \le \frac{\tau}{2}\sum_{j=1}^{\ell}\|f(t_{j-1+\theta})\|^2.$$

Exercise A.4.34 For a triangulation $\mathcal{T}_h$ of $\Omega \subset \mathbb{R}^d$, let $P_{h,0} : L^2(\Omega) \to \mathcal{S}_0^1(\mathcal{T}_h)$ denote the L^2-projection, defined for every $f \in L^2(\Omega)$ by

$$(P_{h,0}f, V) = (f, V)$$

for all $V \in \mathcal{S}_0^1(\mathcal{T}_h)$.

(i) Prove that $P_{h,0}$ defines a bounded linear operator $P_{h,0} : L^2(\Omega) \to L^2(\Omega)$.
(ii) Prove that the discrete Laplacian defines a bijection $-\Delta_h : \mathcal{S}_0^1(\mathcal{T}_h) \to \mathcal{S}_0^1(\mathcal{T}_h)$.
(iii) Show that the Galerkin approximation $u_h \in \mathcal{S}_0^1(\mathcal{T}_h)$ of the Poisson problem with the right-hand side $f \in L^2(\Omega)$ is given by

$$u_h = (-\Delta_h)^{-1}P_{h,0}f.$$

Exercise A.4.35 Let $\mathcal{T}_h$ be a triangulation of $\Omega \subset \mathbb{R}^d$ and consider the semi-discretized heat equation

$$\partial_t U(t) = \Delta_h U(t) - F(t)$$

for $t \in [0, T]$ with initial condition $U(0) = u_0$ and $F : [0, T] \to \mathcal{S}_0^1(\mathcal{T}_h)$. Formulate sufficient conditions for the existence and uniqueness of a solution $U : [0, T] \to \mathcal{S}_0^1(\mathcal{T}_h)$.

Exercise A.4.36 Show that there exists a constant $c > 0$, such that for all $v_h \in \mathscr{S}_0^1(\mathscr{T}_h)$ we have

$$\|\Delta_h v_h\| \le c h_{\min}^{-1} \|\nabla v_h\|$$

for all $v_h \in \mathscr{S}_0^1(\mathscr{T}_h)$ with the minimal mesh-size $h_{\min}$ of $\mathscr{T}_h$.

Exercise A.4.37 For a sequence of triangulations $(\mathscr{T}_h)_{h>0}$, let $(u_h)_{h>0}$ be the corresponding sequence of Galerkin approximations $u_h \in \mathscr{S}_0^1(\mathscr{T}_h)$ of the Poisson problem with the right-hand side $f \in L^2(\Omega)$ and exact solution $u \in H_0^1(\Omega)$.

(i) Prove that for every $h > 0$ we have

$$\|\Delta_h u_h\| \le \|f\|.$$

(ii) Show that

$$\Delta_h u_h \to \Delta u$$

in $L^2(\Omega)$ as $h \to 0$.

Exercise A.4.38

(i) Construct a triangulation with at least eight elements for which the marking of one element requires the refinement of all elements in a bisection strategy.
(ii) Construct a triangulation with four interior nodes, for which none of the interior nodes can be coarsened locally, but all nodes together can be removed to obtain a coarsened triangulation, assuming that the vertices opposite the longest edges are newest vertices.

Exercise A.4.39

(i) Bisect the marked elements in the triangulation shown in the left plot of Fig. A.4 using a minimal number of compatible edge patch bisections.
(ii) Perform as many coarsenings as possible in the triangulation shown in the right plot of Fig. A.4.

Exercise A.4.40 Devise a Crank–Nicolson reconstruction for the case of a right-hand side f that is not piecewise constant but continuously differentiable.

Fig. A.4 Triangulations to be compatibly refined (*left*) and coarsened (*right*)

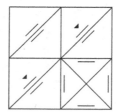 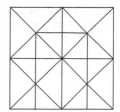

Quiz A.4.4 Decide for each of the following statements whether it is true or false. You should be able to justify your decision.

Weak solutions of the heat equation are not unique in general	
The approximation error for the heat equation is bounded by the operator norm of the residual	
The difference between a function and its elliptic reconstruction can be controlled with estimates for the Poisson problem	
The Crank–Nicolson reconstruction is based on a piecewise quadratic interpolant of the iterates	
Mesh coarsening is realized by reversing the most recent refinements of a given triangulation	

A.4.5 Projects

Project A.4.1 Consider the Poisson problem $-\Delta u = 0$ in the domain

$$\Omega_\gamma = \{x = r(\cos\phi, \sin\phi) : r > 0, 0 < \phi < \gamma\} \cap (-1,1)^2$$

with Dirichlet boundary conditions $u = u_D$ on $\Gamma_D = \partial\Omega_\gamma$ so that the exact solution is given by $u(r,\phi) = r^{\pi/\gamma}\sin(\phi\pi/\gamma)$. Experimentally determine for $\gamma = j\pi/2$, $j = 1,2,3,4$, an optimal grading strength by computing in each case a sequence of approximation errors $\|\nabla(\mathscr{I}_h u - u_h)\|_{L^2(\Omega)}$.

Project A.4.2 Construct a sequence of triangulations $(\mathscr{T}_h)_{h>0}$ to obtain an optimally convergent sequence of approximations of the Poisson problem with the right-hand side $f = 1$ with homogeneous Neumann and Dirichlet boundary conditions on the corresponding parts of $\partial\Omega$ for the domain Ω shown in Fig. A.5.

Project A.4.3 Modify the red-green-blue refinement routine so that a new node is created in marked triangles, i.e., marked triangles are refined as indicated in Fig. A.6.

Fig. A.5 Domain Ω with re-entrant corners and specification of Γ_N

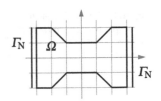

Fig. A.6 Creation of a new node in the interior of an element

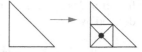

Project A.4.4 Approximate the Poisson problem $-\Delta u = 1$ in the square $\Omega = (-1,1)^2$ and in the L-shaped domain $\Omega = (-1,1)^2 \setminus ([0,1] \times [-1,0])$ with homogeneous Dirichlet boundary conditions on $\partial\Omega$ on sequences of uniform triangulations $(\mathcal{T}_j)_{j=0,1,\ldots}$. For $j = 0,1,\ldots$, let $s_j = \|\nabla u_j\|_{L^2(\Omega)}$ and define the extrapolated values $\tilde{s}_j$ for $j \geq 2$ via

$$\tilde{s}_j = \frac{s_j s_{j-2} - s_{j-1}^2}{s_j - 2s_{j-1} + s_{j-2}}$$

to obtain an accurate approximation of the unknown value $s = \|\nabla u\|_{L^2(\Omega)}$. Use the obtained value to approximate the errors

$$\delta_h^2 = \|\nabla(u - u_h)\|_{L^2(\Omega)}^2 = \|\nabla u_h\|_{L^2(\Omega)}^2 - \|\nabla u\|_{L^2(\Omega)}^2$$

for sequences of uniformly and adaptively refined triangulations and plot the quantities versus numbers of degrees of freedom with a logarithmic scaling for both axes. Determine the experimental rates of convergence.

Project A.4.5 Consider the Poisson problem $-\Delta u = 1$ in the L-shaped domain $\Omega = (-1,1)^2 \setminus ([0,1] \times [-1,0])$ with homogeneous Dirichlet boundary conditions on $\partial\Omega$. For a sequence of uniform triangulations, compute upper bounds for the approximation errors and determine the CPU-time needed to solve each problem. Approximate the problems adaptively, compute error bounds, and compare the CPU-times needed to obtain comparable error bounds. Discuss the benefits of adaptivity on the basis of your results.

Project A.4.6 Let $\mathcal{T}_h$ be a triangulation of Ω and let $\mathcal{T}_{h/2}$ be the triangulation obtained from $\mathcal{T}_h$ by a uniform red refinement. The *h-h/2-estimator* is defined by

$$\eta_{h-h/2}(\mathcal{T}_h) = \|\nabla(u_h - u_{h/2})\|_{L^2(\Omega)},$$

where u_h and $u_{h/2}$ are the finite element approximations corresponding to the triangulations $\mathcal{T}_h$ and $\mathcal{T}_{h/2}$, respectively. Test and compare the error estimator to the residual error estimator for a Poisson problem in a nonconvex domain.

Project A.4.7 Let $u_h \in \mathcal{S}^1(\mathcal{T}_h)$ be a finite element function and define $\mathcal{A}_h[\nabla u_h] = p_h \in \mathcal{S}^1(\mathcal{T}_h)^d$ by

$$p_h(z) = |\omega_z|^{-1} \int_{\omega_z} \nabla u_h \, dx$$

for all nodes $z \in \mathcal{N}_h$ with node patches $\omega_z = \operatorname{supp} \varphi_z$. The *averaging estimator* is defined by

$$\eta_{\mathscr{A}}(\mathscr{T}_h) = \left\| \nabla u_h - \mathscr{A}_h[\nabla u_h] \right\|_{L^2(\Omega)}.$$

Test and compare the error estimator to the residual error estimator for a Poisson problem in a nonconvex domain.

Project A.4.8 Consider the Poisson problem $-\Delta u = f$ in $\Omega = (0,1)^2$, $u|_{\partial\Omega} = 0$, with the function $f \in L^2(\Omega)$ defined by $f = -\Delta u$ for

$$u(x,y) = x(1-x)y(1-y)\arctan\big(50(r(x,y) - 1)\big),$$

$$r(x,y) = ((x - 5/4)^2 + (y + 1/4)^2)^{1/2}.$$

Determine errors and error estimates on sequences of uniformly and adaptively refined triangulations and plot them versus degrees of freedom in a double-logarithmic scaling. Comment on possible benefits of adaptivity.

A.5 Iterative Solution Methods

A.5.1 Multigrid

Exercise A.5.1 Let A_h be the stiffness matrix related to a finite element space $\mathscr{S}_0^1(\mathscr{T}_h)$. Show that the estimate $\operatorname{cond}_2(A_h) \leq ch^{-2}$ is optimal in the sense that there exists a constant $c' > 0$ such that $\operatorname{cond}_2(A_h) \geq c'h^{-2}$. Consider the case $d = 1$ first.

Exercise A.5.2 Let V be a finite-dimensional space, let $(\cdot, \cdot)_V : V \times V \to \mathbb{R}$ be a scalar product on V, and let $a : V \times V \to \mathbb{R}$ be a symmetric bilinear form on V. Show that there exists an orthonormal basis $(v_1, v_2, \ldots, v_n)$ of V, and numbers $\lambda_1, \lambda_2, \ldots, \lambda_n$ such that

$$a(v_j, w) = \lambda_j \langle v_j, w \rangle_V$$

for all $w \in V$.

Exercise A.5.3 Construct an inner product $\langle \cdot, \cdot \rangle$ on $\mathbb{R}^n$ and a symmetric matrix $A \in \mathbb{R}^{n \times n}$ such that A is not symmetric with respect to this inner product, i.e., there exist $V, W \in \mathbb{R}^n$ such that

$$\langle AV, W \rangle \neq \langle V, AW \rangle.$$

Exercise A.5.4 Prove that for every $k \in \mathbb{N}$ we have

$$\max_{t \in [0,1]} t(1-t)^k \leq \frac{1}{ek} \leq k^{-1}.$$

Exercise A.5.5 Let A_h be the system matrix resulting from a P_1-finite element discretization of the elliptic boundary value problem

$$-\Delta u + \alpha u = f \text{ in } \Omega, \quad u|_{\partial\Omega} = 0.$$

Derive an upper bound for the condition number with a precise dependence on the parameter $\alpha \geq 0$.

Exercise A.5.6 Let $A_h \in \mathbb{R}^{n \times n}$ be the finite element stiffness matrix related to the Poisson problem, i.e., for $v_h, w_h \in \mathscr{S}_0^1(\mathscr{T}_h)$ with coefficient vectors $V_h, W_h \in \mathbb{R}^n$, we have

$$V_h^\top A_h W = \int_\Omega \nabla v_h \cdot \nabla w_h \, dx.$$

For a diagonalization $A_h = Q^\top D Q$ with a diagonal matrix $D \in \mathbb{R}^{n \times n}$ and an orthogonal matrix $Q \in \mathbb{R}^{n \times n}$, and $s \in \mathbb{R}$, we define

$$A_h^s = Q^\top D^s Q.$$

(i) Show that the expression $\|v_h\|_s = \|A_h^{s/2} V_h\|$ defines a norm on $\mathscr{S}_0^1(\mathscr{T}_h)$ which coincides with the Euclidean norm if $s = 0$ and with the H^1-norm if $s = 1$.

(ii) Show that for s, r, t with $s = (r+t)/2$, we have $\|v_h\|_s \leq \|v_h\|_r^{1/2} \|v_h\|_t^{1/2}$.

(iii) Show that if $A_h U_h = B_h$, then we have $\|u_h\|_{s+2} = \|b_h\|_s$.

Exercise A.5.7 Let $\mathscr{T}_h$ be a triangulation of Ω, and let $A_h \in \mathbb{R}^{n \times n}$ be the stiffness matrix defined by the space $\mathscr{S}_0^1(\mathscr{T}_h)$. Construct norms $\|\cdot\|_\ell$ and $\|\cdot\|_r$ on $\mathbb{R}^n$ such that for the induced condition number we have $\text{cond}_{\ell r}(A_h) = 1$.

Exercise A.5.8 Let $\mathscr{T}_h$ be a triangulation that is obtained from a triangulation $\mathscr{T}_H$ by a red-green-blue-refinement procedure. Show that there exists a uniquely defined prolongation operator $P : \mathbb{R}^N \to \mathbb{R}^n$, such that $P V_H$ is the coefficient vector of the function $v_H \in \mathscr{S}_0^1(\mathscr{T}_H)$ with respect to the nodal basis in $\mathscr{S}_0^1(\mathscr{T}_h)$, provided that v_H is defined by the coefficient vector $V_H \in \mathbb{R}^N$.

Exercise A.5.9 Let $P : \mathscr{S}_0^1(\mathscr{T}_H) \to \mathscr{S}_0^1(\mathscr{T}_h)$ be the prolongation operator between two nested finite element spaces. Show that the transpose $P^\top$ defines an operator $\mathscr{S}_0^1(\mathscr{T}_h) \to \mathscr{S}_0^1(\mathscr{T}_H)$ that is not the inverse of P and does not coincide with the nodal interpolation on $\mathscr{T}_H$.

Exercise A.5.10 Assume that for norms $\|\cdot\|_X$ and $\|\cdot\|_Y$ we have the smoothing property

$$\|\mathscr{S}^k v_h\|_X \le c_1 h^{-\beta} k^{-\gamma} \|v_h\|_Y,$$

for an iterative method for solving an equation $A_h u_h = b_h$, and the approximation property

$$\|v_h - v_{2h}\|_Y \le c_2 h^{\beta} \|v_h\|_X,$$

for all $v_h \in V_h$, and an appropriate $v_{2h} \in V_{2h}$. Devise and analyze an abstract two-level method for the efficient numerical solution of the equation $A_h u_h = b_h$. Discuss the computational complexity, and apply the framework to the approximation of the Poisson problem.

Quiz A.5.1 Decide for each of the following statements whether it is true or false. You should be able to justify your decision.

The condition number of a finite element stiffness matrix depends on the choice of the basis	
To achieve a reduction to 5% of the initial error, approximately 20 Richardson iterations are necessary	
The smallest eigenvalue of the discrete Laplace operator is of order $\mathcal{O}(h^d)$	
The prolongation operator computes the coefficients of a coarse grid function on a finer grid	
The coarse grid correction c_H is the Galerkin approximation of the Poisson problem $-\Delta u = f$ in Ω, $u|_{\partial\Omega} = 0$	

A.5.2 Domain Decomposition

Exercise A.5.11 Let $f \in L^2(\Omega)$ and $u \in H^2(\Omega) \cap H_0^1(\Omega)$. Show that we have

$$-\Delta u = f \text{ in } \Omega, \quad u|_{\partial\Omega} = 0,$$

if and only if the functions $u_j = u|_{\Omega_j}, j = 1, 2$, satisfy

$$-\Delta u_j = f \text{ in } \Omega_j, \quad u_j|_{\Gamma_j} = 0,$$

for $j = 1, 2$ and

$$u_1 = u_2, \quad \partial_{n_1} u_1 = -\partial_{n_2} u_2,$$

on γ, where $\partial_{n_j} u_j = \nabla u_j \cdot n_j$ on γ with the outer unit normal n_j to $\partial\Omega_j$, i.e., $n_2 = -n_1$.

Exercise A.5.12 Show that if $\Omega = (0,1), f = 0$, and $\Omega_1 = (0,a), \Omega_2 = (a,1)$ for $0 < a < 1/2$, then the Dirichlet–Neumann method converges if and only if $\theta < 1$.

Exercise A.5.13 Prove that every stationary pair (u_1, u_2) for the Dirichlet–Neumann method coincides with the solution $u \in H_0^1(\Omega)$ of the Poisson problem with the right-hand side f, i.e., $u_j = u|_{\Omega_j}$ for $j = 1, 2$.

Exercise A.5.14 Generalize the Dirichlet–Neumann method to partitions with more than two subdomains.

Exercise A.5.15 Let $v \in H^1(\Omega)$ and $\partial\Omega = \Gamma \cup \gamma$, for disjoint sets Γ and γ, and assume that $v|_\Gamma = 0$. Show that for $w \in H^1(\Omega)$ we have

$$-\Delta w = 0 \text{ in } \Omega, \quad w|_\Gamma = 0, \quad w|_\gamma = v|_\gamma$$

if and only if $\|\nabla w\|_{L^2(\Omega)} \leq \|\nabla(w + \phi)\|_{L^2(\Omega)}$ for all $\phi \in H^1(\Omega)$ with $\phi|_\gamma = 0$.

Exercise A.5.16 Let Ω_1, Ω_2 be a nonoverlapping partition of Ω with interface $\gamma = \partial\Omega_1 \cap \partial\Omega_2$.

(i) Construct diffeomorphisms $\Phi_j : \Omega_j \to \omega_j$ for $j = 1, 2$, with sets $\omega_1 \subset \Omega_2$ and $\omega_2 \subset \Omega_1$ such that $\Phi_j(\gamma) = \gamma$ for $j = 1, 2$.

(ii) Show that the expressions $\|\psi\|_j = \|\nabla H_j \psi\|_{L^2(\Omega_j)}$ with the harmonic extension $H_j \psi$ of ψ to $\Omega_j, j = 1, 2$, define equivalent norms on $H_{00}^{1/2}(\gamma)$.

Exercise A.5.17 Let Ω_1, Ω_2 be a nonoverlapping partition of Ω with interface $\gamma = \partial\Omega_1 \cap \partial\Omega_2$, and set $\Gamma_j = \partial\Omega_j \cap \partial\Omega$. Define $T : H_{00}^{1/2}(\gamma) \to H_{00}^{1/2}(\gamma)$ by $T\psi = w_2|_\gamma$, where $w_2 \in H_{\Gamma_2}^1(\Omega_2)$ solves

$$-\Delta w_2 = 0 \text{ in } \Omega_2, \quad w_2|_{\Gamma_2} = 0, \quad \partial_{n_2} w_2 = -\partial_{n_1} H_1 \psi.$$

Show that this is equivalent to

$$a_2(w_2, v_2) = -a_1(H_1 \psi, H_1 v_2|_\gamma)$$

for all $v_2 \in H_{\Gamma_2}^1(\Omega_2)$.

Exercise A.5.18 Compute the iterates $(u_1^k, u_2^k)_{k=0,\dots,5}$ of the alternating Schwarz method for the problem $-u'' = 1, u(0) = u(1) = 0$ for $\Omega_1 = (0, 1/2 + \delta)$ and $\Omega_2 = (1/2 - \delta, 1)$ for $0 < \delta < 1/2$.

Exercise A.5.19 Prove that with the appropriate spaces $V_1, V_2 \subset V$, the alternating Schwarz method can be equivalently defined by the problems of computing $w_1^k \in V_1$ such that

$$a(w_1^k, v_1) = b(v_1) - a(u^k, v_1)$$

for all $v_1 \in V_1$, setting $u^{k+1/2} = u_k + w_1^k$, computing $w_2^k \in V_2$ such that

$$a(w_2^k, v_2) = b(v_2) - a(u^{k+1/2}, v_2)$$

for all $v_2 \in V_2$, and setting $u^{k+1} = u^{k+1/2} + w_2^k$.

Exercise A.5.20 For a Hilbert space V and a subspace $W \subset V$, let $\mathscr{P} : V \to V$ denote the orthogonal projection onto W. Prove that $\mathscr{P} \circ (1 - \mathscr{P}) = 0$, and that the images of $\mathscr{P}$ and $(1 - \mathscr{P})$ are orthogonal subsets of V.

Quiz A.5.2 Decide for each of the following statements whether it is true or false. You should be able to justify your decision.

If $u \in H^1(\Omega)$ and $\overline{\Omega} = \overline{\Omega}_1 \cup \overline{\Omega}_2$, $\gamma = \partial\Omega_1 \cap \partial\Omega_2$, then for $u_j = u\|_{\Omega_j}, j = 1, 2$, we have $u_1\|_\gamma = u_2\|_\gamma$	
If $u \in H^1(\Omega_1 \cup \Omega_2)$, $\gamma = \partial\Omega_1 \cap \partial\Omega_2$, then for $u_j = u\|_{\Omega_j}, j = 1, 2$, we have $u_1\|_\gamma = u_2\|_\gamma$	
if $v \in H^1(\Omega)$, $\gamma \subset \partial\Omega$, then the harmonic extension $w \in H^1(\Omega)$ of $v\|_\gamma$ to Ω with $w\|_{\partial\Omega \setminus \gamma} = 0$ is well defined	
The overlapping Schwarz method converges only if the subdomains are nonfloating	
The Poisson problem in Ω can be decomposed into independent problems on subdomains	

A.5.3   Preconditioning

Exercise A.5.21 Let $A \in \mathbb{R}^{n \times n}$ be symmetric and positive definite and $b \in \mathbb{R}^n$. Assume that $C \in \mathbb{R}^{n \times n}$ and $K \in \mathbb{R}^{n \times n}$ are regular matrices with $C = KK^\top$. Apply the conjugate gradient algorithm to the linear systems of equations $(CA)x = Cb$ and $(K^\top AK)K^{-1}x = K^\top b$ and show that the resulting methods are equivalent to the preconditioned CG algorithm. Compare the cost of the matrix vector products $z \mapsto Cz$ and $z \mapsto (KK^\top)z$.

Exercise A.5.22 Let $\langle \cdot, \cdot \rangle : \mathbb{R}^n \times \mathbb{R}^n \to \mathbb{R}$ be a scalar product on $\mathbb{R}^n$. Show that the minimal and maximal eigenvalues of a symmetric matrix $A \in \mathbb{R}^{n \times n}$ are given as the extrema of the function

$$R : \mathbb{R}^n \setminus \{0\} \to \mathbb{R}, \quad x \mapsto \frac{\langle Ax, x \rangle}{\langle x, x \rangle}.$$

Exercise A.5.23 Let $A, C \in \mathbb{R}^n$ be symmetric regular matrices. Prove that the extremal eigenvalues of CA are the extrema of the mapping

$$R : \mathbb{R}^n \setminus \{0\} \to \mathbb{R}, \quad x \mapsto \frac{Ax \cdot x}{C^{-1}x \cdot x}.$$

Exercise A.5.24 Let $\mathscr{T}_h$ be a triangulation of $\Omega \subset \mathbb{R}^d$ with nodes $(z_1, z_2, \ldots, z_n) = \mathscr{N}_h$. Let M and $\widetilde{M}$ be the matrices in $\mathbb{R}^{n \times n}$ defined by

$$M_{ij} = \int_\Omega \varphi_{z_i} \varphi_{z_j} \, dx, \quad \widetilde{M}_{ij} = \int_\Omega \mathscr{I}_h[\varphi_{z_i} \varphi_{z_j}] \, dx,$$

for $i, j = 1, 2, \ldots, n$, with the nodal interpolation operator $\mathscr{I}_h : C(\overline{\Omega}) \to \mathscr{S}^1(\mathscr{T}_h)$.

(i) Show that M and $\widetilde{M}$ are positive definite and symmetric, and that $\widetilde{M}$ is diagonal.
(ii) Prove that for every $v \in \mathbb{R}^n$ we have

$$v^T M v \leq v^T \widetilde{M} v \leq (d+2) v^T M v.$$

Hint: Prove the inequality on every element $T \in \mathscr{T}_h$ first.
(iii) Show that $C = \widetilde{M}^{-1}$ is an optimal preconditioner for M, i.e.,

$$\mathrm{cond}_2(CM) \leq d + 2,$$

and the evaluation $r \mapsto Cr$ is of complexity $\mathscr{O}(n)$.

Exercise A.5.25 Let $0 < \gamma < 1$ and define $\Gamma_{ij} = \gamma^{|i-j|/2}$ for $i, j = 0, 1, \ldots, L$.

(i) Prove that $\varrho(\Gamma) \leq 1/(1 - \gamma^{1/2})$.
(ii) Show that for vectors $\alpha, \beta \in \mathbb{R}^{L+1}$ we have

$$\gamma^{|i-j|/2} \alpha_i \beta_j \leq \varrho(\Gamma) |\alpha| |\beta|.$$

Exercise A.5.26 Let $A \in \mathbb{R}^{n \times n}$ be regular and define the diagonal matrix $D \in \mathbb{R}^{n \times n}$ by $d_{ii} = \sum_{j=1}^n |a_{ij}|$ for $i = 1, 2, \ldots, n$. Show that for every diagonal matrix $T \in \mathbb{R}^n$, we have

$$\mathrm{cond}_\infty(D^{-1}A) \leq \mathrm{cond}_\infty(TA),$$

where $\mathrm{cond}_\infty(A) = \|A\|_\infty \|A^{-1}\|_\infty$ is the condition number defined by the row sum norm $\|A\|_\infty = \max_{i=1,\ldots,n} \sum_{j=1}^n |a_{ij}|$.

Exercise A.5.27 Derive an upper bound for the computational complexity of the application of the BPX preconditioner, using that the number of levels satisfies $L \sim \log_2(h^{-1})$.

Exercise A.5.28 Let $V_0 \subset V_1 \subset \cdots \subset V_L = V \subset L^2(\Omega)$. Let $v \in V$ and construct functions $w_0, w_1, \ldots, w_L \in V$ with $w_\ell \in V_\ell$ for $\ell = 0, 1, \ldots, L$, such that $w_{\ell+1} = w_{\ell+2} = \cdots = w_L = 0$ of $v \in V_\ell$, and

$$\sum_{\ell=0}^{L} w_\ell = v, \quad \left\| \sum_{\ell=0}^{L} w_\ell \right\|_{L^2(\Omega)}^2 = \sum_{\ell=0}^{L} \|w_\ell\|_{L^2(\Omega)}^2 .$$

Exercise A.5.29 Show that if the assumption about the H^2-regularity of the Poisson problem is omitted, then for the BPX preconditioner $C \in \mathbb{R}^{n \times n}$ we have that

$$\mathrm{cond}_2(CA) \le c \log_2(h^{-1}).$$

Exercise A.5.30 Show that $q : \mathbb{R}^n \to \mathbb{R}$ is a quadratic form, i.e., $q(\lambda v) = \lambda^2 q(v)$ and $q(v + w) + q(v - w) = 2q(v) + 2q(w)$ for all $v, w \in \mathbb{R}^n$ and $\lambda \in \mathbb{R}$, if and only if there exists a symmetric matrix $M \in \mathbb{R}^{n \times n}$ such that $q(v) = v^\top M v$ for all $v \in \mathbb{R}^n$.

Quiz A.5.3 Decide for each of the following statements whether it is true or false. You should be able to justify your decision.

The Gauss–Seidel preconditioner is optimal for diagonal matrices	
The BPX preconditioner is optimal on regular sequences of triangulations if the Poisson problem is H^2-regular	
If a preconditioner is optimal, i.e., $\mathrm{cond}_2(CA) \le c$, then the preconditioned CG algorithm terminates within a finite number of iterations, independently of the problem size	
If the eigenvalues of the symmetric and positive definite matrices A and C satisfy $c_1 \lambda_i \le \mu_i^{-1} \le c_2 \lambda_i$ for $i = 1, 2, \ldots, n$, then $\mathrm{cond}_2(CA) \le c_1/c_2$	
The strengthened Cauchy–Schwarz inequality states that functions v_k, w_ℓ in nested spaces $\mathscr{S}_0^1(\mathscr{T}_k) \subset \mathscr{S}_0^1(\mathscr{T}_\ell)$ are nearly orthogonal	

A.5.4   Projects

Project A.5.1 Let $\mathscr{T}_h$ be a triangulation of $\Omega = (0, 1)^2$. Use the von Mises power method to determine nontrivial functions $u_h^{(1)}, u_h^{(N)} \in \mathscr{S}_0^1(\mathscr{T}_h)$ such that there exist $\lambda_1, \lambda_N \in \mathbb{R}$ with $\lambda_1 < \lambda_N$ such that

$$\int_\Omega \nabla u_h^{(i)} \cdot \nabla v_h \, dx = \lambda_i \int_\Omega u_h^{(i)} v_h \, dx$$

for all $v_h \in \mathscr{S}_0^1(\mathscr{T}_h)$ and $i = 1, N$. Plot the functions for different meshes and plot the ratios $q_h = \lambda_N / \lambda_1$ for a sequence of refined triangulations versus the mesh-size h using a logarithmic scaling, and determine a relation $q_h \sim h^\alpha$.

Project A.5.2 Use a sequence of uniformly refined triangulations $(\mathscr{T}_j)_{j=0,1,...}$ and approximately solve the Poisson problem $-\Delta u = 1$, $u|_{\partial\Omega} = 0$, for $\Omega = (0,1)^d$, $d = 2, 3$, using the backslash operator, a Gauss–Seidel iteration, and the multigrid algorithm. Compare the CPU times and discuss the results.

Project A.5.3 The W-cycle of the multigrid method is defined by applying the multigrid function twice on every level. Explain the name of the method and experimentally compare its performance with the V-cycle. Experimentally determine a sufficient number of post-smoothing steps by comparing the multigrid approximations with the discrete solution obtained with the backslash operator.

Project A.5.4 Consider the Poisson problem $-\Delta u = 1$, $u|_{\partial\Omega} = 0$, in the domain $\Omega = (0,1)^d$, $d = 2, 3$. Use a sequence of uniformly refined triangulations and solve the resulting linear systems of equations of the finite element approximations using the preconditioned conjugate gradient algorithm with Jacobi, equilibration, symmetric Gauss–Seidel, incomplete Cholesky, and BPX preconditioner. Repeat the experiment for a two-dimensional L-shaped domain and discuss your results.

Project A.5.5 Implement the two-level preconditioner and test it for a Poisson problem and a sequence of uniformly refined triangulations.

Project A.5.6 Implement the overlapping Schwarz domain decomposition method, test it for two Poisson problems with different domains $\Omega \subset \mathbb{R}^2$, and determine experimentally the dependence of the number of iterations on the geometry of the overlap region.

Project A.5.7 Generalize the Dirichlet–Neumann method to the case of arbitrarily many nonfloating subdomains and experimentally investigate its convergence for two Poisson problems in $\mathbb{R}^2$ with at least four subdomains.

Project A.5.8 Let $(\Omega_j)_{j=1,2}$ be a nonoverlapping partition of $\Omega \subset \mathbb{R}^2$ with interface γ. For a triangulation $\mathscr{T}_h$ of Ω such that Ω_1 and Ω_2 are matched by unions of triangles in $\mathscr{T}_h$, let $\mathscr{T}_h^{(j)}$ be the induced triangulations of Ω_j, $j = 1, 2$. Let $W_h = \mathscr{S}_0^1(\mathscr{T}_h)|_\gamma$ be the discrete trace space on the interface and defined norms $\|\cdot\|_{j,h}$ for $j = 1, 2$ for $\psi_h \in W_h$ via

$$\|\psi_h\|_{j,h} = \|\nabla u_h^{(j)}\|_{L^2(\Omega_j)},$$

where $u_h^{(j)} \in \mathscr{S}^1(\mathscr{T}_h^{(j)})$ satisfies $u_h^{(j)} = \psi_h$ on γ, $u_h^{(j)}|_{\partial\Omega_j \setminus \gamma} = 0$, and

$$\int_{\Omega_j} \nabla u_h^{(j)} \cdot \nabla v_h \, dx = 0$$

for all $v_h \in \mathscr{S}_0^1(\mathscr{T}_h^{(j)})$. Define an eigenvalue problem to determine mesh-dependent constants λ_1^h, λ_N^h such that

$$\lambda_1^h \|\!|\!|\psi_h|\!|\!|_{1,h} \leq \|\!|\!|\psi_h|\!|\!|_{2,h} \leq \lambda_2^h \|\!|\!|\psi_h|\!|\!|_{1,h}$$

for all $\psi_h \in \mathscr{S}_0^1(\mathscr{T}_h)|_\gamma$. Compute the numbers λ_1^h, λ_N^h for three different partitions of $\Omega = (0, 1)^2$ and three different triangulations in each case.

A.6   Saddle-Point Problems

A.6.1   Discrete Saddle-Point Problems

Exercise A.6.1 Let $M \in \mathbb{R}^{n \times n}$ and $\|\cdot\|_\ell$ be a norm on $\mathbb{R}^n$ with dual norm $\|\cdot\|_{\ell'}$. Show that $k \geq 0$ is the smallest constant such that

$$x^\top M y \leq k \|x\|_\ell \|y\|_\ell$$

if and only if

$$k = \sup_{x \in \mathbb{R}^n} \frac{\|Mx\|_{\ell'}}{\|x\|_\ell}.$$

Exercise A.6.2

(i) Show that $M \in \mathbb{R}^{n \times n}$ is regular if and only if

$$\inf_{v \in \mathbb{R}^n \backslash \{0\}} \sup_{w \in \mathbb{R}^n \backslash \{0\}} \frac{v^\top M w}{\|v\| \|w\|} > 0.$$

(ii) Show that $B \in \mathbb{R}^{m \times n}$ is surjective if and only if

$$\inf_{q \in \mathbb{R}^m \backslash \{0\}} \sup_{v \in \mathbb{R}^n \backslash \{0\}} \frac{q^\top B v}{\|q\| \|v\|} > 0.$$

Exercise A.6.3 Let $f : V \to W$ be a linear mapping between the n-dimensional linear space V and the n-dimensional linear space W. Show that there exist bases $(v_1, v_2, \ldots, v_n)$ for V and $(w_1, w_2, \ldots, w_m)$ for W, such that the matrix $B \in \mathbb{R}^{m \times n}$ representing f in these bases satisfies

$$\ker B = \{v = (v_1, v_2, \ldots, v_n) \in \mathbb{R}^n : v_1 = v_2 = \cdots = v_r = 0\},$$

where $r = \dim \ker B$.

Exercise A.6.4 Systematically investigate the regularity of the saddle-point matrix

$$M = \begin{bmatrix} 0 & 1 & 1 \\ 1 & 0 & -1 \\ 1 & -1 & 0 \end{bmatrix}, \quad M = \begin{bmatrix} 1 & 0 & 1 & 1 \\ 0 & 1 & 1 & 2 \\ 1 & 1 & 0 & 0 \\ 1 & 2 & 0 & 0 \end{bmatrix}, \quad M = \begin{bmatrix} 0 & 1 & 0 & 0 & 1 & 0 \\ 1 & 0 & 0 & 0 & 0 & 1 \\ 0 & 0 & 1 & 3 & 1 & 0 \\ 0 & 0 & 3 & 9 & 0 & 0 \\ 1 & 0 & 1 & 0 & 0 & 0 \\ 0 & 1 & 0 & 0 & 0 & 0 \end{bmatrix}.$$

Exercise A.6.5 Let $A \in \mathbb{R}^{n \times n}$ be positive semidefinite and symmetric, and let $K \subset \mathbb{R}^n$ be a linear subspace. Show that

$$\inf_{v \in K \setminus \{0\}} \sup_{w \in K \setminus \{0\}} \frac{v^T A w}{\|v\| \|w\|} \geq \alpha$$

if and only if $v^T A v \geq \alpha \|v\|^2$ for all $v \in K$.

Exercise A.6.6

(i) For $B \in \mathbb{R}^{n_B \times n_A}$ let B_l denote the restriction of B to $(\ker B)^\perp$. Prove that invertibility of $B_l : \operatorname{Im} B^T \mapsto \mathbb{R}^{n_B}$ is equivalent to the implication

$$B^T z = 0 \quad \Longrightarrow \quad z = 0,$$

(ii) Show that bijectivity of $A \in \mathbb{R}^{n_A \times n_A}$ on $\ker B$ for a matrix $B \in \mathbb{R}^{n_B \times n_A}$ is equivalent to the implication

$$y^T A v = 0 \quad \text{for all } v \in \ker B \quad \Longrightarrow \quad y = 0.$$

Exercise A.6.7 Let $\| \cdot \|_V$ and $\| \cdot \|_Q$ be norms on $\mathbb{R}^{n_A}$ and $\mathbb{R}^{n_B}$ with duals $\| \cdot \|_{V'}$ and $\| \cdot \|_{Q'}$, respectively.

(i) Show that the mapping

$$\| \cdot \|_\ell : \mathbb{R}^{n_A} \times \mathbb{R}^{n_B} \to \mathbb{R}, \quad (y, z) \mapsto \|y\|_V + \|z\|_Q,$$

defines a norm on $\mathbb{R}^{n_A} \times \mathbb{R}^{n_B}$, its dual norm is given by

$$\|(f, g)\|_{\ell'} = \max \left\{ \|f\|_{V'}, \|g\|_{Q'} \right\}$$

for all $(f, g) \in \mathbb{R}^{n_A} \times \mathbb{R}^{n_B}$, and that

$$\|(f, g)\|_{\ell'} \leq \|f\|_{V'} + \|g\|_{Q'} \leq 2 \|(f, g)\|_{\ell'}.$$

(ii) Let $A \in \mathbb{R}^{n_A \times n_A}$ and $B \in \mathbb{R}^{n_B \times n_A}$ be such that there exist constants $k_A, k_B \geq 0$ so that

$$v^\top Ay \leq k_A \|v\|_V \|y\|_V, \quad v^\top Bz \leq k_B \|v\|_V \|z\|_Q$$

for all $v, y \in \mathbb{R}^{n_A}$ and $z \in \mathbb{R}^{n_B}$. Show that for the associated saddle-point matrix M we have with $\| \cdot \|_r = \| \cdot \|_{\ell'}$ that

$$\|M\|_{\ell r} \leq k_A + 2k_B.$$

Exercise A.6.8 Let $\| \cdot \|_V$ and $\| \cdot \|_Q$ be norms on $\mathbb{R}^{n_A}$ and $\mathbb{R}^{n_B}$ with dual norms $\| \cdot \|_{V'}$ and $\| \cdot \|_{Q'}$, respectively, and let the restriction $B_I \in \mathbb{R}^{n_B \times n_B}$ of B to $(\ker B)^\perp$ be regular.

(i) Show that for every $z \in \mathbb{R}^{n_B}$ we have

$$\|z\|_Q = \sup_{s \in \mathbb{R}^{n_B}} \frac{s^\top z}{\|s\|_{Q'}}.$$

(ii) Prove that

$$\|B_I^{-1}\|_{Q'V} = \|(B_I^\top)^{-1}\|_{V'Q}.$$

Exercise A.6.9 Assume that $a : V \times V \to \mathbb{R}$ is a bounded and coercive bilinear form with constants $k_a, \alpha > 0$, $V_h = \mathrm{span}\{v_1, v_2, \ldots, v_n\}$ a finite-dimensional subspace, and $M_{jk} = a(v_j, v_k)$ for $j, k = 1, 2, \ldots, n$.

(i) Show that a norm $\| \cdot \|_\ell$ on $\mathbb{R}^n$ is defined by

$$x \mapsto \left\| \sum_{i=1}^n x_i v_i \right\|_V.$$

(ii) Show that we have

$$\frac{y^\top Mx}{\|x\|_\ell \|y\|_\ell} = \frac{a(u_h, v_h)}{\|u_h\|_V \|v_h\|_V}.$$

(iii) Show that for $\| \cdot \|_r = \| \cdot \|_{\ell'}$ we have $\mathrm{cond}_{\ell r}(M) \leq k_a/\alpha$.

Exercise A.6.10 Let $A \in \mathbb{R}^{n_A \times n_A}$ be symmetric and positive definite and $B \in \mathbb{R}^{n_B \times n_A}$. Prove that the Schur complement $S = BA^{-1}B^\top \in \mathbb{R}^{n_B \times n_B}$ is symmetric and positive definite.

Quiz A.6.1 Decide for each of the following statements whether it is true or false. You should be able to justify your decision.

A necessary condition for the unique solvability of a saddle-point problem is that $n_B \leq n_A$ and B has maximal rank	
If $B^\top$ is injective, then for all $f \in (\ker B)^\perp$ there exists a unique z with $B^\top z = f$	
The inf-sup condition bounds the operator norm of the left-inverse of a matrix	
If B is surjective, then it has a left inverse $B^{-\ell}$	
The symmetric matrix A defines a bijection on the subspace K if and only if A is positive definite on K	

A.6.2 Continuous Saddle-Point Problems

Exercise A.6.11

(i) For $n \in \mathbb{N}$ let $L_n : \mathbb{R}^n \to \mathbb{R}^n$ be defined by $(x_1, x_2, \ldots, x_n) \mapsto (x_1, x_2/2, x_3/3, \ldots, x_n/n)$. Determine the condition number of L_n with respect to the Euclidean norm.

(ii) Let $\ell^2(\mathbb{N})$ be the space of all sequences $x = (x_j)_{j \in \mathbb{N}} \subset \mathbb{R}$ such that $\|x\|^2_{\ell^2(\mathbb{N})} = \sum_{j \in \mathbb{N}} x_j^2 < \infty$. Show that the operator

$$L : \ell^2(\mathbb{N}) \to \ell^2(\mathbb{N}), \quad (x_1, x_2, x_3, \ldots) \mapsto (x_1, x_2/2, x_3/3, \ldots)$$

is bounded, linear, and injective, but $\mathrm{Im}\, L$ is not closed.

Exercise A.6.12 Let $\ell^2(\mathbb{N})$ be the space of all sequences $x = (x_j)_{j \in \mathbb{N}} \subset \mathbb{R}$ such that $\|x\|^2_{\ell^2(\mathbb{N})} = \sum_{j \in \mathbb{N}} x_j^2 < \infty$. Determine which of the following operators $L_j : \ell^2(\mathbb{N}) \to \ell^2(\mathbb{N}), j = 1, 2, \ldots, 4$, satisfy an inf-sup condition:

$$L_1 x = (x_1, 0, x_2, 0, x_3, 0, \ldots),$$
$$L_2 x = (x_1, x_3, x_5, \ldots),$$
$$L_3 x = (x_1 - x_2, x_3 - x_4, \ldots),$$
$$L_4 x = (|x_1|, |x_2|, \ldots).$$

Specify for each operator its adjoint.

Exercise A.6.13 Let X be a Hilbert space and let $a : X \times X \to \mathbb{R}$ be a bounded and coercive bilinear form on X. Show that a satisfies an inf-sup condition, and that a is nondegenerate.

Exercise A.6.14

(i) Show that $p \in H_0^1(\Omega)$ solves the Poisson problem $-\Delta p = g$ in Ω with $p|_{\partial\Omega} = 0$ in the weak sense, if and only if the pair $(u, p) \in L^2(\Omega; \mathbb{R}^d) \times H_0^1(\Omega)$ with $u = \nabla p$ satisfies

$$\int_\Omega u \cdot v \, dx + \int_\Omega u \cdot \nabla p \, dx = 0,$$

$$-\int_\Omega u \cdot \nabla q \, dx \qquad\qquad = -\int_\Omega g q \, dx,$$

for all $(v, q) \in L^2(\Omega; \mathbb{R}^d) \times H_0^1(\Omega)$.

(ii) Verify directly the boundedness, and the inf-sup and nondegeneracy conditions for the associated bilinear form

$$\Gamma\big((u, p), (v, q)\big) = \int_\Omega u \cdot \nabla v \, dx + \int_\Omega u \cdot \nabla p \, dx + \int_\Omega u \cdot \nabla q \, dx.$$

Conclude that the conditions of the generalized Lax–Milgram lemma are satisfied.

Hint: Consider the pair $(v, q) = (u - \nabla p, -2p)$.

Exercise A.6.15 Assume that V and Q are Hilbert spaces, and $a : V \times V \to \mathbb{R}$ and $b : V \times Q \to \mathbb{R}$ are bounded bilinear forms. Assume that the operator

$$L : V \times Q \to V' \times Q', \quad (u, p) \mapsto \big(a(u, \cdot) + b(\cdot, p), b(u, \cdot)\big)$$

is an isomorphism, and there exists $\beta > 0$ such that

$$\inf_{q \in Q \setminus \{0\}} \sup_{v \in V \setminus \{0\}} \frac{b(v, q)}{\|v\|_V \|q\|_Q} \geq \beta.$$

Show that there exists $\alpha > 0$ such that

$$\inf_{v \in \ker B \setminus \{0\}} \sup_{w \in \ker B \setminus \{0\}} \frac{a(v, w)}{\|v\|_V \|w\|_V} \geq \alpha.$$

Exercise A.6.16 Let $a : V \times V \to \mathbb{R}$ be symmetric and positive semidefinite, and $K \subset V$ a closed subspace. Assume that for every $\ell \in K'$, there exists a unique $u \in K$ such that

$$a(u, v) = \ell(v)$$

for all $v \in K$ and such that $\|u\|_V \leq c_K \|\ell\|_{K'}$.

(i) Prove the Cauchy–Schwarz inequality $a(u, v)a(u, v) \leq a(u, u)a(v, v)$.

(ii) Show that a is coercive on K.

Exercise A.6.17 Let the bilinear forms $a : H^1(\Omega) \times H^1(\Omega) \to \mathbb{R}$ and $b : H^1(\Omega) \times \mathbb{R} \to \mathbb{R}$ be defined by

$$a(u, v) = \int_\Omega \nabla u \cdot \nabla v \, dx, \quad b(v, \lambda) = \lambda \int_\Omega v \, dx.$$

(i) Show that for every $\ell \in H^1(\Omega)'$ there exists a uniquely defined pair $(u, \lambda) \in H^1(\Omega) \times \mathbb{R}$ such that

$$a(u, v) + b(v, \lambda) = \ell(v),$$
$$b(u, \mu) \qquad\qquad = 0,$$

for all $(v, \mu) \in H^1(\Omega) \times \mathbb{R}$.

(ii) Show that under appropriate conditions on functions $f \in L^2(\Omega)$ and $g \in L^2(\partial\Omega)$, and an appropriate definition of ℓ, the weak formulation defines a weak solution for the Poisson problem $-\Delta u = f$ in Ω with Neumann boundary condition $\partial_n u = g$ on $\Gamma_N = \partial\Omega$.

Exercise A.6.18

(i) Show that for all $v \in H_0^1(\Omega)$, we have

$$\frac{1}{2} \int_\Omega |\nabla v|^2 \, dx = \sup_{q \in L^2(\Omega; \mathbb{R}^d)} \int_\Omega q \cdot \nabla u \, dx - \frac{1}{2} \int_\Omega |q|^2 \, dx.$$

(ii) Let $f \in L^2(\Omega)$. Derive the optimality conditions for the saddle-point problem

$$\inf_{v \in H_0^1(\Omega)} \sup_{q \in L^2(\Omega; \mathbb{R}^d)} L(v, q), \quad L(v, q) = \int_\Omega q \cdot \nabla v \, dx - \frac{1}{2} \int_\Omega |q|^2 \, dx - \int_\Omega fv \, dx,$$

i.e., compute the derivatives of the mappings $t \mapsto L(v + tw, q)$ and $t \mapsto L(v, q + ts)$ at $t = 0$.

(iii) Determine a partial differential equation that is satisfied by the element u of a saddle point (u, p).

Exercise A.6.19 Let $A \in \mathbb{R}^{n_A \times n_A}$ and $B \in \mathbb{R}^{n_B \times n_A}$ with $n_A, n_B \in \mathbb{N}$ such that $n_A \geq n_B$. Assume that $\dim \operatorname{Im} B = n_B$ and $v^\top A w > 0$ for all $v, w \in \ker A$. Let $C \in \mathbb{R}^{n_B \times n_B}$ be positive semidefinite and $t \geq 0$. Show that the matrix

$$\begin{bmatrix} A & B^\top \\ B & -tC \end{bmatrix}$$

is regular.

Exercise A.6.20

(i) Assume that $\|u\|_V + \|p\|_Q > \delta^{-1}t|p|_c$ and $\|u\|_V > (\beta/(2k_a))\|p\|_Q$. Show that $\|u\|_V + \|q\|_Q + t|q|_c \le (2 + 4k_a/\beta)\|u\|_V$ and for δ sufficiently small so that $\|u\|_V + \|q\|_Q + t|q|_c \le \delta^{-1}\|u\|_V$.

(ii) Show that for $x, y, z \ge 0$ with $x > 0$ and $0 < x \le y + z$, we have $x \le y^2/x + z$.

Quiz A.6.2 Decide for each of the following statements whether it is true or false. You should be able to justify your decision.

If $L : X \to X'$ is self-adjoint, i.e., $L' = L$, then the inf-sup condition implies nondegeneracy
If the image of a bounded linear operator is finite-dimensional, then it is closed
If $X = Y$ and $\Gamma : X \times Y \to \mathbb{R}$ is positive semidefinite and satisfies the conditions of the generalized Lax–Milgram lemma, then Γ is coercive
If L satisfies an inf-sup condition, then L' is surjective
If $L : X \to Y'$ is bijective and satisfies an inf-sup condition, then it is an isomorphism

A.6.3   Approximation of Saddle-Point Problems

Exercise A.6.21 Assume that the bilinear forms $a : V \times V \to \mathbb{R}$ and $b : V \times Q \to \mathbb{R}$, and the families of subspaces $(V_h)_{h>0}$ and $(Q_h)_{h>0}$ satisfy the Babuška–Brezzi conditions. Suppose that for

$$K_h = \{v_h \in V_h : b(v_h, q_h) = 0 \text{ for all } q_h \in Q_h\},$$

we have $K_h \subset \ker B$, where $B : V \to Q'$ is defined by $Bv = b(v, \cdot)$ for all $v \in V$. Show that for the approximation error $u - u_h$, we have

$$\|u - u_h\|_V \le (k_a/\alpha) \inf_{v_h \in K_h} \|u - u_h - v_h\|_V.$$

Exercise A.6.22 For $n \in \mathbb{N}$ and $h = 1/n$, let $\mathscr{T}_h$ be the triangulation of $\Omega = (0, 1)^2$ consisting of halved squares of length side h, with diagonals parallel to the vector $(1, 1)$.

(i) Show that if $v_h \in \mathscr{S}_0^1(\mathscr{T}_h)^2$ satisfies div $v_h = 0$, then we have $v_h = 0$.

(ii) Show that for every $h = 1/n > 0$, there exists a constant $c_h > 0$ such that $\| \operatorname{div} v_h\|_{L^2(\Omega)} \ge c_h \|v_h\|_{L^2(\Omega)}$ for all $v_h \in \mathscr{S}_0^1(\mathscr{T}_h)^2$. Prove that $c_h \le ch$ with a constant $c > 0$ that is independent of h.

(iii) Let $u(x, y) = [\sin(x)\sin(y), \cos(x)\cos(y)]^\top$ for $(x, y) \in T$, where T is for $h > 0$ defined by $T = \operatorname{conv}\{(0, 0), (h, 0), (0, h)\}$. Show that div $u = 0$, div $\mathscr{I}_h u \approx h/2$ for small h, and $|D^2 u| \geq 1/2$. Conclude that $\| \operatorname{div}(u - \mathscr{I}_h u)\|_{L^2(T)} \geq ch\|D^2 u\|_{L^2(T)}$.

Exercise A.6.23 Let $a : V \times V \to \mathbb{R}$ be a bounded, coercive, and symmetric bilinear form, and let $B : V \to Q$ be a bounded linear operator for Hilbert spaces V and Q. Let $0 < t \leq 1$ and $\ell \in V'$, and let $u \in V$ be such that

$$a(u, v) + t^{-2}(Bu, Bv) = \ell(v)$$

for all $v \in V$. For a family of subspaces $(V_h)_{h>0}$, let $(u_h)_{h>0} \subset V$ be the sequence of corresponding Galerkin approximations. Specify and discuss the dependence of the constant in Céa's lemma on the parameter $t > 0$.

Exercise A.6.24 Let $a : V \times V \to \mathbb{R}$ be a coercive, bounded and symmetric bilinear form, and let $b : V \times Q \to \mathbb{R}$ be a bounded bilinear form, so that the conditions of Brezzi's splitting theorem are satisfied. Let $c : Q \times Q \to \mathbb{R}$ be a bounded and positive semidefinite bilinear form, and let $0 < t \ll 1$. Let $(V_h)_{h>0}$ and $(Q_h)_{h>0}$ be families of subspaces, such that the Babuška–Brezzi conditions are satisfied. Use the stability results about perturbed saddle-point problems to derive an error estimate for its numerical approximation that is independent of $0 < t \leq 1$.

Exercise A.6.25 Let $\mathscr{T}_h$ be the triangulation of $\Omega = (0, 1)^2$ with nodes $z_0 = (0, 0)$, $z_1 = (1, 0)$, $z_2 = (1, 1)$, $z_3 = (0, 1)$, and $z_5 = (1, 1)/2$. Construct a function $q_h \in \mathscr{L}^0(\mathscr{T}_h)$ with $\int_\Omega q_h \, dx = 0$, such that div $v_h \neq q_h$ for all $v_h \in \mathscr{S}_0^1(\mathscr{T}_h)^2$, and conclude that these spaces do not lead to an inf-sup condition for

$$b(v_h, q_h) = \int_\Omega q_h \operatorname{div} v_h \, dx.$$

Exercise A.6.26 Specify a Fortin interpolant for the bilinear form

$$b(v, q) = \int_\Omega v \cdot \nabla q \, dx$$

on $V \times Q$ with $V = L^2(\Omega; \mathbb{R}^d)$ and $Q = H_0^1(\Omega)$, and for the subspaces $V_h = \mathscr{L}^0(\mathscr{T}_h)^d$ and $Q_h = \mathscr{S}_0^1(\mathscr{T}_h)$ for a regular family of triangulations $(\mathscr{T}_h)_{h>0}$ of Ω.

Exercise A.6.27 Let $A \in \mathbb{R}^{n_A \times n_A}$, $B \in \mathbb{R}^{n_B \times n_A}$, $f \in \mathbb{R}^{n_A}$, and $g \in \mathbb{R}^{n_B}$. Show that $x \in \mathbb{R}^{n_A}$ satisfies $Bx = g$ and $y^\top A x = y^\top f$ for all $y \in \mathbb{R}^{n_A}$ with $By = 0$, if and only if $Bx = g$ and there exists $\lambda \in \mathbb{R}^{n_B}$ such that $Ax + B^\top \lambda = b$.

Exercise A.6.28 Let $X = Y = H_0^1(\Omega), f \in L^2(\Omega)$, and $X_h = Y_h = \mathscr{S}_0^1(\mathscr{T}_h)$ for a regular family of triangulations $(\mathscr{T}_h)_{h>0}$ of Ω. Let

$$\Gamma(u, v) = \int_\Omega \nabla u \cdot \nabla v \, dx, \quad \ell(v) = \int_\Omega f v \, dx.$$

Prove that the conditions of the generalized Céa lemma are satisfied.

Exercise A.6.29 Let $(\mathcal{T}_h)_{h>0}$ be a regular family of triangulations of Ω and let $g \in L^2(\Omega)$. Let $p \in H_0^1(\Omega)$ and $(\hat{p}_h)_{h>0}$ be the exact solution and its Galerkin approximations of the problem

$$-\Delta p = g \text{ in } \Omega, \quad p|_{\partial\Omega} = 0.$$

For every $h > 0$, let $(u_h, p_h) \in \mathcal{L}^0(\mathcal{T}_h)^d \times \mathcal{S}_0^1(\mathcal{T}_h)$ be the solution of the saddle-point formulation

$$\int_\Omega u_h \cdot v_h \, dx - \int_\Omega v_h \cdot \nabla p_h \, dx = 0,$$

$$\int_\Omega u_h \cdot \nabla q_h \, dx \qquad\qquad = \int_\Omega g q_h \, dx,$$

for all $(v_h, q_h) \in \mathcal{L}^0(\mathcal{T}_h)^d \times \mathcal{S}_0^1(\mathcal{T}_h)$. Show that for every $h > 0$ we have $\hat{p}_h = p_h$.

Exercise A.6.30 Let $\mathcal{T}_h$ be the partition of the interval $(0, 1)$ defined by the nodes $x_j = jh = j/N, j = 0, 1, \ldots, N$, and let $\mathcal{S}^{p,k}(\mathcal{T}_h) \subset C^k([0, 1])$ be the space of spline functions with piecewise polynomial degree $p \geq 0$ and differentiability order $k \geq 0$. Let $\mathcal{S}_0^{r,k}(\mathcal{T}_h)$ be the subspace of functions that vanish for $x \in \{0, 1\}$. Determine pairs (r, k) and (s, ℓ) such that an inf-sup condition

$$\sup_{v_h \in \mathcal{S}_0^{r,k}(\mathcal{T}_h)\backslash\{0\}} \frac{\int_0^1 v_h' q_h \, dx}{\|v_h'\|} \geq \beta_h \|q_h\|$$

holds for all $q_h \in \mathcal{S}^{s,\ell}(\mathcal{T}_h)$ with a positive constant $\beta_h > 0$.

Quiz A.6.3 Decide for each of the following statements whether it is true or false. You should be able to justify your decision.

Unique solvability of $\Gamma(x_h, y_h) = \ell(y_h)$ requires that $\dim X_h = \dim Y_h$	
The inf-sup constant $\gamma \geq 0$ for a bilinear form Γ is bounded from below by the inverse of the continuity constant k_Γ^{-1}	
Uniformity of a discrete inf-sup condition means that the constants γ_h are positive for every $h > 0$	
In the case of the Stokes equation with $V_h = \mathcal{S}_0^1(\mathcal{T}_h)^d$ and $Q_h = \mathcal{L}^0(\mathcal{T}_h)$, we have $K_h = \{v_h \in V_h : \operatorname{div} v_h = 0\}$	
The locking effect refers to a limited flexibility of a finite element space for a particular problem	

A.6.4 Projects

Project A.6.1 Discretize the mixed formulation of the one-dimensional Poisson problem $-p'' = f$ in $\Omega = (0, 1)$, $p(0) = p(1) = 0$, i.e., the problem of determining $(u, p) \in H^1(\Omega) \times L^2(\Omega)$ such that

$$\int_0^1 uv \, dx + \int_0^1 pv' \, dx = 0,$$

$$\int_0^1 qu' \, dx \qquad = -\int_\Omega fq \, dx,$$

for all $(v, q) \in H^1(\Omega) \times L^2(\Omega)$. Use the finite element space $\mathscr{S}^1(\mathscr{T}_h)$ for the approximation of u and the spaces $\mathscr{S}^1(\mathscr{T}_h)$ and $\mathscr{L}^0(\mathscr{T}_h)$ for the approximation of p. Compare the results for the case $f = 1$ with exact solution $p(x) = x(1-x)/2$.

Project A.6.2 Solutions of the Neumann problem $-\Delta u = f$ in Ω, $\partial_n u = g$ on $\Gamma_N = \partial\Omega$ only exist if the compatibility condition

$$\int_\Omega f \, dx + \int_{\Gamma_N} g \, ds = 0$$

is satisfied, and are unique only up to a constant. Determine different finite element approximations using the normalizations

$$u_h(z_0) = 0, \qquad \int_\Omega u_h \, dx = 0,$$

and via minimizing the functional

$$\widetilde{E}_h(v_h) = \frac{1}{2} \int_\Omega |\nabla v_h|^2 \, dx - \int_\Omega f v_h \, dx - \int_{\Gamma_N} g v_h \, ds + \left(\int_\Omega v_h \, dx \right)^2.$$

Implement the three approaches, discuss relations between them, and comment on the linear systems of equations. Test your implementations for the case $\Omega = (0, 1)^2$, $f = 1$, and $g = -1/4$.

Project A.6.3 Consider the Neumann problem $-\Delta u = f$ in Ω, $\partial_n u = g$ on $\Gamma_N = \partial\Omega$, assume that the compatibility condition

$$\int_\Omega f \, dx + \int_{\Gamma_N} g \, ds = 0$$

is satisfied, and let $u \in H^1(\Omega)$ be the unique solution of the problem satisfying

$$\int_\Omega u \, dx = 0.$$

Characterize u as the solution of a saddle-point problem, define a discrete saddle-point problem, and compare its solution using the backslash operator and the Uzawa algorithm for the case $\Omega = (0, 1)^3, f = 1, g = -1/6$.

Project A.6.4 Consider the problem

$$\Delta^2 u = f \quad \text{in } \Omega, \quad u = \Delta u = 0 \quad \text{on } \partial\Omega.$$

Introduce the variable $\xi = -\Delta u$, formulate a saddle-point problem, and approximate it with a $P1$-$P1$ finite element method. Test your problem for the cases $\Omega = (0, 1)^2, f = 1$, and $\Omega = B_1(0)$. Identify the exact solution in the second case using that $\Delta u = \partial_r^2 u + r^{-1}\partial_r u$ for a rotationally symmetric function u.

Project A.6.5 Determine a function $f \in L^2(\Omega; \mathbb{R}^2)$ with $\Omega = (-1, 1)^2$ so that the exact solution $u \in H_0^1(\Omega; \mathbb{R}^2)$ of the variational formulation

$$\int_\Omega \nabla u : \nabla v \, dx + \varepsilon^{-2} \int_\Omega \operatorname{div} u \operatorname{div} v \, dx = \int_\Omega f \cdot v \, dx$$

for all $v \in H_0^1(\Omega; \mathbb{R}^2)$ is for every $\varepsilon > 0$ given by

$$u(x_1, x_2) = \begin{bmatrix} \sin(2\pi x_2) \sin^2(\pi x_1) \\ -\sin(2\pi x_1) \sin^2(\pi x_2) \end{bmatrix}.$$

Investigate the approximation of the problem in $\mathscr{S}_0^1(\mathscr{T}_h)^2$ for sequences of triangulations and the parameters $\varepsilon = 10^{-j}, j = 1, 2$.

Project A.6.6 Implement the discretization of the Stokes problem

$$-\Delta u + \nabla p = f,$$
$$\operatorname{div} u \qquad = 0,$$

with $P1$-finite elements for approximating u and p. Test it for $\Omega = (-1, 1)^2, \Gamma_D = \partial\Omega$, and

$$u(x_1, x_2) = \pi \begin{bmatrix} \sin(2\pi x_2) \sin^2(\pi x_1) \\ -\sin(2\pi x_1) \sin^2(\pi x_2) \end{bmatrix},$$

$$p(x_1, x_2) = \cos(\pi x_1) \sin(\pi x_2).$$

Project A.6.7 For a compact C^2-submanifold $M \subset \mathbb{R}^3$, the nearest-neighbor projection $\pi_M(z)$ of a point $z \in \mathbb{R}^3$ onto M is defined as a point $x \in M$ with $|z - x| = \min_{y \in M} |z - y|$. One can show that there exists an open neighborhood of M in which π_M is well defined. We assume that $M = f^{-1}(\{0\})$ and characterize the nearest-neighbor projection of z as a saddle-point for the functional

$$G(x, \lambda) = \frac{1}{2}|x - z|^2 + \lambda f(x),$$

i.e., a solution (x, λ) of the equation $F(x, \lambda) = 0$ with $F : \mathbb{R}^4 \to \mathbb{R}^4$ defined by

$$F(x, \lambda) = \begin{bmatrix} x - z + \lambda \nabla f(x) = 0 \\ f(x) \end{bmatrix}.$$

Formulate the Newton iteration for the solution of this equation and discuss its well-posedness and convergence. Implement and test it for cases of a sphere and torus.

Project A.6.8 Implement a $P0$-$P1$ method for the primal mixed formulation of the Poisson problem and verify experimentally that it coincides with the standard $P1$ finite element method.

A.7 Mixed and Nonstandard Methods

A.7.1 Mixed Methods for the Poisson Problem

Exercise A.7.1 Prove that the space $H_N(\text{div}; \Omega)$ is a Hilbert space.

Exercise A.7.2 Let $u \in L^2(\Omega; \mathbb{R}^d)$ be such that $u|_{\Omega_i} \in C^1(\overline{\Omega}_i; \mathbb{R}^d)$, $i = 1, 2, \dots, I$, for a partition $(\Omega_i)_{i=1,\dots,I}$ of Ω. Prove that we have $u \in H(\text{div}; \Omega)$ if and only if $u|_{\Omega_i} \cdot n_i = -u|_{\Omega_j} \cdot n_j$ on every interface $\Gamma_{ij} = \partial\Omega_i \cap \partial\Omega_j$ with the outer unit normals n_i and n_j to Ω_i and Ω_j.

Exercise A.7.3 Let $\Omega = \{(x_1, x_2) \in \mathbb{R}^2 : |x| < 1, x_1, x_2 > 0\}$, $S = [0, 1] \times \{0\}$ and

$$u_\varepsilon : \Omega \to \mathbb{R}^2, \quad x \mapsto \frac{[-x_2, x_1]^\top}{\varepsilon + |x|^2}.$$

(i) Show that $u_\varepsilon \in H(\text{div}; \Omega)$ and that there exists $\alpha_\varepsilon \in \mathbb{R}$ such that $\tilde{u}_\varepsilon = \alpha_\varepsilon u_\varepsilon \in H(\text{div}; \Omega)$ and $\|\tilde{u}_\varepsilon\|_{L^2(\Omega)} = 1$.

(ii) Show that $\|\tilde{u}_\varepsilon \cdot n\|_{L^1(S)}$ is unbounded as $\varepsilon \to 0$, and conclude that the trace operator for functions in $H(\mathrm{div}; \Omega)$ is not well-defined as an operator into $L^1(\partial\Omega)$ in general.

(iii) Why is the expression $\int_{\partial\Omega} u_\varepsilon \cdot n \, ds$ well-defined?

Exercise A.7.4 Let $p_D \in H^1(\Omega)$, $g \in L^2(\Omega)$, and $\sigma \in L^2(\Gamma_N)$. Let $p \in H^1(\Omega)$ be the weak solution of the Poisson problem, i.e., we have $p = p_D$ on Γ_D and

$$\int_\Omega \nabla p \cdot \nabla q \, dx = \int_\Omega gq \, dx + \int_{\Gamma_N} \sigma q \, ds$$

for all $q \in H^1_D(\Omega)$. Let $(u, p') \in H(\mathrm{div}; \Omega) \times L^2(\Omega)$ be the solution of the dual mixed formulation of the Poisson problem, i.e., we have $u \cdot n = \sigma$ on Γ_N and

$$\int_\Omega u \cdot v \, dx + \int_\Omega p' \, \mathrm{div}\, v \, dx = \langle v \cdot n, p_D \rangle_{\partial\Omega},$$

$$\int_\Omega q \, \mathrm{div}\, u \, dx = -\int_\Omega gq \, dx,$$

for all $(v, q) \in H_N(\mathrm{div}; \Omega) \times L^2(\Omega)$. Prove that $u = \nabla p$ and $p = p'$ and conclude that the dual mixed formulation is well-posed.

Exercise A.7.5 Let $T \subset \mathbb{R}^d$, $d = 1, 2, 3$, be a nondegenerate simplex. Let $S \subset \partial T$ be a side of T with outer unit normal n_S, and let $z \in T$ be the vertex of T opposite S. Prove that

$$d!|T| = |S|(z - x_S) \cdot n_S$$

for an arbitrary point $x_S \in S$.

Exercise A.7.6 For $v \in H(\mathrm{div}; T)$ and an affine diffeomorphism $\Phi_T : \widehat{T} \to T$, define

$$\hat{v} = (\det D\Phi_T)(D\Phi_T)^{-1}(v \circ \Phi_T).$$

Compute the divergence of $\hat{v}$, and show that for every side $\widehat{S} \subset \partial\widehat{T}$ and $S = \Phi_T(\widehat{S})$ we have

$$\int_S v \cdot n \, ds = \int_{\widehat{S}} \hat{v} \cdot \hat{n} \, d\hat{s}.$$

Exercise A.7.7 Let $v \in H^1(\omega)$ and define $\bar{v} = |\omega|^{-1} \int_\omega v \, dx$.

(i) Show that $\|\bar{v}\|_{L^2(\omega)} \le \|v\|_{L^2(\omega)}$.
(ii) Prove that

$$\|v - \bar{v}\|_{L^2(\omega)} \le c \operatorname{diam}(\omega) \|\nabla v\|_{L^2(\omega)},$$

with a constant $c > 0$ that is independent of v and $\operatorname{diam}(\omega)$.

Exercise A.7.8 Let $T \subset \mathbb{R}^d$ be a simplex, and let $\widehat{T} \subset \mathbb{R}^d$ be a reference simplex. Let S and $\widehat{S}$ be sides of T and $\widehat{T}$, respectively.

(i) Prove that there exists a constant $c > 0$, such that for all $\hat{v} \in H^1(\widehat{T})$ we have

$$\|\hat{v}\|_{L^2(\widehat{S})} \le c \|\hat{v}\|_{H^1(\widehat{T})}.$$

(ii) Prove that there exists a constant $c > 0$, such that for all $v \in H^1(T)$ we have

$$\|v\|_{L^2(S)} \le c\left(h_T^{-1/2} \|v\|_{H^1(T)} + h_T^{1/2} \|\nabla v\|_{L^2(T)}\right).$$

Exercise A.7.9 Assume that $\Omega \subset \mathbb{R}^d$, $\Gamma_D \subset \partial\Omega$, and $\Gamma_N = \partial\Omega \setminus \Gamma_D$ are such that the Poisson problem

$$-\Delta\phi = q \text{ in } \Omega, \quad \partial_n\phi|_{\Gamma_N} = 0, \quad \phi|_{\Gamma_D} = 0$$

is H^2-regular, i.e., there exists $c > 0$ such that $\|D^2\phi\|_{L^2(\Omega)} \le c\|q\|_{L^2(\Omega)}$ for every $q \in L^2(\Omega)$. Use the Fortin interpolant $\mathscr{I}_{\mathscr{R}T} : H^1(\Omega; \mathbb{R}^d) \to \mathscr{R}T_N^0(\mathscr{T}_h)$ to prove the discrete inf-sup condition for the bilinear form b in the dual mixed formulation of the Poisson problem.

Exercise A.7.10 Let $u \in H(\operatorname{div}; \Omega)$. Show that there exists a sequence $(u_\varepsilon)_{\varepsilon>0} \subset C^\infty(\overline{\Omega}; \mathbb{R}^d)$ such that

$$\|u - u_\varepsilon\|_{H(\operatorname{div};\Omega)} \le c_\varepsilon, \quad \|\nabla u_\varepsilon\|_{L^2(\Omega)} \le c\varepsilon^{-1}$$

with $c_\varepsilon \to 0$ as $\varepsilon \to 0$.

Quiz A.7.1 Decide for each of the following statements whether it is true or false. You should be able to justify your decision.

The mapping $(u, q) \mapsto \langle u \cdot n, q \rangle_{\partial\Omega}$ defines a bounded bilinear form on $H(\mathrm{div}; \Omega) \times H^1(\Omega)$	
A piecewise polynomial vector field belongs to $H^1(\Omega; \mathbb{R}^d)$ if its tangential component is continuous across interfaces	
The Raviart–Thomas finite element space consists of all piecewise affine vector fields with continuous normal components across sides	
The divergence operator defines a bijection between the spaces $\mathcal{R}T^0(\mathcal{T}_h)$ and $\mathcal{L}^0(\mathcal{T}_h)$	
There exists a bounded linear Fortin interpolant $\mathcal{I}_{\mathcal{R}T} : H_N(\mathrm{div}; \Omega) \to \mathcal{R}T_N^0(\mathcal{T}_h)$ for the dual mixed Poisson problem	

A.7.2 Approximation of the Stokes System

Exercise A.7.11 Show that the space $L_0^2(\Omega) = \{q \in L^2(\Omega) : \int_\Omega q \, \mathrm{d}x = 0\}$ is a Hilbert space which is isomorphic to the quotient space $L^2(\Omega)/\mathbb{R}$, resulting from identifying functions that coincide up to an additive constant, and which is equipped with the quotient space norm

$$\|q\|_{L^2(\Omega)/\mathbb{R}} = \inf_{c \in \mathbb{R}} \|q - c\|_{L^2(\Omega)}.$$

Exercise A.7.12 Specify the linear system of equations resulting from the discretization of the Stokes system with the MINI-element, and show that the degrees of freedom related to bubble functions can be eliminated by inverting a diagonal matrix.

Exercise A.7.13 Let $\mathcal{T}_h$ be a triangulation of $\Omega \subset \mathbb{R}^d$ with sides $\mathcal{S}_h$. For $q_h \in \mathcal{S}^{1,cr}(\mathcal{T}_h)$ and an inner side $S \in \mathcal{S}_h$, let $[q_h]|_S$ be the jump of q_h across S defined by

$$[q_h(x)] = \lim_{\varepsilon \to 0} \left(q_h(x + \varepsilon n_S) - q_h(x - \varepsilon n_S) \right)$$

for every x in the interior of S.

(i) Show that for every $q_h \in \mathcal{S}^{1,cr}(\mathcal{T}_h)$ and every $S \in \mathcal{S}_h$ we have

$$\int_S [q_h] \, \mathrm{d}s = 0.$$

(ii) Let $u_h \in \mathcal{R}T^0(\mathcal{T}_h)$ and $q_h \in \mathcal{S}^{1,cr}(\mathcal{T}_h)$. Show that

$$\int_\Omega q_h \operatorname{div} u_h \, dx = -\int_\Omega \nabla_\mathcal{T} q_h \cdot u_h \, dx + \int_{\partial\Omega} q_h u_h \cdot n \, ds.$$

Exercise A.7.14 Let $\mathcal{T}_h$ be a triangulation of $\Omega \subset \mathbb{R}^d$.

(i) For $u \in H_0^1(\Omega)$ let $G_h u \in \mathcal{S}_0^1(\mathcal{T}_h)$ be defined by

$$\int_\Omega \nabla G_h u \cdot \nabla v_h \, dx = \int_\Omega \nabla u \cdot \nabla v_h \, dx$$

for all $v_h \in \mathcal{S}_0^1(\mathcal{T}_h)$. Show that $G_h u$ is well defined with $\|\nabla G_h u\|_{L^2(\Omega)} \le \|\nabla u\|_{L^2(\Omega)}$. Prove that if the Poisson problem is H^2-regular in Ω, then we have

$$\|u - G_h u\|_{L^2(\Omega)} \le c_1 h \|\nabla u\|_{L^2(\Omega)}.$$

(ii) For $T \in \mathcal{T}_h$ with vertices $z_1, z_2, \ldots, z_{d+1} \in \mathcal{N}_h$ let $b_T = \varphi_{z_1} \varphi_{z_2} \ldots \varphi_{z_{d+1}}$. Given $v \in L^2(\Omega)$ and $T \in \mathcal{T}_h$, let $\lambda_T \in \mathbb{R}$ be such that

$$\int_T (\lambda_T b_T - v) \, dx = 0$$

and define $R_h v = \sum_{T \in \mathcal{T}_h} \lambda_T b_T$. Show that $\|\nabla R_h v\|_{L^2(\Omega)} \le c_2 h_{\min}^{-1} \|v\|_{L^2(\Omega)}$ for all $v \in L^2(\Omega)$, where $h_{\min} = \min_{T \in \mathcal{T}_h} \operatorname{diam}(T)$.
Hint: Show that $\|\nabla b_{T_{\mathrm{ref}}}\|_{L^2(T_{\mathrm{ref}})} \le c_2' \|b_{T_{\mathrm{ref}}}\|_{L^2(T_{\mathrm{ref}})}$ and use a transformation argument.
For $w \in H_0^1(\Omega; \mathbb{R}^d)$ let $I_F w = G_h w + R_h(w - G_h w)$, where G_h and R_h are applied component-wise. Show that there exists $c > 0$

$$\|\nabla I_F w\|_{L^2(\Omega)} \le c\big(h_{\max}/h_{\min}\big)\|\nabla w\|_{L^2(\Omega)}.$$

(iii) Show that for all $w \in H_0^1(\Omega; \mathbb{R}^d)$ and all $q_h \in \mathcal{S}^1(\mathcal{T}_h)$, we have

$$\int_\Omega q_h \operatorname{div}(w - I_F w) \, dx = 0.$$

(iv) Formulate sufficient conditions for the uniform validity of the inf-sup condition for the spaces $V_h = \mathcal{S}_0^1(\mathcal{T}_h)^d \oplus \mathcal{B}_h(\mathcal{T}_h)^d$ and $Q_h = \mathcal{S}^1(\mathcal{T}_h) \cap L_0^2(\Omega)$ for the approximation of the Stokes system.

Exercise A.7.15

(i) Show that $v_h \mapsto \|\nabla_{\mathscr{T}} v_h\|_{L^2(\Omega)}$ defines a norm on $\mathscr{S}_D^{1,cr}(\mathscr{T}_h)$.

(ii) Let $u_h \in \mathscr{S}_D^{1,cr}(\mathscr{T}_h)$. Show that in general $u_h|_{\Gamma_D} \neq 0$.

(iii) Show that there exists a uniquely defined function $u_h \in \mathscr{S}^{1,cr}(\mathscr{T}_h)$ with $u_h(x_S) = u_D(x_S)$ for all midpoints x_S of sides $S \in \mathscr{S}_h \cap \Gamma_D$ and

$$\int_\Omega \nabla_{\mathscr{T}} u_h \cdot \nabla_{\mathscr{T}} v_h \, dx = \int_\Omega f v_h \, dx + \int_{\Gamma_N} g v_h \, ds$$

for all $v_h \in \mathscr{S}_D^{1,cr}(\mathscr{T}_h)$.

(iv) Show that for all $v_h \in \mathscr{S}_D^1(\mathscr{T}_h)$, we have

$$\int_\Omega \nabla_{\mathscr{T}} (u - u_h) \cdot \nabla (u - u_h - v_h) \, dx = 0,$$

and conclude a best approximation result.

(v) Show that $\mathscr{S}_D^1(\mathscr{T}_h) \subset \mathscr{S}_D^{1,cr}(\mathscr{T}_h)$.

Exercise A.7.16 Let $\Gamma : X \times X \to \mathbb{R}$ be a symmetric bounded bilinear form that satisfies an inf-sup condition. For a subset $X_h \subset X$ and a bounded symmetric positive semidefinite bilinear form $c_h : X_h \times X_h \to \mathbb{R}$ let $\widehat{\Gamma}_h = \Gamma + c_h$. Assume that $\widehat{\Gamma}_h : X_h \times X_h \to \mathbb{R}$ satisfies an inf-sup condition uniformly in $h > 0$. For a given functional $\ell \in X'$, let $x \in X$ and $x_h \in X_h$ be the solutions of

$$\Gamma(x, y) = \ell(y), \quad \widehat{\Gamma}_h(x_h, y_h) = \ell(y_h)$$

for all $y \in X$ and $y_h \in X_h$, respectively. Prove that

$$c^{-1} \|x - x_h\| \leq \inf_{v_h \in X_h} \|x - v_h\|_X + c_h(v_h, v_h)^{1/2}.$$

Exercise A.7.17 For $v \in L^1(\Omega)$, the distributional gradient is the operator

$$\nabla v : C_0^\infty(\Omega; \mathbb{R}^d) \to \mathbb{R}, \quad \phi \mapsto - \int_\Omega v \operatorname{div} \phi \, dx.$$

Show that if $v \in H^1(\Omega)$, then the distributional gradient can be identified with the elementwise weak gradient $\nabla_{\mathscr{T}} v$, but not if $v \in H^1(\mathscr{T}_h)$.

Exercise A.7.18

(i) Show that the Crouzeix–Raviart element is a finite element $(T, \mathscr{P}_1(T), \mathscr{K})$ for an appropriate choice of the functionals $\mathscr{K}$, and determine the corresponding nodal basis functions.

(ii) Show that the space $\mathscr{S}^{1,cr}(\mathscr{T}_h)$ is an affine family.

(iii) Prove that the Crouzeix–Raviart element is not a C^0-element.

Exercise A.7.19 Let $u_h \in \mathscr{S}_0^{1,cr}(\mathscr{T}_h)$ be the Crouzeix–Raviart approximation of the Poisson problem $-\Delta u = f$ in Ω, $u|_{\partial\Omega} = 0$, defined by

$$\int_\Omega \nabla_{\mathscr{T}} u_h \cdot \nabla_{\mathscr{T}} v_h \, dx = \int_\Omega f v_h \, dx$$

for all $v_h \in \mathscr{S}^{1,cr}(\mathscr{T}_h)$. Let $a_h(u_h, v_h)$ and $b_h(v_h)$ be defined by the left- and right-hand sides of the identity. Assume that the Poisson problem is H^2-regular and prove that

$$\sup_{v_h \in \mathscr{S}_D^{1,cr}(\mathscr{T}_h)\setminus\{0\}} \frac{|a_h(u, v_h) - b_h(v_h)|}{\|\nabla_{\mathscr{T}} v_h\|} \le ch\|D^2 u\|_{L^2(\Omega)}.$$

Deduce an error estimate with the second Strang lemma.

Hint: Use that $\int_S \llbracket v_h \rrbracket \, ds = 0$ for every interior side $S \in \mathscr{S}_h$ and $v_h|_{T_\pm}(x) = v_h(x_S) + \nabla v_h|_{T_\pm} \cdot (x - x_S)$ on the neighboring elements $T_\pm$ to S and with $x_S \in S$.

Exercise A.7.20 Assume that $\mathscr{T}_h$ is a quasiuniform triangulation of $\Omega \subset \mathbb{R}^d$, and let $V_h = \mathscr{S}_0^k(\mathscr{T}_h)^d$ and $Q_h = \mathscr{S}^{k-1}(\mathscr{T}_h) \cap L_0^2(\Omega)$, for $k \ge 2$.

(i) Show that

$$\sup_{v_h \in V_h\setminus\{0\}} \frac{\int_\Omega q_h \operatorname{div} v_h \, dx}{\|\nabla v_h\|_{L^2(\Omega)}} \ge c_1 h \frac{\int_\Omega \nabla q_h \cdot v_h \, dx}{\|v_h\|_{L^2(\Omega)}}.$$

(ii) Use the Clément quasi-interpolant to prove that

$$\sup_{v_h \in V_h\setminus\{0\}} \frac{\int_\Omega q_h \operatorname{div} v_h \, dx}{\|\nabla v_h\|_{L^2(\Omega)}} \ge \beta' \|q_h\|_{L^2(\Omega)} - c_2 h \|\nabla q_h\|_{L^2(\Omega)}.$$

(iii) Assume that for every $q_h \in Q_h$, there exists $v_h \in V_h$ such that

$$\int_\Omega \nabla q_h \cdot v_h \, dx \ge c_3 \|\nabla q_h\|_{L^2(\Omega)}^2, \quad \|v_h\|_{L^2(\Omega)} \le c_4 \|\nabla q_h\|_{L^2(\Omega)},$$

and deduce the inf-sup condition for the spaces V_h and Q_h.

(iv) Try to construct $v_h \in V_h$ with the assumed properties.

Quiz A.7.2 Decide for each of the following statements whether it is true or false. You should be able to justify your decision.

The bilinear form b satisfies an inf-sup condition if and only if $-b$ satisfies an inf-sup condition
The MINI-element satisfies an inf-sup condition if $\Omega \subset \mathbb{R}^2$ is convex and $\Gamma_D = \partial\Omega$
The P_1-P_0 method defines a nonconforming stable Stokes element
If a Crouzeix–Raviart function is continuous at two distinct points on a side of a triangle, then it is continuous across the side
We have dim $\mathscr{S}^{1,cr}(\mathscr{T}_h) = |\mathscr{S}_h|$

A.7.3   Convection-Dominated Problems

Exercise A.7.21 Let $\Omega = (0, L_1) \times (0, L_2) \times (-L_3, L_3)$ and let $b = 10\text{m/s}\,[1,\,0,\,0]^T$ be the velocity field of the wind in Ω. Smoke is released from a chimney at $(0, \ell_{ch}, 0)$ with ℓ_{ch} being the height of the chimney, cf. Fig. A.7. Let $c : [0, T] \times \Omega \to \mathbb{R}$ be the smoke concentration with units $[c] = \text{g/m}^3$.

  (i) Explain the following principle:

$$\int_\omega c(t + \delta t, x + b\delta t)\,dx = \int_\omega c(t, x)\,dx - \int_t^{t+\Delta t} \int_{\partial\omega} q(s) \cdot n(s)\,ds$$

   for a control volume $\omega \subset \Omega, t > 0$ and δt sufficiently small. Make use of Fick's law, i.e., $q = -\nu\nabla c$, where $\nu = 1.5 \cdot 10^{-5}\text{m}^2/\text{s}$ is the diffusion coefficient for carbon dioxide in air.
 (ii) Deduce a partial differential equation for c. Non-dimensionalize the equation and explain why the process is convection dominated.
(iii) Derive a two-dimensional simplification by introducing

$$\bar{c}(x_1, x_2) = \int_{-L_3}^{L_3} c(x_1, x_2, x_3)\,dx_3,$$

   and formulate appropriate boundary conditions on $\partial\Omega$.

Fig. A.7 Smoke released from a chimney and transported by a wind field

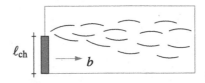

ℓ_{ch}

b

(iv) Justify the assumption $c(t + \delta t, x) = c(t, x)$ for $x \in \Omega$ and $t \in [0, T]$, and deduce a steady state convection-diffusion equation.

Exercise A.7.22 Let $\Omega = (0, 1)^2$ and define $b(x) = (\sin(\phi), -\cos(\phi))$ for $x = r(\cos(\phi), \sin(\phi)) \in \Omega$. Show that div $b = 0$ and construct a solution of the equation $b \cdot \nabla u = 0$ in Ω subject to a boundary condition defined by $u_D(x_1, x_2) = x_2$ imposed on a suitable subset of $\partial\Omega$.

Exercise A.7.23 Let $b \in H_N(\text{div}; \Omega) \cap L^\infty(\Omega; \mathbb{R}^d)$ with div $b = 0$. Show that the bilinear form

$$c(u, v) = \int_\Omega b \cdot (\nabla u) v \, dx$$

is bounded and skew-symmetric on $H_D^1(\Omega)$.

Exercise A.7.24 For $b \in H_N(\text{div}; \Omega) \cap L^\infty(\Omega; \mathbb{R}^d)$ with div $b = 0$ and $u_D \in C(\partial\Omega)$, we consider the equation

$$-\varepsilon \Delta u + b \cdot \nabla u = 0 \text{ in } \Omega, \quad u|_{\partial\Omega} = u_D.$$

(i) Prove by considering $\tilde{u} = \min\{u, c\}$ for an appropriate choice of c that $u(x) \leq \max_{y \in \partial\Omega} u_D(y)$ for $x \in \Omega$.
(ii) Prove $\|D^2 u\|_{L^2(\Omega)} \leq c\varepsilon^{-2}$ provided that the Poisson problem is H^2-regular.

Exercise A.7.25 Compare the error estimates for the standard Galerkin approximation and the streamline-diffusion method.

Exercise A.7.26 For grid points $x_i = i/M$, $i = 0, 1, \ldots, M$, and $a, b \in \mathbb{R}$, consider the scheme

$$-\varepsilon \frac{U_{i-1} - 2U_i + U_{i+1}}{h^2} + \frac{U_{i+1} - U_{i-1}}{2h} = 0, \quad U_0 = a, \quad U_M = b,$$

for $i = 1, 2, \ldots, M-1$ with $h = 1/M$. Rewrite the scheme in the form $\widehat{U}_{i+1} = A\widehat{U}_i$ with a matrix $A \in \mathbb{R}^{2 \times 2}$ and $\widehat{U}_{i+1} = [U_{i+1}, U_i]^\top$, and construct the solution in terms of the eigenvalues of A.

Exercise A.7.27

(i) Devise an upwinding finite difference discretization of the equation

$$-\varepsilon u'' + bu' = 0, \quad u(0) = \alpha, \ u(1) = \beta,$$

and prove a discrete maximum principle.

(ii) Show that the upwinding discretization of the convection term bu' can be interpreted as a symmetric discretization of the modified term $bu' + \delta u''$ with an appropriate parameter δ.

Exercise A.7.28 Assume that the nodes z, y are the endpoints of an inner edge $S = T_1 \cap T_2$ in a Delaunay triangulation. Show that for the associated nodal basis functions $\varphi_z, \varphi_y \in \mathscr{S}^1(\mathscr{T}_h)$, we have

$$\int_\Omega \nabla \varphi_z \cdot \nabla \varphi_y \, dx = -\frac{|m_{T_1} - m_{T_2}|}{|z - y|} = -\frac{1}{2}(\cot \alpha_1 + \cot \alpha_2),$$

where m_1, m_2 are the circumcenters of T_1, T_2, and α_1, α_2 are the inner angles of T_1, T_2 that are opposite S.

Exercise A.7.29

(i) Construct the Voronoi diagram associated with the given points $(x_j)_{j=1,\dots,8} \subset \overline{\Omega} = [0, 5] \times [0, 2]$ shown in Fig. A.8.
(ii) Discuss the regularity of the diagram.
(iii) Construct the Delaunay triangulation of the Voronoi diagram and verify its weak acuteness.

Exercise A.7.30 Let $b \in C^1(\overline{\Omega}; \mathbb{R}^d)$ with $\operatorname{div} b = 0$ and $\varepsilon, \alpha > 0$, and consider the boundary value problem

$$-\varepsilon \Delta u + b \cdot \nabla u + \alpha u = f \text{ in } \Omega, \quad u|_{\partial \Omega} = 0.$$

Show that for the P_1-finite element approximation of the problem, we have the error estimate

$$\varepsilon^{1/2} \|\nabla e\|_{L^2(\Omega)} + \alpha^{1/2} \|e\|_{L^2(\Omega)} \le c\big(\varepsilon^{1/2} + \varepsilon^{-1/2} h \|b\|_{L^\infty(\Omega)} + \alpha^{1/2} h\big) h \|D^2 u\|_{L^2(\Omega)}$$

for $e = u - u_h$. Discuss extreme parameters α, ε for which the Galerkin approximation provides useful approximations.

Fig. A.8 Points $(x_j)_{j=1,\dots,5}$
the define a Voronoi diagram

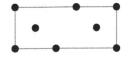

Quiz A.7.3 Decide for each of the following statements whether it is true or false. You should be able to justify your decision.

A convection-diffusion problem can be formulated as an equivalent minimization problem	
Boundary layers occur when a Dirichlet boundary is imposed on an outflow boundary	
The solution of a convection-dominated partial differential equation is uniformly bounded as to $\varepsilon \to 0$ in $H^1(\Omega)$ but not in $H^2(\Omega)$	
The streamline-diffusion method finite element methods by modifying the diffusion term	
The finite volume method allows for a generalization of upwinding techniques to higher-dimensional problems	

A.7.4 Discontinuous Galerkin Methods

Exercise A.7.31 Let $\mathcal{T}_h$ be the triangulation of $\Omega = (-1, 1)^2$ consisting of four halved squares with diagonals parallel to the vector $(1, 1)$. Compute the jumps and averages of the function

$$v(x, y) = \begin{cases} x^2 y, & x \geq 0, y > 0, \\ x(1 - y), & x < 0, y \geq 0, \\ x^3, & x \leq 0, y < 0, \\ 1 - y, & x > 0, y \leq 0. \end{cases}$$

Exercise A.7.32

(i) Prove that if $v \in H^1(\Omega)$, then for every interior side $S \in \mathcal{S}_h$ we have that

$$[\![v]\!]|_S = 0, \quad \{v\}|_S = v|_S.$$

(ii) Show that if $v \in H^2(\mathcal{T}_h)$ such that $\nabla v \in H(\text{div}; \Omega)$, then

$$[\![\nabla v \cdot n_S]\!]|_S = 0,$$

where n_S is extended constantly to a neighborhood of S.

Exercise A.7.33 Let $u_D = \tilde{u}_D$ for $\tilde{u}_D \in H^1(\Omega) \setminus H^2(\Omega)$. Devise a discontinuous Galerkin method for approximating the Poisson problem $-\Delta u = f$ in Ω with $u|_{\partial \Omega} = u_D$.

Exercise A.7.34 Let $\beta_S > 0$ for every $S \in \mathscr{S}_h$. Prove that

$$\|v\|_{\mathrm{dG}}^2 = \|\nabla_{\mathscr{T}} v\|_{L^2(\Omega)}^2 + \sum_{S \in \mathscr{S}_h} \frac{\beta_S}{h_S^\gamma} \int_S |[\![v]\!]|^2 \, ds$$

is a norm on $H^1(\mathscr{T}_h)$.

Exercise A.7.35

(i) Show that the bilinear form $a_{\mathrm{dG}} : \mathscr{S}^{k,\mathrm{dG}}(\mathscr{T}_h) \times \mathscr{S}^{k,\mathrm{dG}}(\mathscr{T}_h) \to \mathbb{R}$ defined by

$$a_{\mathrm{dG}}(u_h, v_h) = \int_\Omega \nabla_{\mathscr{T}} u_h \cdot \nabla_{\mathscr{T}} v_h \, dx + \sum_{S \in \mathscr{S}_h} \int_S \{\nabla u_h \cdot n_S\} [\![v_h]\!] \, ds$$

$$+ \sigma \sum_{S \in \mathscr{S}_h} \int_S \{\nabla v_h \cdot n_S\} [\![u_h]\!] \, ds + \sum_{S \in \mathscr{S}_h} \frac{\beta_S}{h_S^\gamma} \int_S [\![u_h]\!][\![v_h]\!] \, ds$$

is bounded with respect to the norm $\|\!|\cdot|\!\|_{\mathrm{dG}}$.
(ii) Show that a_{dG} is symmetric if and only if $\sigma = 1$.

Exercise A.7.36 Show that $u \in H_0^1(\Omega) \cap H^2(\mathscr{T}_h)$ is a weak solution of the Poisson problem $-\Delta u = f$, if and only if

$$a_{\mathrm{dG}}(u, v) = \ell(v)$$

for all $v \in H^2(\mathscr{T}_h)$ and $\nabla u \in H(\mathrm{div}; \Omega)$.

Exercise A.7.37

(i) Show that for $v \in H^1(\mathscr{T}_h)$ and a side $S \subset \partial T$, we have

$$\|v\|_{L^2(S)} \le c_{\mathrm{Tr}}\big(h_S^{1/2}\|\nabla v\|_{L^2(T)} + h_S^{-1/2}\|v\|_{L^2(T)}\big).$$

(ii) Prove that for $v_h \in \mathscr{S}^{k,\mathrm{dG}}(\mathscr{T}_h)$, we have

$$\|v_h\|_{L^2(S)} \le c_{\mathrm{Tr},k} h_S^{-1/2}\|v_h\|_{L^2(T)},$$

where $c_{\mathrm{Tr},k} \to \infty$ as $k \to \infty$.

Exercise A.7.38 Assume that the Poisson problem with homogeneous Dirichlet conditions on $\partial\Omega$ is H^2-regular. Use the representation

$$\|v\|_{L^2(\Omega)} = \sup_{q \in L^2(\Omega) \setminus \{0\}} \frac{\int_\Omega vq \, dx}{\|q\|_{L^2(\Omega)}}$$

to prove that there exists $c_{P,dG} > 0$ such that

$$\|v_h\|_{L^2(\Omega)} \le c_{P,dG} \|\|v_h\|\|_{dG}$$

for all $v_h \in \mathscr{S}^{1,dG}(\mathscr{T}_h)$.

Exercise A.7.39 Let A_{dG} be the matrix representing the symmetric bilinear form a_{dG} in the basis $\big(\varphi_{T,z} : T \in \mathscr{T}_h, z \in \mathscr{N}_h \cap T\big)$ with $\varphi_{T,z}$ defined with the P_1-hat functions by

$$\varphi_{T,z}(x) = \begin{cases} \varphi_z(x) & \text{if } x \in T, \\ 0 & \text{otherwise.} \end{cases}$$

Show that $\operatorname{cond}_2(A_{dG}) = \mathscr{O}(h^{-1-\gamma})$.

Exercise A.7.40 Let $\mathscr{T}_h$ be a fixed triangulation and let $u_h^\beta \in \mathscr{S}^{1,dG}(\mathscr{T}_h)$ be the solution corresponding to a discontinuous Galerkin method for the Poisson problem with $\beta_s = \beta$ for all $S \in \mathscr{S}_h$. Show that as $\beta \to \infty$, the sequence $(u_h^\beta)_{\beta>0}$ converges to a function $u_h \in \mathscr{S}_0^1(\mathscr{T}_h)$ that is the continuous Galerkin approximation.

Quiz A.7.4 Decide for each of the following statements whether it is true or false. You should be able to justify your decision.

If $v \in H^1(\mathscr{T}_h)$ with $[\![v]\!]\|_S = 0$ for all $S \in \mathscr{S}_h$, then $v \in H_0^1(\Omega)$	
The symmetric interior penalty discontinuous Galerkin method is unconditionally well-posed	
The nonsymmetric interior penalty discontinuous Galerkin method is unconditionally well-posed	
The average $\{v\}\|_S$ is well defined if $v \in H^{1/2}(\mathscr{T}_h)$ and the jump $[\![v]\!]\|_S$ is well defined if $v \in H^{1/2+\varepsilon}(\mathscr{T}_h)$.	
We have $\mathscr{S}_0^1(\mathscr{T}_h) \subset \mathscr{S}^{1,dG}(\mathscr{T}_h)$ and $\mathscr{S}_0^{1,cr}(\mathscr{T}_h) \subset \mathscr{S}^{1,dG}(\mathscr{T}_h)$	

A.7.5 Projects

Project A.7.1 Implement the pressure-stabilized $P1$-$P1$ method for the Stokes problem and investigate the experimental convergence using the model solution

$$u(x,y) = \pi \begin{bmatrix} \sin(2\pi y)\sin^2(\pi x) \\ -\sin(2\pi x)\sin^2(\pi y) \end{bmatrix},$$

$$p(x,y) = \cos(\pi x)\sin(\pi y)$$

in the square $\Omega = (-1, 1)^2$ with homogeneous Dirichlet boundary conditions on $\Gamma_D = \partial\Omega$ for u.

Project A.7.2 Determine experimental convergence rates for approximating the Stokes problem with the nonconforming Crouzeix–Raviart method for the model solution

$$u(x, y) = \pi \begin{bmatrix} \sin(2\pi y) \sin^2(\pi x) \\ -\sin(2\pi x) \sin^2(\pi y) \end{bmatrix},$$

$$p(x, y) = \cos(\pi x) \sin(\pi y)$$

in the square $\Omega = (-1, 1)^2$ with homogeneous Dirichlet boundary conditions on $\Gamma_D = \partial\Omega$ for u.

Project A.7.3 Consider the dual mixed formulation of the Poisson problem and its approximation using the Raviart–Thomas finite element method. Determine experimental convergence rates for both variables on sequences of uniformly refined triangulations for the domain $\Omega = (0, 1)^2$ with exact solution $u(x, y) = \sin(\pi x) \sin(\pi y)$, and the domain $\Omega = (-1, 1)^2 \setminus ([-1, 0] \times [0, 1])$ with the exact solution $u(r, \phi) = r^{2/3} \sin(2\phi/3)$ in polar coordinates.

Project A.7.4 Implement the $P1$-$P0$ finite element method for the dual mixed formulation of the Poisson problem and demonstrate experimentally that it is ill-posed in general. Try to stabilize the method by incorporating an appropriate penalty term.

Project A.7.5 Determine a function $f \in L^2(\Omega; \mathbb{R}^2)$ with $\Omega = (-1, 1)^2$ so that the exact solution $u \in H_0^1(\Omega; \mathbb{R}^2)$ of the variational formulation

$$\int_\Omega \nabla u : \nabla v \, dx + \varepsilon^{-2} \int_\Omega \operatorname{div} u \operatorname{div} v \, dx = \int_\Omega f \cdot v \, dx$$

for all $v \in H_0^1(\Omega; \mathbb{R}^2)$ is given by

$$u(x_1, x_2) = \begin{bmatrix} \sin(2\pi x_2) \sin^2(\pi x_1) \\ -\sin(2\pi x_1) \sin^2(\pi x_2) \end{bmatrix}.$$

Introduce the variable $p = \varepsilon^{-2} \operatorname{div} u$, and rewrite the problem as a saddle-point formulation with penalty term. Discretize it with a nonconforming method and investigate the experimental convergence of approximations for the parameters $\varepsilon = 10^{-j}, j = 1, 2$.

Project A.7.6 A simple mathematical description of the release of smoke from a chimney and its distribution in the environment leads to the convection dominated equation

$$-\nu\Delta c + b \cdot \nabla c = 0$$

for the smoke concentration c, the diffusion coefficient $\nu = 1.5 \cdot 10^{-5} \mathrm{m}^2/\mathrm{s}$ of carbon dioxide in air, and the velocity field $b = [10, 0, 0]^\top \mathrm{m/s}$. Assume that the chimney is 50 m high, and simulate a two-dimensional model reduction of the problem with appropriate boundary conditions up to a height of 200 m and a distance of 1000 m in the direction of the wind. Compare a direct approximation with a stabilized one.

Project A.7.7 Experimentally determine the experimental convergence rate of different discontinuous Galerkin methods for the Poisson problem $-\Delta u = f$ in $\Omega = (0, 1)^2$ with boundary condition $u|_{\partial\Omega} = 0$ on a sequence of uniformly refined triangulations for the exact solution

$$u(x, y) = \sin(\pi x) \sin(\pi y).$$

Investigate also dependence on the parameter γ.

Project A.7.8 Incorporate the treatment of convective terms in implementating the discontinuous Galerkin method and test its performance with meaningful experiments.

A.8 Applications

A.8.1 Linear Elasticity

Exercise A.8.1

(i) Show that $SO(d) = \{Q \in \mathbb{R}^{d \times d} : Q^\top Q = I, \det Q = 1\}$ is a $d(d-1)/2$-dimensional submanifold in $\mathbb{R}^{d \times d}$.

(ii) Prove that the tangent space of $SO(d)$ at the identity matrix is given by $T_I SO(d) = so(d) = \{U \in \mathbb{R}^{d \times d} : U^\top + U = 0\}$.

Exercise A.8.2 Assume that $v \in H^1(\Omega; \mathbb{R}^d)$ for $\Omega \subset \mathbb{R}^d$ with $d = 2, 3$ satisfies $\nabla v(x) \in SO(d)$ for almost every $x \in \Omega$.

(i) Show that $\operatorname{div} \operatorname{Cof} \nabla w = 0$ for $w \in H^1(\Omega; \mathbb{R}^d)$, where $\operatorname{Cof} A = (\det A)A^{-\top}$, and conclude that every component of v is harmonic.

(ii) Show that $\Delta v = 0$ implies $\Delta(|\nabla v|^2 - d) = 2|D^2 v|^2$ and deduce that ∇v is constant.

Exercise A.8.3 Construct a solution $u(x) = Ax$ with a diagonal matrix $A \in \mathbb{R}^{3\times3}$ of the Navier–Lamé equations in the cylinder domain $(-L/2, L/2) \times B_r(0) \subset \mathbb{R}^3$ with $\Gamma_D = \emptyset$, $\Gamma_N = \partial\Omega$, and

$$g(x) = \begin{cases} \pm e_1, & x_1 = \pm L/2, \\ 0, & -L/2 < x_1 < L/2, \end{cases}$$

for $x = (x_1, x_2, x_3) \in \Gamma_N$. Determine the ratio between elongation and radial compression and sketch the solution for different Lamé constants (λ, μ).

Exercise A.8.4 For $\lambda, \mu > 0$ let $\mathbb{C} : \mathbb{R}^{d\times d}_{\text{sym}} \to \mathbb{R}^{d\times d}_{\text{sym}}$ be defined by $\mathbb{C}A = \lambda \operatorname{tr}(A)I + 2\mu A$. Show that $\mathbb{C}$ is invertible with

$$\mathbb{C}^{-1}B = \frac{1}{2\mu}\left(B - \frac{\lambda}{d\lambda + 2\mu} \operatorname{tr}(B)I\right).$$

Exercise A.8.5 We consider the elastic deformation of a solid occupying the domain $\Omega \subset \mathbb{R}^3$.

(i) Assume that the expected behavior of the solid is such that for $\varepsilon = \varepsilon(u) \in \mathbb{R}^{3\times3}$, we have $\varepsilon_{i3} = 0$ for $i = 1, 2, 3$. Derive a simplified two-dimensional model.

(ii) Assume that the expected behavior of the solid is such that for $\sigma = \mathbb{C}\varepsilon(u) \in \mathbb{R}^{3\times3}$, we have $\sigma_{i3} = 0$ for $i = 1, 2, 3$. Derive a simplified two-dimensional model.

(iii) Discuss model situations for which the simplifications apply.

Exercise A.8.6 Assume that $u \in H^1_D(\Omega; \mathbb{R}^d)$ is minimal for

$$I(u) = \frac{1}{2}\int_\Omega \mathbb{C}\varepsilon(u) : \varepsilon(u)\,dx - \int_\Omega f \cdot u\,dx - \int_{\Gamma_N} g \cdot u\,ds.$$

Prove that u is a weak solution of the Navier–Lamé equations.

Exercise A.8.7

(i) Prove that for $v \in C^2(\overline{\Omega}; \mathbb{R}^d)$ we have

$$2\operatorname{div}\varepsilon(v) = \Delta v + \nabla \operatorname{div} v.$$

(ii) Assume that $v(x) = Ax + b$ for $x \in \Omega$ with a skew-symmetric matrix $A \in \mathbb{R}^{d \times d}$ such that $v|_{\Gamma_D} = 0$. Show that $v = 0$.

Exercise A.8.8 Devise and analyze a numerical method for approximating the problem of determining $(\sigma, u) \in L^2(\Omega; \mathbb{R}^{d \times d}_{\mathrm{sym}}) \times H^1_D(\Omega; \mathbb{R}^d)$ such that

$$\int_\Omega \mathbb{C}^{-1}\sigma : \tau \, \mathrm{d}x - \int_\Omega \tau : \varepsilon(u) \, \mathrm{d}x = 0,$$

$$\int_\Omega \sigma : \varepsilon(v) \, \mathrm{d}x = \ell(v),$$

for all $(\tau, v) \in L^2(\Omega; \mathbb{R}^{d \times d}_{\mathrm{sym}}) \times H^1_D(\Omega; \mathbb{R}^d)$. Discuss the dependence of the approximation error on the Lamé constants.

Exercise A.8.9 Consider the Navier–Lamé equations with $\Gamma_D = \partial\Omega$, and let $u_h \in \mathscr{S}^{1,cr}_D(\mathscr{T}_h)^d$ satisfy

$$\mu \int_\Omega \nabla_{\mathscr{T}} u_h : \nabla_{\mathscr{T}} v_h \, \mathrm{d}x + (\mu + \lambda) \int_\Omega \mathrm{div}_{\mathscr{T}} \, u_h \, \mathrm{div}_{\mathscr{T}} \, v_h \, \mathrm{d}x = \int_\Omega f \cdot v_h \, \mathrm{d}x$$

for all $v_h \in \mathscr{S}^{1,cr}_D(\mathscr{T}_h)^d$. Show that the approximations converge to the solution of the Navier–Lamé equations as $h \to 0$.

Exercise A.8.10 Assume that a discrete Korn inequality holds on $\mathscr{S}^{1,cr}_D(\mathscr{T}_h)^d$.

(i) Show that there exists a unique $u_h \in \mathscr{S}^{1,cr}_D(\mathscr{T}_h)^d$ with

$$2\mu \int_\Omega \varepsilon_{\mathscr{T}}(u_h) : \varepsilon_{\mathscr{T}}(v_h) \, \mathrm{d}x + \lambda \int_\Omega \mathrm{div}_{\mathscr{T}} \, u_h \, \mathrm{div}_{\mathscr{T}} \, v_h \, \mathrm{d}x = \ell(v_h)$$

for all $v_h \in \mathscr{S}^{1,cr}_D(\mathscr{T}_h)^d$.

(ii) Assume that the exact solution of the Navier–Lamé equations satisfies $\|u\|_{H^2(\Omega)} + \lambda\|\nabla \mathrm{div}\, u\|_{L^2(\Omega)} \le c_{NL}\|f\|_{L^2(\Omega)}$. Prove that

$$\|\mathbb{C}^{1/2}\varepsilon_{\mathscr{T}}(u - u_h)\|_{L^2(\Omega)} \le c_{cr}\big((2\mu)^{1/2} + \lambda^{-1/2}\big)h\|f\|_{L^2(\Omega)}.$$

Quiz A.8.1 Decide for each of the following statements whether it is true or false. You should be able to justify your decision.

A discrete Korn inequality is used in deriving the error estimate $\|\mathbb{C}^{1/2}\nabla_{\mathscr{S}}(u - u_h)\|_{L^2(\Omega)} \le ch\|D^2u\|_{L^2(\Omega)}$
The sets $\mathbb{R}^{d\times d}_{\text{sym}}$ and $so(d)$ are orthogonal with respect to the scalar product $A : B = \text{tr}\, A^\top B$
For all matrices $A, B \in \mathbb{R}^{d\times d}$ we have $A : B^\top = A^\top : B = B^\top : A$
If the stresses $\sigma = \mathbb{C}\varepsilon(u)$ vanish, then the displacement u is a linearized rigid body motion
Crouzeix–Raviart finite elements avoid incompressibility locking but may lead to ill-posed discrete problems

A.8.2 Plate Bending

Exercise A.8.11 Assume that the Poisson problem is H^2-regular in ω, and let $x_0 \in \bar{\omega}$. Show that there exists a unique weak solution $u \in H^2(\omega)$ of the boundary value problem

$$\Delta^2 u = \delta_{x_0} \text{ in } \omega, \quad u = \nabla u = 0 \text{ on } \partial\omega,$$

where $\delta_{x_0} : H^2(\omega) \to \mathbb{R}$ is defined by $\delta_{x_0}(v) = v(x_0)$ for all $v \in H^2(\omega)$.

Exercise A.8.12

(i) Prove that $V = \{v \in H^2(\omega) : v = \nabla v = 0 \text{ on } \partial\omega\}$ is a closed subspace of $H^2(\omega)$.

(ii) Show that V coincides with the closure of $C_0^\infty(\omega)$ with respect to the Sobolev norm $\|\cdot\|_{H^2(\omega)}$.

Exercise A.8.13 Show that $u \in H^2(\omega) \cap H_0^1(\omega)$ is a minimizer for the Kirchhoff functional

$$I_{Ki}(u) = \frac{\theta}{2}\int_\omega |\Delta u|^2 \, dx + \frac{1-\theta}{2}\int_\omega |D^2u|^2 \, dx - \int_\omega fu \, dx$$

if and only if $a(u, v) = \ell(v)$ for all $v \in H^2(\omega) \cap H_0^1(\omega)$ with

$$a(u, v) = \int_\omega \Delta u \Delta v \, dx + (1 - \theta)\int_\omega 2\partial_1\partial_2 u \partial_1\partial_2 v - \partial_1^2 u \partial_2^2 v - \partial_1^2 v \partial_2^2 u \, dx,$$

$$\ell(v) = \int_\omega fv \, dx.$$

Exercise A.8.14 Show that for $u \in H^4(\omega)$ and $v \in H^2(\omega)$, we have

$$\int_\omega \Delta u \, \Delta v \, dx = \int_\omega \Delta^2 u \, v \, dx - \int_{\partial\omega} (\partial_n \Delta u) \, v \, ds + \int_{\partial\omega} \Delta u \, \partial_n v \, ds.$$

Exercise A.8.15 Let $\phi \in L^2(\partial\omega)$ and assume that

$$\int_{\partial\omega} \phi \partial_n v \, ds = 0$$

for all $v \in H^2(\omega)$. Show that $\phi = 0$ on $\partial\omega$. Assume first that for $x_0 \in \partial\omega$, there exists $\varepsilon > 0$ such that $B_\varepsilon(x_0) \cap \partial\omega \subset \mathbb{R} \times \{0\}$.

Exercise A.8.16 Let $f = 1$ in $\omega = B_1(0) \subset \mathbb{R}^2$, and $\theta \in (0, 1)$. Determine the solutions of

$$\Delta^2 u = f \text{ in } \omega, \quad u = \Delta u = 0 \text{ on } \partial\omega,$$

and

$$\Delta^2 u = f \text{ in } \omega, \quad u = \Delta u + (1 - \theta)\kappa \partial_n u = 0 \text{ on } \partial\omega.$$

Discuss their difference and the approximation of the problems on polygonal domains $\omega_h \subset \omega$.

Exercise A.8.17 Consider the boundary value problem

$$\Delta^2 u = f \text{ in } \omega, \quad u = \Delta u = 0 \text{ on } \partial\omega.$$

Introduce the variable $v = \Delta u$ and formulate and analyze an equivalent saddle-point formulation under appropriate conditions on ω. Discuss error estimates for the approximation of the saddle-point formulation with a low order finite element method.

Exercise A.8.18 Let $u_D = \tilde{u}_D|_{\partial\omega}$ for some $\tilde{u}_D \in H^2(\omega)$ and consider the minimization of the functionals

$$I_1(u) = \frac{1}{2} \int_\omega |\nabla u|^2 \, dx, \quad I_2(u) = \frac{1}{2} \int_\omega |\Delta u|^2 \, dx,$$

subject to $u|_{\partial\omega} = u_D$.

(i) Show that the problems define under appropriate assumptions on u_D surfaces of minimal area and minimal total curvature.
(ii) Derive the Euler–Lagrange equations for both minimization problems.
(iii) Compute the solutions for $\omega = B_1(0)$ and $u_D(\theta) = \sin(\theta)$, $\theta \in [0, 2\pi]$.

Exercise A.8.19 Let $a : H^2(\omega) \times H^2(\omega) \to \mathbb{R}$ be the bilinear form associated with the Kirchhoff bending energy, and let $\ell \in H^2(\omega)'$.

(i) Show that there exists a unique function $u_h \in \mathscr{S}_0^{5,1}(\mathscr{T}_h)$ such that

$$a(u_h, v_h) = \ell(v_h)$$

for all $v_h \in \mathscr{S}_0^{5,1}(\mathscr{T}_h)$.

(ii) Assume that the exact solution of the plate bending problem satisfies $u \in H^4(\omega)$, and prove that we have

$$\|u - u_h\|_{H^2(\omega)} \leq c_{\mathrm{Arg}} h^2 \|u\|_{H^4(\omega)}.$$

Exercise A.8.20 Let $T \subset \mathbb{R}^2$ be a triangle with vertices z_0, z_1, z_2, midpoint x_T and sides S_0, S_1, S_2. Let K_0, K_1, K_2 be the subtriangles with vertex x_T and sides S_0, S_1, S_2, respectively. For $i = 0, 1, 2$ and $\alpha \in \mathbb{N}_0^2$ with $|\alpha| \leq 1$, and $v \in C^1(T)$, define

$$\chi_{i,\alpha}(v) = \partial^\alpha v(z_i), \quad \chi_{i,n}(v) = \nabla v(x_{S_i}) \cdot n_{S_i},$$

with midpoints x_{S_i} and normals n_{S_i} for the sides S_i, $i = 0, 1, 2$. Show that if $v \in C^1(T)$ with $v|_{K_i} \in \mathscr{P}_3(K_i)$, $i = 0, 1, 2$, and

$$\chi_{i,\alpha}(v) = 0, \quad \chi_{i,n}(v) = 0$$

for $i = 0, 1, 2$, and $\alpha \in \mathbb{N}_0^2$ with $|\alpha| \leq 1$, then we have $v = 0$.

Quiz A.8.2 Decide for each of the following statements whether it is true or false. You should be able to justify your decision.

Plate bending refers to a model reduction of linear elasticity corresponding to small thickness $t > 0$	
The bending problem with simple support boundary conditions is unconditionally well-posed	
We have that $\mathscr{S}_0^1(\mathscr{T}_h) \subset H^2(\omega) \cap H_0^1(\omega)$	
If a piecewise polynomial function belongs to $H^2(\omega)$, then its derivatives are continuous	
The Argyris element leads to quadratic convergence rates if $u \in H^4(\omega)$	

A.8.3 Electromagnetism

Exercise A.8.21 Let $w, \phi \in C^1(\overline{\Omega}; \mathbb{R}^3)$.

(i) Show that we have

$$\operatorname{div}(\phi \times w) = w \cdot \operatorname{curl} \phi - \phi \cdot \operatorname{curl} w.$$

(ii) Prove that

$$\int_\Omega w \cdot \operatorname{curl} \phi \, dx = \int_\Omega \operatorname{curl} w \cdot \phi \, dx - \int_{\partial\Omega} (w \wedge n) \cdot \phi \, ds.$$

Exercise A.8.22

(i) Prove that $H_0(\operatorname{curl}; \Omega)$ is a Banach space.
(ii) Show that $\nabla H_0^1(\Omega) \subset H_0(\operatorname{curl}; \Omega)$.
(iii) Prove that for all $v \in H(\operatorname{curl}; \Omega)$ we have $\operatorname{div} \operatorname{curl} v = 0$.

Exercise A.8.23

(i) Show that for $\Omega \subset \mathbb{R}^d$ and $\phi \in C^2(\overline{\Omega}; \mathbb{R}^d)$, we have

$$\operatorname{Curl} \operatorname{curl} \phi = \nabla \operatorname{div} \phi - \Delta\phi,$$

and if $d = 2$, $\operatorname{curl} \operatorname{Curl} \psi = \Delta\psi$ for every $\psi \in C^2(\overline{\Omega})$.
(ii) Show that in the absence of charges and currents, solutions E and B of the Maxwell system in free space are solutions of wave equations with wave speed that coincide with the speed of light. Use that $\varepsilon_0 = 8.854\,187 \cdot 10^{-12} \text{F/m}$ and $\mu_0 = 4\pi \cdot 10^{-7} \text{N/A}^2$.

Exercise A.8.24 Let $F \in H(\operatorname{div}; \Omega)$ with $F \cdot n = 0$ on $\partial\Omega$. Show that there exist functions $\phi \in H^1(\Omega)$ and $G \in H(\operatorname{div}; \Omega)$ with $G \cdot n = 0$ on $\partial\Omega$ and $\operatorname{div} G = 0$, such that

$$F = \nabla\phi + G.$$

Show that the decomposition is L^2-orthogonal.

Exercise A.8.25 Let $G \in C^1(\mathbb{R}^3; \mathbb{R}^3)$. Show that there exists $\phi \in C^1(\mathbb{R}^3)$ such that

$$G = \nabla\phi$$

if and only if $\operatorname{curl} G = 0$.

Exercise A.8.26

(i) Show that the solution operator $(-\Delta)^{-1} : L^2(\Omega) \to L^2(\Omega)$ related to the Poisson problem with homogeneous Dirichlet boundary conditions is bounded, compact, and self-adjoint.

(ii) Show that the boundary value problem

$$-\Delta u - \omega^2 u = f \text{ in } \Omega, \quad u|_{\partial\Omega} = 0,$$

admits for every $f \in L^2(\Omega)$ a unique weak solution, provided that ω^2 does not coincide with an eigenvalue of the operator $-\Delta$.

Exercise A.8.27 Let $\Omega \subset \mathbb{R}^2$. Show that every eigenfunction of the Laplace operator with homogeneous Neumann boundary conditions defines a solution of the Maxwell eigenvalue problem with constraint div $u = 0$.

Exercise A.8.28 Let $v_h \in C(\overline{\Omega}; \mathbb{R}^d)$ be a piecewise polynomial vector field. Show that $v_h \in H_0(\mathrm{curl}; \Omega)$ if and only if the tangential component of v_h is continuous and vanishes on the boundary.

Exercise A.8.29

(i) Prove that we have $\nabla \mathscr{S}_0^1(\mathscr{T}_h) \subset \mathscr{N}ed_T^0(\mathscr{T}_h)$.
(ii) Show that $\mathscr{I}_{Ned} \nabla \phi = \nabla \mathscr{I}_h \phi$ for all $\phi \in C^1(\overline{\Omega})$.

Exercise A.8.30

(i) Let $v \in C^1(\mathbb{R}^3; \mathbb{R}^3)$ with div $v = 0$. Assume that there exists $\psi \in C^1(\mathbb{R}^3; \mathbb{R}^3)$ such that $v = \mathrm{curl}\,\psi$. Show that there exists $\widetilde{\psi} \in C^1(\mathbb{R}^3; \mathbb{R}^3)$ with $\widetilde{\psi}_3 = 0$ and $v = \mathrm{curl}\,\widetilde{\psi}$.
(ii) Let $v(x, y, z) = [x^2, 3xz^2, -2xz]^{\mathsf{T}}$. Construct $\psi \in C^1(\mathbb{R}^3; \mathbb{R}^3)$ such that $v = \mathrm{curl}\,\psi$.

Quiz A.8.3 Decide for each of the following statements whether it is true or false. You should be able to justify your decision.

For $d = 2$ we have $v \wedge n = \det[v, n]$ and for $d = 3$, we have $v \cdot (a \times b) = \det[v, a, b]$	
If $a \in \mathbb{R}^3$ and $a \times e_i = 0$ for $i = 1, 2, 3$, then $a = 0$	
If $v \wedge n = 0$ on $\partial\Omega$, then v is parallel to n on $\partial\Omega$	
We have curl $v_h \in \mathscr{S}^1(\mathscr{T}_h)$ for every $v_h \in \mathscr{R}T^0(\mathscr{T}_h)$.	
We have that $\nabla\varphi_z$ is orthogonal to every side $S \in \mathscr{S}_h$ with $z \notin S$	

A.8.4 Incompressible, Viscous Fluids

Exercise A.8.31 Let $A \in C^1([0,T]; \mathbb{R}^{d \times d})$ and $t_0 \in [0,T]$ be such that $A(t_0)$ is invertible. Show that we have

$$\frac{d}{dt}\Big|_{t=t_0} \det A(t) = \det A(t_0) \operatorname{tr}\left(A(t_0)^{-1} A'(t_0)\right), \quad A'(t) = \frac{d}{dt} A(t)$$

Hint: Use Leibniz's formula to prove the identity for the special case $A(t_0) = I$ first. Consider the function $B(t) = A(t_0)^{-1} A(t)$ to prove the general case.

Exercise A.8.32 Let $\Phi : [0,T] \times \Omega \to \mathbb{R}^d$ be a C^2 mapping, such that $\Phi(t, \cdot)$ defines a diffeomorphism between Ω and $\Omega_t = \Phi(t, \Omega)$ for every $t \in [0,T]$. Assume that $\Phi(0,x) = x$ for all $x \in \Omega$, i.e., $\Omega_0 = \Omega$, and that $J(t,x) = \det D\Phi(t,x) > 0$ for all $(t,x) \in [0,T] \times \Omega$.

(i) For $t \in [0,T]$, $x \in \Omega$, and $y = \Phi(t,x)$ let $v(t,y) = \partial_t \Phi(t,x)$. Show that

$$\partial_t J(t,x) = J(t,x) \operatorname{div} v\big((t, \Phi(t,x))\big).$$

(ii) For $t \in [0,T]$ and $y \in \Omega_t$, let $\varrho(t,y)$ be the mass density of a material occupying the domain Ω_t. The mass of the set $\omega_t = \Phi(t, \omega)$ for $\omega \subset \Omega$ is given by

$$m_{\omega_t} = \int_{\omega_t} \varrho(t, y) \, dy.$$

Assume that mass is conserved to deduce that

$$\partial_t \rho + \operatorname{div}(\rho v) = 0.$$

What can you conclude for incompressible materials, i.e., when $t \mapsto m_{\omega_t}$ is constant for every $\omega \subset \Omega$?

Exercise A.8.33 For a vector field $u : \Omega \to \mathbb{R}^d$, let

$$\sigma = 2\mu \varepsilon(u) + \lambda \operatorname{tr}(u) I - pI.$$

Show that

$$\operatorname{div} \sigma = \mu \Delta u + (\lambda + \mu) \nabla \operatorname{div} u - \nabla p,$$

where the divergence is taken row-wise.

Exercise A.8.34 Prove that the trilinear form $n : \left[H_0^1(\Omega; \mathbb{R}^d)\right]^3 \to \mathbb{R}$,

$$n(z; u, v) = \int_\Omega (z \cdot \nabla u) \cdot v \, dx,$$

is skew-symmetric in the second and third variable, provided that $\operatorname{div} z = 0$.

Exercise A.8.35 Show that a non-dimensionalization of the stationary Navier–Stokes equations with a characteristic length L, e.g., $L = \operatorname{diam}(\Omega)$, a characteristic speed U, e.g., $U = \max_{x \in \Gamma_D} |u_D(x)|$, leads to the system of equations

$$-\frac{1}{R}\Delta\tilde{u} + \tilde{u} \cdot \nabla\tilde{u} + \nabla\tilde{p} = \frac{L}{U^2}\tilde{f}, \quad \operatorname{div}\tilde{u} = 0,$$

where $R = UL/\nu$.

Exercise A.8.36 Let $\tilde{u} \in H_N(\operatorname{div}; \Omega)$ and $(u, p) \in H_N(\operatorname{div}; \Omega) \times L_0^2(\Omega)$ be a weak solution of the system

$$u + \nabla p = \tilde{u}, \quad \operatorname{div} u = 0.$$

Show that u is the L^2-projection of $\tilde{u}$ onto the space $\{v \in H_N(\operatorname{div}; \Omega) : \operatorname{div} v = 0\}$.

Exercise A.8.37 Show that if $\Gamma_D = \partial\Omega$ and $\|f\|_{L^2(\Omega)} \le \nu^2/(c_P c_S^2)$, where $c_P, c_S > 0$ are such that $\|v\|_{L^2(\Omega)} \le c_P \|\nabla v\|_{L^2(\Omega)}$ and $\|v\|_{L^4(\Omega)} \le c_S \|\nabla v\|_{L^2(\Omega)}$ for all $v \in H_0^1(\Omega; \mathbb{R}^d)$, solutions of the stationary Navier–Stokes equations are unique.

Exercise A.8.38

(i) Let $1 \le p, q, r \le \infty$ and $u \in L^p(\Omega)$, $v \in L^q(\Omega)$, and $w \in L^r(\Omega)$. Show that

$$\int_\Omega uvw \, dx \le \|u\|_{L^p(\Omega)} \|v\|_{L^q(\Omega)} \|w\|_{L^r(\Omega)}$$

provided that $1/p + 1/q + 1/r = 1$.

(ii) For which exponents $1 \le p, q, r \le \infty$ and dimensions $1 \le d \le 3$ is the trilinear form

$$n : W^{3,p}(\Omega) \times W^{1,q}(\Omega) \times L^r(\Omega) \to \mathbb{R}, \quad n(u, v, w) = \int_\Omega (\Delta u) v w \, dx,$$

bounded?

Exercise A.8.39 Brouwer's fixed point theorem states that every continuous mapping $f : C \to C$ on a nonempty, convex, and compact set $C \subset \mathbb{R}^n$ has a fixed point.

Prove via contradiction that for every continuous mapping $F : \mathbb{R}^n \to \mathbb{R}^n$ with the property that

$$F(U) \cdot U \geq 0$$

for all $U \in \mathbb{R}^n$ with $|U| \geq R > 0$, there exists $U^* \in \mathbb{R}^n$ with $|U^*| \leq R$ such that $F(U^*) = 0$.

Exercise A.8.40

(i) Use the closed range theorem to show that if $\phi \in H_0^1(\Omega; \mathbb{R}^d)'$ is such that

$$\phi(v) = 0$$

for all $v \in H_0^1(\Omega; \mathbb{R}^d)$ with div $v = 0$, then there exists $p \in L_0^2(\Omega)$ such that

$$\phi(v) = \int_\Omega p \, \mathrm{div}\, v \, dx$$

for all $v \in H_0^1(\Omega; \mathbb{R}^d)$.

(ii) Conclude that it suffices to determine $u \in H_0^1(\Omega; \mathbb{R}^d)$ as the solution of the equation

$$\nu \int_\Omega \nabla u : \nabla v \, dx + \int_\Omega (u \cdot \nabla u) \cdot v \, dx = \int_\Omega f \cdot v \, dx$$

subject to div $u = 0$ and for all $v \in H_0^1(\Omega; \mathbb{R}^d)$ with div $v = 0$, in order to solve the stationary Navier–Stokes equations.

Quiz A.8.4 Decide for each of the following statements whether it is true or false. You should be able to justify your decision.

Solutions of the stationary Navier–Stokes equations are unique	
If $\Gamma_D = \partial\Omega$ and $u_D = 0$, then the Picard iteration is globally convergent	
The vector field $u : \Omega \to \mathbb{R}^d$ in the Navier–Stokes equations determines the displacements of particles in the domain Ω	
If $a : V \times V \to \mathbb{R}$ is coercive and $n : V \times V \to \mathbb{R}$ skew-symmetric, then $a + \lambda v$ is coercive on $V \times V$ for every $\lambda \in \mathbb{R}$	
The Stokes system is a linearization of the Navier–Stokes equations	

A.8.5 Projects

Project A.8.1 Implement the discretization of the Navier–Lamé equations with a
$P1$ finite element method and test your code using the exact solution

$$u(t, x, y) = \begin{bmatrix} \sin(2\pi y) \sin^2(\pi x) \\ -\sin(2\pi x) \sin^2(\pi y) \end{bmatrix},$$

by defining the right-hand side f in $\Omega = (0, 1)^2$ appropriately. Illustrate the failure
of the method in the case of a nearly incompressible material, i.e., investigating the
dependence of the approximation error on $\lambda = 10^j$, $j = 1, 2, \ldots, 6$. Compare this
to the approximation with the stabilized Crouzeix–Raviart method.

Project A.8.2 Consider a cylinder domain Ω of length $\ell = 1$ and radius $r = 1/10$,
i.e., $\Omega = B_r(0) \times (0, \ell) \subset \mathbb{R}^3$ that represents an elastic rod. We assume that the
rod is fixed on the side $\Gamma_D = B_r(0) \times \{0\}$. Undamped vibrations of the rod are then
described by the equation

$$\partial_t^2 u - \operatorname{div} \mathbb{C} \varepsilon(u) = 0 \quad \text{in } (0, T) \times \Omega$$

supplemented with boundary and initial conditions. Devise a weak formulation and
a numerical method for simulating the vibrations and carry out experiments with
different discretization parameters.

Project A.8.3 Use the Argyris element to discretize the Poisson problem with
homogeneous Dirichlet boundary conditions on the unit square. Determine the
experimental convergence rate for the exact solution

$$u(x, y) = \sin(\pi x) \sin(\pi y).$$

Improve the conditioning of the local linear systems of equations by using scaled
monomials $p_{jk}(x_1, x_2) = h_T^{-(j+k)}(x_1 - x_{T,1})^j (x_2 - x_{T,2})^k$.

Project A.8.4 We consider a simply supported plate occupying the domain $\Omega = B_1(0)$ with force $f = 1$ and material parameter $\theta = 1/2$, i.e., the boundary value
problem

$$\Delta^2 u = 1 \quad \text{in } \Omega,$$

$$u = 0 \quad \text{on } \partial\Omega,$$

$$\Delta u - \frac{1}{2} \partial_n u = 0 \quad \text{on } \partial\Omega.$$

Verify that for a rotationally symmetric function, we have $\Delta u = \partial_r^2 u + r^{-1} \partial_r u$,
$\partial_n u = \partial_r u$, and $\Delta^2 u = \left(\partial_r^4 + 2r^{-1}\partial_r^3 - r^{-2}\partial_r^2 + r^{-3}\partial_r\right)u$, so that the exact solution

is given by

$$u(r) = \frac{1}{64}r^4 - \frac{14}{3 \cdot 64}r^2 + \frac{11}{3 \cdot 64}.$$

Approximate the problem on a sequence of polygonal domains $(\Omega_n)_{n=0,1,...}$ and show experimentally that the approximations $(u_n)_{n=0,1,...}$ do not converge to u by comparing $u_n(0)$ for $n = 0, 1, \ldots$ with $u(0)$. Show that in the case of clamped boundary conditions, the exact solution is given by $u(r) = r^4/64 - r^2/32 + 1/64$ and that the problem does not occur.

Project A.8.5 Implement the approximation of the Maxwell equations with a $P1$-finite element method and test its performance on the unit square $\Omega = (0, 1)^2$ with $f = [1, 1]^T$, the L-shaped domain $\Omega = (-1, 1)^2 \setminus (-1, 0) \times (0, 1)$ with $f = [1, 1]^T$, and the ring domain $\Omega = B_1(0) \setminus B_{1/2}(0)$ with $f(x) = \text{Curl} |x|$. Compare approximations qualitatively for mesh-sizes $h \approx 1/20$ to approximations obtained with the Nédélec method for different choices of ω^2.

Project A.8.6 Verify that $\lambda = \pi^2$ is an eigenvalue of the Maxwell operator in the unit square $\Omega = (0, 1)^2$. For a fixed triangulation of Ω with $h \approx 1/100$, investigate the dependence of the approximation error on the difference $|\omega^2 - \lambda|$, by considering $\omega^2 = \lambda + 1/10^j$ for $j = 0, 1, 2, \ldots$ and constructing an appropriate reference solution.

Project A.8.7 Use the Crouzeix–Raviart method and a backward difference quotient to discretize the Stokes flow

$$\partial_t u - \Delta u + \nabla p = f, \quad \text{div } u = 0,$$

in $(0, T) \times \Omega$ with no-slip boundary conditions for u on $\partial\Omega$. Define $f : (0, T) \times \Omega \to \mathbb{R}^2$ so that the exact solution of the problem with $\Omega = (-1, 1)^2$ and $T = 1$ is given by

$$u(t, x, y) = \pi \sin(t) \begin{bmatrix} \sin(2\pi y) \sin^2(\pi x) \\ -\sin(2\pi x) \sin^2(\pi y) \end{bmatrix},$$

$$p(t, x, y) = \sin(t) \cos(\pi x) \sin(\pi y).$$

Use a fixed mesh-size $0 < h \ll 1$ and different step-sizes $\tau > 0$ to investigate the convergence behavior of the errors

$$\max_{k=0,\ldots,K} \|p - p_h^k\|_{L^2(\Omega)}, \quad \max_{k=0,\ldots,K} \|\nabla_{\mathscr{T}}(u - u_h^k)\|_{L^2(\Omega)}$$

as $\tau \to 0$. Compare the convergence behavior to that for approximations obtained with the Chorin projection scheme. Plot the pressure errors at the final time T and discuss your observations.

Project A.8.8 Use the Picard iteration and a Crouzeiz–Raviart discretization to solve the stationary Navier–Stokes equations in the cylinder domain with hole $\Omega = \big((-\ell, \ell) \times B_r(0)\big) \setminus B_{r/2}(0) \subset \mathbb{R}^d$ for $\ell = 3$, $r = 1$, and $d = 2, 3$, for the Dirichlet boundary conditions

$$u(x_1, x_2, \dots, x_d) = \begin{cases} x_2^2 + \cdots + x_d^2 - 1, & x_1 = \pm \ell, \\ 0, & -\ell < x_1 < \ell. \end{cases}$$

Use a mesh-size $h \approx 1/20$ and test the convergence behavior of the iteration for relative viscosities $\nu = 10^{-j}$, $j = 0, 1, \dots, 3$. Replace the Dirichlet boundary condition at $x_1 = \ell$ by a homogeneous Neumann boundary condition and repeat the experiment. Visualize some solutions with `paraview`.

Appendix B
Implementation Aspects

B.1   Basic MATLAB Commands

B.1.1   Matrix Operations

The programming language MATLAB provides various optimized implementations
of matrix operations. Some important commands, whose usage is canonical, are
listed in Table B.1. Various operations such as matrix multiplication can be applied
component-wise by placing a dot in front of the operand, e.g.,

$$\begin{bmatrix} 1 & 2 \\ 3 & 4 \end{bmatrix} \cdot * \begin{bmatrix} 5 & 6 \\ 7 & 8 \end{bmatrix} = \begin{bmatrix} 5 & 12 \\ 21 & 32 \end{bmatrix},$$

whereas the command without the dot gives the result

$$\begin{bmatrix} 1 & 2 \\ 3 & 4 \end{bmatrix} * \begin{bmatrix} 5 & 6 \\ 7 & 8 \end{bmatrix} = \begin{bmatrix} 19 & 22 \\ 43 & 50 \end{bmatrix}.$$

Similarly, functions can be applied component-wise, e.g.,

$$\begin{bmatrix} 1 & 2 \\ 3 & 4 \end{bmatrix} \cdot \hat{\ } 2 = \begin{bmatrix} 1 & 4 \\ 9 & 16 \end{bmatrix}, \quad \cos\left(\begin{bmatrix} 0 & \pi/2 \\ \pi & 2\pi \end{bmatrix}\right) = \begin{bmatrix} 1 & 0 \\ -1 & 1 \end{bmatrix}.$$

Linear systems of equations can be solved with the backslash operator, e.g.,

$$\begin{bmatrix} 2 & 1 \\ 1 & 2 \end{bmatrix} \setminus \begin{bmatrix} 3 \\ 3 \end{bmatrix} = \begin{bmatrix} 1 \\ 1 \end{bmatrix} \quad \Longleftrightarrow \quad \begin{bmatrix} 2 & 1 \\ 1 & 2 \end{bmatrix} * \begin{bmatrix} 1 \\ 1 \end{bmatrix} = \begin{bmatrix} 3 \\ 3 \end{bmatrix}.$$

© Springer International Publishing Switzerland 2016   501
S. Bartels, *Numerical Approximation of Partial Differential Equations*,
Texts in Applied Mathematics 64, DOI 10.1007/978-3-319-32354-1

Table B.1 Elementary matrix constructions and operations

`[a,b,...;x,y,...]`	Definition of a matrix (commas may be omitted)
`[a,b,...], [x;y;...]`	Definition of row and column vectors
`A(i,j), I(j)`	Entry at position (i,j), j-th entry
`a:b, a:step:b`	List of numbers
`A(i,:), A(:,j)`	i-th row and j-th column
`A(I,J)`	Submatrix defined by lists I and J
`ones(m,n), zeros(m,n)`	Matrix with entries one or zero
`A+B, A-B, A*B`	Sum, difference, and product
`A', inv(A), det(A)`	Transpose, inverse, and determinant
`x = A\b`	Solution of a linear system of equations
`eye(n), speye(n)`	Unit and sparse unit $n \times n$ matrix
`A.*B, A./B`	Component-wise multiplication and division
`lu(A), chol(A)`	LU and Cholesky factorization
`eig(A)`	Eigenvectors and eigenvalues
`diag(A), tril(A)`	Diagonal and lower triangular part
`sparse(I,J,X,m,n)`	Creation of a sparse matrix

The backslash operator is flexible and can also be used, e.g., to solve overdetermined or singular systems in an appropriate sense. To solve large linear systems of equations with sparse system matrices, it is important that the matrices are defined correspondingly. In most cases this can be done using the MATLAB command sparse, which generates a matrix by providing the coordinates and values of the relevant entries, e.g.,

$$A = \begin{bmatrix} 1 & 0 & 0 & 0 \\ 2 & 3 & 0 & 0 \\ 0 & 0 & 4 & 0 \\ 0 & 0 & 0 & 5 \end{bmatrix} \iff \begin{array}{l} I = [1,2,2,3,3,4], \\ J = [1,1,2,3,3,4], \\ X = [1,2,3,2,2,5], \\ A = \texttt{sparse(I,J,X,4,4)}. \end{array}$$

Note that in the example the position $(i,j) = (3,3)$ occurs twice with value 2. By convention, values in the lists corresponding to the same position in the matrix are added. This is a crucial feature for the efficient assembly of finite element matrices.

B.1.2 List Manipulation

The manipulation of lists and arrays is frequently used in the implementation of finite element methods, e.g., to extract implicit information about a triangulation from the lists of vertices and elements. The arrays c4n, n4e, Db, and Nb specify the triangulation, e.g., of $\Omega = (0,1)^2$ with $\Gamma_D = [0,1] \times \{0\} \cup \{1\} \times [0,1]$ and

$\Gamma_N = \partial\Omega \setminus \Gamma_D$ into two triangles via

$$c4n = \begin{bmatrix} 0 & 0 \\ 1 & 0 \\ 1 & 1 \\ 0 & 1 \end{bmatrix}, \quad n4e = \begin{bmatrix} 1 & 2 & 3 \\ 1 & 3 & 4 \end{bmatrix}, \quad Db = \begin{bmatrix} 1 & 2 \\ 2 & 3 \end{bmatrix}, \quad Nb = \begin{bmatrix} 3 & 4 \\ 4 & 1 \end{bmatrix}.$$

We then obtain the nodes belonging to the Dirichlet boundary via the command `dNodes = unique(Db)`. Precise information about the sides in the triangulation can be obtained via first arranging all sides of elements in one array. Interior sides then occur twice while boundary sides occur only once. To obtain a list in which all sides of a two-dimensional triangulation only appear once, we use the following commands:

```
all_sides = [n4e(:,[1,2]);n4e(:,[2,3]);n4e(:,[3,1])];
[sides,i,j] = unique(sort(all_sides,2),'rows');
```

In the above example we have

$$\text{all\_sides} = \begin{bmatrix} 1 & 2 \\ 1 & 3 \\ 2 & 3 \\ 3 & 4 \\ 3 & 1 \\ 4 & 1 \end{bmatrix}, \quad \text{sides} = \begin{bmatrix} 1 & 2 \\ 1 & 3 \\ 1 & 4 \\ 2 & 3 \\ 3 & 4 \end{bmatrix}.$$

The output arguments `i` and `j` of the command `unique` provide mappings between the two arrays and thereby specify sides occurring repeatedly. Table B.2 displays further MATLAB commands for manipulating lists.

Table B.2 Elementary list manipulation commands

`sort(A,p)`	Sorts columns or rows of a matrix
`unique(A,'rows')`	Extracts unique rows
`reshape(A,p,q)`	Rearranges entries of A in a $p \times q$ matrix
`A(:)`	Writes columns of A as one column vector
`repmat(A,r,s)`	Builds block matrix with copies of A
`accumarray(N,X,[m,n])`	Creates a matrix by summing entries of X
`length(x), size(A)`	Length and dimensions of arrays
`setdiff(A,B)`	Set-theoretic difference of A and B
`max(x), min(x)`	Minimal and maximal entry
`find(I)`	Indices of nonvanishing entries in I

The right-hand side f in a partial differential equation gives rise to a vector $b \in \mathbb{R}^M$ with entries corresponding to the nodes $(z_1, z_2, \ldots, z_M)$ defined by

$$\int_\Omega f\varphi_{z_i} \, dx \approx \sum_{T \in \mathcal{T}_h : z_i \in T} \frac{|T|}{d+1} f(x_T) = b_i$$

for $i = 1, 2, \ldots, M$. The ℓ-th row of the array n4e specifies those nodes z_i that belong to the element T_ℓ. With the help of the command accumarray the vector b can thus be assembled with the following lines:

```
Z = (1/(d+1))*Vol_T.*f(Mp_T); ZZ = repmat(Z,1,d+1);
b = accumarray(n4e(:),ZZ(:),[nC,1]);
```

Here, Vol_T and Mp_T are arrays that contain the volumes and midpoints of the elements.

B.1.3 Graphics

Finite element functions can be visualized as graphs or color plots with the commands trisurf, trimesh, and tetramesh, e.g., if u is the coefficient vector of a $P1$-finite element function via:

```
trisurf(n4e,c4n(:,1),c4n(:,2),u);
tetramesh(n4e,c4n,u);
```

Other useful commands that plot vector fields as arrows or change the view of an object are listed in Table B.3.

Table B.3 MATLAB commands that generate and manipulate plots and figures

figure	Selects a figure window
plot, plot3	Plots a polygonal curve in $\mathbb{R}^2$ or $\mathbb{R}^3$
loglog	Plot with logarithmic scaling for both axes
legend	Adds a legend to a plot
hold on/off	Plotting of several objects in one figure
trimesh, tetramesh	Displays a triangulation in $\mathbb{R}^2$ or $\mathbb{R}^3$
trisurf	Shows the graph of a $P1$ function ($d = 2$)
quiver, quiver3	Plots a two- or three-dimensional vector field
drawnow, clf	Updates and clears a figure
axis	Sets the axes in a figure including color range
axis on/off	Switches coordinate axes on or off
xlabel, ylabel	Adds labels to axes
colorbar	Displays a color bar
subplot	Shows several plots in one figure
view	Changes the perspective
colormap	Chooses a color scale
clc	Clears the command window

Table B.4 Standard programming commands

`for .. end`	For loop
`while .. end`	While loop
`if .. else .. end`	If-then-else structure
`pause, break`	Pause until key is hit, stop a program
`disp, fprintf`	Displays or prints a quantity
`load, save`	Loads and saves variables
`fopen, fclose`	Opens and closes file to write data
`addpath, rmpath`	Adds and removes a directory path
`Ctrl-C`	Stops a running program

Table B.5 Examples of functions available in MATLAB

`exp, ln, log`	Exponential and logarithms
`sqrt, ^`	Square root and power
`sin, cos, tan, pi`	Trigonometric functions and constant π

B.1.4 Standard Commands

Most standard programming structures and mathematical functions are available in MATLAB. Tables B.4 and B.5 provide an overview over frequently used commands.

B.2 Finite Element Matrix Assembly

B.2.1 Global Loops

The standard assembly of a finite element matrix is based on a global loop over elements in a triangulation, e.g., via

$$s_{zy} = \int_\Omega \nabla \varphi_z \cdot \nabla \varphi_y \, \mathrm{d}x = \sum_{T \in \mathscr{T}_h} |T| \nabla \varphi_z|_T \cdot \nabla \varphi_y|_T.$$

A realization in MATLAB is shown in Fig. B.1. The routine computes the coordinate lists I, J and corresponding entries X of the $P1$-finite element stiffness matrix, which are then used to assemble the stiffness matrix via

$$\texttt{s = sparse(I,J,X,nC,nC);}$$

For the use of the routine, the volumes of elements and elementwise gradients of nodal basis functions are precomputed and stored in the arrays `Vol_T` and `Grads_T` arranged by elements $T_1, T_2, \ldots, T_L$ and local node numbers z_i^ℓ, $i =$

```
function [I,J,X] = fe_matrix_loop(c4n,n4e,Vol_T,Grads_T)
d = size(c4n,2); nE = size(n4e,1);
ctr = 0; ctr_max = (d+1)^2*nE;
I = zeros(ctr_max,1); J = zeros(ctr_max,1);
X = zeros(ctr_max,1);
for j = 1:nE
    grads_T = Grads_T((j-1)*(d+1)+(1:d+1),:);
    vol_T = Vol_T(j);
    for m = 1:d+1
        for n = 1:d+1
            ctr = ctr+1; I(ctr) = n4e(j,m); J(ctr) = n4e(j,n);
            X(ctr) = vol_T*grads_T(m,:)*grads_T(n,:)';
        end
    end
end
end
```

Fig. B.1 Standard assembly of the stiffness matrix in a loop over all elements

$1, 2, \ldots, d + 1$, i.e.,

$$
\text{Vol\_T} = \begin{bmatrix} |T_1| \\ |T_2| \\ \vdots \\ |T_L| \end{bmatrix} \in \mathbb{R}^L, \quad
\text{Grads\_T} = \begin{bmatrix} \nabla \varphi_{z_1^1}|_{T_1} \\ \vdots \\ \nabla \varphi_{z_{d+1}^1}|_{T_1} \\ \nabla \varphi_{z_1^2}|_{T_2} \\ \vdots \\ \nabla \varphi_{z_{d+1}^L}|_{T_L} \end{bmatrix} \in \mathbb{R}^{L(d+1) \times d}.
$$

Although the assembly via a loop over elements is of linear complexity, its practical performance in interpreted programming languages is suboptimal, i.e., its runtime is typically longer than the CPU-time needed for solving the linear system of equations. This is problematic in the case of time-dependent problems, when a system matrix has to be assembled in every time step.

B.2.2 Vectorized Loop

A way to accelerate the computation of the entries of the stiffness matrix is to avoid the global loop over elements and to compute the relevant quantities with matrix operations. The idea is related to the simple observation that e.g.,

$$
\sum_{i=1}^{n} \text{a(i)} * \text{b(i)} = \text{a'} * \text{b} = \text{sum(a.*b)},
$$

```
function [I,J,X] = fe_matrix_vectorized_1(c4n,n4e,Vol_T,Grads_T)
d = size(c4n,2); nE = size(n4e,1); ctr_max = (d+1)^2*nE;
I = zeros(ctr_max,1); J = zeros(ctr_max,1); X = zeros(ctr_max,1);
for m = 1:d+1
    for n = 1:d+1
        idx = ((m-1)*(d+1)+(n-1))*nE+(1:nE);
        vals = Vol_T.*...
            sum(Grads_T(m:d+1:end,:).*Grads_T(n:d+1:end,:),2);
        I(idx) = n4e(:,m); J(idx) = n4e(:,n); X(idx) = vals;
    end
end
```

```
function [I,J,X] = fe_matrix_vectorized_2(c4n,n4e,Vol_T,Grads_T)
d = size(c4n,2); nE = size(n4e,1);
I = repmat(reshape(n4e',(d+1)*nE,1),1,d+1)';
J = repmat(n4e,1,d+1)';
B = reshape(repmat(Grads_T',d+1,1),d,(d+1)^2*nE)';
C = reshape(repmat(reshape(Grads_T',(d+1)*d,nE),d+1,1),...
    d,(d+1)^2*nE)';
rep_Vol_T = repmat(Vol_T,1,(d+1)^2)';
X = rep_Vol_T(:).*sum(B.*C,2);
I = I(:); J = J(:); X = X(:);
```

Fig. B.2 Vectorized computations of the coordinate and entry vectors for the stiffness matrix

where the implementation of the method on the right-hand side is significantly faster than the loop suggested by the left-hand side in a MATLAB implementation. The first code displayed in Fig. B.2 results from the routine shown in Fig. B.1 by eliminating the global loop and writing corresponding operations in vectorized form. In the second code of Fig. B.2, the loops with boundedly many repetitions have also been eliminated. These implementations are significantly faster than the code that uses the loop over elements, which is due to the optimized implementation of matrix operations in MATLAB.

B.2.3 Assembly in C

An alternative to the acceleration via vectorization is the assembly of the matrix entries in the compiled programming language C. To include C code in MATLAB, we use the interface MEX, which allows for a simple variable transfer. Figure B.3 shows a C code that is equivalent to the MATLAB routine shown in Fig. B.1. Its compilation is done in MATLAB using the command

```
mex fe_matrix_mex.c;
```

```c
#include <mex.h>   /* fe_matrix_mex.c */
void lists(double n4e[], double c4n[],
        double Vol_T[], double Grads_T[],
        int nE, int nC, int d,
        double I[], double J[], double X[]){
    int j, m, n, r, idx1, idx2, ctr;
    double val;
    ctr = 0;
    for (j=0; j<nE; j++){
        for (m=0; m<d+1; m++){
            for (n=0; n<d+1; n++){
                I[ctr] = n4e[j+m*nE]; J[ctr] = n4e[j+n*nE];
                val = 0.0;
                for (r=0; r<d; r++){
                    idx1 = j*(d+1)+m+r*(d+1)*nE;
                    idx2 = j*(d+1)+n+r*(d+1)*nE;
                    X[ctr] += Vol_T[j]*Grads_T[idx1]*Grads_T[idx2];
                }
                ctr += 1;
            }
        }
    }
}
void mexFunction(int nlhs, mxArray *plhs[], int nrhs,
        const mxArray *prhs[]){
    double *n4e, *c4n, *Vol_T, *Grads_T;
    int nE, nC, d;
    double *I, *J, *X;
    if (nrhs != 4)
        mexErrMsgTxt("4 input arguments required");
    nC      = mxGetM(prhs[0]);
    d       = mxGetN(prhs[0]);
    nE      = mxGetM(prhs[1]);
    c4n     = mxGetPr(prhs[0]);
    n4e     = mxGetPr(prhs[1]);
    Vol_T   = mxGetPr(prhs[2]);
    Grads_T = mxGetPr(prhs[3]);
    if (nlhs != 3)
        mexErrMsgTxt("3 output arguments required");
    plhs[0] = mxCreateDoubleMatrix(nE*(d+1)*(d+1),1,mxREAL);
    plhs[1] = mxCreateDoubleMatrix(nE*(d+1)*(d+1),1,mxREAL);
    plhs[2] = mxCreateDoubleMatrix(nE*(d+1)*(d+1),1,mxREAL);
    I = mxGetPr(plhs[0]);
    J = mxGetPr(plhs[1]);
    X = mxGetPr(plhs[2]);
    lists(n4e,c4n,Vol_T,Grads_T,nE,nC,d,I,J,X);
}
```

Fig. B.3   Computing the entries of the stiffness matrix in C using MATLAB to C interface MEX

For this the gnu C compiler gcc has to be selected via the MATLAB command mex -setup. The routine can then be called within MATLAB as in the code shown in Fig. B.4.

```
function p1_comparison(d,red,assembly,solver)
[c4n,n4e,Db,Nb] = triang_cube(d); Db = [Db;Nb]; Nb = [];
for j = 1:red
    [c4n,n4e,Db,Nb] = red_refine(c4n,n4e,Db,Nb);
end
nC = size(c4n,1); h = 1/nC^(1/d);
dNodes = unique(Db); fNodes = setdiff(1:nC,dNodes);
[Vol_T,Grads_T,Mp_T] = nodal_basis(c4n,n4e);
Z = (1/(d+1))*Vol_T.*f(Mp_T); ZZ = repmat(Z,1,d+1);
switch assembly
    case 0
        [I,J,X] = fe_matrix_loop(c4n,n4e,Vol_T,Grads_T);
    case 1
        [I,J,X] = fe_matrix_vectorized_1(c4n,n4e,Vol_T,Grads_T);
    case 2
        [I,J,X] = fe_matrix_vectorized_2(c4n,n4e,Vol_T,Grads_T);
    case 3
        [I,J,X] = fe_matrix_mex(c4n,n4e,Vol_T,Grads_T);
end
s = sparse(I,J,X,nC,nC); u = u_D(c4n);
b = accumarray(n4e(:),ZZ(:),[nC,1])-s*u;
s_fN = s(fNodes,fNodes); b_fN = b(fNodes);
switch solver
    case 0
        u(fNodes) = s_fN\b_fN;
    case 1
        K = size(fNodes,2); eps_stop = h;
        C = spdiags(diag(s_fN),0,K,K);
        u(fNodes) = pcg(s_fN,b_fN,eps_stop,K,C);
end
show_p1(c4n,n4e,Db,Nb,u);

function val = f(x); val = ones(size(x,1),1);
function val = u_D(x); val = zeros(size(x,1),1);

function [Vol_T,Grads_T,Mp_T] = nodal_basis(c4n,n4e)
d = size(c4n,2); nE = size(n4e,1);
Grads_T = zeros((d+1)*nE,d);
Vol_T = zeros(nE,1); Mp_T = zeros(nE,d);
for j = 1:nE
    X_T = [ones(1,d+1);c4n(n4e(j,:),:)'];
    Grads_T((j-1)*(d+1)+(1:d+1),:) = X_T\[zeros(1,d);eye(d)];
    Vol_T(j) = det(X_T)/factorial(d);
    Mp_T(j,:) = sum(c4n(n4e(j,:),:),1)/(d+1);
end
```

Fig. B.4 Comparison of four different ways of computing the entries of the $P1$-finite element stiffness matrix and two different solvers for the linear system of equations

B.2.4   Comparison

Tables B.6 and B.7 show a comparison of the runtimes for the different ways to assemble the stiffness matrix for two- and three-dimensional Poisson problems:

$$-\Delta u = 1 \text{ in } \Omega = (0, 1)^d, \quad u = 0 \text{ on } \Gamma_D = \partial\Omega.$$

The experiments were carried out on a standard Desktop (Intel Core i3-3220 CPU, 8 GB RAM). To relate the numbers, we also included the runtimes of different solvers for the linear systems of equations, where A\b refers to the solution via backslash operator and PCG to the use of a preconditioned conjugate gradient algorithm with the diagonal Jacobi preconditioner. A dash indicates that an experiment was not carried out due to memory limitations.

Our conclusions from the experiments are as follows:

- all assembly routines scale linearly with respect to the numbers of degrees of freedom, i.e., when the mesh-size is halved, the runtime increases approximately by a factor 2^d;
- a vectorized assembly reduces the runtime to a few percent of the assembly via a loop in MATLAB and is comparable to the solution with the backslash operator;
- eliminating loops with a mesh-independent number of repetitions does not lead to an additional runtime reduction;

Table B.6 Runtime comparison for different assemblies of the stiffness matrix and solution methods for a two-dimensional Poisson problem; numbers are in seconds

$d = 2, \#\mathcal{N}_h$	M-loop	vec-1	vec-2	C-loop	A\b	PCG
9	0.0007	0.0003	0.0003	0.0001	0.0000	0.0020
25	0.0021	0.0003	0.0003	0.0001	0.0001	0.0024
81	0.0076	0.0003	0.0002	0.0001	0.0001	0.0029
289	0.0148	0.0002	0.0002	0.0001	0.0003	0.0025
1089	0.0590	0.0007	0.0007	0.0002	0.0011	0.0053
4225	0.2411	0.0026	0.0023	0.0006	0.0054	0.0218
16641	0.9608	0.0137	0.0099	0.0028	0.0284	0.1123
66049	3.9347	0.0651	0.0397	0.0107	0.1551	0.8683
263169	15.8410	0.3031	0.2444	0.0571	0.7964	7.8872

Table B.7 Runtime comparison for different assemblies of the stiffness matrix and solution methods for a three-dimensional Poisson problem; numbers are in seconds

$d = 3, \#\mathcal{N}_h$	M-loop	vec-1	vec-2	C-loop	A\b	PCG
27	0.0045	0.0018	0.0017	0.0129	0.0000	0.3710
125	0.0193	0.0004	0.0004	0.0001	0.0003	0.0134
729	0.1516	0.0026	0.0027	0.0006	0.0020	0.0019
4913	1.2204	0.0273	0.0218	0.0044	0.0439	0.0077
35937	10.0983	0.2363	0.1851	0.0345	2.3275	0.0979
274625	–	2.0396	1.6933	0.3291	–	1.7406

- a loop in C reduces the assembly runtime to about ten percent of the solution time;
- the backslash operator provides good results in two-dimensional situations but may fail in three-dimensional situations;
- in the three-dimensional situation, the preconditioned conjugate gradient algorithm with Jacobi preconditioner outperforms the solution via the backslash operator;
- the solution of the linear system of equations is the *bottleneck* of the problem, i.e., dominates the total CPU time, provided that the assembly of the stiffness matrix is done appropriately.

Remark B.1 Precomputation of the elementwise gradients of the nodal basis functions in the routine `nodal_basis.m` in Fig. B.4 can be accelerated by vectorizing the computation or transferring it to C. A useful formula for efficient computation of the gradients is the identity

$$\nabla \varphi_z|_T = -\frac{|S_z|}{d|T|} n_{S_z},$$

for $d = 2, 3$, where S_z is the side of T which is opposite to z and n_{S_z} is the outward unit normal on S_z. The quantity $(d-1)|S_z|n_{S_z}$ can be obtained from a cross-product of edge vectors if $d = 3$, or a rotation by $\pi/2$ of an edge vector if $d = 2$.

B.3   Mesh Generation and Visualization

B.3.1   Mesh Generation

The free MATLAB package `distmesh` provides routines to generate triangulations of domains in $\mathbb{R}^d$. It assumes the domain to be defined by a signed distance function $s_\Omega : \mathbb{R}^d \to \mathbb{R}$ and contained in a box of diameter $2R$, i.e,

$$\Omega = \{x \in \mathbb{R}^d : s_\Omega(x) < 0\} \subset K_R^\infty(0),$$

where $K_R = \{x \in \mathbb{R}^d : \|x\|_\infty \le R\}$. It is assumed that s_Ω grows linearly away from $\partial\Omega$, i.e., that

$$|\nabla s_\Omega| = 1.$$

On $\partial\Omega$ the gradient ∇s_Ω coincides with the outer unit normal to Ω. A triangulation is generated with the command:

```
[c4n,n4e] = distmeshnd(@sdist_Omega,@mesh_density,...
                       h_min,bd_box,fixed_vertices);
```

```
function [c4n,n4e] = gen_triang_ball(d_tmp,r_tmp,h_min,alpha_tmp)
addpath('~/auxiliary/distmesh');
global d r alpha;
d = d_tmp; r = r_tmp; alpha = alpha_tmp;
R = r; fixed = [];
box = [-R*ones(1,d);R*ones(1,d)];
[c4n,n4e] = distmeshnd(@s,@phi,h_min,box,fixed);

function val = s(x)
global r;
val = sqrt(sum(x.^2,2))-r;

function val = phi(x)
global alpha;
val = (1+sqrt(sum(x.^2,2))).^alpha;
```

Fig. B.5 Generation of a graded triangulation of a d-dimensional ball $B_r(0)$

The parameter h_min determines the approximate minimal mesh-size h_{min} of the triangulation. The function mesh_density allows for a grading of the triangulation specified by a function $\phi(x)$ so that

$$\mathrm{diam}(T) \approx \phi(x_T)h_{min}$$

for all triangles or tetrahedra T with midpoint x_T in the generated triangulation $\mathscr{T}_h$. The argument fixed_vertices allows for prescribing vertices in the triangulation.

Example B.1 The generation of a triangulation of the ball $B_r(x) \subset \mathbb{R}^d$ with radius $r > 0$ using the function $s_\Omega(x) = |x| - r$, the bounding box defined by $R = r$, minimal mesh-size h_{min}, and grading function $\phi(x) = (1+|x|)^\alpha$ is shown in Fig. B.5.

The function distmeshnd initially chooses a quasiuniform triangulation of the bounding box $K_R(0)$, discards all vertices that do not belong to Ω, i.e., vertices z for which $s_\Omega(z) > \varepsilon_{tol}$, and then equidistributes and projects the remaining vertices with the help of a repulsive force function that involves the function ϕ. A large class of Lipschitz domains Ω can be obtained as the union or intersection of simple domains and their complements. If Ω_1 and Ω_2 are Lipschitz domains with signed distance functions s_{Ω_1} and s_{Ω_2}, then signed distance functions for the union, intersection, complement, and transformed domain are given by

$$s_{\Omega_1 \cup \Omega_2} = \min\left\{s_{\Omega_1}, s_{\Omega_2}\right\},$$

$$s_{\Omega_1 \cap \Omega_2} = \max\left\{s_{\Omega_1}, s_{\Omega_2}\right\},$$

$$s_{\mathbb{R}^d \setminus \Omega_1} = -s_{\Omega_1},$$

$$s_{\Phi(\Omega_1)} = s_{\Omega_1} \circ \Phi^{-1}.$$

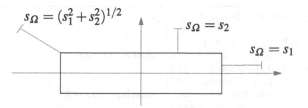

Fig. B.6 Construction of a signed distance function for a parallelepiped

A signed distance function for the parallelepiped $\Omega = \prod_{i=1}^{d} (-\ell_i, \ell_i)$ is obtained from the functions

$$s_i(x) = |x_i| - \ell_i$$

via $s_\Omega = \max_{i=1,2,\dots,d} s_i$. The performance of the routine `distmeshnd` is significantly improved by setting

$$s_\Omega(x) = \begin{cases} \left(\sum_{i=1}^{d} s_i(x)^2 \right)^{1/2}, & s_i(x) > 0 \text{ for } i = 1, 2, \dots, d, \\ \max_{i=1,2,\dots,d} s_i(x), & \text{otherwise}, \end{cases}$$

cf. Fig. B.6. With this definition the function $s_\Omega(x)$ is a proper distance function to Ω in the sense that $s_\Omega(x) = \inf_{y \in \Omega} |x - y|$ for all $x \in \mathbb{R}^d \setminus \Omega$. This function provides a mechanism to compute projections onto $\partial\Omega$. Alternatively, boundary points of Ω can be prescribed as fixed vertices to improve the stability of the mesh generation routine.

Example B.2 Figure B.7 shows a function that generates a triangulation of a cylinder domain of length $2L_c$ and diameter $2r_c$, with a spherical hole of diameter $2r_s$ centered at the origin, i.e.,

$$\Omega = \left((-L_c, L_c) \times B_{r_c}(0) \right) \setminus B_{r_s}(0) \subset \mathbb{R}^d.$$

The subroutines `fix_orientation` and `find_bdy_sides` adjust the ordering of the nodes of triangles or tetrahedra to obtain a positive orientation and identify the boundary sides in the triangulation.

```
function gen_triang_cyl_w_hole(d_tmp)
addpath('~/auxiliary/distmesh');
global d r_sph L_cyl r_cyl;
d = d_tmp; L_cyl = 2; r_cyl = 1; r_sph = 1/2;
R = 2; h_min = 0.1; fixed = [];
box = [-R*ones(1,d);R*ones(1,d)];
[c4n,n4e] = distmeshnd(@s,@phi,h_min,box,fixed);
n4e = fix_orientation(c4n,n4e);
Db = find_bdy_sides(n4e); Nb = [];
str = strcat('save triang_cyl_w_hole_',num2str(d),...
    'd.mat c4n n4e Db Nb');
% eval(str);

function val = s(x)
global d r_sph L_cyl r_cyl;
dist_hor = abs(x(:,1))-L_cyl;
dist_rad = sqrt(sum(x(:,2:d).^2,2))-r_cyl;
dist_cyl = max(dist_hor,dist_rad);
idx = dist_hor>0 & dist_rad>0;
dist_cyl(idx) = sqrt(dist_hor(idx).^2+dist_rad(idx).^2);
dist_compl_sph = r_sph-sqrt(sum(x.^2,2));
val = max(dist_cyl,dist_compl_sph);

function val = phi(x)
global r_sph;
dist_sph = sqrt(sum(x.^2,2))-r_sph;
val = min(dist_sph+1,2);

function bdy = find_bdy_sides(n4e)
d = size(n4e,2)-1;
if d == 2
    all_sides = [n4e(:,[1,2]);n4e(:,[2,3]);n4e(:,[3,1])];
elseif d == 3
    all_sides = [n4e(:,[2,4,3]);n4e(:,[1,3,4]);...
        n4e(:,[1,4,2]);n4e(:,[1,2,3])];
end
[sides,~,j] = unique(sort(all_sides,2),'rows');
valence = accumarray(j(:),1);
bdy = sides(valence==1,:);

function n4e = fix_orientation(c4n,n4e)
global d; nE = size(n4e,1); or_Vol_T = zeros(nE,1);
for j = 1:nE
    X_T = [ones(1,d+1);c4n(n4e(j,:),:)'];
    or_Vol_T(j) = det(X_T)/factorial(d);
end
n4e(or_Vol_T<0,[1,2]) = n4e(or_Vol_T<0,[2,1]);
```

Fig. B.7   Generation of a triangulation of a cylindrical domain with hole

B.3.2   *Visualization*

To generate plots with streamlines of velocity fields or to produce movies from a sequence of plots of finite element functions, the platform-independent and free application `paraview`, cf. Fig. B.8, can be used. It requires the `vtu` file format of the triangulation and finite element functions. Finite element functions can be elementwise constant functions, specified by *cell values*, or continuous elementwise affine functions, specified by *point values*. Functions can be scalar-valued or vector fields. The data file shown in Fig. B.9 was generated with the MATLAB routine `export2vtu.m` displayed in Fig. B.10 for functions and vector fields specified in the following example. All vectorial quantities are embedded into $\mathbb{R}^3$ by appending a trivial component if necessary.

Example B.3 Let $\Omega = (0,1)^2$ and $\mathscr{T}_h$ be the triangulation of Ω consisting of two triangles specified by

$$z_1 = (0,0),\ z_2 = (1,0),\ z_3 = (1,1),\ z_4 = (0,1)$$

and

$$T_1 = \mathrm{conv}\{z_1, z_2, z_3\},\ T_2 = \mathrm{conv}\{z_1, z_3, z_4\}.$$

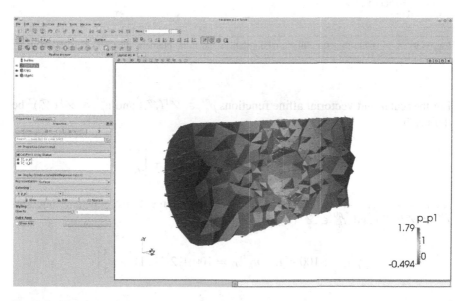

Fig. B.8   Visualizing finite element functions in `paraview` (screenshot)

```
<VTKFile type="UnstructuredGrid" version="0.1">
  <UnstructuredGrid>
    <Piece NumberOfPoints="4" NumberOfCells="2">
      <PointData>
        <DataArray type="Float32" Name="p_p1">
          1.000 2.000 3.000 4.000
        </DataArray>
        <DataArray type="Float32" Name="u_p1" NumberOfComponents="3">
          5.000 6.000 0.000 7.000 8.000 0.000 9.000 10.000 0.000 11.000 12.000 0.000
        </DataArray>
      </PointData>
      <CellData>
        <DataArray type="Float32" Name="p_p0">
          101.000 102.000
        </DataArray>
        <DataArray type="Float32" Name="u_p0" NumberOfComponents="3">
          103.000 104.000 0.000 105.000 106.000 0.000
        </DataArray>
      </CellData>
      <Points>
        <DataArray type="Float32" NumberOfComponents="3">
          0.000 0.000 0.000 1.000 0.000 0.000 1.000 1.000 0.000 0.000 1.000 0.000
        </DataArray>
      </Points>
      <Cells>
        <DataArray type="Int32" Name="connectivity">
          0 1 2 0 2 3
        </DataArray>
        <DataArray type="Int32" Name="offsets">
          3 6
        </DataArray>
        <DataArray type="UInt8" Name="types">
          5 5
        </DataArray>
      </Cells>
    </Piece>
  </UnstructuredGrid>
</VTKFile>
```

Fig. B.9 A vtu file that specifies a triangulation and finite element functions (displayed using Opera)

Let the scalar and vectorial affine functions $p_h^{p1} \in \mathscr{S}^1(\mathscr{T}_h)$ and $u_h^{p1} \in \mathscr{S}^1(\mathscr{T}_h)^2$ be defined by

$$p_h^{p1}(z_i) = i, \quad u_h^{p1}(z_i) = 2(i-1) + \begin{bmatrix} 5 \\ 6 \end{bmatrix}$$

for $i = 1, 2, \ldots, 4$. Let the scalar and vectorial elementwise constant functions $p_h^{p0} \in \mathscr{L}^0(\mathscr{T}_h)$ and $u_h^{p0} \in \mathscr{L}^0(\mathscr{T}_h)^2$ be defined by

$$p_h^{p0}|_{T_i} = 100 + i, \quad u_h^{p0}|_{T_i} = 100 + 2(i-1) + \begin{bmatrix} 3 \\ 4 \end{bmatrix}$$

```
function export2vtu(file,c4n,n4e,p_p1,u_p1,p_p0,u_p0)
[nC,d] = size(c4n); nE = size(n4e,1); type = d^2+1;
if d == 2
    c4n = [c4n,zeros(nC,1)];
    u_p1 = [u_p1(1:2:2*nC),u_p1(2:2:2*nC),zeros(nC,1)];
    u_p0 = [u_p0(1:2:2*nE),u_p0(2:2:2*nE),zeros(nE,1)];
end
fid = fopen([[file '.vtu'],'wt');
fprintf(fid,'<?xml version="1.0"?>\n');
fprintf(fid,'<VTKFile type="UnstructuredGrid" version="0.1">\n');
fprintf(fid,'<UnstructuredGrid>\n');
fprintf(fid,'<Piece NumberOfPoints="%d"',nC);
fprintf(fid,' NumberOfCells="%d">\n',nE);
fprintf(fid,'<PointData>\n');
fprintf(fid,'<DataArray type="Float32" Name="p_p1">\n');
fprintf(fid,'%3.3f\n',p_p1);
fprintf(fid,'</DataArray>\n');
fprintf(fid,'<DataArray type="Float32" Name="u_p1"');
fprintf(fid,' NumberOfComponents="3">\n');
fprintf(fid,'%3.3f\n',u_p1');
fprintf(fid,'</DataArray>\n');
fprintf(fid,'</PointData>\n');
fprintf(fid,'<CellData>\n');
fprintf(fid,'<DataArray type="Float32" Name="p_p0">\n');
fprintf(fid,'%3.3f\n',p_p0);
fprintf(fid,'</DataArray>\n');
fprintf(fid,'<DataArray type="Float32" Name="u_p0"');
fprintf(fid,' NumberOfComponents="3">\n');
fprintf(fid,'%3.3f\n',u_p0');
fprintf(fid,'</DataArray>\n');
fprintf(fid,'</CellData>\n');
fprintf(fid,'<Points>\n');
fprintf(fid,'<DataArray type="Float32" NumberOfComponents="3">\n');
fprintf(fid,'%3.3f\n',c4n');
fprintf(fid,'</DataArray>\n');
fprintf(fid,'</Points>\n');
fprintf(fid,'<Cells>\n');
fprintf(fid,'<DataArray type="Int32" Name="connectivity">\n');
fprintf(fid,'%d\n',(n4e-1)');
fprintf(fid,'</DataArray>\n');
fprintf(fid,'<DataArray type="Int32" Name="offsets">\n');
fprintf(fid,'%d\n',d+1:d+1:(d+1)*nE);
fprintf(fid,'</DataArray>\n');
fprintf(fid,'<DataArray type="UInt8" Name="types">\n');
fprintf(fid,'%d\n',type*ones(nE,1));
fprintf(fid,'</DataArray>\n');
fprintf(fid,'</Cells>\n');
fprintf(fid,'</Piece>\n');
fprintf(fid,'</UnstructuredGrid>\n');
fprintf(fid,'</VTKFile>\n');
fclose(fid);
```

Fig. B.10 Export of a triangulation and scalar and vectorial finite element functions to vtu format

Table B.8 Filters and source functions for visualizing finite element functions in `paraview`

`slice`	Displays cut through domain
`clip`	Displays cross section of domain
`glyph`	Visualizes velocity field
`warp`	Visualizes displacement field
`contour`	Visualizes level sets
`stream`	Displays stream lines
`annotate time`	Displays a time counter
`calculator`	Allows for modifying, e.g., scaling, variables
`text` (source)	Displays text

for $i = 1, 2$. A `vtu` file that specifies the triangulation and the functions is shown in Fig. B.9. The file was generated with the MATLAB routine `export2vtu.m` shown in Fig. B.10 via the following commands:

```
>> c4n = [0 0;1 0;1 1;0 1]; n4e = [1 2 3;1 3 4];
>> p_p1 = [1;2;3;4]; u_p1 = [5;6;7;8;9;10;11;12];
>> p_p0 = [101;102]; u_p0 = [103;104;105;106];
>> file = 'test';
>> export2vtu(file,c4n,n4e,p_p1,u_p1,p_p0,u_p0);
```

B.3.3   Manipulating Plots

The visualization of finite element functions or vector fields in `paraview` is done using filters. Some examples are listed in Table B.8. A sequence of data files can be exported as a movie using the file menu item `Save Animation`; background and text colors can be modified in the menu item `Settings`.

References

Further details on implementing finite element methods in MATLAB can be found in [1–3]. Details about `paraview` are provided in [5], the usage of the mesh generator `distmesh` is explained in [6]. Another powerful mesh generator is `gmsh` which is described in [4].

1. Alberty, J., Carstensen, C., Funken, S.A.: Remarks around 50 lines of Matlab: short finite element implementation. Numer. Algorithms **20**(2–3), 117–137 (1999). URL http://dx.doi.org/10.1023/A:1019155918070
2. Chen, L.: iFEM: an integrated finite element methods package in MATLAB. Tech. rep., University of California at Irvine (2009). URL https://bitbucket.org/ifem/ifem/get/tip.zip

3. Funken, S., Praetorius, D., Wissgott, P.: Efficient implementation of adaptive P1-FEM in Matlab. Comput. Methods Appl. Math. **11**(4), 460–490 (2011). URL http://dx.doi.org/10.2478/cmam-2011-0026
4. Geuzaine, C., Remacle, J.F.: Gmsh Reference Manual (2015)
5. Henderson, A., Ahrens, J., Law, C.: The ParaView Guide (2004)
6. Persson, P.O., Strang, G.: A simple mesh generator in Matlab. SIAM Rev. **46**(2), 329–345 (electronic) (2004). URL http://dx.doi.org/10.1137/S0036144503429121

Appendix C
Notation, Inequalities, Guidelines

C.1 Frequently Used Notation

Real Numbers, Vectors, and Matrices

$\mathbb{Z}, \mathbb{N}, \mathbb{N}_0$	Integers, positive and nonnegative integers
$\mathbb{R}, \mathbb{C}, \mathbb{R}_{\geq 0}$	Real and complex numbers, nonnegative real numbers
$[s,t], (s,t)$	Closed and open interval
$\mathbb{R}^n$	n-dimensional Euclidean vector space
$\mathbb{R}^{m \times n}$	Vector space of m by n matrices
$B_r(x), B_r$	Open ball of radius r centered at x or at the origin
$K_r(x), K_r$	Closed ball of radius r centered at x or at the origin
$A \subset B$	A is a subset of B or $A = B$
a, A	(Column) vector and matrix
$a^\top, A^\top$	Transpose of a vector or matrix
$\lvert \cdot \rvert$	Euclidean length or Frobenius norm
$a \cdot b = a^\top b$	Scalar product of vectors a and b
$A : B$	Inner product of matrices A and B
$\operatorname{tr} A$	Trace of the matrix A
I_L	$L \times L$ identity matrix
$[x,y]^\top, (x,y)$	Vectors with entries x and y
$\begin{bmatrix} x_1 & x_2 \\ y_1 & y_2 \end{bmatrix}$	Matrix with entries x_1, x_2, y_1, y_2
$a \gg b$	a significantly greater than b
$a \approx b$	a approximately equal to b
$a \sim b$	a proportional to b

© Springer International Publishing Switzerland 2016
S. Bartels, *Numerical Approximation of Partial Differential Equations*,
Texts in Applied Mathematics 64, DOI 10.1007/978-3-319-32354-1

Sets and Domains

Ω	Bounded Lipschitz domain in $\mathbb{R}^d$, $d = 2, 3$
$\partial\Omega$	Boundary of the domain Ω
n	Outer unit normal on $\partial\Omega$
Γ_{D}	Dirichlet boundary, closed subset of $\partial\Omega$
Γ_{N}	Neumann boundary, $\Gamma_{\mathrm{N}} = \partial\Omega \setminus \Gamma_{\mathrm{D}}$
$[0, T]$	Time interval

Linear Spaces and Operators

id	Identity operator
ker	Kernel of an operator
X, Y	Banach spaces
$\|\cdot\|_X$	Norm in X
X'	Linear bounded functionals $\phi : X \to \mathbb{R}$
$\langle \phi, x \rangle = \phi(x)$	Duality pairing of $\phi \in X'$ and $x \in X$
$\|\cdot\|_{X'}$	Operator norm in X'
$\mathrm{L}(X, Y)$	Bounded linear operators $A : X \to Y$
$\|\cdot\|_{\mathrm{L}(X,Y)}$	Operator norm in $\mathrm{L}(X, Y)$
A'	Adjoint of $A \in \mathrm{L}(X, Y)$
H	Hilbert space
$(x, y)_H$	Inner product of x and y in a Hilbert space H

Differential Operators

∂_i, ∂_{x_i}, $\frac{\partial}{\partial x_i}$	Partial derivative with respect to the i-th coordinate
∇	Gradient of a function
div	Divergence of a vector field
D, D^2	Total derivative and Hessian of a function
∂_x, ∂_y, ∂_t, ∂^α	Partial derivatives
$\partial_n u = \nabla u \cdot n$	Normal derivative on $\partial\Omega$
u_t	Partial derivative with respect to t
$\varepsilon(u)$	Symmetric gradient of a vector field
Δ	Laplace operator

Function Spaces

$C^k(A)$	k-times continuously differentiable functions
$C_c^\infty(\Omega)$	Compactly supported, smooth functions
$C_0(\Omega)$	Closure of $C_c^\infty(\Omega)$ with respect to maximum norm
$L^p(\Omega)$	Functions whose p-th power is Lebesgue integrable
$W^{k,p}(\Omega)$	k-times weakly differentiable functions
$W_D^{k,p}(\Omega), W_0^{k,p}(\Omega)$	Functions in $W^{k,p}(\Omega;\mathbb{R}^m)$ vanishing on Γ_D or $\partial\Omega$
$H^k(\Omega)$	Hilbert space $W^{k,2}(\Omega)$
$H_N(\mathrm{div};\Omega)$	Vector fields with square integrable divergence
$\|\cdot\|, (\cdot,\cdot)$	Norm and inner product in $L^2(\Omega;\mathbb{R}^m)$
$L_0^2(\Omega)$	Functions in $L^2(\Omega)$ with vanishing integral over Ω

Finite Differences

$\tau, \Delta t, \Delta x, \Delta y$	Step-sizes
$\partial^-, \partial^+, \widehat{\partial}$	Backward, forward, and central difference quotient
d_t	Backward difference quotient in time
$t_k, t_{k+1/2}$	Time steps $k\tau$ and $(k+1/2)\tau$
$u^k, u^{k+1/2}$	Approximations associated with time steps

Finite Element Spaces

$h, h_{\min}$	Maximal and minimal diameter of elements in $\mathcal{T}_h$
h_T, h_S, h_z	Local mesh-sizes
$\mathcal{N}_h$	Nodes that define vertices of elements
$\mathcal{T}_h$	Set of elements that define a triangulation
$\mathcal{S}_h$	Sides of elements in a triangulation
z, E, S, T	Node, edge, side, and element in a triangulation
φ_z	Nodal basis function
ω_z	Patch of a node
$\mathcal{P}_k(T)$	Polynomials of maximal degree k restricted to T
$\mathcal{L}^0(\mathcal{T}_h)$	$\mathcal{T}_h$-elementwise constant functions
$\mathcal{S}^1(\mathcal{T}_h)$	Continuous, $\mathcal{T}_h$-elementwise affine functions
$\mathcal{S}_D^1(\mathcal{T}_h), \mathcal{S}_0^1(\mathcal{T}_h)$	Functions in $\mathcal{S}^1(\mathcal{T}_h)$ vanishing on Γ_D or $\partial\Omega$
$\mathcal{I}_h$	Nodal interpolation operator on $\mathcal{T}_h$
$\mathcal{J}_h$	Clément quasi-interpolant
$[\![\nabla u_h \cdot n_S]\!]$	Jump of the normal component of ∇u_h across S
$\mathcal{S}^{1,cr}(\mathcal{T}_h)$	Crouzeix–Raviart finite element space

Other Notation

$c, C, C', C'', c_1, c_2, \ldots$	Mesh-size independent, generic constants
dx, ds	Volume and surface element for Lebesgue measure
$\overline{A}$	Closure of a set A
$|A|$	Cardinality, volume, surface area, or length of a set A
$\mathrm{diam}(A)$	Diameter of the set A
χ_A	Characteristic function of a set A
δ_{ij}	Kronecker symbol
$\mathscr{O}(t)$, $o(t)$	Landau symbols
$\mathrm{supp} f$	Support of a function f
$\mathscr{C}$	Consistency term

MATLAB Routines

`d, red`	Space dimension, number of uniform refinements
`c4n, n4e`	Lists of coordinates of nodes, and nodes of elements
`Db, Nb`	Lists of sides on Γ_D and Γ_N
`dNodes, fNodes`	Nodes belonging to Γ_D and remaining nodes
`nC, nE, nDb, nNb`	Number of nodes, elements and sides on Γ_D and $\overline{\Gamma}_N$
`s, m, m_lumped`	$P1$ stiffness, mass, and lumped mass matrix
`m_Nb, m_Nb_lumped`	Exact and discrete inner products on Γ_N
`vol_T, mp_T`	Volume and midpoint of an element
`grads_T`	Elementwise gradients of nodal basis functions
`tau`	Step-size
`I, J, X`	Lists to generate a sparse matrix

Special Constants

c_P	Poincaré inequality
c_S	Sobolev inequality
c_{Tr}	Trace inequality
c_Δ	H^2-regularity for Δ
$c_{\mathscr{I}}$	Nodal interpolation
$c_{\mathscr{J}}$	Clément quasi-interpolation
c_{usr}	Uniform shape regularity
c_{inv}	Inverse estimate
α, β, γ	Ellipticity, inf-sup condition
k_a, k_b, κ_Γ	Continuity constants

C.2 Important Inequalities

Young's Inequality For all $a, b \in \mathbb{R}_{\geq 0}$ and $1 < p, q < \infty$ with $1/p + 1/q = 1$ we have that

$$ab \leq \frac{1}{p}a^p + \frac{1}{q}b^q.$$

Cauchy–Schwarz Inequality For all $x, y \in \mathbb{R}^n$ we have that

$$x \cdot y = \sum_{i=1}^n x_i y_i \leq \Big(\sum_{i=1}^n x_i^2\Big)^{1/2}\Big(\sum_{i=1}^n y_i^2\Big)^{1/2} = |x||y|.$$

Triangle Inequality For all $x, y \in \mathbb{R}^n$ we have that

$$|x + y| \leq |x| + |y|.$$

Hölder's Inequality For $f \in L^p(\Omega)$ and $g \in L^q(\Omega)$ and $1 \leq p, q \leq \infty$ with $1/p + 1/q = 1$ we have that

$$\int_\Omega fg \, dx \leq \|f\|_{L^p(\Omega)} \|g\|_{L^q(\Omega)}.$$

Minkowski's Inequality For $f, g \in L^p(\Omega)$ with $1 \leq p \leq \infty$ we have that

$$\|f + g\|_{L^p(\Omega)} \leq \|f\|_{L^p(\Omega)} + \|g\|_{L^p(\Omega)}.$$

Poincaré (or Friedrichs) Inequality For $v \in W^{1,p}(\Omega)$ with $1 \leq p \leq \infty$ and $v|_{\Gamma_D} = 0$ or $\int_\Omega v \, dx = 0$ we have that

$$\|v\|_{L^p(\Omega)} \leq c_P \|\nabla v\|_{L^p(\Omega)}.$$

Trace Inequality For $v \in W^{1,p}(\Omega)$ and $1 \leq p \leq \infty$ we have that

$$\|v\|_{L^p(\partial\Omega)} \leq c_{\mathrm{Tr}} \|v\|_{W^{1,p}(\Omega)}.$$

Sobolev Inequalities For $v \in W^{1,p}(\Omega)$ with $1 \leq p < d$ and $1 \leq q \leq p^*$ for $p^* = dp/(d-p)$ we have that

$$\|v\|_{L^q(\Omega)} \leq c_S \|v\|_{W^{1,p}(\Omega)}.$$

Nodal Interpolation Estimate If $v \in W^{2,p}(\Omega)$ with $p \geq 2$ we have that

$$\|v - \mathscr{I}_h v\|_{L^p(\Omega)} + h\|\nabla(v - \mathscr{I}_h v)\|_{L^p(\Omega)} \leq c_{\mathscr{I}} h^2 \|D^2 v\|_{L^p(\Omega)}.$$

Uniform Shape Regularity For all $h > 0$ and all $T \in \mathscr{T}_h$ we have that

$$h_T \leq c_{\mathrm{usr}} \varrho_T.$$

Inverse Estimate For all $v_h \in \mathscr{S}^1(\mathscr{T}_h)$ and $1 \leq p \leq \infty$ we have that

$$\|\nabla v_h\|_{L^p(\Omega)} \leq c_{\mathrm{inv}} h_{\mathrm{min}}^{-1} \|v_h\|_{L^p(\Omega)}.$$

Discrete Norm Equivalence For all $v_h \in \mathscr{S}^1(\mathscr{T}_h)$ we have that

$$c_{\mathrm{eq}}^{-1} \|v_h\|_{L^p(\Omega)} \leq \left(\sum_{z \in \mathscr{N}_h} h_z^d |v_h(z)|^p \right)^{1/p} \leq c_{\mathrm{eq}} \|v_h\|_{L^p(\Omega)}.$$

H^2-Regularity If $\Omega \subset \mathbb{R}^d$ is convex, then for all $v \in H^2(\Omega) \cap H_0^1(\Omega)$ we have that

$$\|D^2 v\|_{L^2(\Omega)} \leq c_\Delta \|\Delta v\|_{L^2(\Omega)}.$$

C.3   Guidelines for Discretizing Differential Equations

The following aspects should be taken into account when developing a numerical scheme for approximating the solution of a partial differential equation.

1. Is the problem under consideration *well-posed* and what are typical properties of solutions, e.g., occurrence of characteristics, validity of a maximum principle, regularity properties, energy conservation principles?
2. Is the discretization *consistent*, i.e., does an interpolant of the exact solution nearly satisfy the numerical scheme?

3. Is the scheme *stable*, i.e., do numerical solutions remain bounded in a meaningful norm as discretization parameters tend to zero, e.g., is a discrete maximum principle or energy conservation principle satisfied?

4. Can the numerical solution be efficiently computed and is the discrete problem *well-conditioned*? What is a good stopping criterion for iterative methods? It is necessary for the problem to have the same number of degrees of freedom as conditions that characterize solutions.

5. Do numerical solutions reflect the typical behavior of the physical process with a reasonable discretization fineness, e.g., transport along characteristics?

6. Which terms should be discretized *implicitly* and which ones *explicitly*? Low-order terms, which involve lower-order derivatives, can often be treated explicitly which simplifies computations, while the explicit discretization of highest order derivatives typically leads to restrictive stability conditions.

7. If the discretization is not well-posed, *stabilizing terms* such as $-\varepsilon\Delta u$ can be introduced. This requires relating ε to the mesh-size and typically leads to reduced convergence rates.

8. How do *nonconformities* such as implementation aspects affect approximations, e.g., the use of quadrature or domain approximation?

9. Does an appropriate *reformulation* of the problem allow for more stable and accurate schemes, e.g., integration is more stable than differentiation.

10. Has the implementation been *debugged*, e.g., by solving a simple discretization by hand, checking enumerations of sides and vertices, and testing the correctness of matrices.

Index

© Springer International Publishing Switzerland 2016
S. Bartels, *Numerical Approximation of Partial Differential Equations*,
Texts in Applied Mathematics 64, DOI 10.1007/978-3-319-32354-1

Backcover

Finite element methods for approximating partial differential equations have reached a high degree of maturity, and are an indispensable tool in science and technology. This textbook aims at providing a thorough introduction to the construction, analysis, and implementation of finite element methods for model problems arising in continuum mechanics. The first part of the book discusses elementary properties of linear partial differential equations along with their basic numerical approximation, the functional-analytical framework for rigorously establishing existence of solutions, and the construction and analysis of basic finite element methods. The second part is devoted to the optimal adaptive approximation of singularities and the fast iterative solution of linear systems of equations arising from finite element discretizations. In the third part, the mathematical framework for analyzing and discretizing saddle-point problems is formulated, corresponding finite element methods are analyzed, and particular applications including incompressible elasticity, thin elastic objects, electromagnetism, and fluid mechanics are addressed. The book includes theoretical problems and practical projects for all chapters, an introduction to the implementation of finite element methods, and model implementations of most devised schemes.

© Springer International Publishing Switzerland 2016   535
S. Bartels, *Numerical Approximation of Partial Differential Equations*,
Texts in Applied Mathematics 64, DOI 10.1007/978-3-319-32354-1

Printed in the United States
By Bookmasters